T0191784

THE FRONTIERS COLLECTION

THE FRONTIERS COLLECTION

The books in this collection are devoted to challenging and open problems at the forefront of modern science, including related philosophical debates. In contrast to typical research monographs, however, they strive to present their topics in a manner accessible also to scientifically literate non-specialists wishing to gain insight into the deeper implications and fascinating questions involved. Taken as a whole, the series reflects the need for a fundamental and interdisciplinary approach to modern science. Furthermore, it is intended to encourage active scientists in all areas to ponder over important and perhaps controversial issues beyond their own speciality. Extending from quantum physics and relativity to entropy, consciousness and complex systems—the Frontiers Collection will inspire readers to push back the frontiers of their own knowledge.

More information about this series at http://www.springer.com/series/5342

For a full list of published titles, please see back of book or springer.com/series/5342

Reinhold Bertlmann · Anton Zeilinger
Editors

QUANTUM [UN]SPEAKABLES II

Half a Century of Bell's Theorem

Editors
Reinhold Bertlmann
Faculty of Physics
University of Vienna
Vienna
Austria

Anton Zeilinger
Faculty of Physics, Vienna Center for Quantum Science and Technology (VCQ)
University of Vienna
Vienna
Austria

and

Institute for Quantum Optics and Quantum Information (IQOQI)
Austrian Academy of Sciences
Vienna
Austria

ISSN 1612-3018 ISSN 2197-6619 (electronic)
THE FRONTIERS COLLECTION
ISBN 978-3-319-81785-9 ISBN 978-3-319-38987-5 (eBook)
DOI 10.1007/978-3-319-38987-5

Printed on acid-free paper

This Springer imprint is published by Springer Nature
The registered company is Springer International Publishing AG Switzerland

Preface

John Stewart Bell was without doubt one of the most influential scientists of the second half of the twentieth century. His scope of the research was very broad. Starting in accelerator physics in the 1950s at Harwell and Malvern, he soon turned to particle physics. In particular, after moving to CERN in 1960, he considered particle physics as a “job” he got paid for, but he certainly was enthusiastic about doing it. Parallel to his “job”, he worked continuously on his “hobby”, the foundations of quantum mechanics, the field he is most famous for, until his untimely death on 1 October 1990.

In all these areas, Bell made outstanding contributions. Let us mention just a few.

Already Bell’s Ph.D. thesis included a fundamental paper “Time reversal in field theory”. In that work, he proved independently from Gerhart Lüders and Wolfgang Pauli the celebrated CPT theorem, a basic symmetry of Nature that leaves a physical system unchanged under the joint action of charge conjugation C, parity inversion P and time reversal T.

Bell’s most far-reaching contribution to particle physics, developed together with Roman Jackiw, was the discovery of the so-called Adler–Bell–Jackiw anomaly, which is responsible for the decay of the pion into two photons. It turned out to be the key to a deeper understanding of quantum field theory.

In accelerator physics, Bell wrote several papers, alone or in collaboration with William Walkinshaw at Harwell, mostly on how to focus a bunch of electrons or protons in a linear accelerator. At CERN, he collaborated with his wife Mary who was working in the Accelerator Research Group. Together they published several papers, for example on electron cooling in storage rings.

A particularly attractive work, in our opinion, was Bell’s combination of the Unruh effect of quantum field theory with accelerator physics. The idea was to use the polarization of accelerated electrons as a thermometer that measures the temperature of the blackbody radiation experienced by the electrons. The results, small but measureable, were published together with Jon Leinaas.

Of course, John Bell is most famous for his contributions to the foundations of quantum mechanics. This topic attracted his interest already in the late 1940s while he was a student at Queen's University Belfast, got stimulated in the 1950s by Bohm's reinterpretation of quantum mechanics and culminated in 1964 when he was on sabbatical in the USA. There, he wrote his two seminal papers. The first one (but published secondly due to a delay in the publishing journal) was "*On the problem of hidden variables in quantum mechanics*", where he discovered that non-contextual hidden variable theories are in conflict with quantum mechanics. The second one, "*On the Einstein-Podolsky-Rosen paradox*", contained the celebrated Bell inequality, or what is called Bell's theorem, stating that any local realistic theory disagrees with quantum mechanics.

Thus, the year 2014 marked the 50th anniversary of Bell's theorem, one of the most significant developments in quantum theory. For us, it was immediately clear that we had to organize some kind of celebration. Discussing this idea with our colleagues, we received such a huge and enthusiastic response that our initial intention finally resulted in the conference "*Quantum [Un]Speakables II: Half a Century of Bell's Theorem*", which took place June 18th–22nd at the University of Vienna. About 400 scientists of the quantum foundations community attended. We were also very happy and felt privileged that Mary Bell took the effort to come as a *Guest of Honour* and to speak at the opening. The major part of the contributions to the conference is collected in this book.

As is well known by now, when John Bell started to work on the foundations of quantum mechanics, there was hardly any interest in such topics. Even worse, working on foundations was not considered to be a proper topic for a physicist. The first who had the courage to carry out an experiment on Bell inequalities was John Clauser in the 1970s; he had to struggle enormously to get the resources for doing the experiment. The situation began to change after the experiments of Aspect in the 1980s. Slowly, the community began to realize that there was something essential to Bell's theorem. The third generation of Bell experiments commenced in the 1990s and has extended into the twenty-first century. It has taken advantage of new technologies, such as spontaneous parametric downconversion, which is an effective way to create entangled photons. Also, more recently, it became possible to create entanglement in other systems, such as atoms or ions in traps or superconducting devices. In such experiments, the case against local realism, the viewpoint excluded by Bell's theorem, and for quantum mechanics became stronger and stronger, and more and more loopholes for the experiments were closed.

Also in the 1990s, the field of applications of entangled states and of Bell's theorem opened up. This was signified by experiments on quantum teleportation, quantum cryptography, long-distance quantum communication and the realization of some of the basic entanglement-based concepts in quantum computation.

Today, Bell's theorem and the underlying physics of entangled states have become cornerstones of the evolving technology of quantum information. Violation of Bell's inequality has become a litmus test for the realization of quantum entanglement in the laboratory. It has become part of the common understanding that a loophole-free Bell experiment is the final and definitive demonstration that quantum cryptography can be unconditionally secure. Also, entanglement swapping, the teleportation of an entangled state, is central for quantum repeaters, which are expected to be the backbone of a future worldwide quantum internet. Furthermore, Bell's theorem, as a fundamental contradiction between local realism and quantum mechanics, has been extended to higher dimensions and multiparticle systems.

The number of citations of Bell inequalities over the last decades is shown in Fig. 1.

Nowadays, physicists agree that John Bell would have definitely received the Nobel Prize for his outstanding contributions to the foundations of quantum mechanics if he had lived longer. This was, for instance, expressed explicitly by Daniel Greenberger in an interview given at the conference *Quantum [Un]Speakables II* in Vienna:

> Of course, people more and more appreciate John Bell's beautiful work. He was essentially starting the field, his work was totally seminal, and if he were alive he certainly would have won the Nobel Prize!

We also want to mention that Bell was not only an outstanding scientist with a sharp and clear view of Nature, but also a man of honest character and high morals. The late Abner Shimony expressed his appreciation for Bell, which we fully share, in the following way:

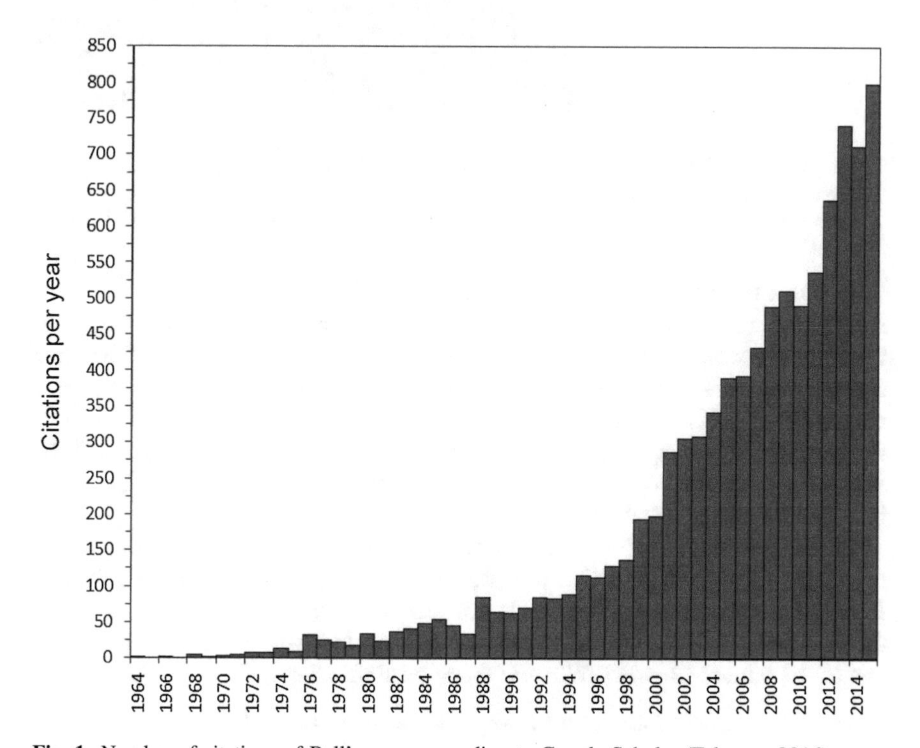

Fig. 1 Number of citations of Bell's paper according to Google Scholar (February 2016)

Fig. 2 The Belfast City Council named a street "*Bells Theorem Crescent*" in the Titanic Quarter of Belfast to honour its eminent scientist John Stewart Bell. *Photo* Joan Whitaker

> His [Bell's] passion for understanding, uncompromising honesty, simplicity of lifestyle and demeanour, dignity, courtesy, generosity to other scientists, and passion for social justice were combined into a character that was inspiring to all who had the privilege to be acquainted with him.

In 2014, the city of Belfast honoured John Bell as "*one of Northern Ireland's most eminent scientists*". The City Council named a street after his theorem "*Bells Theorem Crescent*" in the Titanic Quarter of Belfast, bending their rule of avoiding streets being named after individuals (see Fig. 2). Today, the Technical College, renamed Belfast Metropolitan College, that Bell attended is situated in that location. Furthermore, the Naughton Gallery at Queen's University Belfast organized the exhibition "*Action at a distance: The life and legacy of John Stewart Bell*" combined with lectures about Bell at the university.

Finally, we would like to mention that the late Walter Thirring, Austria's most prominent theoretical physicist, who was a member of the International Advisory Board of our conference series, in his last years developed a deep interest in Bell's theorem and published several papers about it together with Heide Narnhofer and one of the editors (R.A.B.). When working on Bell's ideas, Thirring said:

> I have to apologize to John Bell that I recognized the significance of Bell's theorem only so late.

Thirring's original German phrasing was:

Ich muss John Bell Abbitte leisten, dass ich erst so spät die Bedeutung von Bell's Theorem erkannt habe.

This collection of articles is based to some extent on presentations made at the conference "Quantum [Un]Speakables II" in Vienna. However, where possible, contributors have made an effort to write at a level accessible to non-specialists and have also updated and expanded their texts as necessary. We are confident that the result will be of interest to graduate students and researchers in quantum theory, specifically in the conceptual foundations of quantum mechanics. But it will be also of value to philosophers and historians of science working in this field, as well as providing stimulating reading for many scientifically literate persons from other fields.

The current volume would not exist without our editor at Springer, Angela Lahee. Her unequivocal support right from the beginning and her continuing feedback and guidance were invaluable for the making of this book.

The organization of the conference in Vienna would have been impossible without the financial support of the Austrian Academy of Sciences, the SFB Foundations and Applications of Quantum Science, the Vienna Center for Quantum Science and Technology (VCQ), the Science and Research Funding of the City of Vienna, and the University of Vienna. We are very grateful to Andrea Aglibut who managed the whole organization of the conference with great commitment and charm and thank her for supporting the editing process of the present book. We are grateful to the numerous students who helped us with the organization, in particular to Bernhard Wittmann and Robert Fickler who were responsible for the technical run. Last but not least, our thanks go to the Austrian Central Library for Physics, in particular to Gerlinde Fritz and Daniel Winkler for the video documentation and to Rudi Handl for taking a complete photographic record of the event.

Vienna, Austria

Reinhold Bertlmann
Anton Zeilinger

Contents

Contributors

Antonio Acín ICFO–Institut de Ciencies Fotoniques, Barcelona Institute of Science and Technology, Castelldefels, Barcelona, Spain; ICREA–Institucio Catalana de Recerca i Estudis Avançats, Barcelona, Spain

Mary Bell European Organization for Nuclear Research (CERN), Geneva, Switzerland

Reinhold Bertlmann Faculty of Physics, University of Vienna, Vienna, Austria

Dagmar Bruß Heinrich-Heine-University Düsseldorf, Düsseldorf, Germany

Časlav Brukner Vienna Center for Quantum Science and Technology (VCQ), Vienna, Austria; Institute of Quantum Optics and Quantum Information (IQOQI), Austrian Academy of Sciences, Vienna, Austria; Faculty of Physics, University of Vienna, Vienna, Austria

Jeffrey Bub Philosophy Department, Institute for Physical Science and Technology, Joint Center for Quantum Information and Computer Science, University of Maryland, College Park, MD, USA

Adán Cabello Departamento de Física Aplicada II, Universidad de Sevilla, Sevilla, Spain

Yuan Cao Shanghai Branch, National Laboratory for Physical Sciences at Microscale and Department of Modern Physics, University of Science and Technology of China, Shanghai, China; CAS Centre for Excellence and Synergetic Innovation Centre in Quantum Information and Quantum Physics, University of Science and Technology of China, Hefei, Anhui, China

Eric G. Cavalcanti School of Physics, University of Sydney, Sydney, NSW, Australia

Bradley G. Christensen Department of Physics, University of Illinois at Urbana-Champaign, Urbana, IL, USA

John F. Clauser J. F. Clauser and Associates, Walnut Creek, CA, USA

J.I. de Vicente Departamento de Matemáticas, Universidad Carlos III de Madrid, Madrid, Spain

Michael Epping Heinrich-Heine-University Düsseldorf, Düsseldorf, Germany

Nicolas Gisin Group of Applied Physics, University of Geneva, Geneva, Switzerland

Marissa Giustina Institute for Quantum Optics and Quantum Information (IQOQI), Austrian Academy of Sciences, Vienna, Austria; Vienna Center for Quantum Science and Technology (VCQ), Vienna, Austria; Faculty of Physics, University of Vienna, Vienna, Austria

Daniel M. Greenberger City Collegeof New York (CCNY), City University of New York (CUNY), New York, NY, USA

Otfried Gühne Faculty IV for Science and Technology, University of Siegen, Siegen, Germany

Lucien Hardy Perimeter Institute for Theoretical Physics, Waterloo, ON, Canada

Yuji Hasegawa Atominstitut, Vienna University of Technology, Vienna, Austria

Beatrix Hiesmayr Faculty of Physics, University of Vienna, Vienna, Austria

Michael Horne Department of Physics and Astronomy, Stonehill College, Easton, MA, USA

Hermann Kampermann Heinrich-Heine-University Düsseldorf, Düsseldorf, Germany

Matthias Kleinmann Department of Theoretical Physics, University of the Basque Country UPV/EHU, Bilbao, Spain

Simon B. Kochen Mathematics Department, Princeton University, Princeton, NJ, USA

B. Kraus Institute for Theoretical Physics, University of Innsbruck, Innsbruck, Austria

Paul G. Kwiat Department of Physics, University of Illinois at Urbana-Champaign, Urbana, IL, USA

Jan-Åke Larsson Institutionen för systemteknik, Linköpings Universitet, Linköping, Sweden

N. David Mermin Laboratory of Atomic and Solid State Physics, Cornell University, Ithaca, NY, USA

Tobias Moroder Faculty IV for Science and Technology, University of Siegen, Siegen, Germany

Miguel Navascués Institute for Quantum Optics and Quantum Information (IQOQI), Austrian Academy of Sciences, Vienna, Austria

Jian-Wei Pan Shanghai Branch, National Laboratory for Physical Sciences at Microscale and Department of Modern Physics, University of Science and Technology of China, Shanghai, China; CAS Centre for Excellence and Synergetic Innovation Centre in Quantum Information and Quantum Physics, University of Science and Technology of China, Hefei, Anhui, China

Cheng-Zhi Peng Shanghai Branch, National Laboratory for Physical Sciences at Microscale and Department of Modern Physics, University of Science and Technology of China, Shanghai, China; CAS Centre for Excellence and Synergetic Innovation Centre in Quantum Information and Quantum Physics, University of Science and Technology of China, Hefei, Anhui, China

Helmut Rauch Atominstitut, Vienna University of Technology, Vienna, Austria

Terence Rudolph Faculty of Natural Sciences, Imperial College London, London, UK

Valerio Scarani Centre for Quantum Technologies and Department of Physics, National University of Singapore, City Area, Singapore

C. Spee Institute for Theoretical Physics, University of Innsbruck, Innsbruck, Austria

Gregor Weihs Institute for Experimental Physics, University of Innsbruck, Innsbruck, Austria

Andrew Whitaker School of Mathematics and Physics, Queen's University Belfast, Belfast, UK

Howard M. Wiseman Centre for Quantum Dynamics, Griffith University, Brisbane, QLD, Australia

Anton Zeilinger Vienna Center for Quantum Science and Technology (VCQ), Vienna, Austria; Institute for Quantum Optics and Quantum Information (IQOQI), Austrian Academy of Sciences, Vienna, Austria; Faculty of Physics, University of Vienna, Vienna, Austria

Qiang Zhang Shanghai Branch, National Laboratory for Physical Sciences at Microscale and Department of Modern Physics, University of Science and Technology of China, Shanghai, China; CAS Centre for Excellence and Synergetic Innovation Centre in Quantum Information and Quantum Physics, University of Science and Technology of China, Hefei, Anhui, China

Marek Żukowski Institute of Theoretical Physics and Astrophysics, University of Gdánsk, Gdánsk, Poland

Part I
John Stewart Bell—The Man

Chapter 1
Address to Participants at Quantum [Un]Speakables II

Mary Bell

I thank everyone for coming to the meeting. J.S.B. would have been amazed to see so many people here.

I first saw John in the autumn of 1949 when he first arrived at the Theory Division of A.E.R.E. After a few months he agreed to go to Malvern where a small group (already part of A.E.R.E.) under Bill Walkinshaw was working on short electron accelerators. Later on, I joined the group. At that time, only desk calculators were available, so it was a different world from now. As you may have read, he always referred to himself as a 'quantum engineer'. He must have become aware of a class difference between physicists and so-called engineers. Recently, I have seen a number of such references. In the book 'Quantum Enigma' by Bruce Rosenblum and Fred Kuttner, they express great surprise that, after graduation, John took a job involving an 'almost engineering role'. In fact, he liked the work a lot. However, not everyone has the same attitude. Later, when we went to Stanford, we were amused at immigration. The officer didn't think much of John's description as a 'theoretical physicist', but was very impressed by mine as 'mathematical engineer'.

Later, the Malvern group moved to Harwell to join the rest of A.E.R.E. The Theory Division had a number of groups, but we all had tea or coffee together, and it was there that John and Franz Mandl had many friendly arguments about quantum mechanics, arguments which they both enjoyed.

In 1953, he was offered a year at a University on his usual salary. In the end he chose Birmingham where Rudolf Peierls was professor. He saw a lot of the late Paul Matthews there. The first part of his thesis was the T.C.P. theorem. On his return to Harwell he joined the Nuclear Physics Group of the Theoretical Physics Division, at that time headed by the late Tony Skyrme. In 1960 we moved to CERN, as John thought that he would like to work in high-energy physics.

M. Bell (✉)
European Organization for Nuclear Research (CERN), CH-1211 Geneva 23, Switzerland

R. Bertlmann and A. Zeilinger (eds.), *Quantum [Un]Speakables II*,
The Frontiers Collection, DOI 10.1007/978-3-319-38987-5_1

As he always said, the quantum mechanics foundations was a hobby. Most of the time he was thinking about other things. As he had a wide knowledge people often consulted him. At the end of 1963 we went for a year to Stanford. It was there that the inequality was thought out during a week-end.

There was not a lot of interest at first. Most people took the Tini Veltman attitude. As he writes in his book "Facts and Mysteries in Elementary Particle Physics", Tini asked John: "Why are you doing this? Does it make the slightest difference in the calculations such as I am doing?" To which Tini says John answered: "You are right, but are you not interested and curious about the interpretation?"

Gradually, more people began to take an interest, many of them here today. It also attracted a number of non-physicists. John always said that CERN was like a railway station with many passers-by. On one occasion when he arrived in the morning, he found his blackboard filled with questions. A stranger wished to write a paper with him. It was to consist of the questions, with the answers supplied by John.

Of course, he had many visits from well-known physicists, many of whom I am pleased to see here today.

Chapter 2
John Stewart Bell, Quantum Information and Quantum Information Theory

Andrew Whitaker

It is traditional to take virtually for granted [1, 2] that John Bell's work on the foundations of quantum theory led fairly directly to the founding of the discipline of quantum information theory, and thus it is natural to give Bell credit perhaps for the very existence of this subject.

This tradition obviously provides a massive boost to anyone, in the present or in the future, who has the task of describing Bell's importance and demonstrating his stature to non-scientists as well as to scientists. Anybody not interested in such arcane matters as determinism, locality and realism may be excited by his having provided the means of breaking codes, of running safe and efficient elections, or even of teleportation.

Of course looking at events historically there is every reason to take the influence of Bell on quantum information theory as obvious and beyond question. Many of the people involved in foundational studies moved on seamlessly to work on quantum information, though they were joined there, particularly in the study of quantum computation, by many with little genuine interest or understanding of Bell's work, and Mermin [3] has pointed out in his introduction to quantum computation how little physics is needed to work in this area.

Experimental and theoretical techniques designed for one or other of the areas of quantum foundations and quantum information theory were often capable of being adapted to be used in work on the other. For example experimental methods used in the study of quantum teleportation [4] were the basis of those used to demonstrate the existence of the GHZ states [5].

And of course it is also natural to bring up the matter of entanglement. It is surely fair to take note of Bell's importance in the full realisation of the significance of entanglement. Its importance was probably first pointed out by EPR and was stressed by Schrödinger, particularly in his famous statement that it was 'not *one*

A. Whitaker (✉)
School of Mathematics and Physics, Queen's University Belfast, Belfast BT7 1NN, UK
e-mail: a.whitaker@qub.ac.uk

R. Bertlmann and A. Zeilinger (eds.), *Quantum [Un]Speakables II*,
The Frontiers Collection, DOI 10.1007/978-3-319-38987-5_2

but *the* characteristic feature of quantum mechanics, the one that enforces the entire departure from classical thought' [6].

Subsequently, however, it was not discussed in a substantial way until Bell used it in his important discoveries [7, 8]. Then, of course, and almost entirely as a result of his influence and his theorems, it became central in at least all the standard examples met in both quantum foundational studies and quantum information theory. It must surely be fair to give him credit for that. So we may feel quite happy in talking about Bell's major part in the development of quantum information theory. But it is still slightly different to claim that his work was its main theoretical stimulus or underpinning.

We may remember David Deutsch, who was, at least in many people's views, the genuine founder of the study of quantum computation [9]. His important insight, obvious enough in retrospect, was that, not only must there be a subject of quantum computation, with its own theoretical basis, but that, at least from a rigorous point of view, we do not have classical and quantum computation on an equal footing, any more than classical and quantum physics are on an equal footing.

Rather quantum physics and quantum computation are fundamental, and classical physics and classical computation, are again, in principle, merely good approximations, often extremely good approximations, to the quantum versions.

Thus Deutsch declared that the classical theory of computation, unchallenged for more than fifty years, was obsolete, and it was necessary to rewrite the Church-Turing argument to meet quantum-mechanical requirements [10]. (Turing, Deutsch said, had restricted his argument to paper, thinking, or at least acting as though he thought, that paper was classical, but it wasn't—it was quantum.)

Incidentally Mermin [3] has effectively made a reply to Deutsch's point about the primacy of the quantum computer. To say that a quantum computer is one that obeys the laws of quantum mechanics, he says, is a temptation to be resisted. It would imply that any laptop or even a mainframe computer is a quantum computer, and that is just not how we think of things. Rather: 'A quantum computer is one whose operation exploits certain very special transformations of its internal state'.

For the present argument, the important and obvious point is that Deutsch's argument had nothing to do with the work of Bell. Rather in terms of analysis of quantum theory it had a close, perhaps a symbiotic, relationship with one of Deutsch's other core beliefs, that in many worlds [10, 11]. For Deutsch, the existence of many worlds could be read straight off from quantum theory by any unprejudiced student, but, if that argument was questioned, he could argue that they certainly explained the possible vast speed-up of a quantum computer—the calculations were being carried out in the many worlds; this is known, of course, as *parallelism* [9, 10].

Deutsch's claim that quantum computation relies on many worlds and parallelism has been strongly criticised by Steane [12], who argues that a quantum computer requires only a single universe, and that is misleading to argue that quantum computers perform more operations than those allowed by a single universe.

Steane himself has suggested that the real source of speed-up in quantum computation is entanglement. As we have already said entanglement is indeed present, and indeed seemingly centrally important, in the examples normally considered, so it seems to be a very natural suggestion. However various arguments have shown that quantum computation may proceed with the usual speed-up without any entanglement at all [13–15].

As well as parallelism and entanglement, the combination of superposition and interference has been suggested, but certainly does not *guarantee* any speed-up as compared to classical computation.

Another contender is contextuality, which was essentially Bell's response [7] to the demonstration by Kochen and Specker [16] that without contextuality hidden variables could not exist. Contextuality says that measurement depends on context—the result of measuring physical quantity A depends on whether it is measured together with physical quantity B or C. Again it may be relevant for the speed-up in quantum computation in some cases [17].

Indeed every attempt to isolate 'the' ingredient of effective quantum computation, essential in every case, seems doomed to failure. Overall none of these factors seem to be either necessary or sufficient for quantum computational speed-up. It may, in the end, be best just to say that the power of quantum computation is a result of a fusion of every aspect of quantum theory, different elements of quantum computation relying on different aspects of quantum theory [18].

While on the question of the centrality of entanglement in quantum information theory, we will remember, of course, that one of the central methods of quantum cryptography, BB84, of course, at least in its basic form [19], does not rely at all on entanglement. Indeed it requires only the most elementary aspects of quantum theory, well-known in fact from the 1920s.

In fact when we analyse the extent to which John Bell's work led to quantum information theory, we have two big difficulties. First, as we have seen, it is difficult or impossible to understand what the essential features of quantum information theory actually are. But secondly, over and above the actual mathematical logic of Bell's inequalities, there are very different understandings of what the conclusions of the experiments will be, once all loopholes are removed. Some, perhaps most, believe they show that local realism is impossible [20]. Some believe that non-locality has actually been demonstrated [21, 22]. Still others just say they show that Copenhagen was right all along [23].

However first we will try to discover Bell's actual views on the idea of information. To what extent might he have found palatable the views of those, such as Brukner and Zeilinger [24], Vedral [25], Wheeler [26], Lloyd [27], and many others, who stress the primacy of information? While those mentioned have a wide range of views, they might be summed up by saying that they regard information as the fundamental quantity in the universe, and that using this idea to derive the basic laws of physics makes clear the reason for their quantum nature.

At first sight it is unlikely that Bell would have given much support to this set of ideas, as it seems that his only comment on information comes in his famous or notorious paper *Against "measurement"*, originally presented at the Erice conference *62 years of uncertainty* in August 1989, and published in *Physics World* in August 1990 [28], sadly at about the time of Bell's sudden death.

This paper, of course, was a diatribe against a list of words that Bell claimed were used illegitimately to 'explain' the results of quantum mechanical experiments. Most of the words were criticised either for implying an artificial division of the world between measuring and measured systems, with the intention of being cavalier about interactions across the divide, or alternatively for defying precise definition. Of course 'measurement' was, in his view, the 'worst', and most of the rest of the paper is used to savage various attempts to use this word to explain how the results of experiments are produced, those of Dirac [29], Landau and Lifshitz [30], Gottfried [31] and van Kampen [32].

All of the words except one—'system', 'apparatus', 'environment', 'microscopic, 'macroscopic', 'reversible', 'irreversible', 'observable' and 'measurement' itself clearly relate to Bell's bugbear about conventional approaches to quantum theory—the 'shifty split' between apparatus and system. Bell had been suspicious of this division from his earliest days as a student of quantum theory, and had been actively hostile to it for many years, so it was scarcely surprising that all these words appeared on his hit list [33, 34].

Much more surprising was the inclusion of the word 'information'. The only comment is: 'Information? *Whose* information? Information about *what*?' The inclusion of this word is indeed surprising because there is little other evidence that Bell had been particularly concerned with the use or misuse of this word, or its (possible) synonym, 'knowledge'.

There is no doubt that these words had often been used in a way that was quite capable of annoying Bell. They were often used, by quite a variety of writers, to provide what seemed to be an easy 'explanation' of the conceptual problems of quantum theory, but which in fact explained nothing and avoided all the real problems.

Perhaps the simplest misunderstanding is just to assume that *all* a measurement or an experiment does is to provide information about a property of an observed system that we may regard as existing before, during and after the measurement process. There is every temptation to regard the actual system just as conventionally or classically as we wish, with all observables having precise values at all times—in a sense the measured system is not itself really part of the quantum world, which is *just* the information.

That may perhaps apply just when we gather information for the first time. Collapse of wave-function occurs when we have some prior knowledge but perform an observation or experiment to bring our knowledge up to date. Naturally our information changes. So in this approach collapse appears to be altogether a straightforward process, merely representing the alterations in our brain when we take in factual information. Collapse need not imply any change of any type in the observed system.

Of course there may be complications. It may be that we initially know the value of s_x and come to know the value of s_y. We must then recognise that our knowledge of s_x is defunct. But a knowledge or information interpretation certainly does not necessitate, but may often encourage, the belief that all properties of a system (such as s_x and s_y) *have* precise values at all times even though our information about these values is necessarily limited.

Bell may well have seen such interpretations claiming all the conceptual gains of hidden variables without accepting any of the accompanying difficulties of such theories, the labour in their creation, or, of course, the struggle in analysing and making sense of their own properties.

While it has been said that Bell had no particular target in mind, it may be suggested that his criticism was rather obliquely on the practically universal belief that quantum theory is about the results of measurements, rather than what actually 'exists', and just a little less obliquely on the well-known views of John von Neumann and Eugene Wigner.

For von Neumann, [35] final collapse was in the 'abstract ego' or perhaps mind of the observer, not too far from talking in terms of knowledge. Bell was, of course, a great admirer of Wigner, but he did make Wigner's idea of consciousness [36] performing the required collapse of wave-function one of his three 'romantic', and hence in Bell's view 'bad' worlds of quantum mechanics. (The kind of interpretation that brought in the same type of stochastic terms mathematically and so in what Bell took to be a more professional or 'unromantic' way, and was thus a 'good' world became exemplified in GRW [37].) Thus Wigner may also have been in his sights in this paper.

Another suggestion is that of a paper of Cavalcanti and Wiseman [38], that studies two of Bell's papers in which he analyses his theorem in a little more depth than when it was originally presented. These are 'The theory of local beables' [39] from 1976, and 'La novelle cuisine' [40], published after his death in 1990.

In these papers, having reached the general existence of non-locality, Bell asks whether this implies that 'we' can signal faster than light. He produces an argument showing that this is not possible, and his argument in itself is not much different from that produced by others. He divides his 'beables' into two classes, 'controllables' and 'uncontrollables'. Controllables may send or receive signals, but uncontrollables may only receive them. What he calls an 'exercise' in quantum mechanics shows that a change in a controllable variable cannot result in a change in a spacelike separated region.

However his words show a lot less enthusiasm for the analysis and the very idea of signal locality. To give a proper answer to the question, he says, or in other words to discuss signal locality, actually requires at least a schematic theory of 'what "we" can do', or in other words 'a fragment of a theory of human beings'. Clearly he is unhappy about the use of such anthropocentric ideas as 'controllability' and, in the background, 'information'.

In the later paper he questions whether 'no signalling faster than light' can be an expression of the fundamental causal structure of theoretical physics, but he rejects the idea. 'No signalling', he says, should really be expressed as 'We cannot signal faster than light', which, he says, immediately provokes the question: 'Who do we think "we" are?'

Do 'we' include chemists or just physicists?
Plants or just animals?
Calculators or just computers?

The 'we' who can 'signal', he says are the same 'we' who can manipulate 'external fields', and, in particular, the same 'we' who take 'measurements'.

So in this paper, written at about the same time as 'Against "measurement"', we do seem to have reached the closest connection between 'information' and the other words on Bell's banned list.

Of Bell's questions, the first—'Whose information?' may have genuinely been a request for a coherent answer, one which was actually to be supplied over the next years. Probably more likely is that it was intended to be pointing out that Bell considered to be an obvious inconsistency of the idea—surely different people must have different amounts of information. As we shall see, this was not necessarily a defect of the theory.

The other question—'Knowledge about what?'—perhaps brought out Bell's main frustration. Information, he assumed, must be about something, in which case, why not discuss what it is actually about? In other words return the discussion to atoms, molecules, electrons and discuss, for example, how they behave at a measurement, and which if their properties may have values simultaneously. Bell may have felt that talk of information or knowledge may not have actually been wrong, but rather unhelpful; it may have failed to distinguish between things that we are prohibited from knowing by the laws of quantum theory, and those that we could know but have not bothered to find out!

In terms of the later development of quantum information theory it may be remarked that there are perhaps two different definitions of information being used. What we may call information^1 is by definition telling you about something—it has some content. Thus a parent might see their child's scribble—it has no meaning, no information. When the parent is told that it is in fact a picture of him or her, it immediately becomes information^1. Bell, it must be assumed, was thinking of information^1. Information as in information theory, classical or quantum, is information^2.

Bell's paper was regarded as a polemic, and it was not surprising that quite a few replies were sent to *Physics World*, among them letters from Gottfried [41] and van Kampen [42] defending their arguments, and from Squires [43] supporting Bell.

The most interesting was from Peierls [44], which managed to include, in a totally non-contrived manner, both of Bell's targets, measurement *and* information. Peierls was a great admirer of Bell. He had given Bell his first chance to enter genuinely mainstream physics by advocating his move at Harwell to a division devoted to tackling such problems as quantum field theory and particle physics [34], and, for the rest of Bell's career, Peierls had probably been split between admiration of his mathematical ability and his honesty, and horror at his constant attacks on the conventional interpretations of quantum theory [45]. A conspiracy theorist might think that Bell had deliberately included the attack on 'information' to allow Peierls to give full rein to his beliefs.

Peierls regarded himself as a complete believer in Bohr's views, to the extent that he rejected the term 'Copenhagen interpretation'. For him, using this term implied that there were several interpretations of quantum theory, of which Copenhagen was just one. For Peierls, there was only one interpretation, so if you say 'Copenhagen interpretation' you really mean 'quantum mechanics' [46].

Yet Peierls' reply to Bell, titled 'In defence of "measurement"', seemed to be very different from what would come from the mouth of Bohr or Heisenberg, being, as stated above, in terms of knowledge, indeed very much along the lines sketched above in a rather cavalier way but with care taken to avoid the obvious problems. In our previous terms, Peierls' knowledge was presumably still information^1.

If our knowledge is complete, by which Peierls meant the greatest that could be allowed by the laws of quantum theory, in particular the uncertainty principle, we may represent this knowledge with a wave-function. However for less knowledge we must use a density-matrix. Uncontrolled disturbances may reduce our knowledge. Measurement may increase it. If we start with the wave-function case and gain new information, some of the previous information must be lost, and so on.

When there is a change of knowledge, Peierls says that the density-matrix must obviously change, but this is not a physical process so we should certainly not expect the change to follow the Schrödinger equation. This argument does indeed seem an excellent way of giving some explanation of von Neumann's poser of the contrast between type 2 processes, processes outside of measurement, which follow the Schrödinger equation, and type 1 processes, measurements, which follow completely different dynamics such as collapse.

On the first of Bell's questions—whose information?—rather than regarding the question as a means of ridiculing the whole idea, Peierls produced a intelligent and convincing answer. His basic point is that the knowledge of different observers must not contradict each other. Contradiction would occur if one observer 'knew' that s_z was +½, while another 'knew' that it was −½, but it would also occur if one 'knew' that s_z was +½ while another 'knew' that s_y was +½. Mermin [47] says that Peierls uses two conditions. A strong one is that the density-matrices of the two observers must commute, while a weaker one is just that the product of the two density-matrices must not be zero.

Mermin also points out that Peierls does not answer Bell's other question—information about what? He may have felt that the whole point of the Bohr approach was that one need not, could not and should not answer it. It might be surmised, though, that if an observer *knows* that s_z is +½, then even an orthodox interpretation would admit that there is a system with s_z equal to +½. It would probably prefer, though, not to comment on the values of s_y or s_x.

Mermin reports that he was initially on Bell's side in his clash with Peierls, but his view was changed by sustained interaction with those involved in quantum computation, who were convinced that quantum theory was 'self-evidently and unproblematically' a theory of information. In our previous terms, this is, of course, information^2.

Like Peierls, Mermin was keen to answer the question—Whose information? By demonstrating a subtlety of entanglement, he was able to demonstrate a weakness in Peierls' strong condition and to suggest improvements, thus again demonstrating that, if Bell thought the question showed up the weakness of the whole position, he was definitely wide of the mark.

However Mermin rejected altogether Bell's other question—Information about what? He described this as a fundamentally metaphysical question and considered that it should not distract 'tough-minded physicists'. It is not possible, he says, to discover whether information is about something objective or just about other information, and one certainly should not waste time trying to do so. Of course this is information^2.

And of course once one recognises the primacy of information, the whole argument from the quantum to information may be reversed. Rather one argues from information to the physical Universe. For Brukner and Zeilinger [24], the obvious quantisation of information is the cause of the quantisation we see in the Universe. For Vedral [25], information is the only entity on which we may base our most fundamental theories; for example evolution is *purely* the inheritance of information with occasional changes of the basic units, the genes. For Wheeler [26] too, the concept of information may unlock some of the most basic mysteries of the Universe. For Lloyd [27], the Universe is just a quantum computer, and what it computes is just its own behaviour. Smolin [48] believed that quantum information is a possible alternative to string theory as an attempt to solve the most basic problems of physics. And so on.

Now let us return to Bell. It must seem bizarre to allocate credit to him for this development, when his only contribution consisted of seven words, of which four were the same—'information', and also when he seemed to end up on the wrong side of the argument. Yet it may also be said that Bell did not believe in wasting words. He took an issue that had maybe been under the radar for half a century, challenged some basic assumptions, and asked precisely the telling questions, the questions that would take others so far, even if they took them to regions which would have surprised him.

I now want to go back to the beginning of quantum computation and the work of Richard Feynman. If you take the founder of quantum computation as David Deutsch, as I have done, it is natural, following Brown [49], to think of Feynman's earlier work as rather analogous to 'the old quantum theory', the period between 1900 and 1925 when many important results were obtained but without any rigorous foundation to the work.

Feynman published two important papers, the first, a conference paper from 1981, published in 1982 under the title 'Simulating physics with computers' [50], and the second, 'Quantum mechanical computers' [51], published in 1985.

In the first he asks a number of important questions about simulations. In each case the simulation must be exact, and the computational work required must increase only in proportion to the size of the system being studied, not exponentially.

First he shows that a classical system may simulate a classical system, but it cannot simulate a quantum system. He then asks if it can simulate the quantum system probabilistically. The answer is again—no, but in this case Feynman has to provide a detailed argument. He examines in some detail an EPR system, and calculates the probabilities all the way through. Everything works out well—with the exception that some probabilities used in the course of the analysis, not in the answers, have to be negative. Feynman now demonstrates why negative probabilities cannot be avoided, and he effectively works through a proof of Bell's theorem.

Tony Hey, who has edited both Feynman's own work in this area (with Robin Allen) [52], and also more recent papers by his collaborators [53], comments [53] that 'Only Feynman could discuss "hidden variables", the Einstein-Podolsky-Rosen paradox and produce a proof of Bell's Theorem, without mentioning John Bell'!

Hey appears to assume that Feynman had encountered Bell's work, but had perhaps forgotten the name of the originator and worked through the analysis himself.

It must be said that Feynman did have form for this kind of sloppiness. Most significantly, in 1957 George Sudarshan, as a research student being supervised by Robert Marshak, had demonstrated in great detail that the structure of the weak interaction was of a V-A type (vector minus axial vector). However Feynman and Murray Gell-Mann were privately informed of this result and thought about the matter themselves, but forgetting their source of information, published the result in their own names and gained priority over Sudarshan and Marshak. Feynman did write later that: 'The V-A theory was discovered by Sudarshan and Marshak, and published by Feynman and Gell-Mann.' [54–56].

But it is, of course, quite possible that Feynman did produce the Bell-type analysis independently of Bell. Gottfried and Mermin [57] do say that the actual analysis is 'extraordinarily elementary'. Presumably it is having the motivation to think about the matters that is requires a special intellect [57, 58], and Bell and Feynman would both come into that category.

It may be stressed then that for Feynman it is Bell's theorem that makes going beyond classical computers inevitable, and he discusses very briefly the possibility of quantum computers or 'universal quantum simulators'.

However more interestingly he remarks that he often has fun trying to squeeze the difficulty of quantum mechanics into a smaller and smaller place—to isolate the essential difficulty, so as to give the possibility of analysing it is detail. He feels that he has located it in the contrast between two numbers—one required by quantum theory, the other the demand of classical theory—a direct result of a Bell type of analysis.

Thus the significance of Bell's work, according to Feynman, can scarcely be over-exaggerated. It is the core element of quantum theory!

Let us briefly turn to Quantum Key Distribution (QKD). It is well-known that Nicolas Gisin, Antonio Acín and co-workers have produced much detailed analysis of the part that the Bell Inequalities play in many aspects and many variants of QKD.

Here I just want to pick up on one important point that they make [59]. It would perhaps normally be said that QKD relies on the fact that, if Alice and Bob are sufficiently entangled, then Eve is effectively factorised out. Yet it could be the case that Alice and Bob share a space of higher dimension that is separable, and thus the method for QKD becomes insecure. This is easily shown to be the case for BB84.

Usual proofs of the security of QKD rely on entanglement theory, and thus they assume that Alice and Bob share knowledge of the fixed dimension of the Hilbert space of their system. If this assumption, which is in fact rather arbitrary, cannot be made, QKD must involve the violation of some Bell inequality. Thus yet again we see the crucial role played by Bell's theorem.

We are seeing that, while quantum theory is obviously different from classical theory in many ways, in many cases at least the core of the difference or the essential discrimination is just Bell's inequality or a result of Bell's inequality.

We may think of Holland's comment [60] in his book on Bohmian mechanics. He was answering a complaint that Bohm theory was trying to restore classicality to quantum theory. His reply was that Bohm theory fully recognised the great differences between quantum and classical theory. In fact, he said, it gave a possibility of discussing classical and quantum mechanics in the same language, but not of reducing one to the other. But it might be said that the task could not be a total success, precisely because of non-locality, or in other words yet again a result of Bell's theorem.

We might well contemplate adjusting Schrödinger's comment mentioned at the beginning of this discussion to obtain the rather striking statement that it is not (just) entanglement but actually Bell's theorem and its implications that are 'not one but the characteristic trait of quantum mechanics, the one that enforces the entire departure from classical thought.'

References

1. R.A. Bertlmann, A. Zeilinger (eds.), *Quantum [Un]speakables: From Bell to Quantum Information* (Springer, Berlin, 2002)
2. A. Whitaker, *The New Quantum Age: From Bell's Theorem to Quantum Computation and Teleportation* (Oxford University Press, Oxford, 2012)
3. N.D. Mermin, *Quantum Computer Science: An Introduction* (Cambridge University Press, Cambridge, 2007)
4. D. Bouwmeester, J.-W. Pan, K. Mattle, M. Eibl, H. Weinfurter and A. Zeilinger, experimental quantum teleportation. Nature **390**, 575 (1997)
5. D. Bouwmeester, J.-W. Pan, M. Daniell, H. Weinfurter, A. Zeilinger, Observation of three-photon Greenberger-Horne-Zeilinger entanglement. Phys. Rev. Lett. **82**, 1345 (1999)
6. E. Schrödinger, Die gegenwärtige Situation in der Quantenmechanik. Naturwissenschaften **23**, 807, 823, 844 (1935); translated by J.D. Trimmer as The present situation in quantum mechanics, in *Quantum Theory and Measurement*, ed. by J.A. Wheeler, W.H. Zurek (Princeton University Press, Princeton, 1983), p. 152
7. J.S. Bell, On the problem of hidden variables in quantum mechanics. Rev. Mod. Phys. **38**, 447 (1966)

8. J.S. Bell, On the Einstein-Podolsky-Rosen paradox. Physics **1**, 195 (1964)
9. D. Deutsch, Quantum theory, the Church-Turing principle and the universal quantum computer. Proc. R. Soc. A **400**, 97 (1985)
10. D. Deutsch, *The Fabric of Reality* (Allen Lane, London, 1997)
11. D. Deutsch, Quantum theory as a universal physical theory. Int. J. Theor. Phys. **24**, 1 (1985)
12. A.M. Steane, A quantum computer only needs one universe. Stud. History Philos. Modern Phys. **34**, 469 (2003)
13. R. Jozsa, N. Linden, On the role of entanglement in quantum computational speed-up. Proc. R. Soc. A **459**, 2011 (2003)
14. B.P. Lanyon, M. Barbieri, M.P. Almeida, A.G. White, Experimental quantum computing without entanglement. Phys. Rev. Lett. **101**, 20051 (2008)
15. M. Van den Nest, Universal quantum computation with little entanglement. Phys. Rev. Lett. **110**, 060504 (2013)
16. S. Kochen, E.P. Specker, The problem of hidden variables in quantum mechanics. J. Math. and Mech. **17**, 59 (1967)
17. M. Howard, J. Wallman, V. Veitch, J. Emerson, Contextuality supplies the 'magic' for a quantum computation. Nature **510**, 351 (2014)
18. P. Ball, Questioning quantum speed. Phys. World **27**(1), 38 (2014)
19. C.H. Bennett and G. Brassard, Quantum cryptography: public-key distribution and coin tossing, in *Proceedings of 1984 IEEE International Conference on Computers, Systems and Signal Processing* (IEEE, New York, 1984), pp. 175–179
20. T. Scheidl, R. Ursin, J. Kofler, S. Ramelow, X.-S. Ma, T. Herbst, L. Ratschbacher, A. Fedrizzi, N.K. Langford, T. Jennewein, A. Zeilinger, Violation of local realism with freedom of choice. Proc. Natl. Acad. Sci. **107**, 19708 (2010)
21. T. Norsen, Against 'realism'. Found. Phys. **37**, 311 (2007)
22. N. Gisin, Non-realism: deep thought or a soft option? Found. Phys. **42**, 80 (2012)
23. J. Hebor, *The Standard Conception as Genuine Quantum Realism* (University Press of Southern Denmark, Odense, 2005)
24. Č. Brukner and A. Zeilinger, Quantum physics as physics of information, in *Quo Vadis Quantum Mechanics?*, ed. by A. Elitzur, S. Dolev, N. Kolenda (Berlin, Springer, 2005)
25. V. Vedral, *Decoding Reality: The Universe as Quantum Information* (Oxford University Press, Oxford, 2010)
26. J.A. Wheeler, *Geons, Black Holes and Quantum Foam: A Life in Physics* (Norton, New York, 1998)
27. S. Lloyd, *Programming the Universe: A Quantum Computer Scientist Takes on the Cosmos* (Knopf, New York, 2006)
28. J.S. Bell, Against 'measurement'. Phys. World **3**(8), 33 (1990)
29. P.A.M. Dirac, *Quantum Mechanics*, 4th edn. (Oxford University Press, Oxford, 1948)
30. L.D. Landau, E.M. Lifshitz, *Quantum Mechanics*, 3rd edn. (Pergamon, Oxford, 1977)
31. K. Gottfried, *Quantum Mechanics* (Benjamin, New York, 1966)
32. N.G. van Kampen, Ten theorems about quantum mechanical measurements. Phys. A **153**, 97 (1988)
33. J. Bernstein, *Quantum Profiles* (Princeton University Press, Princeton, 1991)
34. A. Whitaker, *John Stewart Bell and Twentieth-Century Physics: Vision and Integrity* (Oxford University Press, Oxford, 2016)
35. J. von Neumann, *Mathematical Foundations of Quantum Mechanics* (Princeton University Press, Princeton, 1955)
36. E. Wigner, Remarks on the mind-body problem, in *The Scientist Speculates*, ed. by I.J. Good (Heinemann, London, 1961), pp. 284–302
37. G.-C. Ghirardi, A. Rimini, T. Weber, Unified dynamics for microscopic and macroscopic systems. Phys. Rev. D **34**, 470 (1986)
38. E.G. Cavalcanti, H.M. Wiseman, Bell nonlocality, signal locality and unpredictability (or what Bohr could have told Einstein at Solvay had he known about Bell experiments). Found. Phys. **42**, 1329 (2012)

39. J.S. Bell, The theory of local beables. Epistemol. Lett. **9**(1976); Dialectica **39**, 86 (1985)
40. J.S. Bell, La novella cuisine, in *Between Science and Technology*, ed. by A. Sarlemijn, P. Kroes (North-Holland, Amsterdam, 1990), pp. 97–115
41. K. Gottfried, Does quantum mechanics carry the seeds of its own destruction? Phys. World **4** (10), 34 (1991)
42. N.G. van Kampen, Quantum criticism. Phys. World **3**(10), 20 (1990); and Mystery of quantum measurement. Phys. World **4**(12), 16 (1991)
43. E.J. Squires, Quantum challenge. Phys. World **5**(1), 18 (1992)
44. R. Peierls, In defence of 'measurement'. Phys. World **4**(1), 19 (1991)
45. R. Peierls, Bell's early work. Europhys. News **22**, 65 (1991)
46. P.C.W. Davies, J.R. Brown, *The Ghost in the Atom* (Cambridge University Press, Cambridge, 1986); the interview with Peierls is on pp. 70–82
47. N.D. Mermin, Whose knowledge?, in Ref. 1, pp. 271–280
48. L. Smolin, *The Trouble with Physics: The Rise of String Theory, the Fall of a Science and What Comes Next* (Penguin, London, 2008)
49. J.R. Brown, *Quest for the Quantum Computer* (Simon and Schuster, Riverside, New York, 2001); hardback: *Minds, Machines and the Multiverse: The Quest for the Quantum Computer* (2000)
50. R.P. Feynman, Simulating physics with computers. Int. J. Theor. Phys. **21**, 467 (1982); also in Ref. 53, pp. 133–53
51. R.P. Feynman, Quantum mechanical computers. Opt. News **11**(2), 11 (1985); also in Ref. 52, pp. 185–211
52. R.P. Feynman, *Feynman Lectures on Computation*, ed. A.J.G. Hey and R.W. Allen (Addison-Wesley, Reading, Mass, 1996)
53. A.J.P. Hey (ed.), *Feynman and Computation* (Perseus, Reading, Mass, 1999)
54. J. Gleick, *Genius: The Life and Science of Richard Feynman* (Vintage, New York, 1992)
55. J. Mehra, *The Beat of a Different Drum: The Life and Science of Richard Feynman* (Clarendon Oxford, 1994)
56. S. Glashow, Message for the Sudarshan symposium, Sudarshan: seven science quests 2006. J. Phys: Conf. Ser. **196**, 011003 (2009)
57. K. Gottfried, N.D. Mermin, John Bell and the moral aspect of quantum mechanics. Europhys. News **22**, 67 (1991)
58. R. Jackiw, A. Shimony, The depth and breadth of John Bell's physics. Phys. Perspect. **4**, 78 (2002)
59. A. Acin, N. Gisin, L. Masanes, From Bell's theorem to secure quantum key distribution. Phys. Rev. Lett. **97**, 120405 (2006)
60. P. Holland, *The Quantum Theory of Motion* (Cambridge University Press, Cambridge, 1993)

Chapter 3
Bell's Universe: A Personal Recollection

Reinhold Bertlmann

Abstract My collaboration and friendship with John Bell is recollected. I will explain his outstanding contributions in particle physics, in accelerator physics, and his joint work with Mary Bell. Mary's work in accelerator physics is also summarized. I recall our quantum debates, mention some personal reminiscences, and give my personal view on Bell's fundamental work on quantum theory, in particular, on the concept of contextuality and nonlocality of quantum physics. Finally, I describe the huge influence Bell had on my own work, in particular on entanglement and Bell inequalities in particle physics and their experimental verification, and on mathematical physics, where some geometric aspects of the quantum states are illustrated.

Keywords Bell inequalities · Nonlocality · Contextuality · Entanglement · Factorization algebra · Geometry

PACS number: 03.65.Ud · 03.65.Aa · 02.10.Yn · 03.67.Mn

Collaboration with John Bell

In April 1978 I moved from Vienna to Geneva to start with my Austrian Fellowship at CERN's Theory Division. Already in one of the first weeks, after one of the Theoretical Seminars, when all newcomers had a welcome tea in the Common Room, I got acquainted with John Stewart Bell. I remember when he approached me straightaway, *"I'm John Bell, who are you?"* I answered a bit shy, *"I am Reinhold Bertlmann from Vienna, Austria"*. *"What are you working on?"* was his next question and I replied *"Quarkonium..."*, which was already a *magic* word since we immediately fell into a lively discussion about bound states of quark-antiquark

---

Dedicated to Mary Bell, John's lifelong companion.

R. Bertlmann (✉)
Faculty of Physics, University of Vienna, 1090 Vienna, Austria
e-mail: Reinhold.Bertlmann@univie.ac.at

R. Bertlmann and A. Zeilinger (eds.), *Quantum [Un]Speakables II*,
The Frontiers Collection, DOI 10.1007/978-3-319-38987-5_3

Fig. 3.1 First encounter with John Bell at CERN in April 1978. Cartoon: ©Reinhold A. Bertlmann.

systems, a very popular subject at that time, which continued in front of the blackboard in his office. A fruitful collaboration and warm friendship began (Fig. 3.1).

The first problem we attacked was how to understand the production of hadrons (strong interacting particles) in e^+e^- collisions. The experiments showed the following feature: At low energies there occurred pumps, the resonances, in the hadronic cross-section whereas at high energies the cross-section became quite flat or asymptotically smooth. Hadrons consist of quarks and antiquarks thus the e^+e^- collisions actually produce quark-antiquark ($q\bar{q}$) pairs. The idea was that at high energies, which corresponds to short distances, the $q\bar{q}$ pairs behave as quasi-free particles providing such the flat cross-section. However, at low energies, where the quarks can penetrate into larger distances they are confined and generate bound states—called quarkonium—which show up as resonances.

Our starting point was the idea of *duality* that stated that smearing each resonance in energy already appropriately matches the corresponding result of the asymptotic cross-section determined by the short-distance interaction [1, 2], an idea that can be traced back to a work of Sakurai [3]. Theoretically, it can be understood in the following way. Allowing for an energy spread means—via the uncertainty relation—that we focus on short times. But for short times the corresponding wave does not spread far enough to feel the details of the long distances, the confining potential. So this part can be neglected and the wave function at the origin of the bound state, which determines the leptonic width or area of a resonance, matches the averaged quasi-free $q\bar{q}$ pair. However, if we want to push the idea of duality even further in order to become sensitive to the position of the bound state in the mass spectrum then we have to include into the wave function the contributions of larger distances, confinement.

How could we include confinement in our duality concept? Our starting point was the vacuum polarization tensor $\Pi(q^2)$, the vacuum expectation value of the time ordered product of two quark currents in quantum field theory, quantum chromo dynamics

$$i\int dx\, e^{iqx}\langle\Omega|T\, j_\mu(x) j_\nu(0)|\Omega\rangle \;=\; \Pi(q^2)(q_\mu q_\nu \,-\, q^2 g_{\mu\nu})\,. \tag{1}$$

This quantity was proportional to the hadronic cross-section, where at low energy the bound states, the resonances, occurred. More precisely, the imaginary part of the vacuum polarization function (the forward scattering amplitude) was related to the total cross-section via the optical theorem

$$\mathrm{Im}\,\Pi(E) \;\sim\; \sigma_{\mathrm{tot}}(E)\,, \tag{2}$$

and calculable within perturbation theory with help of Feynman diagrams, the loop diagrams depicted in Fig. 3.2. At that time the Russian group, Shifman-Vainshtein-Zakharov (SVZ) [4], claimed that the so-called gluon condensate $\langle\frac{\alpha_s}{\pi}GG\rangle$, the vacuum expectation value of two gluon fields, would be responsible for the

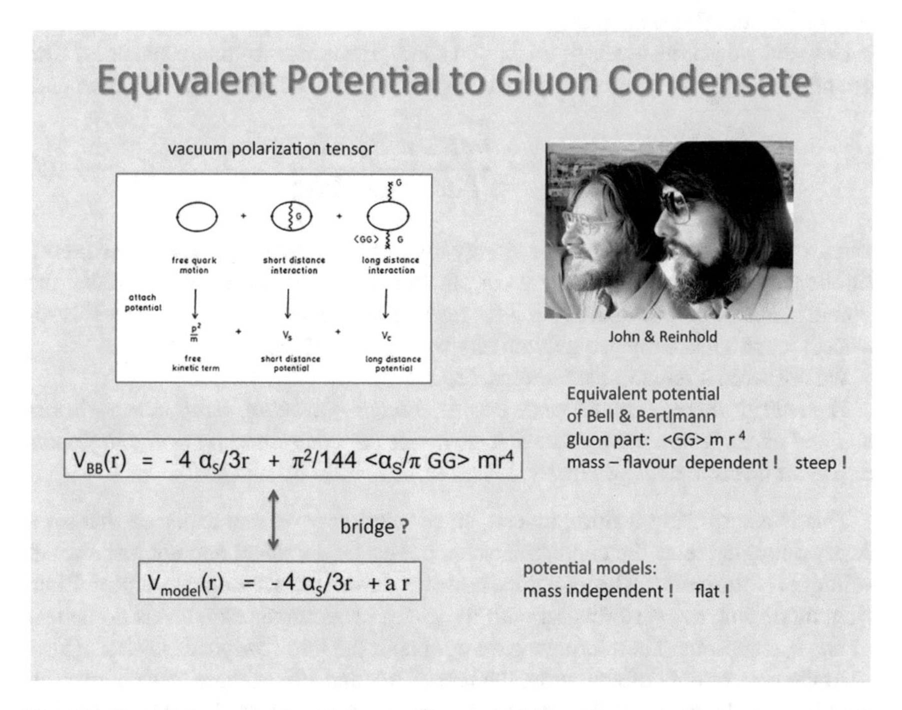

Fig. 3.2 Perturbative expansion (Feynman diagrams) of the vacuum polarization tensor in QCD including the gluon condensate, and the *equivalent potential of Bell and Bertlmann* to this expansion. Foto: © Renate Bertlmann.

influence of confinement (represented by the third loop diagram in Fig. 3.2). This idea we wanted to examine further.

Approximating quantum field theory by potential theory, we could calculate both the perturbative *and* the exact expression. For the energy smearing we had chosen an exponential as weight function, which was called a *moment* by the mathematicians

$$M(\tau) \;=\; \int dE \; e^{-E\tau} \; \mathrm{Im}\,\Pi(E) \;=\; \frac{3}{8m^2} \, \langle \vec{x} = 0 | e^{-H\tau} | \vec{x} = 0 \rangle \; . \tag{3}$$

In this case, where we had rediscovered the τ dependent Green's function at $\vec{x} = 0$, the procedure corresponded to perturbation theory of a Hamiltonian H with respect to an imaginary time τ. This I found quite fascinating. The actual calculation provided the following result

$$M(\tau) \;=\; \frac{3}{8m^2} \, 4\pi \, (\frac{m}{4\pi\tau})^{\frac{3}{2}} \left\{ 1 \,+\, \frac{4}{3}\alpha_s \sqrt{\pi m} \; \tau^{\frac{1}{2}} \;-\; \frac{4\pi^2}{288m} \langle \frac{\alpha_s}{\pi} GG \rangle \; \tau^3 \right\} . \tag{4}$$

The leading term corresponded to the free motion of the quarks (first diagram in Fig. 3.2); it was perturbed by the α_s term, representing the short distance interaction (second diagram), and by the gluon condensate $\langle \frac{\alpha_s}{\pi} GG \rangle$ term, responsible for the longer distances (third diagram).

How did we get the levels of the bound states, the masses of the resonances? The ground state level we could extract by using the logarithmic derivative of a moment,

$$R(\tau) \;=\; -\frac{d}{d\tau} \, \log M(\tau) \;=\; \frac{\int dE \; E \; e^{-E\tau} \, \mathrm{Im}\,\Pi(E)}{\int dE \; e^{-E\tau} \, \mathrm{Im}\,\Pi(E)} \quad \overset{\tau \longrightarrow \infty}{\longrightarrow} \quad E_1 \; , \tag{5}$$

which approached the ground state energy for large (imaginary) times since the contributions of the higher levels were cut off. In this way we were able to predict the ground states of charmonium (the J/ψ resonances) and of bottonium (the Υ resonances) to a high accuracy, quantitatively within 10 % [5–8].

We observed a remarkable *balance*, see Fig. 3.3:

The energy average can be made coarse enough—involving small times—for the modified perturbation theory to work, while on the other hand fine enough for the individual levels to emerge clearly.

This is a surprising feature, indeed, since intuitively we had expected that for a clearly emerging level the confinement force must be dominant and not just a small additional perturbation. The moments, however, forced us to re-educate our intuition, when modifying the perturbation with the gluon condensate term, levels do appear for *magical* reasons. Therefore we gave our paper the title *"Magic Moments"* [5].

At the time of our collaboration, the late 1970 s and 1980s, there was no internet, no quick email exchange. The way we interacted, when I was absent from CERN, was via letters written by hand. Of course, this communication took some time, the writing itself, the search for stamps, the walk to the post office, etc. In retrospect these

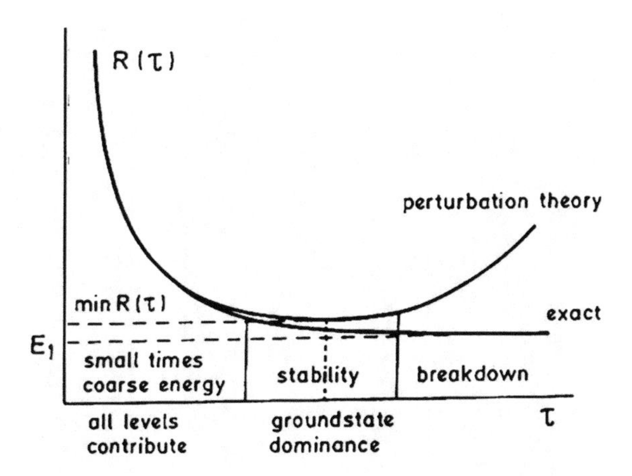

Fig. 3.3 The ratio of moments $R(\tau)$ is depicted qualitatively. There occurs a balance between the short and long distance contributions in the ratio of moments. The stability of the expansion determined by the Feynman diagrams of Fig. 3.2 approximates quite accurately the exact ground state level.

letters were beautiful documents expressing not only our scientific thoughts but also our personal attitudes, our characters, how we had investigated a physical subject and how we had presented our scientific work, what to include and what to leave out. As a typical example, I would like to show a letter of John (see Fig. 3.4), which he wrote to me in November 1983 [9], while I was staying in Vienna. The letter was written in connection with the preparation of our paper on the *"SVZ moments for charmonium and potential model"* [8], which we were going to publish. It shows quite nicely our struggle for an accurate and clear presentation.

Since John and I were working within potential theory, which functioned remarkably well [8], it was quite natural for us to ask whether one can attach a potential to the occurrence of the gluon condensate. Indeed, we found ways to do this [10, 11].

One way was to work within the moments, which regularize the divergence of the long-distance part of the gluon propagator, the gluon condensate contribution. In this case a static, nonrelativistic potential containing the gluon condensate can be extracted, which is called in the literature the *equivalent potential of Bell and Bertlmann* [10]

$$V_{\mathrm{BB}}(r) \;=\; -\frac{4}{3}\frac{\alpha_{\mathrm{s}}}{r} \;+\; \frac{\pi^2}{144}\langle\frac{\alpha_{\mathrm{s}}}{\pi}GG\rangle\, m\, r^4\;, \tag{6}$$

where α_{s} is the QCD coupling constant, m the quark mass and $\langle\frac{\alpha_{\mathrm{s}}}{\pi}GG\rangle$ the gluon condensate.

The short-distance part was the well-known Coulomb potential, whereas the long-distance component, the gluon condensate contribution, emerged as a quartic potential $m\,r^4$ and is mass-, i.e., flavour-dependent. In this respect it differed considerably from the familiar mass-independent, rather flat potential models [12–15]. However, for a final comparison with potential models one has to go further and take into account the higher order fluctuations [16].

I very well remember one of our afternoon rituals in our collaboration. John, a true Irishman, always had to drink a *4 o'clock tea*, and this we regularly practiced

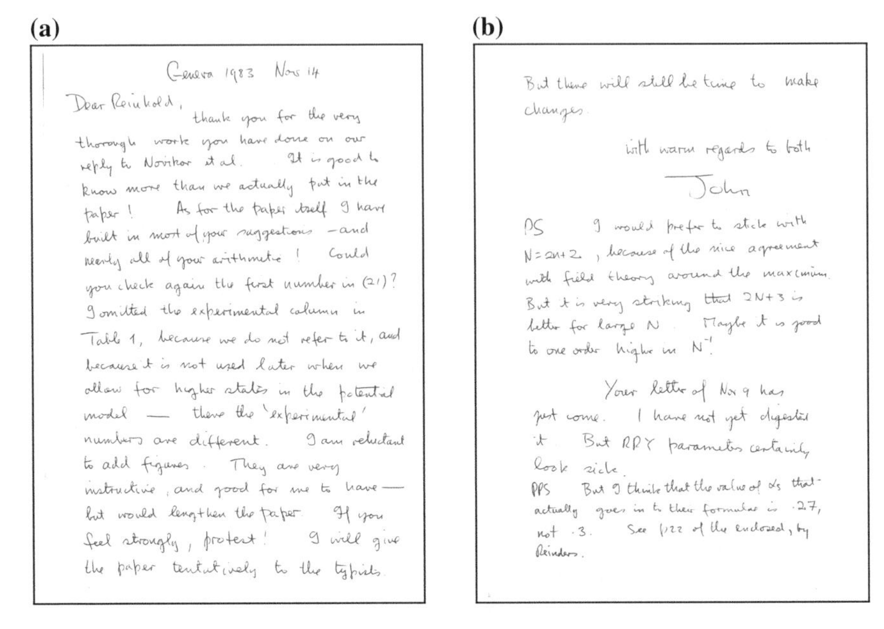

(a)

Geneva 1983 Nov 14

Dear Reinhold,
thank you for the very thorough work you have done on our reply to Novikov et al. It is good to know more than we actually put in the paper! As for the paper itself I have built in most of your suggestions — and nearly all of your arithmetic! Could you check again the first number in (21)? I omitted the experimental column in Table 1, because we do not refer to it, and because it is not used later when we allow for higher states in the potential model — there the 'experimental' numbers are different. I am reluctant to add figures. They are very instructive, and good for me to have — but would lengthen the paper. If you feel strongly, protest! I will give the paper tentatively to the typists.

(b)

But there will still be time to make changes.

with warm regards to both

John

PS I would prefer to stick with $N = 2n + 2$, because of the nice agreement with field theory around the maximum. But it is very striking that $2N + 3$ is better for large N. Maybe it is good to one order higher in N^{-1}!

Your letter of Nov 9 has just come. I have not yet digested it. But RRY parameters certainly look sick.

PPS But I think that the value of α_s that actually goes in to their formulae is .27, not .3. See p22 of the enclosed, by Reinders.

Fig. 3.4 **a** Page 1 of the handwritten letter of John Bell to Reinhold Bertlmann from 1983 [9], in connection with the preparation of their paper on the *"SVZ moments for charmonium and potential model"* [8]. **b** Page 2 of the letter.

Fig. 3.5 Bertlmann (*left*) and Bell (*right*) choosing the *right* sort of tea at Bell's home in 1980. Foto: © Renate Bertlmann.

in the CERN Cafeteria or at John's home, when we were working there. Choosing the right sort of tea was quite a ceremony, see Fig. 3.5. Then, in this relaxed tea-atmosphere, we talked not only about physics but also about politics, philosophy, and when we were joined by my artist wife Renate, we three also had heated debates about modern art.

Next we studied very heavy quarkonium systems, which described e.g. bottonium (the Υ resonances). There John and I found another way to extract a potential from the gluon condensate effect [11]. In that case the low-lying bound states, because of their small size, were dominated by the Coulomb potential and the condensate effect, an external colour field representing the gluon, could be added as a small perturbation. Leutwyler [17] and Voloshin [18] had considered such a colour-electric Stark effect and calculated the energy spectrum for all quantum numbers n and l. John and myself, on the other hand, were able to construct a gluonic potential, which by perturbing the Coulomb states provided the energy spectrum of Leutwyler-Voloshin

$$V_{\text{gluon}}(r) = \frac{4\pi^2}{81\beta}\langle\frac{\alpha_s}{\pi}GG\rangle\left(r^3 - \frac{304}{81}\frac{r^2}{m\beta} + \frac{53}{10}\frac{r}{m^2\beta^2} - \frac{113}{100}\frac{1}{m^3\beta^3}\right) \quad \text{with} \quad \beta = \frac{4\alpha_s}{3}\,. \quad (7)$$

The leading term in potential (7), the infinite mass limit, had a cubic r^3 dependence and was therefore mass-independent. But for finite masses there were further corrections necessary proportional to $\frac{r^2}{m}, \frac{r}{m^2}, \frac{constant}{m^3}$ such that the potential became flavour-dependent again.

While our equivalent potential (6) was constructed to reproduce the gluon condensate shifts in the moments à la Shifman, Vainshtein and Zakharov [4], the potential (7) was designed to reproduce gluon condensate level shifts à la Leutwyler [17] and Voloshin [18] in hydrogen-like heavy $q\bar{q}$ systems. The two potentials differed because they were fitted to different aspects of quantum field theory—and potential theory is, of course, not field theory.

In conclusion, whereas the moment procedure including the gluon condensate worked surprisingly well for predicting the ground state levels of quark-antiquark bound states, no adequate bridge was found between a field theory containing the gluon condensate, quantum chromodynamics, on one side, and popular potential models on the other. For an overview of this field I would like to refer to Ref. [19].

John Bell—the Particle Physicist

Bell graduated with First-Class Honours in Experimental Physics in 1948 at Queen's University Belfast, where the senior staff members of the Physics Department were Karl Emeleus and Robert Sloane. He spent an additional year at the University and obtained a second degree in Mathematical Physics, where his teacher was the famous crystallographer Peter Paul Ewald. Subsequently, in 1949, he got a position the Atomic Energy Research Establishment (AERE) at Harwell, Oxfordshire, but was soon sent to the Telecommunications Research Establishment (TRE) at Malvern, Worcestershire. There he began to work in accelerator physics, see Sect. "John Bell—the Accelerator Physicist". In 1951, the accelerator group at Malvern moved to the Atomic Energy Research Establishment (AERE) at Harwell, Oxfordshire.

In 1953, one of the few persons who had got that privilege, John was selected to spend some time at a university to perform a PhD Thesis while keeping his usual AERE salary. He had eventually chosen the University of Birmingham, where Rudolf Peierls was the head of Theoretical Physics. There he could study quantum field theory having much contact with Paul T. Matthews. John succeeded already as a student to write a fundamental work, namely his PhD Thesis. It consisted of two parts: *"Time reversal in field theory"* and *"Some functional methods in field theory"*. In the first part, published in 1955 in the Proceedings of the Royal Society [20], he established the celebrated *CPT* Theorem (*C*, the charge conjugation operator, which replaces particles by anti-particles; *P*, the parity operator, which performs a reflection; and *T*, which carries out a time reversal). The theorem states that any Lorentz invariant, local quantum field theory with a Hermitian Hamiltonian must be *CPT* symmetric. It is commonly believed that Nature is *CPT* invariant, no *CPT* violating phenomena have been found so far. For a long time Lüders [21] and Pauli [22], who proved the theorem a little ahead of Bell, received all the credit but nowadays Bell's 'elementary' derivation, being more accessible than the formal mathematical arguments of Lüders and Pauli, is also rightly recognized.

After his PhD, about 1955, John turned to nuclear and particle physics, a growing field at that time. At AERE he closely worked with Tony Skyrme, the head of the Nuclear Physics Group, whose soliton model for nucleons, called skyrmions, became well-known later on. Together they investigated the *"Nuclear spin-orbit coupling"* [23], the *"Magnetic moments of nuclei and the nuclear many body problem"* [24, 25] and the *"Anomalous magnetic moments of the nucleons"* [26]. His mastery in field theory he also showed in a paper about a *"Variational method in field theory"* [27], where he found a new form of Skyrme's variational principle for the one-nucleon propagator.

John also contributed various papers to the many-body theory of nucleons [28–31] and together with Euan Squires he derived an effective one-body potential, the *"Formal optical model"* for the scattering of a particle incident on a complex target [32, 33].

An other one of John's colleagues at Harwell was Franz Mandl. They harmoniously worked together in particle physics, for instance, on the *"Polarization-asymmetry equality"* [34] for elastic scattering of spin $\frac{1}{2}$ particles off unpolarized targets, but concerning the interpretation of quantum mechanics Mandl was John's strong opponent. At that time David Bohm's papers on the *"Interpretation of quantum theory in terms of hidden variables"* [35] had appeared, which were a *"revelation"* for John. After John had reflected upon the papers he also gave a talk about them in the Theory Division and, as Mary Bell remembered, there were many interruptions from Mandl *"with whom he had many fierce arguments"* [36].

In 1960, John together with Mary joined CERN (Conseil Européen pour la Recherche Nucléaire), the European Organization for Nuclear Research which attracted the preeminent scientists, and he worked there until the rest of his life. His interest was both in phenomenological aspects of particle physics, where he frequently interacted with experimentalists, and in more formal, mathematical features of the theory, which often had no relation to experiment. As I could experience, John always was open to discuss and study any topic in physics, no matter how speculative it was. He liked to test his thoughts by basic examples. *"Always test your general reasoning against simple models!"* was his maxim.

In the 1960s, the theory of weak interactions became a hot topic. John also entered in this field and collaborated with several physicists. With his experimental colleague Jack Steinberger, who received the Nobel Prize in 1988 *"for the discovery of the muon neutrino"*, John has written the influential review on *"Weak interaction of kaons"* [37]. The so-called *Bell-Steinberger unitarity relations* for the kaon decay amplitudes belong still to the standard achievements in this area.

Most important, I think, was John's scientific exchange with his colleague and friend Martinus Veltman, just called with his nickname 'Tini', who was a Fellow at CERN. They had a fruitful and each other appreciative collaboration in particle physics, however, concerning John's quantum work Tini was, as most people at that time, quite reluctant to recognize its value (see Veltman's book [38] or Bell's essay [39]).

John and Tini started to investigate the carriers of the weak interactions, the intermediate bosons or W-bosons, which were just hypothetical particles at that time. But they had faith in their existence, encouraged and assisted the experimentalists at CERN to find these particles, and published important papers on the W-boson production by neutrinos on nuclei [40–42], which served as a basis for the neutrino experiments. As we know, it took two decades to discover the W-bosons, due to their big mass.

They also had numerous discussions about the characteristics of a quantum field theory, the issue of symmetries turned up and the feature of currents in a quantized theory. In these discussions already the seed was planted that led John finally to his most important discovery in particle physics. In case of a modern quantum field theory the *gauge symmetry* formed the basis for the so-called *gauge theory*, i.e., the Lagrangian and the basic equations of the theory were invariant with respect to possible gauge transformations. To each generator of the transformation corresponded a gauge field whose quanta were called gauge bosons, the W-bosons in case of weak interactions. Veltman successfully pursued these ideas further with his former student Gerald't Hooft and both were awarded with the Nobel Prize in 1999 for *"elucidating the quantum structure of electroweak interactions"*.

However, such a quantized theory for the weak interaction, what we call nowadays *standard model*, the unification of the electromagnetic and weak interactions, was not yet developed in the 1960s and physicists probed different ideas. Gell-Mann's *current algebra*, where instead of the conventional fields the currents were considered, was quite popular. Using plausible ideas from group theory Murray Gell-Mann postulated a canonical structure for the commutators of the current components which were involved in the physical process [43]. Conserved vector current (CVC) and partially conserved axial current (PCAC) were assumed. Physical processes could be studied by calculating the corresponding matrix elements of the currents. Amusingly, Gell-Mann published his current algebra, the *"Symmetry group of vector and axial-vector currents"* [43], in the now-defunct journal *Physics*, in which Bell just some pages behind published his seminal *"On the Einstein Podolsky Rosen paradox"* paper [44]. Due to the success of current algebra many particle physicists began to research its foundations and applications. So did John, and it is within this area that he made his outstanding contribution to particle physics.

In quantum field theory infinities occur and quantities like currents must be renormalized. It was not clear at that time whether the postulated relations of the canonical commutators for the current components survived a proper renormalization procedure. John illuminated this problem by studying an *"Equal-time commutator in a solvable model"* [45], the unrealistic but completely solvable Lee model. He demonstrated that indeed the canonical commutation relations must be taken with care since in a certain case, in the calculation of a related sum rule, the canonical values do not agree with the summation of the explicitly calculated amplitudes. On the other hand, relying on a work by Veltman about gauge invariance of sum rules [46], John showed in *"Current algebra and gauge invariance"* [47] that the desired commutation relations of the currents are achieved if gauge invariance is imposed in the matrix elements of interest.

Next, trusting current algebra and PCAC an analysis revealed that the decay of the eta meson into three pions, $\eta \to 3\pi$, is forbidden [48, 49], even though the decay is experimentally seen. Moreover, with the same assumptions the calculation of the decay of the neutral pion into two photons, $\pi^0 \to \gamma\gamma$, yielded a vanishing result [50, 51], again in contradiction to experiment.

These features were generally considered as shortcomings of the otherwise successful theory of current algebra. But John, dissatisfied with incompleteness and deficiency of a theory, always kept these current algebra defects in his mind. When Roman Jackiw, a postdoc from MIT visiting CERN in 1967–1968, asked John for a research problem he suggested to analyze the failure of current algebra in the $\pi^0 \to \gamma\gamma$ decay (see Jackiw's contribution to Ref. [52]). The result then turned out, as we know now, to be John's most far-reaching contribution to particle physics and his most-quoted paper.

But how to tackle the problem? Interestingly, John's colleague Steinberger with whom he had frequently contact, in a discussion during a coffee break in the CERN cafeteria pointed the right way. In 1949, Steinberger [53] had already calculated in his PhD a Feynman diagram, a triangle diagram with two vector current vertices and one axial vertex (see Fig. 3.6), in an at that time fashionable pion—nucleon model in order to describe the decay $\pi^0 \to \gamma\gamma$. He obtained a nonvanishing result that,

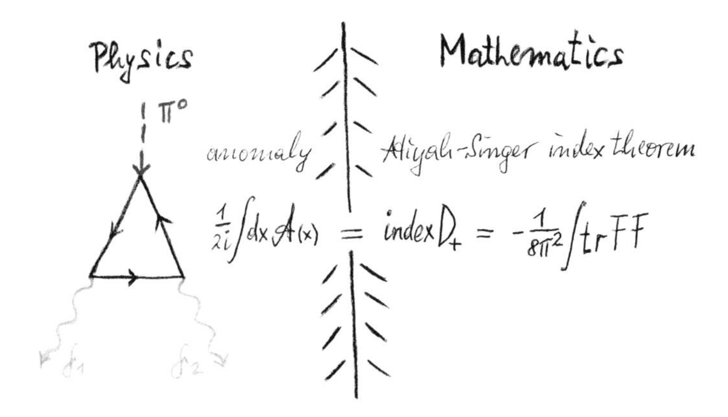

Fig. 3.6 An interplay of physics and mathematics: The Abelian anomaly is responsible for the decay $\pi^0 \to \gamma\gamma$. It is represented by the triangle diagram with two vector current vertices that couple to the two photons and one axial vertex linking to the π^0. The anomaly is related to the Atiyah-Singer index theorem in topology.

moreover, agreed well with experiment. John and Jackiw recognized immediately that Steinberger's procedure could be pursued in the sigma model (σ-model) [54], a field theory based on current algebra and PCAC. So they calculated the pion decay amplitude directly and found that the σ-model did not satisfy the requirements of PCAC, the effective coupling constant of the decay did not vanish. Their paper was entitled *"A PCAC puzzle: $\pi^0 \to \gamma\gamma$ in the σ-model"* [55].

Independently, in the same year, Stephan L. Adler from the Institute of Advanced Study in Princeton investigated the axial-vector vertex in spinor electrodynamics [56]. He found that the axial Ward identity failed in case of the triangle diagram, which led him to modify PCAC by an extra term, the *anomaly* as he phrased it. When applying this modified PCAC equation to calculate the pion decay rate Adler obtained an excellent agreement with experiment, when the fermions propagating in the triangle diagram are interpreted as quarks with their fractional charges, which occur in three species, *colours* as we know now. Thus the *modified PCAC equation*

$$\partial^\mu j_\mu^{5(3)}(x) = f_\pi m_\pi^2 \phi_\pi^{(3)}(x) + \mathcal{A} \,, \tag{8}$$

states that axial current $j_\mu^{5\,a} = \bar{\psi}\gamma_\mu\gamma_5 \frac{\sigma^a}{2}\psi$ is not conserved in massless limit of the theory but represents the celebrated *ABJ anomaly* (in honour of Adler, Bell and Jackiw)

$$\mathcal{A} = \frac{e^2}{16\pi^2}\,\varepsilon^{\mu\nu\alpha\beta} F_{\mu\nu} F_{\alpha\beta} \,, \tag{9}$$

where $F_{\mu\nu}$ is the electromagnetic field strength tensor.

Now the ice was broken, it turned out that the anomaly was not just a pathology of the quantization procedure but opened the door to a deeper understanding of quantum field theory. A new era of field theory research began (for an overview see Ref. [57]).

Quantum field theories with non-Abelian fields had been studied subsequently and anomalies were found there. In terms of differential geometry the anomalies could be formulated very concisely:

Singlet anomaly (corresponding to the Abelian- or ABJ anomaly with $e = 1$)

$$\mathcal{A} = d * j^5 = \frac{1}{4\pi^2}\,\mathrm{Tr}\, FF = \frac{1}{4\pi^2}\, d\,\mathrm{Tr}(AdA + \frac{2}{3}A^3) \,, \tag{10}$$

non-Abelian anomaly

$$G^a[A] = -(D * j)^a = \pm\frac{1}{24\pi^2}\,\mathrm{Tr}\, T^a d\,(AdA + \frac{1}{2}A^3) \,, \tag{11}$$

with $A = A_\mu^a T^a\, dx^\mu$ the non-Abelian 1-form (or connection, geometrically), $F = \frac{1}{2} F_{\mu\nu}^a T^a\, dx^\mu \wedge dx^\nu$ the field strength 2-form (or curvature), D the covariant derivative and T^a the generators of the gauge group. The signs $\pm$ corresponded to positive or negative chiral fields. Expression (11) was determined by a simple equation, the Wess-Zumino consistency condition

$$s\,G(v,A) \;=\; s\int v^a G^a[A] \;=\; 0\,, \tag{12}$$

where $v = v^a T^a$ denoted the Faddeev-Popov ghost and s the BRST operator with $s^2 = 0$, a gauge variation with respect to the gauge fields *and* the Faddeev-Popov ghosts (for literature, see the book [57]).

Particularly interesting was the connection of the anomaly to topology in mathematics. Several authors, among them Roman Jackiw, had discovered that the singlet anomaly was determined by the distinguished Atiyah-Singer index theorem (see Fig. 3.6). The reason was that the anomaly could be expressed by a sum of eigenfunctions of the Dirac operator, where only the zero-modes of a given chirality (n_+, n_-) survived

$$\frac{1}{2i}\int dx\,\mathcal{A}(x) \;=\; \int dx\sum_n \varphi_n^\dagger(x)\gamma_5\,\varphi_n(x) \;=\; n_+ \,-\, n_- \;=\; \mathrm{index}\,D_+\,. \tag{13}$$

The difference of the chirality zero modes represented the index of the Weyl operator D_+ for positive chirality, which was expressed via the Atiyah-Singer index theorem by a Chern character

$$\mathrm{index}\,D_+ \;=\; -\frac{1}{8\pi^2}\int \mathrm{Tr}\,FF\,. \tag{14}$$

Furthermore, when gravitation was considered as a gauge theory, where the gauges were the general coordinate transformations or the rotations in the tangent frame, then the classical conservation law of the energy-momentum tensor could be broken in the quantum case, an Einstein- or Lorentz anomaly occurred (see Refs. [57, 58]).

Thus quantum anomalies play a vital role in all quantum theories and it is their double feature which makes them so important for physics. On one hand, anomalies are needed to describe certain experimental facts but, on the other hand, they must be avoided since they violate a classical conservation law and signal the breakdown of gauge invariance, which ruins the consistency of the theory. This avoidance of the anomaly, which may be possible, leads to severe constraints on the physical content of the theory. For example, the *standard model* of electro-weak interaction is constructed such that no anomalies occur (actually, they canceled each other), which has led to the prediction of the top quark that has been discovered much later on.

It is interesting that John did not participate actively in these further developments of the anomaly, whereas Jackiw, Adler and many other physicists did. Roman Jackiw once asked me in a letter (from May 17th, 1996 [59]) why this might be the case:

"I was very interested in your 'Preface' (of the book [57]) *where you reminisce about conversations with Bell on anomalies a decade ago. I was very happy to read about this, because I always felt a certain surprise (disappointment) that he did not take an active role in subsequent developments, after our paper was published. Did you ever learn why this was so?"*

My assessment of John's missing engagement in the further developments I expressed in a return letter to Jackiw (from July 21st, 1996 [60]) as follows:

"At first side it seems indeed surprising that John Bell did not take an active role in the further developments of anomalies after your common paper which had such a great influence on QFT. He never mentioned explicitly a reason for that. But my impression is that one aspect of his character or of his attitude towards physics was to find out the fundamental weaknesses of a theory. It was his criticism which has led him to important discoveries. This was so in his famous works on quantum mechanics and also I experienced this attitude in our collaboration. After having found the crucial point (error, ...) he was not so interested any more in working out the further details.

In the case of the anomalies John got interested again in the middle of the 80ties since he was puzzled by the fact that several types of anomalies are linked in different dimensions (descent equations). In this connection John always spoke with high respect of your works."

Of course, after the anomaly paper John did further important works. Let me just mention a few.

Together with Bell and de Rafael [61] he calculated an upper bound on the hadronic contribution to the anomalous magnetic moment of the muon, which is satisfied by today's accepted value. Nevertheless, the topic is still of high interest due to the present discrepancy between experimental- and theoretical value.

More phenomenological work, also in view of the experiments at CERN, John carried out in collaboration with Christopher LLewellyn Smith who later on became Director General of CERN (1994–1998) and was knighted in 2001. They studied several effects in neutrino—nucleus interactions [62, 63].

More formal, mathematical topics on *hadronic symmetry classification schemes, Melosh transformations, and all that* ..., John presented in his Lectures Notes of the Schladming Winter School 1974 [64], Austria, and in further papers in collaboration with Hey [65] and Bell and Ruegg [66, 67].

About John's last interests, QCD, gluon condensate and potential models, I have reported already in the previous Section; that was the topic I had the great joy to collaborate with him.

John received several prestigious awards, among them the Dirac Medal of the Institute of Physics (1988), the Dannie Heinemann Prize of the American Physical Society (1989), and the Hughes Medal of the Royal Society (1989):

"For his outstanding contributions to our understanding of the structure and interpretation of quantum theory, in particular demonstrating the unique nature of its predictions."

Moreover, in 1988 he received honorary degrees from both Queen's University Belfast and Trinity College Dublin. John obtained these honours, as Mary Bell emphasized, mainly for his contributions in particle physics. I personally think that the discovery of the anomaly had such an enormous impact on several branches in physics that sooner or later John, if he had lived longer, would have been awarded the Nobel Prize together with Jackiw and Adler.

John Bell—the Accelerator Physicist

As already mentioned John started his scientific career at Malvern in 1949. There he joined the group of William Walkinshaw and was mainly concerned with the theory of particle accelerators. Walkinshaw highly appreciated Bell's abilities (recorded in Ref. [68]):

"Here was a young man of high caliber who soon showed his independence on choice of project, with a special liking for particle dynamics. His mathematical talent was superb and elegant."

John began with the study of a dielectric-loaded LINAC (Linear Accelerator) for electrons [69, 70], which at the beginning were used for medical purposes and later on for basic science. In 1951 Walkinshaw's group moved to Harwell and was one of those groups that established the Theory Division of AERE. In the course of the setting-up of CERN they also began to investigate proton accelerators. John himself contributed to *"Scattering-"* and *"Phase debunching by focussing foils in a proton linear accelerator"* [71, 72]. Although at that time John's works appeared as internal reports they were highly appreciated and read by the accelerator community, in particular, his report on *"Basic algebra of the strong focussing system"* [73] received much attention. Just 2 works out of 21 papers (see Bell's collected works in Ref. [74]), the topic was on strong focussing, John's special interest, have been published in journals [75, 76]. In fact, an other work about *"Linear accelerator phase oscillations"* [77], John was pleased with, he submitted to a journal and got back a 'typical' referee report, as Mary Bell remembered [36]: *"One referee said it was too short, the other that it was too long."* So John gave up, he was leaving the accelerator field anyhow and turned to particle physics.

In the beginning of the 1980s John became interested again in accelerator physics. It was the time of the construction of the SPS (Super Proton Synchrotron) and LEAR (Low Energy Antiproton Ring) at CERN. There he collaborated with Mary on *electron cooling, quantum beamstrahlung and bremsstrahlung*, which will be reported in Sect. "Joint Works of the Bells".

A particularly attractive work, in my opinion, was Bell's combination of the Unruh effect of quantum field theory with accelerator physics. According to Unruh, a uniformly accelerated observer through the electromagnetic vacuum will experience a black body radiation at a temperature proportional to the acceleration [78]. There is a close connection between the Unruh effect and the Hawking radiation of a black hole, which can be seen already in the temperature formula. From the Unruh temperature T_U for a given acceleration a

$$k T_U \;=\; \frac{\hbar\, a}{2\pi c} \quad \longrightarrow \quad \frac{\hbar\, \kappa}{2\pi c} \;=\; \frac{\hbar\, c^3}{8\pi G M} \;=\; k T_H \tag{15}$$

follows directly the Hawking T_H temperature [79] for the *surface gravity* $\kappa = \frac{c^2}{2R_S}$ of a black hole with Schwarzschild radius $R_S = \frac{2GM}{c^2}$.

John discussed this subject with Jon Leinaas, a Norwegian CERN Fellow in the early 1980s, who got interested in the Unruh effect. Starting from a paper of Jackson [80] about the spin polarization effect of electrons circulating in a storage ring, they tried to find a connection to the Unruh effect. Their idea was to consider accelerated electrons as detectors, with the spin degree of freedom used to measure the temperature of the radiation. Indeed, what John and Leinaas found was that the spin depolarization effect of the electrons in a storage ring is closely related to the thermal effect of linearly accelerated electrons [81]. The effect is small but in agreement with the measured values at a storage ring. However, there are complications due to the circular motion (Thomas precession), a resonance occurs, etc. ... For this reason *"the measurements cannot be considered as a direct demonstration of the (circular) Unruh effect. Therefore a measurement of the vertical fluctuations would be of interest, as a more direct demonstration of this effect. However, these fluctuations are small, and it is not clear whether it would be possible to separate this effect from other perturbations in the orbit."* [82]. These subtle points show their great effort and achievement to test a quite sophisticated theory in an advanced experiment of accelerator physics. A discussion of the Unruh effect for the case of extended thermometers is given in Ref. [83].

John and Mary Bell

John and Mary met for the first time in 1949 when John came to Harwell, where Mary had worked on reactors since 1947. At 1949/1950, both moved for a year to Malvern, first John then Mary followed, and they returned again to Harwell. At that time John was still without a beard as one could see from an excursion of the couple to Stonehenge in the 1950s, see Fig. 3.7. They obviously enjoyed both their scientific and private life. In 1954, they were married and pursued their careers together. In course of their life they collaborated several times on issues of accelerator physics.

In 1960, both moved to CERN and lived in Geneva. End of 1963, they took a sabbatical year staying in the USA. There John had time to work on his *'hobby'*, the foundations of quantum mechanics, and it was at SLAC where he had written his celebrated *'inequality paper'*.

When I think of John I always remember both, John and Mary, the couple. In lasting memory are the many pleasant events, the lunches, walks, ..., Renate and myself spent together with the Bells: For instance, the lunch in the former *'Haas Haus Restaurant'* in Vienna in 1982, see Fig. 3.8, when John was the distinguished *'Schrödinger Guest Professor'* at our Institute of Theoretical Physics; Or the beautiful walks in the Calanque de Port d'Alon in 1983, where we even could arrange a typically British weather for John and Mary, which was not so easy to get in South of France, see Fig. 3.9. I was at that time a Visiting Professor at the University of Marseille, Luminy, and the Bells visited us in Bandol; Or, finally, our last reunion in Bures sur Yvette in April 1990, when I stayed at the CNRS (Centre Nationale de la Recherche Scientific).

Fig. 3.7 Mary and John Bell at Stonehenge in the 1950s. Foto: © Mary Bell.

Fig. 3.8 Mary and John Bell in the former *'Haas Haus Restaurant'* in Vienna in 1982. Foto: © Renate Bertlmann.

Mary always was present, what John nicely had expressed in the preface of his book *Speakable and Unspeakable in Quantum Mechanics* [84], when he wrote the moving tribute to Mary:

Fig. 3.9 Reinhold, John and Mary on a walk in the Calanque de Port d'Alon, South of France, 1983. Foto: © Renate Bertlmann.

"... I here renew very especially my warm thanks to Mary Bell. When I look through these pages again I see her everywhere."

Mary Bell

Mary, with maiden name Ross, originates from a Scottish family resident in Glasgow. The parents were Alexander Munro Ross, a commercial manager in a ship-building firm, and Catherine Brown Wotherspoon, an elementary school teacher. Her father was quite enthusiastic to send Mary and her two older sisters to university.

As a teenager Mary attended the *Hyndland Secondary School* which was situated at a two minutes walk from where she lived. There she spent all her school life except for the year 1939–1940, when her father had sent her to the *Kingussie High School* in Iverness-shire to stay with his two sisters. This was to elude the bombing in Glasgow at the beginning of World War II. There Mary's talents as a storywriter had been already discovered, for instance, her amusing agent short-story *"The Secret Service Agent"* had been printed in the *Kingussie Secondary School Magazine* [85].

Mary returned to Hyndland for the year 1940–1941. The Hyndland Secondary School catered for *all* types of pupils, academic and non-academic, but in different classes. Mary was quite fortunate that this school had a strong science department. Two of the headmasters were science teachers: There was an older building, but also a fine airy new building with a number of laboratories, art rooms, and an apartment for domestic science. Hyndland School was without fees, in contrast to the *'Girls' High School'* (in the centre of the city) which had fees. Hyndland had pupils, boys *and* girls, from a wide area. There was an examination which all pupils (at all schools) had to take on a given day for getting money for the Girls' High School, which was considered as a good school. Mary told me in a letter [86]:

"Unfortunately, I won this, but my father insisted that I stay at the local school, instead of travelling into the town each day. He was right. The Girls' High School had only botany among the science subjects (no physics)."

At Hyndland, science was a strong subject. Mary even obtained a prize for *"General Excellence in Dynamics"* in the season 1938–1939, a phrase John found quite amusing, and she still remembers learning Newton's Laws of Motion.

Mary also won an open *'Bursary Competition for the University'*, in particular for a female studying science, and got enough money to pay the university fees. *"But money for fees was not really a problem"* as she said.

The university course was four years in *Mathematics and Natural Philosophy* (Natural Philosophy was the former term for Physics). Due to World War II many students were called up after two years and could qualify for a wartime degree. Mary was allowed to stay three years because of the need for demonstrators. Eventually in 1944, she was called up and sent to TRE in Malvern, which was the big government establishment for radio navigation, radar, infra-red detection and related subjects. After a year the war ended and she returned to the University. When she graduated in 1947, it was quite natural for her to apply for AERE at Harwell.

At that time, there were only a few female scientists in the Theory Division, about 3 of 20 persons as Mary remembers. Science was not considered as an appropriate profession for a woman. Thus it was quite exceptional, in my opinion, that Mary had chosen already as a young girl a scientific subject for her career and not just a *'girl subject'*. When I once asked her why, she replied in her beautiful handwriting [87]:

"I always liked to solve little problems, even when I knew only arithmetic."

So already at an early stage her vision for a career was science and not so much being just a 'house wife', may be also her good scientific education in the Hyndland School was quite encouraging for her.

Mary's appointment at Harwell started in February 1947. The research was on nuclear reactors but soon it broadened including nuclear physics, accelerators and related topics. The head of the theory division was the well-known Klaus Fuchs, a left-wing refugee from Germany with British citizenship, who had worked before on the *'Manhatten Project'* in Los Alamos and on the *'Tube Alloys'* programme, the British atomic bomb project. Fuchs had close contact with his compatriot Rudolf Peierls (later on Sir Rudolf) who was a consultant to AERE and had much influence there.

"At the beginning the work at Harwell was not very interesting", Mary remembered [87]. But then Fuchs told her *"to move to more interesting problems."* She was instructed to work on treating the control rods in a fast fission reactor. Some calculations she published in the paper: *"Slow neutron absorption cross-section of the elements"* [88]. In fact, she had to carry out a special perturbation calculation

devised by Klaus Fuchs. Soon after this work was completed Fuchs confessed, in 1950, that he was a spy (out of conviction) supplying information from the American, British, Canadian atomic bomb project to the Soviet Union. He was sentenced to 14 years imprisonment, got arrested but released in 1959 and emigrated to the former German Democratic Republic.

As already mentioned, in 1950, Mary and John were sent to TRE in Malvern to join the accelerator group. After a year working there, for a second time, Mary and all the other physicists moved back to Harwell to the Theory Division which was then headed by Brian Flowers (later on Lord Flowers). In the next years Mary worked there on several subjects of accelerator physics. A fairly complete list of papers, mainly AERE reports, are given by the Refs. [88–104]. Some works, whose topics I personally find very interesting, are:

- *"Series impedance of double-ridged wave guides."* [90]
- *"Focussing system for the 600 MeV proton linear accelerator."* [96]
- *"Nonlinear equations of motion in the synchrotron."* [97]
- *"Injection into the 7 GeV synchrotron."* [102]

After working for a decade at Harwell John and Mary felt somehow ready for a change, mainly since John had been attracted by particle physics. In 1960, they moved to CERN, the big European center for particle physics, John to the Theory Division and Mary to the Accelerator Research Group.

At CERN Mary wrote a lot of computer programmes representing the orbits of the circulating particles in different accelerator machines. These orbits, of course, had to be extremely accurate. In the 1970s, when particle accelerators for higher energy (some GeV region) were planed and constructed the question of cooling particles in a storage ring became a hot issue. Two methods were debated at CERN for further use in the accelerators in order to reduce the large phase space spread of an ion, proton or antiproton beam in the ring: *'Electron cooling'* proposed by Gersh Itskovich Budker [105] and successfully applied at the Novosibirsk storage ring [106] and *'stochastic cooling'* invented by Simon van der Meer at CERN.

Mary was studying electron cooling (a picture of her at about that time can be seen in Fig. 3.10) which is based on repeated interactions of protons circulating in the storage ring with a dense and cold electron beam. She published a paper on *"Electron cooling with magnetic field"* [107], where she discussed the basic equations for the slowing of a proton in an electron gas with magnetic field. She was also part of a working group headed by Frank Kienen, which studied the cooling of protons by means of electrons in the ICE (Initial Cooling Experiment) storage ring in order to test the feasibility of a high luminosity proton—antiproton collider. Kienen often had discussions with John. It raised John's interest again in accelerator physics and he began to collaborate with Mary on these topics, which will be reported in Sect. "John and Mary Bell". The results of the ICE measurements of the momentum spread, beam profile, beam lifetime, etc. ..., were quite in agreement with theory and were published in a common paper [108], where also Carlo Rubbia (appointed to Director-

Fig. 3.10 Mary Bell in 1982. Foto: © Renate Bertlmann.

General of CERN in 1989) and Simon van der Meer contributed. Further details of the cooling method in the ICE were published in Ref. [109] and Mary had to prepare the paper for publication. As she remembers [36]:

"This involved a lot of condensing to make the paper a reasonable size, and John kindly helped me. He particularly liked to be acknowledged at the end for 'helping with the typing'."

Also stochastic cooling was explored in the ICE storage ring and in these tests it turned out that van der Meer's stochastic cooling method was sufficient to be implemented in the newly constructed SPS (Super Proton Synchrotron) to collide protons and antiprotons in the same ring. Rubbia and van der Meer were the leading figures in constructing the SPS and detected in 1983 the W and Z bosons. In 1984, they received the Nobel Prize *"for their decisive contributions to the large project, which led to the discovery of the field particles W and Z, communicators of weak interactions."* Nevertheless, electron cooling was used further on in the redesigned ring LEAR (Low Energy Antiproton Ring) to decelerate and store antiprotons.

Joint Works of the Bells

When John stayed at Malvern also Mary joined the accelerator group of Walkinshaw. Having previously worked on nuclear reactors she now switched to accelerator physics. Clearly, Mary and John had many discussions on accelerator issues, and back again at Harwell they wrote a common paper in 1952 (where Mary still signed with her maiden name Mary Ross) about: *"Heating of focussing foils by a proton beam"* [92]. Since John was leaving the accelerator field and turned to particle physics there was a long pause of collaboration of about 30 years. John's interest in accelerator physics got revived about 1980 at CERN when Mary was working on electron cooling.

John and Mary's first common paper at CERN was devoted to *"Electron cooling in storage rings"* [110], where they calculated the effect of *'flattening'* of the electron velocity distribution, which meant that the longitudinal velocity spread was suppressed, to increase the rate of cooling of small betatron oscillations. This paper they devoted to the nuclear and accelerator physicist Yuri Orlov who was imprisoned at that time in the Soviet Union for his human rights activism, freed later on and deported to the USA. Such an act of solidarity was quite typical for the Bells.

A paper on a similar issue: *"Capture of cooling electrons by cool protons"* [111] followed. There Mary and John presented formulae for the capture of low-energy electrons by stationary protons using Maxwellian and flattened electron velocity distributions. The latter was more appropriate for the electron beams used in the accelerator proton beam cooling experiments. They found out that the flattening increased the capture rate by a factor of about two. Also similar formulae for the capture of antiprotons by protons were mentioned.

In a further paper *"Radiation damping and Lagrange invariants"* [112] Mary and John proposed a general formula for the damping of small oscillations about closed orbits. It was applied to derive the results for the effects of classical radiation damping on storage ring orbits.

Next, Mary and John turned to *'beamstrahlung'* which is the radiation of the whole beam of charged particles in a storage ring or LINAC, when the beam interacts with the electromagnetic field of the other beam. In their paper *"Quantum beamstrahlung"* [113] they showed that the well-known Blankenbecler-Drell formula [114] for the bremsstrahlung energy loss of a relativistic electron passing through the field of a cylindric charge agrees with the popular *'Russian formula'* [115–118] if the spin-flip contributions were added to Blankenbecler-Drell approach.

In *"End effects in quantum beamstrahlung"* [119] Mary and John also included so-called *'end effects'* in their beamstrahlung calculations. These effects occurred at the ends of a sharply bounded cylindrical charge bunch. Their conclusion was that end effects were indeed negligible when the bunch length is large compared to its typical quantum length.

In their last joint work *"Quantum bremsstrahlung in almost uniform fields"* [120] Mary and John studied inhomogeneity effects in quantum bremsstrahlung in the extreme quantum limit, in the case of a weakly nonuniform deflecting field.

Out of the Blue

At CERN Bell was a kind of *'Oracle'* for particle physics, consulted by many colleagues who wanted to get his approval for their ideas. Of course, I had heard that he was also a leading figure in quantum mechanics, specifically, in quantum foundations. But nobody could actually explain to me his work in this quantum area, neither at CERN nor anywhere else. The standard answer was: *"He discovered some 'relation' whose consequence was that quantum mechanics turned out alright. But we knew that anyway, so don't worry."* And I didn't. John, on the other hand, never mentioned his quantum works to me in the first years of our collaboration. Why? This I understood later on, John was reluctant to push somebody into a field that was quite unpopular at that time.

At the end of summer 1980, I returned for some time to my home institute in Vienna to continue our collaboration on the gluon condensate potential from there. At that time, there was no internet, and it was a common practice to send preprints (typed manuscripts) of the work, prior to publication, to all main physics institutions in the world. Also we in Vienna had such a preprint shelf where each week the new incoming preprints were exhibited.

One day, on the 15th of September, I was sitting in the Institute's computer room, handling my computer cards, when my colleague Gerhard Ecker, who was in charge of receiving the preprints, rushed in waving a preprint in his hands (see Fig. 3.11). He shouted, *"Reinhold look—now you're famous!"* I hardly could believe my eyes

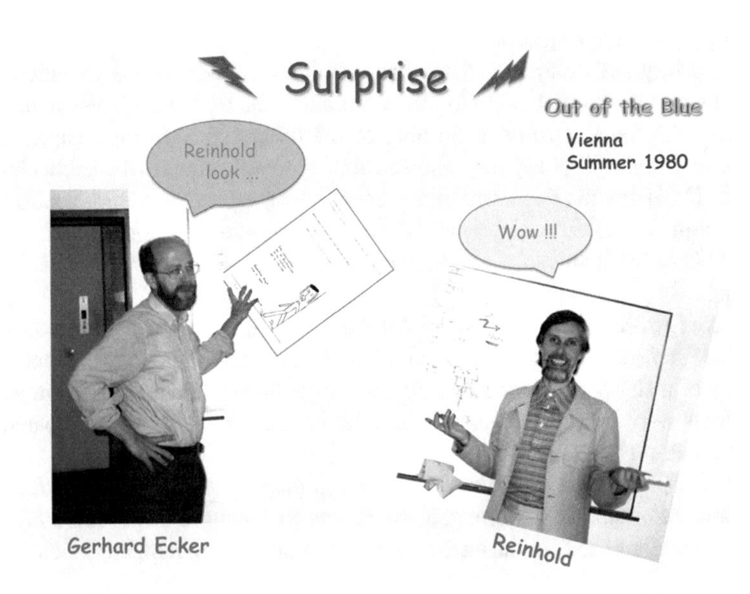

Fig. 3.11 Gerhard Ecker standing in front of Reinhold Bertlmann holds a preprint in his hands with the title *"Bertlmann's socks and the nature of reality"* [121]. Fotos: © Renate Bertlmann.

Fig. 3.12 Cartoon in the CERN preprint Ref.TH.2926-CERN *"Bertlmann's socks and the nature of reality"* of John Bell from 18th July 1980 [121]. The article is based on an invited lecture John Bell has given at le Colloque sur les *"Implications conceptuelles de la physique quantique"*, organisé par la Foundation Hugot du Collège de France, le 17 juin 1980, published in Journal de Physique [123].

as I read and reread the title of a paper by Bell [121]: *"Bertlmann's socks and the nature of reality"*.

I was totally excited. Reading the first page my heart stood still:

"The philosopher in the street, who has not suffered a course in quantum mechanics, is quite unimpressed by Einstein-Podolsky-Rosen correlations [122]. *He can point to many examples of similar correlations in every day life. The case of Bertlmann's socks is often cited. Dr. Bertlmann likes to wear two socks of different colours. Which colour he will have on a given foot on a given day is quite unpredictable. But when you see* (Fig. 3.12) *that the first sock is pink you can be already sure that the second sock will not be pink. Observation of the first, and experience of Bertlmann, gives immediate information about the second. There is no accounting for tastes, but apart from that there is no mystery here. And is not the EPR business just the same ?"*

Seeing the cartoon John has sketched by himself (see Fig. 3.12), showing me with my odd socks, nearly knocked me down. All this came so unexpectedly—I had not the slightest idea that John had noticed my habits of wearing socks of different colours, a habit I cultivated since my early student days, my special *'generation-68'* protest. This article pushed me *instantaneously* into the quantum debate, which changed my life. Since then *'Bertlmann's socks'* had developed a life of its own. You can find Bertlmann's socks everywhere on the internet, in popular science debates, and even in the fields of literature and art.

Now the time had come for diving into the quantum world of John, to understand why the *"EPR business"* was *not* just the same as *"Bertlmann's socks"*, and to appreciate his profound insight. It was John who pushed the rather philosophical Einstein-Bohr discussions of the 1930s about realism and incompleteness of quantum mechanics onto physical grounds. His axiom of *locality* or *separability* was the essential ingredient of a *hidden variable theory* and illuminated the physical difference between all such hidden variable theories and the predictions of quantum mechanics. Due to *'Bell's Theorem'* we can distinguish *experimentally* between quantum mechanics and all local realistic theories with hidden variables. I was impressed by the clarity and depth of John's thoughts. From this time on we had fruitful discussions about the foundations of quantum mechanics and this was a great fortune and honour for me. It was just about the time when Aspect [124] finished his time-flip experiments on Bell inequalities and the whole field began to attract the increasing interest of physicists. For me a new world opened up—the universe of John Bell—and caught my interest and fascination for the rest of my life.

Bell and the Quantum

John never was satisfied with the interpretations of quantum mechanics. Already as a student at Queen's University Belfast he disliked the so-called *Copenhagen Interpretation* with its distinction between the quantum world and the classical world. He wondered, *"Where does the quantum world stop and the classical world begin?"* He wanted to get rid of that division.

For him it was clear that hidden variable theories, where quantum particles do *have* definite properties governed by hidden variables, would be appropriate to reformulate quantum theory. *"Everything has definite properties!"* I remember John saying.

Contextuality

Hidden variable theories (HVT) as well as quantum mechanics describe an ensemble of individual systems. Whereas in QM the orthodox (Copenhagen) doctrine tells us that measured properties, e.g. the spin of a particle, have no definite values before measurement, the HVT in contrast postulate that the properties of individual systems do *have* pre-existing values revealed by the act of measurement.

Given a set of observables $\{A, B, C, ...\}$ then a hidden variable theory assigns to each individual system a set of values (eigenvalues) $\{v(A), v(B), v(C), ...\}$ to the corresponding observables [125]. The hidden variable theory provides a rule how to distribute the values over all individual systems of an ensemble. Such states, specified by the quantum mechanical state vector *and* by the additional hidden variable, are called *dispersion-free*.

If a functional relation, $f(A, B, C, \ldots) = 0$, is satisfied by a set of mutually commuting observables $A, B, C, \ldots$, then the same relation must hold for the values in the individual systems, $f\big(v(A), v(B), v(C), \ldots\big) = 0$. Amazingly, just by relying on above two conditions a contradiction can be constructed, a so-called *No-Go Theorem*.

Bell started his investigation *"On the problem of hidden variables in quantum mechanics"* [126] in 1964[1] by criticizing John von Neumann who had given already in 1932 a proof [127] that dispersion-free states, and thus hidden variables, are incompatible with quantum mechanics. What was the criticism? Consider three operators A, B, C with condition $C = A + B$, then the correspondingly attached values also have to satisfy $v(C) = v(A) + v(B)$, since the operators A, B are supposed to commute.

Von Neumann, however, also imposed the additivity property for *noncommuting* operators. *"This is wrong!"* John grumbled and illustrated his dictum, before giving a general proof, with the example of spin measurement. Measuring the spin operator σ_x for a magnetic particle requires a suitably oriented Stern-Gerlach apparatus. The measurements of σ_y and $(\sigma_x + \sigma_y)$ demand different orientations. Since the operators cannot be measured simultaneously, there is no reason for imposing additivity. Of course, for the quantum mechanical expectation values we have additivity in the mean $\langle\psi|A + B|\psi\rangle = \langle\psi|A|\psi\rangle + \langle\psi|B|\psi\rangle$, irrespective whether A, B commute or not.

Interestingly, already in 1935 the mathematician and philosopher Hermann [128] raised her objection to von Neumann's assumption but she was totally ignored. Also Kochen and Specker [129], when reading von Neumann's proof in 1961, had their doubts about the additivity for noncommuting operators.

John being aware of Gleason's Theorem [130], which was not explicitly addressed to HVT but aimed instead to reduce the axioms for QM, established the following corollary [126], more directed to HVT:

Corollary 1 (Bell's Corollary) *Consider a state space $\mathcal{V}$. If $\dim \mathcal{V} > 2$ then the additivity requirement for expectation values of commuting operators cannot be met for dispersion-free states.*

Corollary 1 states that for $dim\mathcal{V} > 2$ it is in general impossible to assign a definite value for each observable in each individual quantum system. Thus Bell pointed to another class of hidden variable models, where the results may depend on different settings of the apparatus. Such models are called *contextual* and may agree with quantum mechanics. Corollary 1, on the other hand, states that all *noncontextual* HVT are in conflict with QM (for $dim > 2$). Hence the essential feature for the difference between HVT and QM is *contextuality*.

In 1967, Simon Kochen and Ernst Specker published their famous paper on *"The problem of hidden variables in quantum mechanics"* [131], where they established their No-Go Theorem that noncontextual hidden variable theories are incompatible with quantum mechanics.

---

[1] Due to a delay in the Editorial Office of Review of Modern Physics the paper was published not until 1966.

Theorem 1 (Kochen-Specker Theorem) *In a Hilbert space $\mathcal{H}$ of $\dim\mathcal{H} > 2$ it is impossible to assign values to all physical observables while simultaneously preserving the functional relations between them.*

Since then, contextuality has become an important issue in the research of quantum systems (see, e.g., Refs. [132–138], and references therein).

Nonlocality

The starting point of John's quantum studies was Bohm's [35] reinterpretation of quantum theory as a deterministic, realistic theory with hidden variables. Although Bohm's work was not at all appreciated by the physics community, neither by Einstein nor by Pauli, John was very much impressed by Bohm's work and often remarked, *"I saw the impossible thing done"*. To me John continued, *"In every quantum mechanics course you should learn Bohm's model!"*

John examined Bohm's model quite carefully and analyzed a system of two particles with spin $\frac{1}{2}$ [126], interacting with the external magnetic fields $\vec{B}$ of two magnets that analyze the spins. The argumentation goes as follows. The hidden variables in this two-particle system are two vectors $\vec{X}_1$ and $\vec{X}_2$ which yield the results for position measurements. The variables are supposed to be distributed in configuration space with probability density

$$\rho(\vec{X}_1, \vec{X}_2) = \sum_{i,j} |\psi_{ij}(\vec{X}_1, \vec{X}_2)|^2 \,, \tag{16}$$

which describes the quantum mechanical state; ψ_{ij} is the solution of the Schrödinger equation. The position operators, the hidden variables, for the two-particle system then vary in time according to ($\hbar = 1, 2m = 1$)

$$\begin{aligned}
\frac{d\vec{X}_1}{dt} &= \frac{1}{\rho(\vec{X}_1, \vec{X}_2, t)} \,\mathrm{Im} \sum_{i,j} \psi_{ij}^*(\vec{X}_1, \vec{X}_2, t) \frac{\partial}{\partial \vec{X}_1} \psi_{ij}(\vec{X}_1, \vec{X}_2, t) \\
\frac{d\vec{X}_2}{dt} &= \frac{1}{\rho(\vec{X}_1, \vec{X}_2, t)} \,\mathrm{Im} \sum_{i,j} \psi_{ij}^*(\vec{X}_1, \vec{X}_2, t) \frac{\partial}{\partial \vec{X}_2} \psi_{ij}(\vec{X}_1, \vec{X}_2, t) \,.
\end{aligned} \tag{17}$$

The strange feature now is that the trajectory equations (17) for the operators, the hidden variables, have a highly nonlocal character. Only in case of a factorizable wave function for the quantum system

$$\psi_{ij}(\vec{X}_1, \vec{X}_2, t) = \eta_i(\vec{X}_1, t) \cdot \chi_j(\vec{X}_2, t) \,, \tag{18}$$

the trajectories decouple

$$\frac{d\vec{X}_1}{dt} = \frac{1}{\sum_i |\eta_i(\vec{X}_1,t)|^2} \, \mathrm{Im} \sum_i \eta_i^*(\vec{X}_1,t) \frac{\partial}{\partial \vec{X}_1} \eta_i(\vec{X}_1,t)$$

$$\frac{d\vec{X}_2}{dt} = \frac{1}{\sum_j |\chi_j(\vec{X}_2,t)|^2} \, \mathrm{Im} \sum_j \chi_j^*(\vec{X}_2,t) \frac{\partial}{\partial \vec{X}_2} \chi_j(\vec{X}_2,t) , \qquad (19)$$

and the trajectories of $\vec{X}_1$ and $\vec{X}_2$ are determined separately by involving the magnetic fields $\vec{B}(\vec{X}_1)$ and $\vec{B}(\vec{X}_2)$ respectively. However, in general, this is not the case. The trajectory of particle 1 depends in a complicated way on the trajectory and wave function of particle 2 , and thus on the analyzing magnetic field acting on particle 2 , no matter how remote the particles are. Therefore, as John remarked [126]: *"In this theory an explicit causal mechanism exists whereby the disposition of one piece of apparatus affects the results obtained with a distant piece."*

John, realizing the importance of this *nonlocal feature*, wondered if it was just a defect of this particular hidden variable model, or is it somehow intrinsic in a hidden variable theory reproducing quantum mechanics. After playing around a bit to find a local account for the quantum results he could construct an impossibility proof, a *Bell inequality*.

Bell Inequalities

In his paper *"On the Einstein-Podolsky-Rosen paradox"* [44] John reconsidered the at that time totally disregarded paper of Albert Einstein, Boris Podolsky and Nathan Rosen (EPR) [122]. Therein the authors argued that quantum mechanics is an *incomplete* theory and that it should be supplemented by additional parameters, the hidden variables. These additional variables would restore causality and locality in the theory. What was John's essence when considering Bohm's spin version [139] of EPR?

Let us analyze such a Bohm-EPR setup, where a pair of spin $\frac{1}{2}$ particles is produced in a spin singlet state and propagates freely into opposite directions (see Fig. 3.13). The spin measurement on one side, called Alice, performed by a Stern-Gerlach magnet along some direction $\vec{a}$ is described by the operator $\vec{\sigma}_A \cdot \vec{a}$ and yields the values ± 1 . Since we can predict in advance the result of $\vec{\sigma}_B \cdot \vec{b}$ on the other side, Bob's side, the result must be predetermined. This predetermination we specify by the additional variable λ . In such an extended theory we denote the measurement result of Alice and Bob by

$$A(\vec{a},\lambda) = \pm 1, 0 \qquad \text{and} \qquad B(\vec{b},\lambda) = \pm 1, 0 \, . \qquad (20)$$

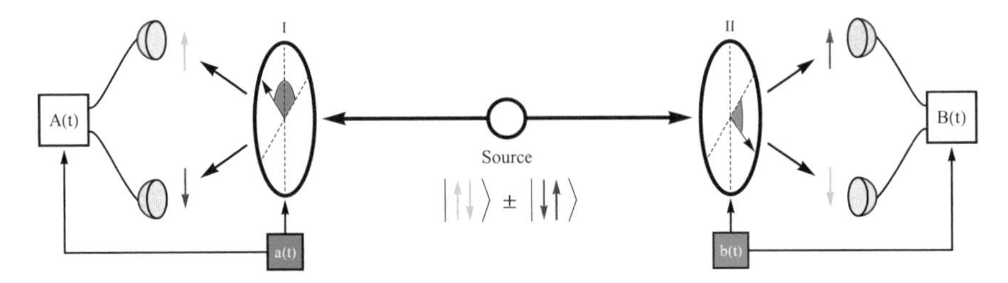

Fig. 3.13 In a Bohm-EPR setup a pair of spin $\frac{1}{2}$ particles, prepared in a spin singlet state, propagates freely in opposite directions to the measuring stations called Alice and Bob. Alice measures the spin in direction $\vec{a}$, whereas Bob measures simultaneously in direction $\vec{b}$.

We also include 0 for imperfect measurements to be more general, i.e., what we actually require is

$$|A| \leq 1 \qquad \text{and} \qquad |B| \leq 1 \,. \tag{21}$$

Then the expectation value of the joint spin measurement of Alice and Bob is

$$E(\vec{a},\vec{b}) \;=\; \int d\lambda\, \rho(\lambda)\, A(\vec{a},\lambda) \cdot B(\vec{b},\lambda)\,. \tag{22}$$

The choice of the product in expectation value (22), where A does not depend on Bob's settings, and B does not depend on Alice's setting, is called *'Bell's Locality Hypothesis'*. It is the obvious definition of a physicist as an engineer, and *must not* be confused with other locality definitions, like local interactions or locality in quantum field theory.

The function $\rho(\lambda)$ represents some probability distribution for the variable λ, and does not depend on the measurement settings $\vec{a}$ and $\vec{b}$ which can be chosen truly free or random. This is essential! The distribution is normalized $\int d\lambda\, \rho(\lambda) \;=\; 1\,.$

Now it is quite easy to derive *Bell's original inequality* [44] by assuming perfect (anti-)correlation $E(\vec{a},\vec{a}) \;=\; -1$

$$S_{\text{Bell}} \;:=\; |E(\vec{a},\vec{b}) \;-\; E(\vec{a},\vec{b}^{\,\prime})| \;-\; E(\vec{b}^{\,\prime},\vec{b}) \;\leq\; 1\,. \tag{23}$$

The quantum mechanical expectation value for the joint measurement when the system is in the spin singlet state $|\psi^-\rangle \;=\; \frac{1}{\sqrt{2}}(|\Uparrow\rangle \otimes |\Downarrow\rangle \;-\; |\Downarrow\rangle \otimes |\Uparrow\rangle)$, also called Bell state, is given by

$$\begin{aligned} E(\vec{a},\vec{b}) \;&=\; \langle\psi^-|\, \vec{a}\cdot\vec{\sigma}_A \otimes \vec{b}\cdot\vec{\sigma}_B\, |\psi^-\rangle \\ &=\; -\,\vec{a}\cdot\vec{b} \;=\; -\,\cos(\alpha-\beta)\,, \end{aligned} \tag{24}$$

where α, β are the angles of the orientations in Alice's and Bob's parallel planes. Inserting expectation value (24) then inequality (23) is violated maximally for the choice of $(\alpha,\beta,\beta^{\prime}) \;=\; (0, 2\frac{\pi}{3}, \frac{\pi}{3})$, the *'Bell angles'*

$$S_{\text{Bell}}^{\text{QM}} = \frac{3}{2} = 1.5 > 1 . \tag{25}$$

Well adapted to experiment is an other inequality, the familiar *CHSH inequality*, named after Clauser, Horne, Shimony, and Hold who published it in 1969 [140]

$$S_{\text{CHSH}} := |E(\vec{a},\vec{b}) - E(\vec{a},\vec{b}')| + |E(\vec{a}',\vec{b}) + E(\vec{a}',\vec{b}')| \le 2 . \tag{26}$$

As we know, in case of quantum mechanics (24) the CHSH inequality (26) is violated maximally

$$S_{\text{CHSH}}^{\text{QM}} = 2\sqrt{2} = 2.828 > 2 , \tag{27}$$

for the choice of the Bell angles $(\alpha, \beta, \alpha', \beta') = (0, \frac{\pi}{4}, 2\frac{\pi}{4}, 3\frac{\pi}{4})$. Inequality (26) has been tested experimentally, e.g., by Zeilinger's group [141, 142] by using entangled photons in the Bell state $|\psi^-\rangle$. In the photon case, however, the expectation value (24) changes to $E(\vec{a},\vec{b}) = -\cos 2(\alpha - \beta)$, i.e., the Bell angles become a factor of 2 smaller as compared to the spin case.

Other types of Bell inequalities, often used in experiments, are:

Firstly, *Wigner's inequality* derived by Eugene P. Wigner in 1970 [143]. He focused on probabilities which are proportional to the number of clicks in a detector. In terms of probabilities P for the joint measurements the expectation value can be expressed by

$$E(\vec{a},\vec{b}) = P(\vec{a} \Uparrow, \vec{b} \Uparrow) + P(\vec{a} \Downarrow, \vec{b} \Downarrow) - P(\vec{a} \Uparrow, \vec{b} \Downarrow) - P(\vec{a} \Downarrow, \vec{b} \Uparrow) , \tag{28}$$

and assuming that $P(\vec{a} \Uparrow, \vec{b} \Uparrow) \equiv P(\vec{a} \Downarrow, \vec{b} \Downarrow)$ and $P(\vec{a} \Uparrow, \vec{b} \Downarrow) \equiv P(\vec{a} \Downarrow, \vec{b} \Uparrow)$ together with $\sum P = 1$ the expectation value becomes

$$E(\vec{a},\vec{b}) = 4\,P(\vec{a} \Uparrow, \vec{b} \Uparrow) - 1 . \tag{29}$$

Inserting expression (29) into Bell's inequality (23) yields *Wigner's inequality* for the joint probabilities, where Alice measures spin $\Uparrow$ in direction $\vec{a}$ and Bob also $\Uparrow$ in direction $\vec{b}$ (we drop from now on the spin notation $\Uparrow$ in the formulae)

$$P(\vec{a},\vec{b}) \le P(\vec{a},\vec{b}') + P(\vec{b}',\vec{b}) , \tag{30}$$

or rewritten

$$S_{\text{Wigner}} := P(\vec{a},\vec{b}) - P(\vec{a},\vec{b}') - P(\vec{b}',\vec{b}) \le 0 . \tag{31}$$

For the Bell state $|\psi^-\rangle$ the quantum mechanical probability gives

$$P(\vec{a},\vec{b}) = |\big(\langle \vec{a} \Uparrow | \otimes \langle \vec{b} \Uparrow |\big)|\psi^-\rangle|^2 = \frac{1}{2} \sin^2 \frac{1}{2}(\alpha - \beta) , \tag{32}$$

and leads to a maximal violation of inequality (31)

$$S^{\mathrm{QM}}_{\mathrm{Wigner}} = \frac{1}{8} = 0.125 > 0\,, \tag{33}$$

for $(\alpha, \beta, \beta') = (0, 2\frac{\pi}{3}, \frac{\pi}{3})$, the same choice as in Bell's original inequality.

Secondly, the *Clauser-Horne inequality* of 1974 [144]. It relies on weaker assumptions and is very well suited for photon experiments with absorptive analyzers. Clauser and Horne work with relative counting rates, i.e., number of registered particles in the detectors. More precisely, the quantity $N(\vec{a}, \vec{b})$ is the rate of simultaneous events, coincidence rate, in the photon detectors of Alice and Bob after the photons passed the corresponding polarizers in direction $\vec{a}$ or $\vec{b}$ respectively. The relative rate $N(\vec{a}, \vec{b})/N = P(\vec{a}, \vec{b})$, where N represents all events when the polarizers are removed, corresponds in the limit of infinitely many events, which is practically the case, to the joint probability $P(\vec{a}, \vec{b})$. If one polarizer is removed, say on Bob's side, then expression $N_A(\vec{a})/N = P_A(\vec{a})$ stands for the single probability at Alice's (or the correspondingly at Bob's) side.

Starting from a pure algebraic inequality $-XY \leq x_1 y_1 - x_1 y_2 + x_2 y_1 + x_2 y_2 - Y x_2 - X y_1 = S \leq 0$ for numbers $0 \leq x_1, x_2 \leq X$ and $0 \leq y_1, y_2 \leq Y$, it is now easy to derive the corresponding inequality for probabilities, which is the *Clauser-Horne inequality*

$$-1 \leq P(\vec{a}, \vec{b}) - P(\vec{a}, \vec{b}') + P(\vec{a}', \vec{b}) + P(\vec{a}', \vec{b}') - P_A(\vec{a}') - P_B(\vec{b}) := S_{\mathrm{CH}} \leq 0\,. \tag{34}$$

Inequality (34) has been used by Aspect in his time-flip experiment [124]. The two-photon state produced was the symmetrical Bell state $|\phi^+\rangle = \frac{1}{\sqrt{2}}(|R\rangle \otimes |L\rangle + |L\rangle \otimes |R\rangle) = \frac{1}{\sqrt{2}}(|H\rangle \otimes |H\rangle + |V\rangle \otimes |V\rangle)$, where $|R\rangle, |L\rangle$ denote the right and left handed circularly polarized photons and $|H\rangle, |V\rangle$ the horizontally and vertically polarized ones.

In case of $|\phi^+\rangle$ entangled photons the quantum mechanical probability to detect a linear polarized photon with an angle α on Alice's side, and simultaneously an other linear polarized one with angle β on Bob's side, is given by

$$P(\vec{a}, \vec{b}) = \left|\left[\left(\langle H|\cos\alpha + \langle V|\sin\alpha\right) \otimes \left(\langle H|\cos\beta + \langle V|\sin\beta\right)\right]|\phi^+\rangle\right|^2 = \frac{1}{2}\cos^2(\alpha - \beta)\,. \tag{35}$$

Choosing now for the Bell angles $(\alpha, \beta, \alpha', \beta') = (0, \frac{\pi}{8}, 2\frac{\pi}{8}, 3\frac{\pi}{8})$ the quantum mechanical probabilities (35) violate the Clauser-Horne inequality (34) maximally

$$S^{\mathrm{QM}}_{\mathrm{CH}} = \frac{\sqrt{2} - 1}{2} = 0.207 > 0\,. \tag{36}$$

The violation of the above discussed Bell inequalities is expressed by the following theorem:

Theorem 2 (Bell's Theorem 1964) *In a certain experimental situation* all *local realistic theories are incompatible with quantum mechanics !*

For a thorough discussion we refer to Ref. [145] and further literature can be found in the review article [146, 147].

Conclusions:

I remember very well, when I had derived Bell's inequality (23) for the first time, I was totally astonished and fascinated that quantum mechanics contradicted an inequality that relied on such general and quite 'natural' assumptions. It was impressive to see how John could turn the pure philosophical debate of Einstein and Bohr into exact mathematical terms. And, this formulation could be tested experimentally!

What are the conclusions? In all Bell inequalities the essential ingredient is Bell's locality hypothesis, Eq. (22), i.e., Einstein's vision of reality and Bell's concept of locality, therefore we have to conclude as expressed in Theorem 2:

Local realistic theories are incompatible with quantum mechanics !

Bell in his seminal work [44] realized the far reaching consequences of a realistic theory as an extension to quantum mechanics and expressed it in the following way:

"In a theory in which parameters are added to quantum mechanics to determine the results of individual measurements, without changing the statistical predictions, there must be a mechanism whereby the setting of one measuring device can influence the reading of another instrument, however remote. Moreover, the signal involved must propagate instantaneously, so that such a theory could not be Lorentz invariant."

He continued and stressed the crucial point in such EPR-type experiments: *"Experiments ..., in which the settings are changed during the flight of the particles, are crucial."*

Thus it is of utmost importance *not* to allow some mutual report by the exchange of signals with velocity less than or equal to that of light.

Historical Experiments

First Generation Experiments of the Seventies:

After John had published his paper about the inequality there was practically no interest in this field. It was the *'dark era'* of the foundations of quantum mechanics. Pauli's opinion was often cited [148]:

"One should no more rack one's brain about the problem of whether something one cannot know anything about exists at all, than about the ancient question of how many angels are able to sit on the point of a needle."

The first who got interested in the subject was John Clauser, a young graduate student from Columbia University, in the late sixties. When he studied Bell's inequality paper [44] that contained a bound for all hidden variable theories, he was astounded by its result. As a true experimentalist he wanted to see the experimental evidence for it. So he planed to do the experiment. However, experiments of this type were not appreciated at that time. When Clauser had an appointment with Richard Feynman at Caltech to discuss an experimental EPR configuration for testing the predictions of QM, he immediately threw him out of his office saying [149]:

"Well, when you have found an error in quantum-theory's experimental predictions, come back then, and we can discuss your problem with it."

But, fortunately, Clauser remained stubborn, he belonged to the revolting generation, and prepared the experiment. He sent an Abstract to the Spring Meeting of the American Physical Society proposing the experiment [150]. Soon afterwards, Abner Shimony called and told him that he and his student Michael Horne had the same ideas. So they joined and wrote together with Richard Holt, a PhD student of Francis Pipkin from Harvard, the famous CHSH paper [140], where they proposed an inequality that was well adapted to experiments.

Clauser carried out the experiment in 1972 together with Stuart Freedman [151], a graduate student at Berkeley, who received his PhD with this experiment. As pointed out in the CHSH paper [140], pairs of photons emitted in an atomic radiative cascade would be suitable for a Bell inequality test. Clauser and Freedman chose Calcium atoms pumped by lasers, where the excited atoms emitted the desired photon pairs. The signals were very weak at that time, a measurement lasted for about 200 hours. For comparison with theory a very practical inequality was used, which was derived by Freedman [152]. The outcome of the experiment is well known, they obtained a clear violation of the Bell inequality very much in accordance with QM.

In 1976, at Houston Edward Fry and his student Randall Thompson set up an experiment by using mercury atoms. As in Clauser's experiment the correlated photons were produced in a radiative cascade from by lasers excited atomic levels. Due to the much better signals with improved lasers they could collect enough data already in 80 min. The result was in excellent agreement with QM, the Bell inequality was violated by 4 standard deviations [153].

Thus at that time, it was already convincing that hidden variable theories did not work but quantum mechanics was correct. However, the experiments were not perfect yet, the analyzers were static, only a small amount of photon pairs were registered, etc. There still existed several loopholes, the detection efficiency or fair sampling loophole, and the communication absence or freedom of choice loophole, just to mention some important ones. To close these loopholes was the challenge of the future experiments.

Second Generation Experiments of the Eighties:

In the late 1970s and beginning of the 1980s, the general atmosphere in the physics community was still such:

"Quantum mechanics works very well, so don't worry!"

I remember, in 1980 I stayed for some time at the Rockefeller University. There I met Abraham Pais, an outstanding particle physicist, with whom I had several stimulating discussions. He had published a bestseller *"Subtle is the Lord: The Science and the Life of Albert Einstein"* [154], where he described very thoroughly all the works of Einstein. However, the EPR paper was, in my opinion, treated a bit poor and not with his usual enthusiasm for Einstein. Pais summarized (on p.456 of his book [154]):

"The content of this paper has been referred to on occasion as the Einstein-Podolsky-Rosen paradox. It should be stressed that this paper contains neither a paradox nor any flaw of logic. It simply concludes that objective reality is incompatible with the assumptions that quantum mechanics is complete. This conclusion has not affected subsequent developments in physics, and it is doubtful that it ever will."

Having read this I felt somehow unease about his EPR assessment. So I asked him frankly: *"You don't appreciate the EPR paper?"* And with an impish smile Pais responded: *"The EPR paper was the only slip Einstein made."* How wrong can be sometimes the judgement and prophecy of a physical work!

Alain Aspect, on the other hand, when reading Bell's inequality paper [44], was so strongly impressed that he immediately decided to do his *"thèse d'état"* on this fascinating topic. He visited John Bell at CERN to discuss his proposal. John's first question was, as Alain told me, *"Do you have a permanent position?"* Only after Aspect's positive answer the discussion could begin. Aspect's goal was to include variable analyzers.

Aspect and his collaborators performed a whole series of experiments [124, 155–158] with an improved design and approached step by step the 'ideal' setup configuration. As Clauser, they chose a radiative cascade in calcium that emitted photon pairs in the Bell state $|\phi^+\rangle$. For comparison with theory the Clauser-Horne inequality (34) was used, which was significantly violated in each experiment.

In the final time-flip experiment [124] together with Jean Dalibard and Gérard Roger a clever acoustic-optical switch mechanism was incorporated. It worked such that the switching time between the polarizers, as well as the lifetime of the photon cascade, was much smaller than the time of flight of the photon pair from the source to the analyzers. That implied a space-like separation of the event intervals. However, the time flipping mechanism was still not ideal, i.e., truly random, but *"quasi-periodic"*, as they called it. The mean for two runs which lasted about 2 h yielded the result $S_{\mathrm{CH}}^{\mathrm{exp}} = 0.101 \pm 0.020$ in very good agreement with the quantum mechanical prediction $S_{\mathrm{CH}}^{\mathrm{QM}} = 0.113 \pm 0.005$ that had been adapted for the experiment (recall the ideal value is $S_{\mathrm{CH}}^{\mathrm{QM}} = 0.207$ (36)).

This time-flip experiment of Aspect received much attention in the physics community and also in popular science, and Alain was the best apologist. In my opinion, it caused a turning point, the physics community began to realize that there was something essential in it. The research started and flourished into a new direction, into what is called nowadays quantum information and quantum communication [159, 160].

Third Generation Experiments of the Nineties and Beyond:

In the 1990s, the spirit towards foundations in quantum mechanics totally changed since quantum information gained increasing interest, Bell inequalities and quantum entanglement were the basis.

Meanwhile, the technical facilities improved considerably too, the electronics and the lasers. Most important was the invention of a new source for creating two entangled photons. That was spontaneous parametric down conversion, where a nonlinear crystal was pumped with a laser and the pump photon was converted into two photons that propagated vertically and horizontally polarized on two different cones. In the overlap region they were entangled.

Such an EPR source was used by Anton Zeilinger and his group, when they performed their Bell experiment in 1998 [141]. Zeilinger's student Gregor Weihs obtained his PhD with this experiment [142] (see Fig. 3.14). Their challenging goal was to construct an ultra-fast and truly random setting of the analyzers at each side of Alice and Bob, such that strict Einstein locality—no mutual influence between the two observers Alice and Bob—was achieved in the experiment. The data were compared with the CHSH inequality (26) and the experimental result was: $S^{\text{exp}}_{\text{CHSH}} = 2.73 \pm 0.02$, which corresponded to a violation of the inequality of 30 standard deviations. Due to this high efficiency photon source the measurement could be performed already in 3–4 min. It was *the* experiment that truly included the vital time factor, John Bell insisted upon so strongly.

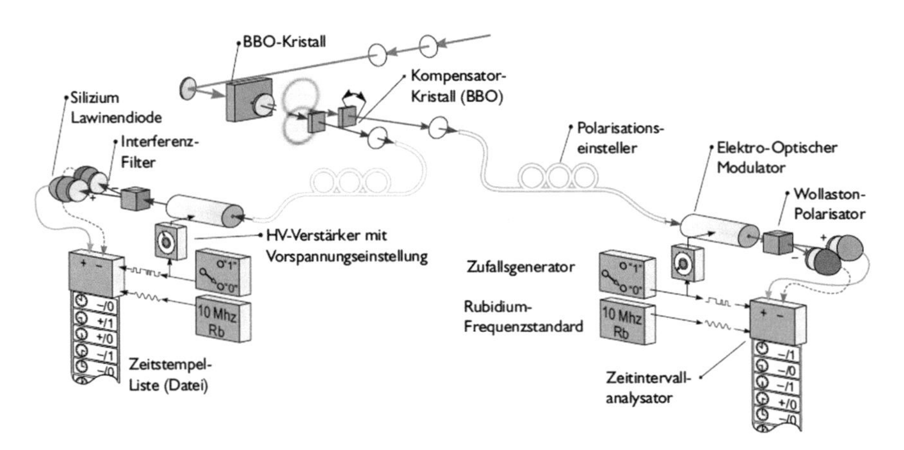

Fig. 3.14 The timing experiment of Weihs et al. [141]. The EPR source is a so-called BBO crystal pumped by a laser, the outgoing photons are vertically and horizontally polarized on two different cones and in the overlap region they are entangled. This entangled photons are led separately via optical fibres to the measurement stations Alice and Bob. During the photon propagation the orientations of polarizations are changed by an electro-optic modulator which is driven by a truly random number generator, on each side. In this way the strict Einstein locality condition—no mutual influence between the two observers Alice and Bob—is achieved in the experiment. The figure is taken from Ref. [142], © Gregor Weihs.

About the same time other groups investigated energy correlated photon pairs to test Bell inequalities [161, 162]. A record was set by the group of Gisin et al. [163], by using energy-time entangled photon pairs in optical fibres. They managed to separate their observers Alice and Bob by more than 10 Km and could show that this distance had practically no effect on the entanglement of the photons. The investigated Bell inequalities had been violated by 16 standard deviations.

Fascinating experiments on quantum teleportation [164, 165] and quantum cryptography [166, 167] followed.

Then a race started in achieving records of entanglement based long distance quantum communication. The vision was to install a global network, in particular via satellites or the International Space Station, that provided an access to secure communication via quantum cryptography at any location.

It was again Zeilinger's group that pushed the distance limits further and further. Firstly, in an open air experiment in the City of Vienna over a distance of 7.8 km the group [168] could violate a CHSH inequality (26) by more than 13 standard deviations. Secondly, this is presently the world record, the group [169] successfully carried out an open air Bell experiment over 144 km between the two Canary Islands La Palma and Tenerife.

In search of closing loopholes a recent Bell experiment of the group [170] closed the fair-sampling loophole, i.e., their results of violating an inequality à la Eberhard [171] were valid without assuming that the sample of measured photons accurately represented the entire ensemble.

Another loophole, the detection efficiency loophole, could be closed with ion traps. Working with ions the group Rowe et al. [172] tested a Bell inequality with perfect detection efficiency.

Finally, I also want to refer to Bell inequality tests of the group of Rauch et al. [173–175]. These neutron interferometer experiments were of particular interest since in this case the quantum correlations were explored in the degrees of freedom of a single particle, the neutron. Physically, it meant that rather contextuality was tested than nonlocality in space.

It is quite interesting and amusing to see the development of Bell experiments in the history of time. Beginning in the 1970s, where one had to overcome huge technical difficulties and the enormous resistance of the physics community, the development ended in the 2010s in such a way that a Bell experiment belonged already to the standard educational program *"Laboratory Quantum Optics"* for the students at the Faculty of Physics of the University of Vienna. It would have been nice if John Bell could have seen that!

For a detailed description of all Bell-type experiments the reader may consult *Quantum [Un]speakables* [160].

Some Memories and John's Quantum Legacy

When I think back again and recall the sometimes lively discussions I had with John about quantum mechanics and its interpretations, in particular, about the meaning of contextuality and nonlocality, then interestingly, John was never so much concerned about contextuality and its implications. Whereas, I always thought that contextuality was *the* important quantum feature and had a profound rooting in Nature, and I am still convinced that some day it will have technical applications.

John, on the other hand, was deeply disturbed by the nonlocality feature of quantum mechanics since for him it was equivalent to a *"breaking of Lorentz invariance"* in an extended theory for quantum mechanics, what he hardly could accept. He often remarked: *"It's a great puzzle to me ... behind the scenes something is going faster than the speed of light."*

At the end of his Bertlmann's socks paper John expressed again his concern [123]: *"It may be that we have to admit that causal influences do go faster than light. The role of Lorentz invariance of a completed theory would then be very problematic. An 'ether' would be the cheapest solution. But the unobservability of this ether would be disturbing. So would be the impossibility of 'messages' faster than light, which follows from ordinary relativistic quantum mechanics in so far as it is unambiguous and adequate for procedures we can actually perform. The exact elucidation of concepts like 'message' and 'we', would be a formidable challenge."*

When John gave a talk on the foundational issues, there often arose a great tension between him and the audience, especially, about the item of nonlocality. People didn't want to listen, didn't want to accept what John was saying, John was like *'a lone voice in the wilderness'*. Once, in the late 1980s, he gave a talk at the ETH Zürich and I asked him afterwards, *"How was it?"* John replied clenching his fists, *"I could beat them!"*

On a summer afternoon in 1987, John and I were sitting outside in the garden of the CERN cafeteria, drinking our late *4 o'clock tea*, and talked as so often about the implications of nonlocality. In this tea-atmosphere I spontaneously said: *"John, you deserve the Nobel prize for your theorem."* John, for a moment puzzled, replied quite strictly: *"No, I don't. ... it's like a null experiment, and you don't get the Nobel prize for a null experiment. ... for me, there are Nobel rules as well, it's hard to make the case that my inequality benefits mankind."* I countered: *"I disagree with you! It's not a null result. You have proved something new, nonlocality! And for that I think you deserve the Nobel Prize."* John, although feeling somehow pleased, raised slowly his arms, shrugged his shoulders and mumbled sadly: *"Who cares about this nonlocality?"*

In John's last paper *"La nouvelle cuisine"*, published in 1990 [176] (and see his collected quantum works [177]), he still remained profoundly concerned with this nonlocal structure of Nature. The paper was based on a talk he gave at the University of Hamburg, in 1988, about the topic: *"What cannot go faster than light?"*. Somebody with Hanseatic humor added to the announcement by hand: *"John Bell, for example!"* (see Fig. 3.15). This remark made John thinking, what exactly that meant, his whole body or just his legs, his cells or molecules, atoms, electrons ... Was it meant that none of his electrons go faster than light?

In our modern view of Nature the concepts of a classical theory changed, the sharp location of objects had been dissolved by the fuzzyness of the wave function or by

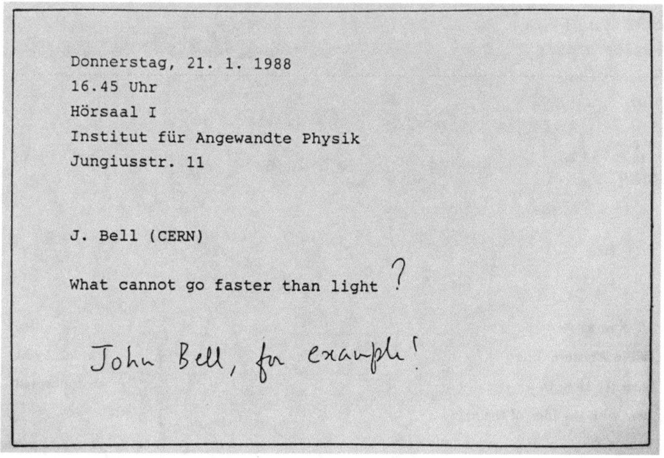

Donnerstag, 21. 1. 1988
16.45 Uhr
Hörsaal I
Institut für Angewandte Physik
Jungiusstr. 11

J. Bell (CERN)

What cannot go faster than light?

John Bell, for example!

Fig. 3.15 Announcement of John Bell's talk: *"What cannot go faster than light?"* at the University of Hamburg in 1988. Somebody with Hanseatic humor added to the announcement by hand: *"John Bell, for example!"*.

the fluctuations in quantum field theory. As John remarked: *"The concept 'velocity of an electron' is now unproblematic only when not thought about it."*

Finally, John discussed *"Cause and effect"* in this paper. As Einstein [178] already pointed out, if an effect follows its cause faster than the propagation of light, then there exists an inertial frame where the effect happens before the cause. Such a thing was unacceptable for both, Einstein and Bell. Therefore, sticking to *"no signals faster than light"* John defined locally causal theories and demonstrated, via an EPR-Bell type experiment, that *"ordinary quantum mechanics is not locally causal"*, or more precisely, *"quantum mechanics cannot be embedded into a locally causal theory"*. It was essential in his argumentation that the measurement settings $\vec{a}$ at Alice's side and $\vec{b}$ at Bob's side could be chosen totally free, i.e., at random. *"But still, we cannot signal faster than light"* John noted at the end.

Let me finally cite John's point of view concerning the existence, the realism of Nature, John in his own words, taken from an interview he gave in the late 1980s.[2]

John's confession:

"Oh, I'm a realist and I think that idealism is a kind of ... it's a kind of ... I think it's an artificial position which scientists fall into when they discuss the meaning of their subject and they find that they don't know what it means. I think that in actual daily practice all scientists are realists, they believe that the world is really there, that it is not a creation of their mind. They feel that there are things there to be discovered, not a world to be invented but a world to be discovered. So I think that realism is a natural position for a scientist and in this debate about the meaning of quantum mechanics I do not know any good arguments against realism."

[2]The whole interview with John Bell can be seen on a DVD available at the Austrian Central Library for Physics and Chemistry, Vienna.

Fig. 3.16 Sketch of my conclusions in the paper *"Bell's theorem and the nature of reality"*, which I dedicated John Bell in 1988 on occasion of his 60th birthday [179]. Cartoon: © Reinhold A. Bertlmann.

John was totally convinced that *realism* is the right position of a scientist. He believed that the experimental results are predetermined and not just induced by the measurement process. Even more, in John's EPR analysis reality is not assumed but inferred! Otherwise (without realism), he said, *"It's a mystery if looking at one sock makes the sock pink and the other one not-pink at the same time."*

So he did hold on the hidden variable program continuously, and was not discouraged by the outcome of the EPR-Bell experiments but rather puzzled. For him *"The situation was very intriguing that at the foundation of all that impressive success* (of quantum mechanics) *there are these great doubts"*, as he once remarked.

I got back at John for *"Bertlmann's socks"* in paper *"Bell's theorem and the nature of reality"* [179] that I dedicated to him in 1988 on occasion of his 60th birthday. I sketched my conclusions in a cartoon, shown here as Fig. 3.16. John, as a strict teetotaler, was very much amused by my illustration, since the *spooky, nonlocal ghost* emerged from a bottle of *Bell's Whisky*, a brand that *really* did exist.

In the 1990s, Bell's huge impact on the developments of quantum information became widely appreciated. It is known that in 1990 (the year of John's unexpected death) John was appointed for the Nobel Prize. Nowadays physicists agree that John would have definitely received the prize for his outstanding contributions to the foundations of quantum mechanics if he had lived longer. For instance, Daniel Greenberger expressed it explicitly in an interview given at the Conference *Quantum [Un]Speakables II* in Vienna [180]:

"Of course, people more and more appreciate John Bells beautiful work. He was essentially starting the field, his work was totally seminal, and if he were alive he certainly would have won the Nobel Prize!"

Turn to Quantum Mechanics

A year after John's death Franco Selleri, Professor at the Universitá degli Studi di Bari Aldo Moro, and his crew organized the Conference *"Bell's Theorem and the Foundations of Modern Physics"* (7–10 October, 1991, Cesena). It was an international conference in memory of John Bell. Many leading scientists of the field participated and gave talks to honour posthumously John Bell. Among them where: Jeffrey Bub, James Cushing, Bernard d'Espagnat, Giancarlo Ghirardi, Daniel Greenberger, Max Jammer, Leonard Mandel, Sir Roger Penrose, Franco Selleri, Euan Squires, Anton Zeilinger, myself and many others.

There I met Anton Zeilinger, he was one of the two people who responded to my *'Bell-revenge'* paper [179]. Anton just became a Professor at the University of Innsbruck and was establishing a quantum group. At that time I still worked in particle physics, but we found an overlapping interest in the foundations of quantum mechanics. Since we both were fascinated by this topic we thought that it would be a good idea to educate the young Austrian generation in this field. So we intensified the contact and exchange of our universities, which, in 1994, resulted in founding officially the Joint Seminar *"Foundations of Quantum Mechanics"* between the universities of Vienna and Innsbruck. For the meetings Anton and his group came to Vienna, and alternatingly, the Vienna students were travelling by train to Innsbruck, see Fig. 3.17.

This Seminar immediately became very popular among the Viennese students since Anton's group reported on fascinating experiments carried out in Innsbruck, and the experiments were performed by young scientists, in fact by students, what impressed our students very much. Also the quite informal, familiar character of the Seminar—we always served coffee and cake—added to the success of the Seminar. One event is unforgettable, when Alois Mair, a student of Anton's group reported on the first experiment of *"Entangled states of orbital angular momentum of photons"*(published in Ref. [181]). In such states the phase surface of the wave resembled a screw in direction of the wave propagation. At the filters it impressively looked like a doughnut. After the talk we had a big doughnut party, where 70 *real* doughnuts, delivered by the nearby bakery *"Ritz"*, had been served. Meanwhile the Seminar belongs to the regular student educational programme of our Faculty.

Of course, after Anton's move from Innsbruck to Vienna in 1999, we further intensified our collaboration in several areas, for instance, in organizing conferences like *"Quantum [Un]Speakables"* in 2000 in commemoration of John Bell, *"Quantum [Un]Speakables II"* in 2014 to celebrate 50 years of Bell's Theorem, see Fig. 3.17. In 2014 there were celebrations of Bell's quantum achievements all over the world, also Queen's University Belfast, John's home university, organized an exhibition *"Action*

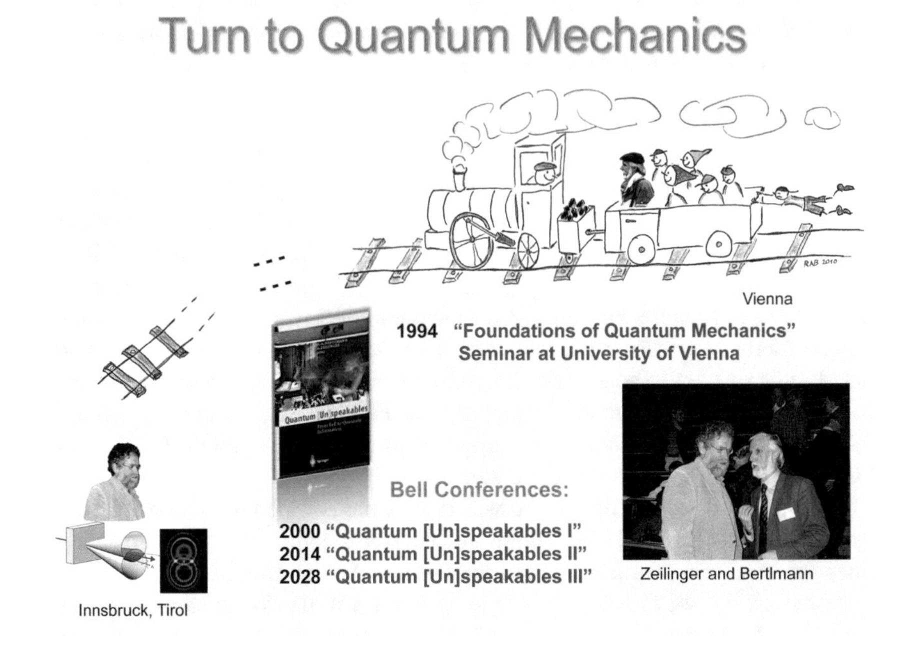

Fig. 3.17 Activities of Anton Zeilinger and Reinhold Bertlmann: Establishing in 1994 *"Foundations of Quantum Mechanics"*, a Joint Seminar between the Universities of Vienna and Innsbruck. Organizing the Conferences *"Quantum [Un]Speakables I"* 2000, *"Quantum [Un]Speakables II"* 2014, and *"Quantum [Un]Speakables III"* 2028. Cartoon: © Reinhold A. Bertlmann, Foto: © Renate Bertlmann.

at a Distance" in The Naughton Gallery at Queen's [182] and the Belfast City Council named a street *"Bell's Theorem Crescent"* in Belfast's Titanic Quater to honour John Bell as *"One of the Northern Ireland's most eminent scientists"*. Finally, for 2028 a third conference *"Quantum [Un]Speakables III"* had been announced to commemorate Bell's 100th birthday [183], see Fig. 3.17.

This fascination for the foundations of quantum mechanics also stimulated my research interest in this field and I began to study this topic in the area I was familiar with, that was particle- and mathematical physics. About this research I want to report next.

Entanglement in Particle Physics

Decoherence of Entangled Particle–Antiparticle Systems

The second person who responded to my *'Bell-revenge'* paper [179] was Walter Grimus, a distinguished particle physicist of our Institute in Vienna. He was an expert in *strangeness systems*, the K-mesons, and in *beauty systems*, the B-mesons. So it was quite natural that we discussed the phenomena of quantum information, the peculiar quantum correlations of bipartite and multipartite systems, within these systems in particle physics.

The difference to the photon systems, discussed so far, was that particle systems had entirely different and additional properties, which the photons did not have. First of all, the investigated particles were very massive, they decayed into other particles, they oscillated between their flavour content, i.e., between their particle and antiparticle nature, and they could regenerate, e.g., once the short-lived kaon state had decayed it could be regenerated from the surviving long-lived component. In addition, they possessed internal symmetries, like the *CP* symmetry (charge conjugation and parity), which turned out to be essential. For these reasons, I think, that it was, and still is, of great importance to investigate such systems, particularly, with regard to the EPR-Bell quantum correlations.

Experimentally, the particle–antiparticle systems that were generated in the huge particle accelerators were already entangled due to conservations laws. For example, the $K^0\bar{K}^0$ system was produced at the Φ resonance in the e^+e^- machine DAΦNE at Frascati, for a sketch see Fig. 3.18, and the $B^0\bar{B}^0$ system at the $\Upsilon(4S)$ resonance in machines like DORIS II (Doppel Ring Speicher) at DESY (Deutsches Elektronen Synchrotron), in CESR (Cornell Electron Storage Ring) at Cornell, or nowadays in the KEK B-factory in Japan.

The first system Walter and I investigated was the $B^0\bar{B}^0$ system since there existed already usable data from DORIS II and CESR. We found that the $B^0\bar{B}^0$ state generated in the decay of the resonance $\Upsilon(4S)$ at 10.6 GeV is very well suited to perform tests of the EPR correlations over macroscopic distances. Using measurements of the ratio R = (*No. like-sign dilepton events*)/(*No. opposite-sign dilepton events*) we could show that already presently existing data strongly favoured the contribution of the interference term to R, as it was required by the rules of quantum mechanics [184, 185].

The next system we explored was the $K^0\bar{K}^0$ system, where data was available from the CPLEAR experiment at CERN [186]. It was precisely the time, when a

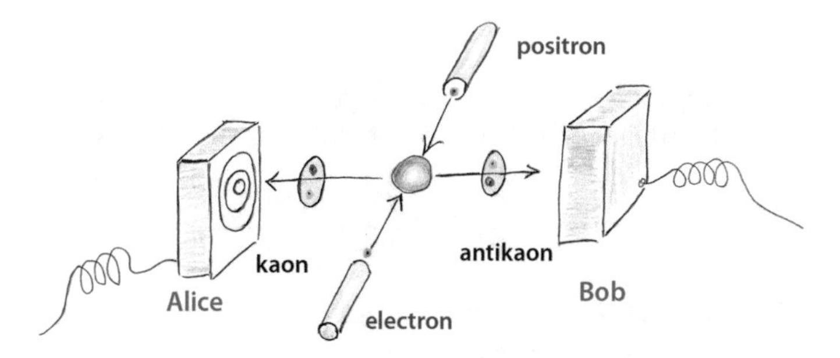

Fig. 3.18 Entanglement of matter and antimatter created in an accelerator of particle physics. Cartoon: © Reinhold A. Bertlmann.

young enthusiastic graduate student, with name Beatrix Hiesmayr, approached me searching for a Diploma Thesis. I thought, that's a good topic for her, so she joined and it was the starting point for fruitful and still ongoing collaboration.

Our aim was to show, if a pair of particle–antiparticle had been created by any kind of interaction in an entangled state, the two-particle wave function retained its non-separable character even if the particles were space-like separated over large macroscopic distances (about 9 cm). To describe quantitatively spontaneous factorization and decoherence of the $K^0\bar{K}^0$ system we modified the quantum-mechanical interference term of the entangled 2-kaon state by a multiplication with the term $(1-\zeta)$. Thus we changed the quantum mechanical expression in such a way that the effective decoherence parameter ζ quantified the deviation from quantum mechanics (corresponding to $\zeta = 0$) and provided a measure for the distance of the total system from its total decoherence or spontaneous factorization ($\zeta = 1$).

The relevant quantities we had to calculate were the probabilities to measure *like-strangeness* and *unlike-strangeness* events at a time t_l on the left side and t_r on the right side. Starting from a produced asymmetric Bell state $|\psi^-\rangle = \frac{1}{\sqrt{2}}(|K_S\rangle \otimes |K_L\rangle - |K_L\rangle \otimes |K_S\rangle)$, where the short-lived kaon K_S and the long-lived kaon K_L played the role of spin up $\Uparrow$ and spin down $\Downarrow$, we found for the *like-strangeness probability* [187] (see also reviews [188, 189])

$$
\begin{aligned}
P(K^0, t_l; K^0, t_r) &= ||\langle K^0|_l \otimes \langle K^0|_r \; |\psi^-(t_l, t_r)\rangle||^2 \\
\longrightarrow \quad P_\zeta(K^0, t_l; K^0, t_r) &= \frac{1}{2}\Big\{ e^{-\Gamma_S t_l - \Gamma_L t_r} |\langle K^0|K_S\rangle_l|^2 \; |\langle K^0|K_L\rangle_r|^2 \\
&+ e^{-\Gamma_L t_l - \Gamma_S t_r} |\langle K^0|K_L\rangle_l|^2 \; |\langle K^0|K_S\rangle_r|^2 \; - \; 2 \; \underbrace{(1-\zeta)}_{\text{modification}} \; e^{-\Gamma(t_l+t_r)} \\
&\times \mathrm{Re}\big\{ \langle K^0|K_S\rangle_l^* \langle K^0|K_L\rangle_r^* \langle K^0|K_L\rangle_l \langle K^0|K_S\rangle_r \; e^{-i\Delta m \Delta t} \big\} \Big\} \\
&= \frac{1}{8}\Big\{ e^{-\Gamma_S t_l - \Gamma_L t_r} + e^{-\Gamma_L t_l - \Gamma_S t_r} \; - \; 2 \; \underbrace{(1-\zeta)}_{\text{modification}} \; e^{-\Gamma(t_l+t_r)} \cos(\Delta m \Delta t) \Big\},
\end{aligned}
\tag{37}
$$

and the *unlike-strangeness probability* just changed the sign of the interference term.

The quantity directly sensitive to the interference term was the asymmetry

$$
A(t_r, t_l) = \frac{P_{\text{unlike}}(t_r, t_l) - P_{\text{like}}(t_r, t_l)}{P_{\text{unlike}}(t_r, t_l) + P_{\text{like}}(t_r, t_l)}, \tag{38}
$$

that had been measured in the CPLEAR experiment. The comparison of the theoretical expression for the given basis $\{K_S, K_L\}$ [187]

$$A_{\zeta}^{K_L K_S}(t_r, t_l) = (1-\zeta) A^{\mathrm{QM}}(t_r, t_l) \qquad \text{with} \qquad A^{\mathrm{QM}}(t_r, t_l) = \frac{\cos(\Delta m\,\Delta t)}{\cosh(\frac{1}{2}\Delta\Gamma\Delta t)}, \tag{39}$$

where $\Delta m = m_L - m_S$, $\Delta\Gamma = \Gamma_L - \Gamma_S$ and $\Delta t = t_r - t_l$, with the CPLEAR data [186] restricted the decoherence parameter ζ to the interval [187]

$$\zeta = 0.13^{+0.16}_{-0.15}\,. \tag{40}$$

This result confirmed nicely in a quantitative way the existence of entangled massive particles over macroscopic distances (9 cm).

When concentrating on a specific decay process of the kaon system $\phi \to K_S K_L \to \pi^+\pi^-\pi^+\pi^-$ the quantitative estimate of the ζ could be improved by orders of magnitude [190]

$$\zeta = 0.003 \pm 0.018_{\mathrm{stat}} \pm 0.006_{\mathrm{sys}}\,. \tag{41}$$

We also found out that the decoherence parameter ζ, which we had introduced by hand, also had a deeper physical basis. It was related to the decoherence strength λ of a Lindblad [191] and Gorini-Kossakowski-Sudarshan [192] master equation for the density matrix ρ of the total quantum system

$$\frac{d\rho}{dt} = -iH\rho + i\rho H^{\dagger} - D[\rho]\,. \tag{42}$$

The *dissipator* $D[\rho]$ was chosen as

$$D[\rho] = \lambda\left(P_1\rho P_2 + P_2\rho P_1\right) = \frac{\lambda}{2}\sum_{j=1,2}\left[P_j, [P_j, \rho]\right], \tag{43}$$

with the projectors $P_j = |e_j\rangle\langle e_j|$ ($j = 1, 2$) onto the states $|e_1\rangle = |K_S\rangle_l \otimes |K_L\rangle_r$ and $|e_2\rangle = |K_L\rangle_l \otimes |K_S\rangle_r$. Then the connection was [188, 193]

$$\zeta(t) = 1 - e^{-\lambda t}\,. \tag{44}$$

The parameter λ representing the strength of the interaction of the system with its environment had to be considered as the more fundamental one.

The increase of decoherence of the initially totally entangled $K^0\bar{K}^0$ system as time evolves means on the other hand a decrease of entanglement of the system, an entanglement loss. Interestingly, there is a direct relation between the decoherence parameter ζ, or λ, which quantifies the spontaneous factorization of the wave function, and the entanglement loss of the system that is defined via the entropy. The amount of entanglement is defined by the familiar measures: *Entanglement of formation E* [194] or *Concurrence C* [195–197].

Then we have the following proposition:

Proposition 1 (Bertlmann-Durstberger-Hiesmayr [198]) *The entanglement loss* $(1 - C)$ *or* $(1 - E)$ *equals the amount of decoherence:*

$$1 - C\big(\rho(t)\big) \;=\; \zeta(t)\,, \tag{45}$$

$$1 - E\big(\rho(t)\big) \;\doteq\; \frac{1}{\ln 2}\,\zeta(t) \;\doteq\; \frac{\lambda}{\ln 2}\,t\,. \tag{46}$$

Bell Inequalities in Particle Physics

Together with Beatrix I also turned to the topic of Bell inequalities in particle physics. The typical feature of these particle systems, e.g., of a kaon–antikaon system, is that the joint expectation value of a measurement at Alice's and Bob's detectors depends on both, on the flavour content, which corresponds to a quasi-spin property, and on the time of the measurement, once the system is created. More precisely, the expectation value for the combined measurement $E(k_a, t_a; k_b, t_b)$ is a function of the flavour k_a measured on the left side at a time t_a and on a (possibly different) k_b on the right side at t_b. Relying on the usual argumentation for Bell inequalities we could derive the following *Bell–CHSH inequality* [199] (see Ref. [188], for an overview in this field)

$$|E(k_a, t_a; k_b, t_b) - E(k_a, t_a; k_{b'}, t_{b'})| \\ + |E(k_{a'}, t_{a'}; k_b, t_b) + E(k_{a'}, t_{a'}; k_{b'}, t_{b'})| \;\leq\; 2\,, \tag{47}$$

which expressed both the freedom of choice in time *and* in flavour. Identifying $E(k_a, t_a; k_b, t_b) \equiv E(\vec{a}, \vec{b})$ we are back at the inequality (26) for the spin–$\frac{1}{2}$ case.

Therefore we may choose in a Bell inequality:

(I) *varying the flavour (quasi-spin) or fixing the time,*
(II) *fixing the flavour (quasi-spin) or varying the time.*

However, the experimental test of Bell inequalities in particle physics is much more intricate than in photon physics. Active measurements have to be carried out, but they are difficult to achieve. Usually, the measurements are passive since they happen through the decays of the particles, for a detailed analysis see Ref. [200].

Let me mention two important cases.

Case I: By varying the flavour content in the particle–antiparticle system a Wigner-type inequality, like Eq. (30), can be established for the kaon system [201]

$$P(K_S, \bar{K}^0) \;\leq\; P(K_S, K_1^0) \;+\; P(K_1^0, \bar{K}^0)\,. \tag{48}$$

Although inequality (48) cannot be tested directly, the *CP*-conserving kaon state K_1^0 does not exist in Nature, it can be converted into a Bell inequality for *CP* violation when studying the *leptonic charge asymmetry*

$$\delta = \frac{\Gamma(K_L \to \pi^- l^+ \nu_l) - \Gamma(K_L \to \pi^+ l^- \bar{\nu}_l)}{\Gamma(K_L \to \pi^- l^+ \nu_l) + \Gamma(K_L \to \pi^+ l^- \bar{\nu}_l)} \quad \text{with} \quad l = \mu, e \,, \tag{49}$$

where l represents either a muon or an electron.

Then Beatrix, Walter and myself [202] could convert the Wigner-type inequality (48) into the inequality

$$\delta \leq 0 \,, \tag{50}$$

for the measurable leptonic charge asymmetry which is proportional to *CP* violation. However, inequality (50) is in contradiction to the experimental value

$$\delta_{\text{exp}} = (3.27 \pm 0.12) \cdot 10^{-3} \,. \tag{51}$$

In fact, considering further Bell inequalities [202] restricts the asymmetry to $\delta = 0$, which means strict *CP* violation.

In conclusion, the premises of local realistic theories are *only* compatible with strict *CP* violation in $K^0\bar{K}^0$ mixing. Conversely, *CP* violation in $K^0\bar{K}^0$ mixing always leads to a *violation* of a Bell inequality. In this way, $\delta \neq 0$ is a manifestation of the entanglement of the considered state. I also want to remark that in case of Bell inequality (48), since it is considered at $t = 0$, it is rather contextuality than nonlocality that is tested. This connection between the violation of a Bell inequality and the violation of an internal symmetry of a particle is quite remarkable and must have a deeper meaning, and probably will occur for other symmetries as well.

Case II: When fixing the flavour of the kaons and varying the time of the measurements, it turns out that due to the fast decay compared to the slow oscillation, which increases the mixedness of the total system, a Bell inequality is not violated anymore by quantum mechanics [199]. However, Beatrix and a group of experimentalists and theorists [203] succeeded to establish a generalized Bell inequality for the $K^0\bar{K}^0$ system, which is violated by quantum mechanics in certain measurable time regions. In this case hidden variable theories are excluded. For such an experiment the preparations at DAΦNE for the KLOE-2 detector are in progress [204].

I also want to draw attention to possible experiments that test Bell inequalities by inserting a regenerator, that is a piece of matter, into the kaon beam [205–209]. These experiments are of particular interest since regeneration, a typical quantum feature of the K meson, is directly related to a Bell inequality.

Furthermore, a Bell test for quite a different system, a hyperon system like the $\Lambda\bar{\Lambda}$ system, has been studied [210, 211] and is experimentally planned by the FLAIR collaboration, Darmstadt.

Last but not least, tests of local realism in the decay of a charmed particle into entangled vector mesons should be mentioned as well [212–214].

These direct Bell-type tests of our basic concepts about matter are of utmost importance since there is always a slim chance of an unexpected result, despite the fact of the general success of quantum mechanics.

I think John Bell, who was both, a particle- and a quantum-physicist, had been pleased seeing the developments of this kind of experiments.

As a particle physicist I certainly was very much interested in studying the entanglement features in a relativistic setting. I was again lucky that an enthusiastic graduate student, Nicolai Friis whom I knew very well from my university lectures, approached me for a Diploma Thesis and showed his strong interest in the connection of quantum mechanics and relativity.

The first topic Nicolai and I investigated was *"Relativistic entanglement of two massive particles"* [215], see Fig. 3.19. We described the spin and momentum degrees of freedom of a system of two massive spin–$\frac{1}{2}$ particles as a 4 qubit system, one qubit for each of the two momenta and each of the two spins. Of course, relativistically spin and momentum of a particle were not independent of each other, what we had to take into account. Then we explicitly showed how the entanglement changed between different partitions of the qubits, when considered by different inertial observers [215]. Although the two particle entanglement corresponding to a partition into Alice's and Bob's subsystems was, as often stated in the literature, invariant under Lorentz boosts, the entanglement with respect to other partitions of the Hilbert space on the other hand, was not. It certainly did depend on the chosen inertial frame and on the initial state considered. This surprising feature we could understand clearly. The change of entanglement arose, because a Lorentz boost on the momenta of the particles caused a Wigner rotation of the spin, which in certain cases entangled the spin- with the momentum states. We systematically investigated the situation for different classes of initial spin states and different partitions of the 4 qubit space.

Furthermore, we studied the behaviour of Bell inequalities for different observers and demonstrated how the maximally possible degree of violation, using the Pauli-Lubanski spin observable, could be recovered by any inertial observer, when Lorentz transforming both the states *and* the observables such that each observer will measure the same expectation value if the correct measurement directions were chosen.

As a next step it was quite natural for Nicolai and me to consider non-inertial particle systems and to study the entanglement features there. For example, two observers shared a bipartite entangled state, where one observer was moving with uniform acceleration. There occurred an entanglement degradation in such a state by the accelerated motion, which was commonly attributed to the thermalization due to the Unruh effect. Whereas for bosonic modes the entanglement vanished in the infinite acceleration limit [216] there still remained a non-zero residual entanglement in case of (anti-) fermionic modes [217]. So we asked ourself if this residual entanglement could be used for quantum information tasks. The criterion for it was the inspection of a Bell inequality. The result published with our colleagues was *"Residual entanglement of accelerated fermions is not nonlocal"* [218].

The statement, more precisely, was the following: Two observers share a maximal entangled state of two fermions, where the entanglement decreases for increas-

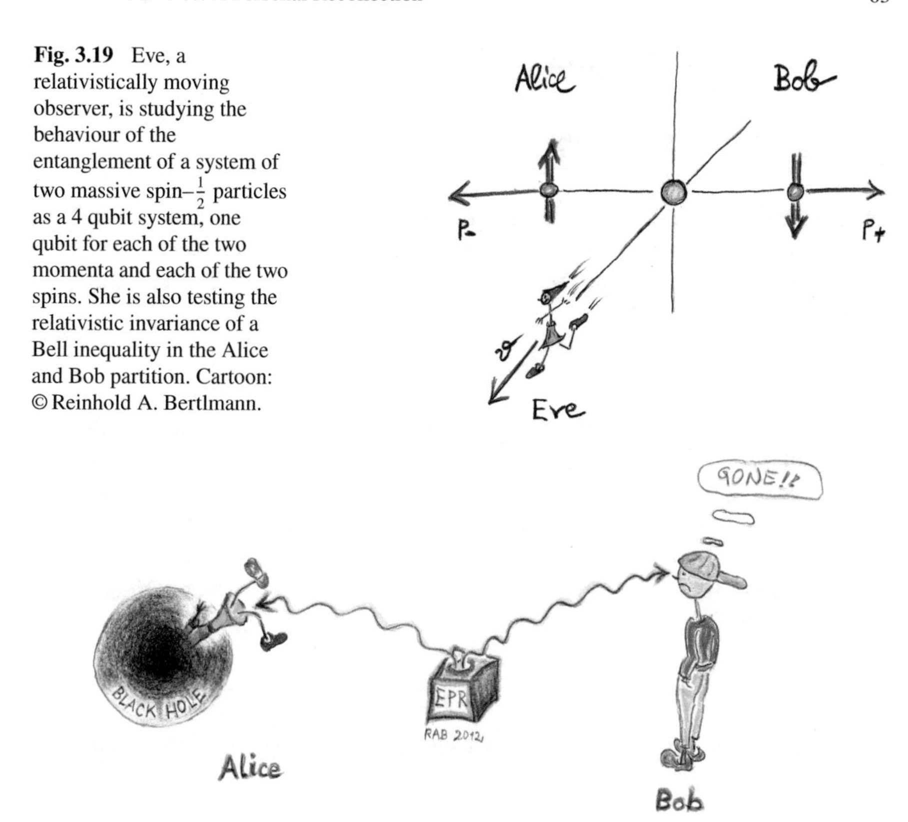

Fig. 3.19 Eve, a relativistically moving observer, is studying the behaviour of the entanglement of a system of two massive spin–$\frac{1}{2}$ particles as a 4 qubit system, one qubit for each of the two momenta and each of the two spins. She is also testing the relativistic invariance of a Bell inequality in the Alice and Bob partition. Cartoon: © Reinhold A. Bertlmann.

Fig. 3.20 Relativistic entanglement: Alice, being an accelerated observer when falling into a black hole, cannot communicate on a quantum information level (via an EPR pair) with an observer, Bob, who is resting near the horizon. *Thus quantum mechanics cannot overcome relativity!* Cartoon: © Reinhold A. Bertlmann.

ing acceleration. The surviving entanglement, in the infinitely accelerated frame, however, cannot be used to violate the CHSH inequality, which is the optimal Bell inequality for this situation. Therefore no quantum information tasks using these correlations can be performed!

This is especially important not only for the results in the infinite acceleration limit but also if we identify this limit with a black hole situation, where an observer is freely falling and another observer is resting arbitrarily close to the event horizon. Alice, when falling into a black hole, cannot communicate on a quantum information level with an observer who is resting near the horizon, see Fig. 3.20. Therefore we have to conclude:

Quantum mechanics cannot overcome relativity !

Entanglement in Mathematical Physics

Entanglement and Bell Inequalities

In mathematical physics the quantum states are described by density matrices. Then all quantum states can be classified into *separable* or *entangled* states. The *set of separable states* is defined by the convex (and compact) hull of product states

$$S = \left\{ \rho = \sum_i p_i \, \rho_i^A \otimes \rho_i^B \mid 0 \leq p_i \leq 1 \, , \sum_i p_i = 1 \right\} . \tag{52}$$

A state is called *entangled* if it is not separable, i.e., $\rho_{\text{ent}} \in S^c$ where S^c denotes the complement of S, with $S \cup S^c = \widetilde{\mathcal{H}} \subset L(\mathcal{H})$ and $\widetilde{\mathcal{H}} = \widetilde{\mathcal{H}}_A \otimes \widetilde{\mathcal{H}}_B$ represents the Hilbert-Schmidt space of linear operators $L(\mathcal{H})$ on the finite dimensional bipartite Hilbert space $\mathcal{H} = \mathcal{H}_A \otimes \mathcal{H}_B$ of Alice and Bob, with dimension $D = d_A \times d_B$. For our discussion of qubits $d_A = d_B = 2$.

In terms of density matrices the CHSH inequality (26) can be rewritten in the following way

$$\langle \rho | \mathcal{B}_{\text{CHSH}} \rangle \; = \; \text{Tr} \, \rho \mathcal{B}_{\text{CHSH}} \; \leq \; 2 \, , \tag{53}$$

for all local states ρ, where the CHSH-Bell operator in case of qubits is expressed by

$$\mathcal{B}_{\text{CHSH}} \; = \; \vec{a} \cdot \vec{\sigma}_A \otimes (\vec{b} - \vec{b}\,') \cdot \vec{\sigma}_B \; + \; \vec{a}\,' \cdot \vec{\sigma}_A \otimes (\vec{b} + \vec{b}\,') \cdot \vec{\sigma}_B \, . \tag{54}$$

Rewriting inequality (53) gives

$$\langle \rho | \, 2 \cdot \mathbb{1} \; - \; \mathcal{B}_{\text{CHSH}} \rangle \; \geq \; 0 \, . \tag{55}$$

If we choose, however, the entangled Bell state $\rho^- = |\psi^-\rangle\langle\psi^-|$ the inner product changes the sign

$$\langle \rho^- | \, 2 \cdot \mathbb{1} \; - \; \mathcal{B}_{\text{CHSH}} \rangle \; < \; 0 \, . \tag{56}$$

Now we can ask, is the inner product (56) negative for *all* entangled states? The answer is *yes* for all pure entangled states [219], i.e., there exist measurement directions for which the CHSH inequality is violated. For mixed states, however, the situation is much more subtle (see, e.g., Sect. 5 of Ref. [220]). Werner [221] discovered that a certain family of bipartite mixed states, which remained entangled, produced an outcome that admitted a local hidden variable model for projective measurements, that is, it satisfies all possible Bell inequalities.

This feature is nicely demonstrated by the so-called *Werner states*

$$\rho_{\text{Werner}} = \alpha \rho^- + \frac{1-\alpha}{4} \mathbb{1}_4 = \frac{1}{4} \left(\mathbb{1} \otimes \mathbb{1} - \alpha \sigma_i \otimes \sigma_i \right) \equiv \rho_\alpha , \tag{57}$$

written in terms of the Bloch decomposition with the parameter values $\alpha \in [0, 1]$.

The region of separability is determined by the so-called *PPT criterion* (positive partial transposition) of Peres [222] and the Horodecki family [223]. Given a general density matrix ρ in Hilbert-Schmidt space $\widetilde{\mathcal{H}} = \widetilde{\mathcal{H}}_A \otimes \widetilde{\mathcal{H}}_B$ in its Bloch decomposition form

$$\rho = \frac{1}{4} \left(\mathbb{1} \otimes \mathbb{1} + r_i \sigma_i \otimes \mathbb{1} + u_i \mathbb{1} \otimes \sigma_i + t_{ij}\, \sigma_i \otimes \sigma_j \right) , \tag{58}$$

then a *partial transposition* is defined by the operator T acting in a subspace $\widetilde{\mathcal{H}}_A$ or $\widetilde{\mathcal{H}}_B$ and transposing there the off-diagonal elements of the Pauli matrices: $T\,(\sigma^i)_{kl} = (\sigma^i)_{lk}$. If and only if, in 2×2 and 2×3 dimensions, a state remains positive under partial transposition then the state is separable. In higher dimensions the PPT criterion is only necessary but not sufficient for separability.

In case of the Werner states (57) the partial transposition provides the following result: the states are separable for $\alpha \leq \frac{1}{3}$ and entangled for $\alpha > \frac{1}{3}$.

In order to find the states violating a Bell inequality an other theorem of the Horodecki family [224] is very powerful since we do not have to check all measurement directions $\vec{a}$ and $\vec{b}$. There one has to consider the square root of the two larger eigenvalues t_1^2, t_2^2 of the product of the t-matrices $(t_{ij})^T(t_{ij})$. If $\mathcal{B}^{\max} = \frac{1}{2} \max_{\mathcal{B}} \operatorname{Tr} \rho\, \mathcal{B}_{\text{CHSH}} = \sqrt{t_1^2 + t_2^2} > 1$ then the CHSH inequality is maximal violated.

In case of the Werner states we can read off the coefficient matrix directly from the Bloch decomposition (57), which yields the maximal violation of the CHSH inequality by $\mathcal{B}^{\max} = \sqrt{2\alpha^2} > 1$. Thus, for all $\alpha > 1/\sqrt{2}$ the CHSH-Bell inequality is violated.

Entanglement Witness Inequality

I remember, when Anton and I organized the conference *"Quantum [Un]Speakables 2000"* Walter Thirring, our Doyen of Theoretical Physics, participated very actively. We had many discussions about Bell inequalities and their physical meaning. These resulted in an enjoyable collaboration and series of works together with Heide Narnhofer, a prominent mathematical physicist.

In *"A geometric picture of entanglement and Bell inequalities"* [225] we asked ourselves how to 'detect' entanglement and to discriminate it from all separable quantum states. A Bell operator given by expression (54) was obviously not appropriate to find all entangled states. In order to locate entanglement accurately a different operator had to be constructed. This was a Hermitian operator, the so-called *entanglement witness* A, that detected the entanglement of a state ρ_{ent} via an *entanglement witness inequality*. So we arrived at the following theorem.

Theorem 3 (Entanglement Witness Theorem [223, 225, 226])
A state ρ_{ent} is entangled if and only if there is a Hermitian operator A—the entanglement witness—such that

$$\begin{aligned} \langle \rho_{\text{ent}} | A \rangle \;=\; Tr\, \rho_{\text{ent}} A \;&<\; 0\,, \\ \langle \rho | A \rangle = Tr\, \rho A \;&\geq\; 0 \qquad \forall \rho \in S\,, \end{aligned} \tag{59}$$

where S denotes the set of all separable states.

An entanglement witness is called *optimal*, and denoted by A_{opt}, if apart from Eq. (59) there exists a separable state $\rho_0 \in S$ such that

$$\left\langle \rho_0 | A_{\text{opt}} \right\rangle \;=\; 0\,. \tag{60}$$

The operator A_{opt} defines a tangent plane to the convex set of separable states S (52), as illustrated in Fig. 3.21. Such an A_{opt} always exists due to the Hahn-Banach Theorem and the convexity of S.

On the other hand, with help of the Hilbert-Schmidt norm we can define the *Hilbert-Schmidt distance* between two arbitrary states ρ_1 and ρ_2

$$d_{\text{HS}}(\rho_1, \rho_2) \;=\; \|\rho_1 - \rho_2\| \;=\; \sqrt{< \rho_1 - \rho_2 | \rho_1 - \rho_2 >} \;=\; \sqrt{\text{Tr}\,(\rho_1 - \rho_2)^{\dagger}(\rho_1 - \rho_2)}\,. \tag{61}$$

We view the minimal distance of an entangled state ρ_{ent} to the set of separable states, the *Hilbert-Schmidt measure*

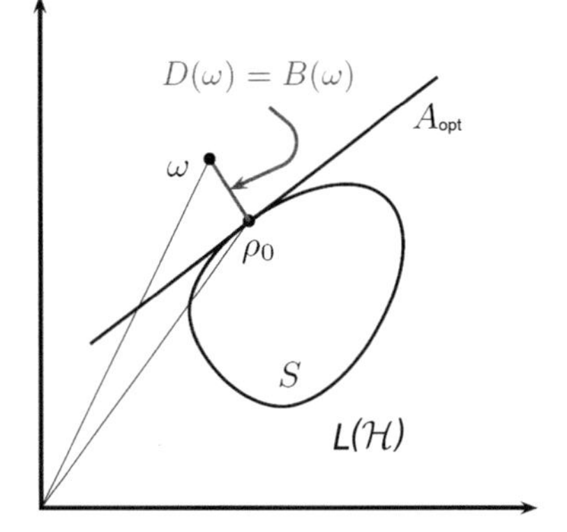

Fig. 3.21 Illustration of the Bertlmann-Narnhofer-Thirring Theorem: $D(\omega) = B(\omega)$, the minimal distance D of the entangled state ω to the set of separable states S in the Hilbert-Schmidt space is equal to the maximal violation B of the entanglement witness inequality. A_{opt} represents the optimal entanglement witness.

$$D(\rho_{\text{ent}}) := \min_{\rho \in S} \|\rho - \rho_{\text{ent}}\| = \|\rho_0 - \rho_{\text{ent}}\| , \tag{62}$$

where ρ_0 denotes the nearest separable state, as a *measure for entanglement.*

What we then discovered was an interesting connection between the Hilbert-Schmidt measure and the entanglement witness inequality. Let us rewrite entanglement witness inequality (59)

$$\langle \rho | A \rangle - \langle \rho_{\text{ent}} | A \rangle \geq 0 \qquad \forall \rho \in S , \tag{63}$$

and define the maximal violation of inequality (63) as follows (ρ and A are still free at our disposal):

Definition 1 (*Maximal violation of the entanglement witness inequality* [225])

$$B(\rho_{\text{ent}}) = \max_A \left(\min_{\rho \in S} \langle \rho | A \rangle - \langle \rho_{\text{ent}} | A \rangle \right) . \tag{64}$$

The minimum is taken over all separable states and maximum over all possible entanglement witnesses A, suitably normalized. Then there holds the following theorem:

Theorem 4 (Bertlmann-Narnhofer-Thirring Theorem [225])

$$(a) \ B(\rho_{\text{ent}}) = D(\rho_{\text{ent}}) , \tag{65}$$

(b) The maximal violation of the entanglement witness inequality is achieved when $\rho \to \rho_0$ *and* $A \to A_{\text{opt}}$ *, then the optimal entanglement witness is given by*

$$A_{\text{opt}} = \frac{\rho_0 - \rho_{\text{ent}} - \langle \rho_0 | \rho_0 - \rho_{\text{ent}} \rangle \mathbb{1}}{\|\rho_0 - \rho_{\text{ent}}\|} . \tag{66}$$

In words:

The maximal violation of the entanglement witness inequality is equal to the Hilbert-Schmidt measure !

As we regard the Hilbert-Schmidt measure (62) as a measure for entanglement, it means that the amount of entanglement is given by the amount of violation (64) of the witness inequality. This is a remarkable result that we have illustrated in Fig. 3.21. Furthermore, the optimal entanglement witness is given explicitly by expression (66). However, we have to know the nearest separable state ρ_0, which is easy to find in low dimensions, but not in higher ones. Nevertheless, there exists an approximation procedure to approach ρ_0 [227]. For a review, see Ref. [147].

For example, in case of Alice and Bob the Werner states are given by ρ_α (57), and the Bell state $\rho^- = |\psi^-\rangle\langle\psi^-|$ by choosing the parameter value $\alpha = 1$, i.e., $\rho^- = \rho_{\alpha=1}$.

The nearest separable state is easily found

$$\rho_0 = \frac{1}{4}\left(\mathbb{1} \otimes \mathbb{1} - \frac{1}{3}\sigma_i \otimes \sigma_i\right), \tag{67}$$

yielding the Hilbert-Schmidt measure

$$D(\rho_\alpha) = \|\rho_0 - \rho_\alpha\| = \frac{\sqrt{3}}{2}(\alpha - \frac{1}{3}). \tag{68}$$

The optimal entanglement witness we calculate from expression (66)

$$A_{\mathrm{opt}} = \frac{1}{2\sqrt{3}}\left(\mathbb{1} \otimes \mathbb{1} + \sigma_i \otimes \sigma_i\right), \tag{69}$$

and the maximal violation of the entanglement witness inequality from Eq. (64)

$$B(\rho_\alpha) = -\left\langle \rho_\alpha | A_{\mathrm{opt}} \right\rangle = \frac{\sqrt{3}}{2}(\alpha - \frac{1}{3}). \tag{70}$$

Clearly, both results (70) and (68) coincide as required by Theorem 4.

Geometry of Quantum States: Entanglement Versus Separability

Let us next turn to a geometrical description of the quantum states, how they are distributed in the Hilbert-Schmidt space of the density matrices. The quantum states for a two-qubit system, the case of Alice and Bob, have very nice geometric features in the Hilbert-Schmidt space, more precisely, in the spin–spin space. Quite generally a quantum state can be decomposed as in Eq. (58), where the last term, the spin–spin term, is the important one to characterize entanglement. If we parameterize the spin-spin space by

$$\rho = \frac{1}{4}\left(\mathbb{1} \otimes \mathbb{1} + \sum_i c_i \sigma_i \otimes \sigma_i\right), \tag{71}$$

the Bell states have the coefficients $c_i = \pm 1$.

Due to the positivity of the density matrix the four Bell states $\psi^-, \psi^+, \phi^-, \phi^+$ set up a simplex, a tetrahedron, in this spin–spin space [225, 228, 229], as illustrated by the two figures in Fig. 3.22. The separable states, given by the PPT criterion, form an octahedron which lies inside, and the maximal mixed state $\frac{1}{4}\mathbb{1}_4 = \frac{1}{4}(\mathbb{1} \otimes \mathbb{1})$ is placed at origin. The entangled states are located in the remaining cones. The local states, on the other hand, satisfying a CHSH-Bell inequality lie within the parachutes, the dark-yellow surfaces in the tetrahedrons of Fig. 3.22 [230]. They are determined by the Horodecki Theorem [224] and contain all separable but also a large amount of mixed entangled states. The Werner states (57), the red line in Fig. 3.22b from the origin to the maximal entangled Bell state ψ^-, show nicely how the states change from maximal mixed and separable to local, mixed entangled, and finally to nonlocal states, ending at ψ^- which is pure and maximal entangled.

There is an important point I became aware of in my discussions with Walter and Heide. For a given quantum state we have the free choice of how to factorize the algebra of a density matrix, which implies either entanglement or separability of the quantum state. Only with respect to such a factorization it makes sense to talk about entanglement or separability. For instance, quantum teleportation precisely relies on this fact that we can think of different factorizations in which entanglement is localized with respect to the measurements that are carried out. Thus we may choose! Via global unitary transformations we can switch from one factorization to the other, where in one factorization the quantum state appears entangled, however, in the other not. These discussions we published in the paper *"Entanglement or separability: The choice of how to factorize the algebra of a density matrix"* [230], where also Philipp Köhler, one of my last Diploma students, joined.

We realized that there is *"total democracy between the different factorizations"* [230], no partition has ontologically a superior status over any other one (see also Ref. [231]). For an experimentalist, however, a certain factorization is preferred and is clearly fixed by the set-up. Consequently, entanglement or separability of a quantum state depends on our choice of factorizing the algebra of the corresponding density matrix, where this choice is suggested either by the set-up of the experiment or by the convenience for the theoretical discussion. This was our basic message.

For pure states the status is quite clear. Any state can be factorized such that it appears separable up to being maximally entangled depending on the factorization. We can prove the following theorems [230]:

Theorem 5 (Factorization algebra) *For any pure state ρ one can find a factorization $M^D = \mathcal{A}_1 \otimes \mathcal{A}_2$ such that ρ is separable with respect to this factorization and an other factorization $M^D = \mathcal{B}_1 \otimes \mathcal{B}_2$ where ρ appears to be maximally entangled.*

The extension to mixed states requires some restrictions, as can be seen from the tracial state $\frac{1}{D}\mathbb{1}_D$ which is separable for any factorization.

Theorem 6 (Factorization in mixed states) *For any mixed state ρ one can find a factorization $M^D = \mathcal{A}_1 \otimes \mathcal{A}_2$ such that ρ is separable with respect to this factorization. An other factorization $M^D = \mathcal{B}_1 \otimes \mathcal{B}_2$ where ρ appears to be entangled exists only beyond a certain bound of mixedness.*

It is interesting to search for this bound, for those states which are separable with respect to all possible factorizations of the composite system into subsystems $\mathcal{A}_1 \otimes \mathcal{A}_2$. This is the case if $\rho_{\mathrm{U}} = U \rho\, U^\dagger$ remains separable for any unitary transformation U. Such states are called *absolutely separable states* [232–234], the tracial state being the prototype. In this connection the *maximal ball* of states around the tracial state $\frac{1}{d^2}\mathbb{1}_{d^2}$ with a general radius $r = \frac{1}{d^2-1}$ of constant mixedness is considered, which can be inscribed into the separable states. This radius is given in terms of the Hilbert-Schmidt distance

$$d(\rho, \mathbb{1}_{d^2}) = \left\| \rho - \frac{1}{d^2}\mathbb{1}_{d^2} \right\| = \sqrt{\operatorname{Tr}\left(\rho - \frac{1}{d^2}\mathbb{1}_{d^2}\right)^2}. \qquad (72)$$

Theorem 7 (Absolute separability of the Kuś-Życzkowski ball [232]) *All states belonging to the maximal ball which can be inscribed into the set of mixed states for a bipartite system are not only separable but also absolutely separable.*

The maximal ball of absolutely separable states we have illustrated by the green shaded ball in Fig. 3.22a.

As illustration of Theorem 7 let us choose the following separable state

$$\rho_{\mathrm{N}} = |\psi_{\mathrm{N}}\rangle\langle\psi_{\mathrm{N}}| = \frac{1}{4}\left(\mathbb{1}\otimes\mathbb{1} + \sigma_x\otimes\sigma_x\right). \qquad (73)$$

It is placed at the corner of the double pyramid of separable states (see Fig. 3.22a) and has the smallest possible mixedness or largest purity. The following unitary transformation [230]

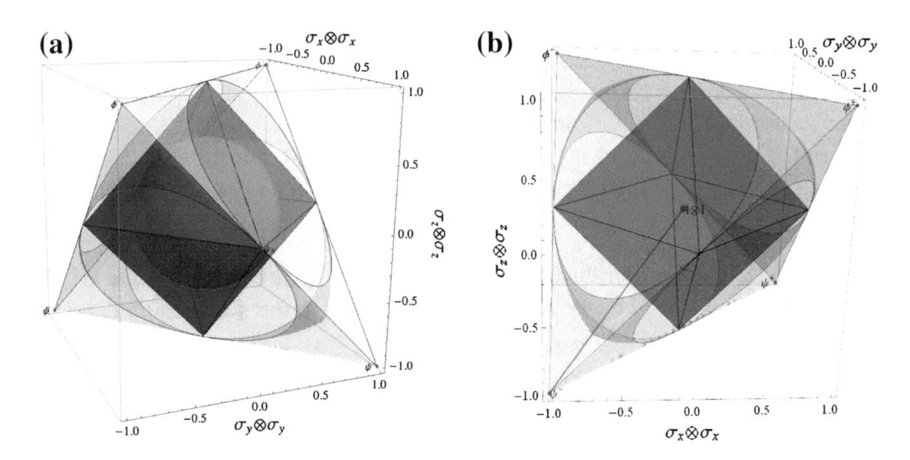

Fig. 3.22 **a** Tetrahedron of physical states in 2×2 dimensions spanned by the four Bell states $\psi^+, \psi^-, \phi^+, \phi^-$: The separable states form the blue double pyramid and the entangled states are located in the remaining tetrahedron cones. The unitary invariant Kuś-Życzkowski ball (shaded in *green*), the maximal ball of absolutely separable states, is located within the double pyramid and the maximal mixture $\frac{1}{4}\mathbb{1}_4$ is at the origin. Outside the ball at the corner of the double pyramid is the state ψ_{N}, the separable (but not absolutely separable) state with maximal purity. The local states according to a Bell inequality lie within the dark-yellow surfaces containing all separable but also some entangled states. **b** Tetrahedron with the illustration of the Werner states (*red line* from the origin to the maximal entangled Bell state ψ^-) that pass through all regions of separability, locality and entanglement.

$$U = \frac{1}{4}\left((2+\sqrt{2})\,\mathbb{1} \otimes \mathbb{1} + i\sqrt{2}\,(\sigma_x \otimes \sigma_y + \sigma_y \otimes \sigma_x) - (2-\sqrt{2})\,\sigma_z \otimes \sigma_z\right) \tag{74}$$

transforms the state ρ_{N} (73) into

$$\rho_{\mathrm{U}} = U \rho_{\mathrm{N}} U^{\dagger} = \frac{1}{4}\left(\mathbb{1} \otimes \mathbb{1} + \frac{1}{2}(\sigma_z \otimes \mathbb{1} + \mathbb{1} \otimes \sigma_z) + \frac{1}{2}(\sigma_x \otimes \sigma_x + \sigma_y \otimes \sigma_y)\right). \tag{75}$$

However, due to the occurrence of the term $(\sigma_z \otimes \mathbb{1} + \mathbb{1} \otimes \sigma_z)$ the transformation U (74) leads to a quantum state that is located outside of the set of Weyl states which are pictured in Fig. 3.22. This new state ρ_{U} (75) is not positive any more under partial transposition, $\rho_{\mathrm{U}}^{\mathrm{PT}} \ngeq 0$, where $\rho_{\mathrm{U}}^{\mathrm{PT}} = (\mathbb{1} \otimes T_{\mathrm{B}})\,\rho_{\mathrm{U}}$ and T_{B} means partial transposition on Bob's subspace. Therefore, due to the Peres-Horodecki criterion the state ρ_{U} (75) is entangled and has a concurrence $C = \frac{1}{2}$. Transformation U (74) is already optimal, i.e., it entangles ρ_{N} maximally [230].

It is also quite instructive to illustrate Theorem 5 by a specific example. General quantum states are expressed by Eq. (58) and separable states can be decomposed into

$$\rho_{\mathrm{sep}} = \frac{1}{4}\left(\mathbb{1} \otimes \mathbb{1} + r_i \sigma_i \otimes \mathbb{1} + u_i \mathbb{1} \otimes \sigma_i + r_i u_j\, \sigma_i \otimes \sigma_j\right), \tag{76}$$

with $\vec{r}^{\,2} = \vec{u}^{\,2} = 1$. A specific separable state is

$$\rho_{\Uparrow\Downarrow} = |\Uparrow\rangle \otimes |\Downarrow\rangle\langle\Uparrow| \otimes \langle\Downarrow| = \frac{1}{4}\left(\mathbb{1} \otimes \mathbb{1} + \sigma_z \otimes \mathbb{1} - \mathbb{1} \otimes \sigma_z - \sigma_z \otimes \sigma_z\right). \tag{77}$$

Let us start with the maximal entangled Bell state

$$\rho^{-} = |\psi^{-}\rangle\langle\psi^{-}| = \frac{1}{4}\left(\mathbb{1} \otimes \mathbb{1} - \vec{\sigma} \otimes \vec{\sigma}\right). \tag{78}$$

Its optimal entanglement witness

$$A_{\mathrm{opt}}^{\rho^{-}} = \frac{1}{2\sqrt{3}}\left(\mathbb{1} \otimes \mathbb{1} + \vec{\sigma} \otimes \vec{\sigma}\right), \tag{79}$$

provides the entanglement witness inequality

$$\begin{aligned} \left\langle \rho^{-} | A_{\mathrm{opt}}^{\rho^{-}} \right\rangle &= \mathrm{Tr}\,\rho^{-} A_{\mathrm{opt}}^{\rho^{-}} = -\frac{1}{\sqrt{3}} < 0, \\ \left\langle \rho_{\mathrm{sep}} | A_{\mathrm{opt}}^{\rho^{-}} \right\rangle &= \mathrm{Tr}\,\rho_{\mathrm{sep}} A_{\mathrm{opt}}^{\rho^{-}} = \frac{1}{2\sqrt{3}}(1+\cos\delta) \geq 0 \qquad \forall \rho \in S, \end{aligned} \tag{80}$$

where δ represents the angle between the unit vectors $\vec{r}$ and $\vec{u}$.

Then there exists a global unitary matrix

$$U = \frac{1}{\sqrt{2}} \left(\mathbb{1} \otimes \mathbb{1} + i \sigma_x \otimes \sigma_y \right) , \tag{81}$$

which transforms the Bell state ρ^- into the separable state $\rho_{\Uparrow\Downarrow}$

$$U \rho^- U^\dagger = \frac{1}{4} \left(\mathbb{1} \otimes \mathbb{1} + \sigma_z \otimes \mathbb{1} - \mathbb{1} \otimes \sigma_z - \sigma_z \otimes \sigma_z \right) \equiv \rho_{\Uparrow\Downarrow} , \tag{82}$$

$$\left\langle U \rho^- U^\dagger | A_{\text{opt}}^{\rho^-} \right\rangle = \text{Tr}\, U \rho^- U^\dagger A_{\text{opt}}^{\rho^-} = 0 , \tag{83}$$

i.e., separability with respect to the algebra $\{\sigma_i \otimes \sigma_j\}$. Thus the transformed state $U \rho^- U^\dagger$ represents a separable pure state as claimed in Theorem 5 and geometrically it has the Hilbert-Schmidt distance

$$d(\rho^-) = \left\| U \rho^- U^\dagger - \rho^- \right\| = 1 , \tag{84}$$

to the state ρ^-. This distance represents the amount of entanglement of ρ^-.

Transforming on the other hand also the entanglement witness, i.e., choosing a different algebra,

$$U A_{\text{opt}}^{\rho^-} U^\dagger = \frac{1}{4} \left(\mathbb{1} \otimes \mathbb{1} - \sigma_z \otimes \mathbb{1} + \mathbb{1} \otimes \sigma_z + \sigma_z \otimes \sigma_z \right) , \tag{85}$$

we then get for the entanglement witness inequality

$$\left\langle U \rho^- U^\dagger | U A_{\text{opt}}^{\rho^-} U^\dagger \right\rangle = \left\langle \rho^- | A_{\text{opt}}^{\rho^-} \right\rangle = -\frac{1}{\sqrt{3}} < 0 , \tag{86}$$

and the transformed state is entangled again with respect to the other algebra factorization $\{\sigma_i \otimes \mathbb{1}, \mathbb{1} \otimes \sigma_j, \sigma_i \otimes \sigma_j\}$. It demonstrates nicely the content of Theorem 5. It can be seen as an analogy to choosing either the Schrödinger picture or the Heisenberg picture in the characterization of the quantum states.

In an other collaboration Walter, Heide and myself studied *"The time-ordering dependence of measurements in teleportation"* [235], where the phenomenon of *"delayed-choice entanglement swapping"* could be traced back to the commutativity of the projection operators that were involved in the corresponding measurement process. We also proposed an experimental set-up which depended on the order of successive measurements corresponding to noncommutative projection operators.

Finally, I would like to mention a recent collaboration with Beatrix and Gabriele Uchida, an expert in scientific computing, where we investigated *"Entangled Entanglement: The Geometry of Greenberger-Horne-Zeilinger States"* [236]. The familiar Greenberger-Horne-Zeilinger (GHZ) states could be rewritten by entangling the Bell states for two qubits with a state of the third qubit, which was named *"entan-*

gled entanglement" [237]. We showed that in a constructive way we could obtain all 8 independent GHZ states that formed the simplex of entangled entanglement, the *magic simplex*. The construction procedure allowed a generalization to higher dimensions both, in the degrees of freedom (considering qudits) as well as in the number of particles (considering *n*-partite states). Such bases of GHZ-type states exhibited a cyclic geometry, a *Merry Go Round*, relevant for experimental and theoretical quantum information applications. We also discussed the inherent symmetries and the regions of (genuine) multi-partite entanglement within the simplex.

As Time Goes by ..

This Article is devoted to the memory of John Stewart Bell, the outstanding scientist and man with honest character and high moral. I had the great fortune to be close to him, to enjoy the fruitful collaboration and warm friendship. My aim was to show the large scope of Bell's Universe, Bell's deep insight into Nature, by describing his superb contributions in particle physics, accelerator physics and quantum physics.

Fig. 3.23 Renate (*left picture*) and Reinhold Bertlmann (*right picture*) having fun during a dinner with John and Mary Bell in Bertlmann's apartment in Geneva in 1982. Fotos: © Renate Bertlmann.

John himself had no preference for his works in the different fields, he was just as pleased with his particle physics papers as with his accelerator papers or with his quantum mechanics papers.

Of course, John had an enormous impact on my own research, in fact, on my whole life. He opened my eyes for a sharp and clear view of Nature paired with honesty and modesty, and for the beauty in scientific thinking. With the taste of Bell's Universe I could enjoy the many collaborations I had in my life in the field of particle physics, mathematical physics and quantum physics.

My essay would not be complete without reporting on Mary Bell, John's wife, who was a committed physicist as well. In my memory are always both, John and Mary. Renate and myself have spent a pleasant time with the Bells and we also had great fun together, as can be seen on Fig. 3.23.

I really feel privileged and thankful for the time I could spend with John and I would like to end with the French saying:

"Le temps passe et le souvenir reste."

Acknowledgments First of all, I am grateful to Mary Bell for the generosity of sharing her memories of her scientific career and her life with John. Many thanks, of course, to all my collaborators and friends, to Beatrix Hiesmayr, Nicolai Friis, Walter Grimus, Philipp Köhler, Heide Narnhofer, Gabriele Uchida and Anton Zeilinger, for the exciting and joyful discussions we had together. Last but not least, I would like to thank Renate Bertlmann for her lovingly company and continuous support in all that years.

References

1. J.S. Bell, R.A. Bertlmann, Z. Phys. C **4**, 11 (1980)
2. R.A. Bertlmann, Acta Phys. Aust. **53**, 305 (1981)
3. J.J. Sakurai, Phys. Lett. B **46**, 207 (1973)
4. M.A. Shifman, V.I. Vainshtein, V.I. Zakharov, Nucl. Phys. B **147**, 385, 448, 519 (1979)
5. J.S. Bell, R.A. Bertlmann, Nucl. Phys. B **177**, 218 (1981)
6. R.A. Bertlmann, Phys. Lett. B **106**, 336 (1981)
7. R.A. Bertlmann, Nucl. Phys. B **204**, 387 (1982)
8. J.S. Bell, R.A. Bertlmann, Phys. Lett. B **137**, 107 (1984)
9. J.S. Bell, letter to R.A. Bertlmann, 14th Nov 1983
10. J.S. Bell, R.A. Bertlmann, Nucl. Phys. B **187**, 285 (1981)
11. R.A. Bertlmann, J.S. Bell, Nucl. Phys. B **227**, 435 (1983)
12. C. Quigg, J.L. Rosner, Phys. Rep. **56**, 167 (1979)
13. M. Krammer, H. Krasemann, Acta Phys. Aust. Suppl. **21**, 259 (1979)
14. H. Grosse, A. Martin, Phys. Rep. **60**, 341 (1980)
15. W. Lucha, F.F. Schöberl, D. Gromes, Phys. Rep. **200**, 127 (1991)
16. R.A. Bertlmann, Phys. Lett. B **148**, 177 (1984)
17. H. Leutwyler, Phys. Lett. B **98**, 447 (1981)
18. M.B. Voloshin, Nucl. Phys. B **154**, 365 (1979)
19. R.A. Bertlmann, Nucl. Phys. B (Proc. Suppl.) **23**, 307 (1991)
20. J.S. Bell, Proc. R. Soc. A **231**, 479 (1955)
21. G. Lüders, Dan. Mat. Fys. Medd. **28**, 5 (1954)

22. W. Pauli, in *Niels Bohr and the Development of Physics*, ed. by W. Pauli, L. Rosenfeld, V. Weisskopf. (McGraw-Hill, New York and Pergamon Press Ltd., London, 1955)
23. J.S. Bell, T.H.R. Skyrme, Phil. Mag. **1**, 1055 (1956)
24. J.S. Bell, R.J. Eden, T.H.R. Skyrme, Nucl. Phys. **2**, 586 (1956)
25. J.S. Bell, Nucl. Phys. **4**, 295 (1957)
26. J.S. Bell, T.H.R. Skyrme, Proc. R. Soc. A **242**, 129 (1957)
27. J.S. Bell, Proc. R. Soc. A **242**, 122 (1957)
28. J.S. Bell, Nucl. Phys. **5**, 167 (1957)
29. J.S. Bell, Proc. Phys. Soc. **73**, 118 (1959)
30. J.S. Bell, Nucl. Phys. **12**, 117 (1959)
31. J.S. Bell, R.J. Blin-Stoyle, Nucl. Phys. **6**, 87 (1958)
32. J.S. Bell, E.J. Squires, Phys. Rev. Lett. **3**, 96 (1959)
33. J.S. Bell, Formal theory of the optical model, in *Lectures on the many body problem*, ed. by E.R. Caianello (Academic Press, New York, 1962)
34. J.S. Bell, F. Mandl, Proc. Phys. Soc. **71**, 272, 867 (1959)
35. D. Bohm, Phys. Rev. **85**, 166 (1952). ibid p. 180 (1952)
36. M. Bell, Some Reminiscences, in *Quantum [Un]Speakables*, eds. by R.A. Bertlmann, A. Zeilinger (Springer, 2002), p. 3
37. J.S. Bell, J. Steinberger, in *Proceedings of the Oxford International Conference on Elementary Particles*, p. 195 (1965)
38. M. Veltman, *Facts and Mysteries in Elementary Particle Physics* (World Scientific, Paperback, 2003), p. 212
39. M. Bell, *Wellcome address to Quantum [Un]Speakables II* (2014). contribution to this book
40. J.S. Bell, J. Løvseth, M. Veltman, in *Proceedings of the Siena International Conference on Elementary Particles*, p. 584 (1963)
41. J.S. Bell, M. Veltman, Phys. Lett. **5**, 94 (1963)
42. J.S. Bell, M. Veltman, Phys. Lett. **5**, 151 (1963)
43. M. Gell-Mann, Physics **1**, 63 (1964)
44. J.S. Bell, Physics **1**, 195 (1964)
45. J.S. Bell, Nuovo Cimento **47**, 616 (1967)
46. M. Veltman, Phys. Rev. Lett. **17**, 553 (1966)
47. J.S. Bell, Nuovo Cimento **50**, 129 (1967)
48. D.G. Sutherland, Phys. Lett. **23**, 384 (1966)
49. J.S. Bell, D.G. Sutherland, Nucl. Phys. B **4**, 315 (1968)
50. D.G. Sutherland, Nucl. Phys. B **2**, 433 (1967)
51. M. Veltman, Proc. R. Soc. A **301**, 107 (1967)
52. R. Jackiw, A. Shimony, Phys. Perspect. **4**, 78 (2004). arXiv:physics/0105046
53. J. Steinberger, Phys. Rev. **76**, 1180 (1949)
54. M. Gell-Mann, M. Lévy, Nuovo Cimento **16**, 705 (1960)
55. J.S. Bell, R. Jackiw, Nuovo Cimento **60**, 47 (1969)
56. S.L. Adler, Phys. Rev. **177**, 2426 (1969)
57. R.A. Bertlmann, *Anomalies in Quantum Field Theory* (Oxford University Press, Paperback, 2000)
58. R.A. Bertlmann, E. Kohlprath, Ann. Phys. **288**, 137 (2001)
59. R. Jackiw, letter to R.A. Bertlmann, 17th May 1996
60. R.A. Bertlmann, letter to R. Jackiw, 21st July 1996
61. J.S. Bell, E. de Rafael, Nucl. Phys. B **11**, 611 (1969)
62. J.S. Bell, C.H.L. Smith, Nucl. Phys. B **24**, 285 (1970)
63. J.S. Bell, C.H.L. Smith, Nucl. Phys. B **28**, 317 (1971)
64. J.S. Bell, Acta Phys. Austriaca Suppl. **XIII**, 395 (1974)
65. J.S. Bell, A.J.G. Hey, Phys. Lett. **51B**, 365 (1974)
66. J.S. Bell, H. Ruegg, Nucl. Phys. B **93**, 12 (1975)
67. J.S. Bell, H. Ruegg, Nucl. Phys. B **104**, 245 (1976)
68. P.G. Burke, I.C. Percival, John Stewart Bell. Biogr. Mems. Fellows R. Soc. Lond. **45**, 1 (1999)

69. J.S. Bell, W. Walkinshaw, A variational approach to disc loaded waveguides. AERE G/R **680** (1950)
70. J.S. Bell, W. Walkinshaw, Review of theory of dielectric loaded linear accelerators. AERE G/R **544** (1950)
71. J.S. Bell, Scattering by focussing foils in a proton linear accelerators. AERE T/M **62** (1952)
72. J.S. Bell, Phase debunching by focussing foils in a proton linear accelerators. AERE T/M **68** (1952)
73. J.S. Bell, Basic algebra of the strong focussing system. AERE T/R **1114** (1953)
74. M. Bell, K. Gottfried, M. Veltman (eds.), *Quantum Mechanics, High Energy Physics and Accelerators, Selected Papers of John S. Bell* (World Scientific, 1995)
75. J.S. Bell, Nature **171**, 167 (1953)
76. J.S. Bell, Proc. Phys. Soc. B **66**, 802 (1953)
77. J.S. Bell, Linear accelerator phase oscillations. AERE T/M **114** (1954)
78. W.G. Unruh, Phys. Rev. D **14**, 870 (1976)
79. S. Hawking, Nature **248**, 30 (1974). Commun. Math. Phys. **43**, 199 (1975)
80. J.D. Jackson, Rev. Mod. Phys. **48**, 417 (1976)
81. J.S. Bell, J.M. Leinaas, Nucl. Phys. B **212**, 131 (1983)
82. J.S. Bell, J.M. Leinaas, Nucl. Phys. B **284**, 488 (1987)
83. J.S. Bell, R.J. Hughes, J.M. Leinaas, Z. Phys. C **28**, 75 (1985)
84. J.S. Bell, *Speakable and Unspeakable in Quantum Mechanics* (Cambridge University Press, 1987)
85. M. Bell, The secret service agent, in *Kingussie Secondary School Magazine*, p. 5 (1939–1940)
86. M. Bell, letter to R.A. Bertlmann, 16th Feb 2015
87. M. Bell, letter to R.A. Bertlmann, 23rd May 2013
88. M. Ross, Slow neutron absorption cross-section of elements. Rep. Prog. Phys. **XII**, 291 (1949)
89. M. Ross, W. Walkinshaw, Effect of corrugation pitch on performance of M.R.C. linear accelerator. AERE T/M **55** (1952)
90. M. Ross, Series impedance of double ridged wave guides. AERE T/M **58** (1952)
91. M. Ross, Note on the prebunching for proton linear accelerator. AERE T/M **66** (1952)
92. J.S. Bell, M. Ross, Heating of focussing foils by a proton beam. AERE T/M **70** (1952)
93. M. Ross, A strong focussing electrostatic lens. AERE T/M **83** (1953)
94. M. Bell, Shunt impedance of rod-loaded wave guides. AERE T/M **99** (1954)
95. M. Bell, A note on gaps at proton energies of 10 and 50 MeV in the proton linear accelerator. AERE T/M **101** (1954)
96. M. Bell, Focussing system for the 600 MeV proton linear accelerator. AERE T/M **112** (1954)
97. M. Bell, Nonlinear equations of motion in the synchrotron. AERE T/M **125** (1955)
98. M. Bell, Focussing in the proton linear accelerator. II, AERE T/M **128** (1955)
99. M. Bell, Perturbation formulae for Hill's equation. AERE T/M **139** (1956)
100. M. Bell, Twist resonances. AERE T/R **2077** (1956)
101. M. Bell, An analytic treatment of nonlinear effects in the spiral ridged cyclotron. AERE T/R **2209** (1957)
102. M. Bell, Injection into the 7 GeV synchrotron. AERE T/M **155** (1957)
103. M. Bell, A note on Mathieu functions. Proc. Glasg. Math. Assoc. **III**, Part III, 132 (1957)
104. M. Bell, The effect of a second harmonic in a 4-ridged spiral cyclotron, AERE R-3326 (1960)
105. G.I. Budker, At. Energy **22**, 346 (1967)
106. G.I. Budker, N.S. Dikansky, V.I. Kudelainen, I.N. Meshkov, V.V. Parkhomchuk, D.V. Pestrikov, A.N. Skrinsky, B.N. Sukhina, Part. Accel. **7**, 197 (1976)
107. M. Bell, Part. Accel. **10**, 101 (1980)
108. M. Bell et al., Phys. Lett. **87B**, 275 (1979)
109. M. Bell, J. Chaney, H. Herr, F. Krienen, P. Møller-Petersen, G. Petrucci, Nucl. Instrum. Methods **190**, 237 (1981)
110. J.S. Bell, M. Bell, Part. Accel. **11**, 233 (1981)
111. M. Bell, J.S. Bell, Part. Accel. **12**, 49 (1982)
112. M. Bell, J.S. Bell, Part. Accel. **13**, 13 (1983)

113. M. Bell, J.S. Bell, Part. Accel. **22**, 301 (1988)
114. R. Bankenbecler, S.D. Drell, Phys. Rev. D **36**, 277 (1987)
115. A.A. Sokolov, I.M. Ternov, *Synchrotron Radiation*, Pergamon (1971)
116. A.I. Nikishov, V.I. Ritus, Sov. Phys. JETP **25**, 1135 (1967)
117. A.N. Matveev, ZETF **31**, 479 (1956). Sov. Phys. JETP **4**, 409 (1957)
118. V.N. Baier, V.M. Katkov, Sov. Phys. JETP **26**, 854 (1968). Sov. Phys. JETP **28**, 807 (1969)
119. M. Bell, J.S. Bell, Part. Accel. **24**, 1 (1988)
120. M. Bell, J.S. Bell, Nucl. Instrum. Methods Phys. Res. A **275**, 258 (1989)
121. J.S. Bell, Bertlmann's socks and the nature of reality, CERN preprint Ref.TH.2926-CERN, 18 July 1980
122. A. Einstein, B. Podolsky, N. Rosen, Phys. Rev. **47**, 777 (1935)
123. J.S. Bell, J. de Phys., Colloque C2, Suppl. 3, Tome 42, C2-41 (1981)
124. A. Aspect, J. Dalibard, G. Roger, Phys. Rev. Lett. **49**, 1804 (1982)
125. N.D. Mermin, Rev. Mod. Phys. **65**, 803 (1993)
126. J.S. Bell, Rev. Mod. Phys. **38**, 447 (1966)
127. J. von Neumann, *Mathematische Grundlagen der Quantenmechanik* (Springer, Berlin 1932). Princeton University Press, English translation (1955)
128. G. Hermann, Die naturphilosophischen Grundlagen der Quantenmechanik. Abhandlungen der Fries'schen Schule **6**, 69 (1935)
129. J. Conway, S.B. Kochen, The geometry of quantum paradoxes, in *Quantum [Un]speakables*, eds. by R.A. Bertlmann, A. Zeilinger (Springer, 2002), p. 257
130. A.M. Gleason, J. Math. Mech. **6**, 885 (1957)
131. S.B. Kochen, E. Specker, J. Math. Mech. **17**, 59 (1967)
132. E. Amselem, M. Bourennane, C. Budroni, A. Cabello, O. Gühne, M. Kleinmann, J.-Å. Larsson, M. Wieśniak, Phys. Rev. Lett. **110**, 078901 (2013)
133. G. Cañas, S. Etcheverry, E.S. Gómez, C. Saavedra, G.B. Xavier, G. Lima, A. Cabello, Phys. Rev. A **90**, 012119 (2014)
134. A. Cabello, S. Severini, A. Winter, Phys. Rev. Lett. **112**, 040401 (2014)
135. A. Cabello, P. Badziag, M.T. Cunha, Phys. Rev. Lett. **111**, 180404 (2013)
136. O. Gühne, C. Budroni, A. Cabello, M. Kleinmann, J.-Å. Larsson, Phys. Rev. A **89**, 062107 (2014)
137. A. Acín, T. Fritz, A. Leverrier, A.B. Sainz, Commun. Math. Phys. **334**, 533 (2015)
138. A. Cabello, The unspeakable why, talk at *Quantum [Un]Speakables II* (2014). Contribution in this book
139. D. Bohm, Y. Aharonov, Phys. Rev. **108**, 1070 (1957)
140. J.F. Clauser, M.A. Horne, A. Shimony, R.A. Holt, Phys. Rev. Lett. **23**, 880 (1969)
141. G. Weihs, T. Jennewein, C. Simon, H. Weinfurter, A. Zeilinger, Phys. Rev. Lett. **81**, 5039 (1998)
142. G. Weihs, Ein Experiment zum Test der Bellschen Ungleichung unter Einstein Lokalität, Ph.D. Thesis, University of Innsbruck (1998)
143. E.P. Wigner, Am. J. Phys. **38**, 1005 (1970)
144. J.F. Clauser, M.A. Horne, Phys. Rev. D **10**, 526 (1974)
145. H.M. Wiseman, E.G. Cavalcanti, Causarum Investigatio and the Two Bells Theorems of John Bell, talk at *Quantum [Un]Speakables II* (2014). Contribution in this book
146. N. Brunner, D. Cavalcanti, S. Pironio, V. Scarani, S. Wehner, Rev. Mod. Phys. **86**, 419 (2014)
147. O. Gühne, G. Tóth, Phys. Rep. **474**, 1 (2009)
148. W. Pauli, M. Born, 31st March 1954, in *Wolfgang Pauli, Scientific Correspondence with Bohr, Einstein, Heisenberg*, vol. IV, ed. by von K. Meyenn (Springer, 1999), Part II, pp. 1953–1954
149. J.F. Clauser, Early history of Bell's theorem, in *Quantum [Un]speakables*, eds. by R.A. Bertlmann, A. Zeilinger (Springer, 2002), p. 61
150. J.F. Clauser, Bull. Am. Phys. Soc. **14**, 578 (1969)
151. S.J. Freedman, J.F. Clauser, Phys. Rev. Lett. **28**, 938 (1972)
152. S.J. Freedman, LBL Rep. **191** (1972). Lawrence Berkeley National Laboratory
153. E.S. Fry, R.C. Thompson, Phys. Rev. Lett. **37**, 465 (1976)

154. A. Pais, *Subtle is the Lord, The Science and Life of Albert Einstein* (Oxford University Press, 1982)
155. A. Aspect, Phys. Rev. D **14**, 1944 (1976)
156. A. Aspect, P. Grangier, G. Roger, Phys. Rev. Lett. **47**, 460 (1981)
157. A. Aspect, P. Grangier, G. Roger, Phys. Rev. Lett. **49**, 91 (1982)
158. A. Aspect, P. Grangier, Lett. Nuovo Cimento **43**, 345 (1985)
159. M.A. Nielsen, I.L. Chuang, *Quantum Computation and Quantum Information* (Cambridge, 2000)
160. R.A. Bertlmann, A. Zeilinger (eds.), *Quantum [Un]speakables* (Springer, 2002)
161. J. Brendel, E. Mohler, W. Martienssen, Eur. Phys. Lett. **20**, 575 (1992)
162. P.R. Tapster, J.G. Rarity, P.C.M. Owens, Phys. Rev. Lett. **73**, 1923 (1994)
163. W. Tittel, J. Brendel, H. Zbinden, N. Gisin, Phys. Rev. Lett. **81**, 3563 (1998)
164. D. Bouwmeester, J.-W. Pan, K. Mattle, M. Eibl, H. Weinfurter, A. Zeilinger, Nature **390**, 575 (1997)
165. X.-S. Ma, T. Herbst, T. Scheidl, D. Wang, S. Kropatschek, W. Naylor, B. Wittmann, A. Mech, J. Kofler, E. Anisimova, V. Makarov, T. Jennewein, R. Ursin, A. Zeilinger, Nature **489**, 269 (2012)
166. T. Jennewein, C. Simon, G. Weihs, H. Weinfurter, A. Zeilinger, Phys. Rev. Lett. **84**, 4729 (2000)
167. N. Gisin, G. Ribordy, W. Tittel, H. Zbinden, Rev. Mod. Phys. **74**, 145 (2002)
168. K.J. Resch, M. Lindenthal, B. Blauensteiner, H.J. Böhm, A. Fedrizzi, C. Kurtsiefer, A. Poppe, T. Schmitt-Manderbach, M. Taraba, R. Ursin, P. Walther, H. Weier, H. Weinfurter, A. Zeilinger, Opt. Express **13**, 202 (2005)
169. R. Ursin, F. Tiefenbacher, T. Schmitt-Manderbach, H. Weier, T. Scheidl, M. Lindenthal, B. Blauensteiner, T. Jennewein, J. Perdigues, P. Trojek, B. Ömer, M. Fürst, M. Meyenburg, J. Rarity, Z. Sodnik, C. Barbieri, H. Weinfurter, A. Zeilinger, Nat. Phys. **3**, 481 (2007)
170. M. Giustina, A. Mech, S. Ramelow, B. Wittmann, J. Kofler, J. Beyer, A. Lita, B. Calkins, T. Gerrits, S.W. Nam, R. Ursin, A. Zeilinger, Nature **497**, 227 (2013)
171. P.H. Eberhard, Phys. Rev. A **47**, 747 (1993)
172. M.A. Rowe, D. Kielpinsky, V. Meyer, C.A. Sackett, W.M. Itano, D.J. Wineland, Nature **409**, 791 (2001)
173. Y. Hasegawa, R. Loidl, G. Badurek, M. Baron, H. Rauch, Nature **425**, 45 (2003)
174. Y. Hasegawa, K. Durstberger-Rennhofer, S. Sponar, H. Rauch, *Kochen-Specker theorem studied with neutron interferometer* (2010). arXiv:1004.2836
175. H. Bartosik, J. Klepp, C. Schmitzer, S. Sponar, A. Cabello, H. Rauch, Y. Hasegawa, Phys. Rev. Lett. **103**, 040403 (2009)
176. J.S. Bell, La nouvelle cuisine, in *Between Science and Technology*, eds. by A. Sarlemijn, P. Kroes (Elsevier Science Publishers, 1990)
177. J.S. Bell, *Speakable and Unspeakable in Quantum Mechanics*, 2nd edn. (Cambridge University Press, 2004)
178. A. Einstein, Ann. Phys. **23**, 371 (1907)
179. R.A. Bertlmann, Bell's theorem and the nature of reality, preprint University of Vienna, 1988. Found. Phys. **20**, 1191 (1990)
180. D. Greenberger, Interview given at *Quantum [Un]Speakables II* (Vienna, 2014). https://vimeo.com/113838670
181. A. Mair, A. Vaziri, G. Weihs, A. Zeilinger, Nature **412**, 313 (2001)
182. Catalogue of the exhibition *Action at a Distance: The Life and Legacy of John Stewart Bell*, The Naughton Gallery, Queen's University Belfast, 2014; Video of the exhibition. http://www.mediator.qub.ac.uk/ms/BellExb.mp4
183. Announcement of *Quantum [Un]Speakables III*. https://www.youtube.com/watch?v=BgUtWihTvDM
184. R.A. Bertlmann, W. Grimus, Phys. Lett. B **392**, 426 (1997)
185. R.A. Bertlmann, W. Grimus, Phys. Rev. D **58**, 034014 (1998)
186. A. Apostolakis et al., CPLEAR Collection. Phys. Lett. B **422**, 339 (1998)

187. R.A. Bertlmann, W. Grimus, B.C. Hiesmayr, Phys. Rev. D **60**, 114032 (1999)
188. R.A. Bertlmann, Entanglement, Bell inequalities and decoherence in particle physics, in *Quantum Coherence: From Quarks to Solids*, vol. 689, eds. by W. Pötz, J. Fabian, U. Hohensteiner. Lecture Notes in Physics (Springer, 2006), pp. 1–45
189. C. Schuler, *Entanglement, Bell inequalities and decoherence in neutral K-meson systems*, Bachelor Thesis, University of Vienna (2014)
190. G. Amelino-Camelia et al., Eur. Phys. J. C **68**, 619 (2010)
191. G. Lindblad, Commun. Math. Phys. **48**, 119 (1976)
192. V. Gorini, A. Kossakowski, E.C.G. Sudarshan, J. Math. Phys. **17**, 821 (1976)
193. R.A. Bertlmann, W. Grimus, Phys. Rev. D **64**, 056004 (2001)
194. C.H. Bennett, D.P. DiVincenzo, J.A. Smolin, W.K. Wootters, Phys. Rev. A **54**, 3824 (1996)
195. S. Hill, W.K. Wootters, Phys. Rev. Lett. **78**, 5022 (1997)
196. W.K. Wootters, Phys. Rev. Lett. **80**, 2245 (1998)
197. W.K. Wootters, Quantum Inf. Comput. **1**, 27 (2001)
198. R.A. Bertlmann, K. Durstberger, B.C. Hiesmayr, Phys. Rev. A **68**, 012111 (2003)
199. R.A. Bertlmann, B.C. Hiesmayr, Phys. Rev. A **63**, 062112 (2001)
200. R.A. Bertlmann, A. Bramon, G. Garbarino, B.C. Hiesmayr, Phys. Lett. A **332**, 355 (2004)
201. F. Uchiyama, Phys. Lett. A **231**, 295 (1997)
202. R.A. Bertlmann, W. Grimus, B.C. Hiesmayr, Phys. Lett. A **289**, 21 (2001)
203. B.C. Hiesmayr, A. Di Domenico, C. Curceanu, A. Gabriel, M. Huber, J.-A. Larsson, P. Moskal, Eur. Phys. J. C **72**, 1856 (2012)
204. A. Di Domenico, privat communication to R.A. Bertlmann (2014)
205. A. Bramon, R. Escribano, G. Garbarino, Found. Phys. **36**, 563 (2006)
206. A. Bramon, R. Escribano, G. Garbarino, *A review on Bell inequality tests with neutral kaons*, ed. by A. Di Domenico. Frascati Physics Series, vol. XLIII, p. 217 (2007)
207. A. Bramon, G. Garbarino, Phys. Rev. Lett. **88**, 040403 (2002)
208. A. Bramon, G. Garbarino, Phys. Rev. Lett. **89**, 160401 (2002)
209. H. Tataroglu, *Nichtlokale Korrelation in Kaonischen Systemen*, Diploma Thesis, University of Vienna (2009)
210. B.C. Hiesmayr, *Weak Interaction Processes: Which Quantum Information is revealed?* (2014). arXiv:1410.1707
211. B.C. Hiesmayr, *Bringing Bells Theorem Back to the Domain of Particle Physics and Cosmology* (2015). Contribution to this book
212. J.-L. Li, C.-F. Qiao, Sci. China G **53**, 870 (2010)
213. Y.-B. Ding, J.-L. Li, C.-F. Qiao, High Energy Phys. Nucl. Phys. **31**, 1086 (2007)
214. J.-L. Li, C.-F. Qiao, Phys. Lett. A **373**, 4311 (2009)
215. N. Friis, R.A. Bertlmann, M. Huber, B.C. Hiesmayr, Phys. Rev. A **81**, 042114 (2010)
216. I. Fuentes-Schuller, R.B. Mann, Phys. Rev. Lett. **95**, 120404 (2005)
217. D.E. Bruschi, J. Louko, E. Martín-Martínez, A. Dragan, I. Fuentes, Phys. Rev. A **82**, 042332 (2010)
218. N. Friis, P. Köhler, E. Martín-Martínez, R.A. Bertlmann, Phys. Rev. A **84**, 062111 (2011)
219. N. Gisin, Phys. Lett. A **154**, 201 (1991)
220. R.A. Bertlmann, J. Phys. A: Math. Theor. **47**, 424007 (2014)
221. R.F. Werner, Phys. Rev. A **40**, 4277 (1989)
222. A. Peres, Phys. Rev. Lett. **77**, 1413 (1996)
223. M. Horodecki, P. Horodecki, R. Horodecki, Phys. Lett. A **223**, 1 (1996)
224. R. Horodecki, P. Horodecki, M. Horodecki, Phys. Lett. A **200**, 340 (1995)
225. R.A. Bertlmann, H. Narnhofer, W. Thirring, Phys. Rev. A **66**, 032319 (2002)
226. B.M. Terhal, Phys. Lett. A **271**, 319 (2000)
227. R.A. Bertlmann, P. Krammer, Ann. Phys. **324**, 1388 (2009)
228. K.G.H. Vollbrecht, R.F. Werner, Phys. Rev. A **64**, 062307 (2000)
229. R. Horodecki, M. Horodecki, Phys. Rev. A **54**, 1838 (1996)
230. W. Thirring, R.A. Bertlmann, P. Köhler, H. Narnhofer, Eur. Phys. J. D **64**, 181 (2011)
231. P. Zanardi, Phys. Rev. Lett. **87**, 077901 (2001)

232. M. Kuś, K. Życzkowski, Phys. Rev. A **63**, 032307 (2001)
233. K. Życzkowski, I. Bengtsson, *An Introduction to Quantum Entanglement: a Geometric Approach*. arXiv:quant-ph/0606228
234. I. Bengtsson, K. Życzkowski, *Geometry of Quantum states: An Introduction to Quantum Entanglement* (Cambridge University Press, 2006)
235. R.A. Bertlmann, H. Narnhofer, W. Thirring, Eur. Phys. J. D **67**, 62 (2013)
236. G. Uchida, R.A. Bertlmann, B.C. Hiesmayr, Phys. Lett. A **379**, 2698 (2015)
237. G. Krenn, A. Zeilinger, Phys. Rev. A **54**, 1793 (1996)

Part II
Bell's Theorem—Fundamental Issues

Chapter 4
Why QBism Is Not the Copenhagen Interpretation and What John Bell Might Have Thought of It

N. David Mermin

Abstract Christopher Fuchs and Rüdiger Schack have developed a way of understanding science, which, among other things, resolves many of the conceptual puzzles of quantum mechanics that have vexed people for the past nine decades. They call it QBism. I speculate on how John Bell might have reacted to QBism, and I explain the many ways in which QBism differs importantly from the orthodox ways of thinking about quantum mechanics associated with the term "Copenhagen interpretation."

Our students learn quantum mechanics the way they learn to ride bicycles (both very valuable accomplishments) without really knowing what they are doing.

—J. S. Bell, letter to R. E. Peierls, 20/8/1980

I think we invent concepts, like "particle" or "Professor Peierls", to make the immediate sense of data more intelligible.

—J. S. Bell, letter to R. E. Peierls, 24/2/1983

I have the impression as I write this, that a moment ago I heard the bell of the tea trolley. But I am not sure because I was concentrating on what I was writing... The ideal instantaneous measurements of the textbooks are not precisely realized

---

Dedicated to my friend and Cornell colleague Geoffrey Chester (1928–2014), who for over 50 years enjoyed my more controversial enthusiasms, while always insisting that I keep my feet firmly on the ground.

N.D. Mermin (✉)
Laboratory of Atomic and Solid State Physics, Cornell University,
Ithaca, NY 14853-2501, USA
e-mail: ndm4@cornell.edu

R. Bertlmann and A. Zeilinger (eds.), *Quantum [Un]Speakables II*,
The Frontiers Collection, DOI 10.1007/978-3-319-38987-5_4

anywhere anytime, and more or less realized, more or less all the time, more or less everywhere.

—J. S. Bell, letter to R. E. Peierls, 28/1/1981[1]

For the past decade and a half Christopher Fuchs and Rüdiger Schack (originally in collaboration with Carlton Caves) have been developing a new way to think about quantum mechanics. Fuchs and Schack have called it QBism.[2] Their term originally stood for "quantum Bayesianism". But QBism is a way of thinking about science quite generally, not just quantum physics,[3] and it is pertinent even when probabilistic judgments, and therefore "Bayesianism", play no role at all. I nevertheless retain the term QBism, both to acknowledge the history behind it, and because a secondary meaning remains apt in the broader context: QBism is as big a break with 20th century ways of thinking about science as Cubism was with 19th century ways of thinking about art.

QBism maintains that my understanding of the world rests entirely on the experiences that the world has induced in me throughout the course of my life. Nothing beyond my personal experience underlies the picture that I have formed of my own external world.[4] This is a statement of empiricism. But it is empiricism taken more seriously than most scientists are willing to do.

To state that my understanding of the world rests on my experience is not to say that my world exists only within my head, as recent popularizations of QBism have wrongly asserted.[5] Among the ingredients from which I construct my picture of my external world are the impact of that world on my experience, when it responds to the actions that I take on it. When I act on my world, I generally have no control over how it acts back on me.

Nor does QBism maintain that each of us is free to construct our own private worlds. Facile charges of solipsism miss the point. My experience of you leads me to hypothesize that you are a being very much like myself, with your own private experience. This is as firm a belief as any I have. I could not function without it.

---

[1] *Selected Correspondence of Rudolf Peierls*, v. 2, Sabine Lee [ed], World Sci., 2009. I have the impression (confirmed at the conference) that all three of these quotations are unfamiliar even to those who, like me, have devoured almost everything John Bell ever wrote about quantum foundations.

[2] C. A. Fuchs and R. Schack, Rev. Mod. Phys. **85**, 1693–1714 (2013).

[3] When the QBist view of science is used to solve classical puzzles I have suggested calling it CBism; N. D. Mermin, Nature, **507**, 421–423, March 27, 2014.

[4] For "my", "me," "I", you can read appropriate versions of "each of us"; the singular personal pronoun is less awkward. But unadorned "our", "us", and "we" are dangerously ambiguous. In QBism the first person plural always means each of us individually; it never means all of us collectively, unless this is spelled out. Part of the 90-year confusion at the foundations of quantum mechanics can be attributed to the unacknowledged ambiguity of the first-person plural pronouns and the carelessness with which they are almost always used.

[5] H. C. von Baeyer, Scientific American **308**, 46–51, June 2013; M. Chalmers, New Scientist, 32–35, May 10, 2014. I believe that in both cases these gross distortions were the fault of overly intrusive copy editors and headline writers, who did not understand the manuscripts they were trying to improve.

If asked to assign this hypothesis a probability I would choose $p = 1$.[6] Although I have no direct personal access to your own experience, an important component of my private experience is the impact on me of your efforts to communicate, in speech or writing, your verbal representations of your own experience. Science is a collaborative human effort to find, through our individual actions on the world and our verbal communications with each other, a model for what is common to all of our privately constructed external worlds. Conversations, conferences, research papers, and books are an essential part of the scientific process.

Fuchs himself may be partly responsible for the silly accusations about solipsism. One of his favorite slogans about QBism is[7] "Quantum mechanics is a single-user theory", sometimes abbreviated to[8] "Me, me, me!" This invites the *s*-word. I hurled it at him myself the first time I came upon such slogans. Although susceptible to misinterpretation, they are important reminders that any application of quantum mechanics must ultimately be understood to be undertaken by a particular person[9] to help her make sense of her own particular experience. They were never intended to mean that there cannot be many different users of quantum mechanics. Nor do they require any particular user to exclude from her own experience what she has heard or read about the private experience of others.

Those who reject QBism—currently a large majority of the physicists who know anything about it—reify the common external world we have all negotiated with each other, removing from the story any reference to the origins of our common world in the private experiences we try to share with each other through language. For all practical purposes reification is a sound strategy. It would be hard to live our daily private or professional scientific lives if we insisted on constantly tracing every aspect of our external world back to its sources in our own private personal experience. My reification of the concepts I invent, to make my immediate sense of data more intelligible, is a useful tool of day-to-day living.

But when subtle conceptual issues are at stake, related to certain notoriously murky scientific concepts like quantum states, then we can no longer refuse to acknowledge that our scientific pictures of the world rest on the private experiences of individual scientists. The most famous investigator Vienna has ever produced, who worked just a short walk from the lecture hall for this conference, put it concisely: "A world constitution that takes no account of the mental apparatus by which we perceive it is an empty abstraction." This was said not by Ludwig Boltzmann, not by Erwin Schrödinger, and not even by Anton Zeilinger. It was said by Sigmund Freud,[10] just down the hill at Berggasse 19. He was writing about religion, but his remark applies equally well to science.

---

[6]I have more to say about $p = 1$ below.

[7]Christopher A. Fuchs, arXiv:1003.5182.

[8]Christopher A. Fuchs, arXiv:1405.2390, especially pp. 546–549.

[9]Generally named Alice.

[10]*The Future of an Illusion*, 1927, concluding paragraph.

After he returned to Vienna in the early 1960s, Schrödinger repeatedly made much the same point,[11] somewhat less concisely than Freud: "The scientist subconsciously, almost inadvertently simplifies his problem of understanding Nature by disregarding or cutting out of the picture to be constructed, himself, his own personality, the subject of cognizance." In expressing these views in the 1960s he rarely mentions quantum mechanics. Only thirty years earlier, in a letter to Sommerfeld, does he explicitly tie this view to quantum mechanics, and even then, he allows that it applies to science much more broadly: "Quantum mechanics forbids statements about what really exists—statements about the object. It deals only with the object-subject relation. Even though this holds, after all, for any description of nature, it evidently holds in quantum mechanics in a much more radical sense."[12] We were rather successful excluding the subject from classical physics (but not completely (Footnote 3)). Quantum physics finally forced (or should have forced) us to think harder about the importance of the object-subject relation.

Niels Bohr, whose views on the meaning of quantum mechanics Schrödinger rejected, also delivered some remarkably QBist-sounding pronouncements, though by "experience" he almost certainly meant the objective readings of large classical instruments and not the personal experience of a particular user of quantum mechanics: "In our description of nature the purpose is not to disclose the real essence of the phenomena but only to track down, so far as it is possible, relations between the manifold aspects of our experience."[13] Thirty years later he was saying pretty much the same thing: "Physics is to be regarded not so much as the study of something a priori given, but as the development of methods for ordering and surveying human experience."[14] Bohr and Schrödinger are not the only dissenting pair who might have found some common ground in QBism.

The fact that each of us has a view of our world that rests entirely on our private personal experience has little bearing on how we actually use our scientific concepts to deal with the world. But it is central to the philosophical concerns of quantum foundational studies. Failing to recognize the foundational importance of personal experience creates illusory puzzles or paradoxes. At their most pernicious, such puzzles motivate unnecessary efforts to reformulate in more complicated ways—or even to change the observational content of—theories which have been entirely successful for all practical purposes.

This talk is not addressed to those who take (often without acknowledging it) an idealistic or Platonic position in their philosophical meditations on the nature of quantum mechanics. They will never be comfortable with QBism. My talk is intended primarily for the growing minority of philosophically minded physicists

---

[11]*Nature and the Greeks, Science and Humanism,* Cambridge (1996), p. 92. See also *Mind and Matter* and *My View of the World*.

[12]Schrödinger to Sommerfeld, 11 December, 1931, in *Schödingers Briefwechsel zur Wellenmechanik und zum Katzenparadoxon*, Springer Verlag, 2011.

[13]Niels Bohr, 1929. In *Atomic Theory and the Description of Nature*, Cambridge (1934), p. 18.

[14]Niels Bohr, 1961. In *Essays 1958–1962 on Atomic Physics and Human Knowledge*, Ox Bow Press, Woodbridge, CT (1987), p. 10.

who, far from rejecting QBism, are starting to maintain that there is nothing very new in it.[15] I am thinking of those who maintain that QBism is nothing more than the Copenhagen interpretation.

I may be partly to blame for this misunderstanding. I have used the above quotations from Bohr in several recent essays about QBism, because QBism provides a context in which these quotations finally make unambiguous sense. While they made sense for Bohr too, it was not a QBist kind of sense, and I very much doubt that people gave them a QBist reading. Similarly, my quotation from Freud does not mean that QBism should be identified with psychoanalysis, and the three epigraphs from John Bell at the head of this text should not be taken to mean that I believe QBism had already been put forth by Bell in the early 1980s. The quotations from Bell's letters to Peierls are only to suggest that John Bell, who strenuously and elegantly identified what is incoherent in Copenhagen, might not have dismissed QBism as categorically. There are many important ways in which QBism is profoundly different from Copenhagen, and from any other way of thinking about quantum mechanics that I know of. If you are oblivious to these differences, then you have missed the point of QBism.

The primary reason people wrongly identify QBism with Copenhagen is that QBism, like most varieties of Copenhagen, takes the quantum state of a system to be not an objective property of that system, but a mathematical tool for thinking about the system.[16] In contrast, in many of the major non-standard interpretations—many worlds, Bohmian mechanics, and spontaneous collapse theories—the quantum state of a system is very much an objective property of that system.[17] Even people who reject all these heresies and claim to hold standard views of quantum mechanics, are often careless about reifying quantum states. Some claim, for example, that quantum states were evolving (and even collapsing) in the early universe, long before anybody existed to assign such states. But the models of the early universe to which we assign quantum states are models that we construct to account for contemporary astrophysical data. In the absence of such data, we would not have come up with the models. As Rudolf Peierls remarked, "If there is a part of the Universe, or a period in its history, which is not capable of influencing present-day events directly or indirectly, then indeed there would be no sense in applying quantum mechanics to it."[18]

A fundamental difference between QBism and any flavor of Copenhagen, is that QBism explicitly introduces each *user* of quantum mechanics into the story, together

[15]I count this as progress. The four stages of acceptance of a radical new idea are: (1) It's nonsense; (2) It's well known; (3) It's trivial; (4) I thought of it first. I'm encouraged to find that stage (2) is now well underway.

[16]Heisenberg and Peierls are quite clear about this. Bohr may well have believed it but never spelled it out as explicitly. Landau and Lifshitz, on the other hand, are so determined to eliminate any trace of humanity from the story that I suspect their flavor of Copenhagen might reject the view of quantum states as mathematical tools.

[17]In consistent histories, which has a Copenhagen tinge, its quantum state can be a true property of a system, but only relative to a "framework".

[18]R. E. Peierls, Physics World January 1991, 19–20.

with the world external to that user. Since every user is different, dividing the world differently into external and internal, every application of quantum mechanics to the world must ultimately refer, if only implicitly, to a particular user. But every version of Copenhagen takes a view of the world that makes no reference to the particular user who is trying to make sense of that world.

Fuchs and Schack prefer the term "agent" to "user". "Agent" serves to emphasize that the user takes actions on her world and experiences the consequences of her actions. I prefer the term "user" to emphasize Fuchs' and Schack's equally important point that science is a user's manual. Its purpose is to help each of us make sense of our private experience induced in us by the world outside of us.

It is crucial to note from the beginning that "user" does not mean a generic body of users. It means a particular individual person, who is making use of science to bring coherence to her own private perceptions. I can be a "user". You can be a "user". But we are not jointly a user, because my internal personal experience is inaccessible to you except insofar as I attempt to represent it to you verbally, and vice-versa. Science is about the interface between the experience of any particular person and the subset of the world that is external to that particular user.[19] This is unlike anything in any version of Copenhagen.[20] It is central to the QBist understanding of science.

The notion that science is a tool that each of us can apply to our own private body of personal experience is explicitly renounced by the Landau-Lifshitz version of Copenhagen. The opening pages of their *Quantum Mechanics*[21] declare that "It must be most decidedly emphasized that we are here not discussing a process of measurement in which the physicist-observer takes part." They explicitly deny the user any role whatever in the story. To emphasize this they add "By measurement, in quantum mechanics, we understand any process of interaction between classical and quantum objects, *occurring apart from and independently of any observer.*" [My italics.] In the second quotation Landau and Lifshitz have, from a QBist point of view, replaced each different member of the set of possible users by one and the same set of "classical objects". Their insistence on eliminating human users from the story, both individually and collectively, leads them to declare that "It is in principle impossible…to formulate the basic concepts of quantum mechanics without using classical mechanics." Here they make two big mistakes: they replace the experiences of each user with "classical mechanics", and they confound the diverse experiences of many different users into that single abstract entity.

Bohr seems not as averse as Landau and Lifshitz[22] to letting scientists into the story, but they come in only as proprietors of a single large, *classical* measurement apparatus. All versions of Copenhagen objectify each of the diverse family of users of science into a single common piece of apparatus. Doing this obliterates the fun-

---

[19]See in this regard my remarks above about the dangers of the first-person plural.

[20]And unlike any other way of thinking about quantum mechanics.

[21]Translated into English by John Bell, who was therefore intimately acquainted with it.

[22]But Peierls identifies their positions, referring to "the view of Landau and Lifshitz (and therefore of Bohr)" in his *Physics World* article. He disagrees with all of them, saying that it is incorrect to require the apparatus to obey classical physics.

damental QBist fact that a quantum-mechanical description is always relative to the particular user of quantum mechanics who provides that description. Replacing that user with an apparatus introduces the notoriously ill-defined "shifty split" of the world into quantum and classical, that John Bell so elegantly and correctly deplored.

Bell's split is shifty in two respects. Its character is not fixed. It can be the Landau-Lifshitz split between "classical" and "quantum". But sometimes it is a split between "macroscopic and microscopic". Or between "irreversible" and "reversible." The split is also shifty because its location can freely be moved along the path between whatever poles have been used to characterize it.

There is also a split in QBism, but it is specific to each user. That it shifts from user to user is the full extent to which the split is "shifty". For any particular user there is nothing shifty about it: the split is between that user's directly perceived internal experience, and the external world that that user infers from her experience.

Closely related to its systematic suppression of the user, is the central role in Copenhagen of "measurement" and the Copenhagen view of the "outcome" of a measurement. In all versions of Copenhagen a measurement is an interaction between a quantum system and a "measurement apparatus". Depending on the version of Copenhagen, the measurement apparatus could belong to a "classical" domain beyond the scope of quantum mechanics, or it could itself be given a quantum mechanical description. But in any version of Copenhagen the *outcome* of a measurement is some strictly classical information produced by the measurement apparatus as a number on a digital display, or the position of an ordinary pointer, or a number printed on a piece of paper, or a hole punched somewhere along a long tape—something like that. Words like "macroscopic" or "irreversible" are used at this stage to indicate the objective, substantial, non-quantum character of the outcome of a measurement.

In QBism, on the other hand, a measurement can be *any* action taken by *any* user on her external world. The outcome of the measurement is the *experience* the world induces back in that particular user, through its response to her action. The QBist view of measurement includes Copenhagen measurements as a special case, in which the action is carried out with the aid of a measurement apparatus and the user's experience consists of her perceiving the display, the pointer, the marks on the paper, or the hole in the tape produced by that apparatus. But a QBist "measurement" is much broader. Users are making measurements more or less all the time more or less everywhere. Every action on her world by every user constitutes a measurement, and her experience of the world's reaction is its outcome. Physics is not limited to the outcomes of "piddling" laboratory tests, as Bell complained about Copenhagen.[23]

In contrast to the Copenhagen interpretation (or any other interpretation I am aware of), in QBism the outcome of a measurement is special to the user taking the action—a private internal experience of that user. The user can attempt to communicate that experience verbally to other users, who may hear[24] her words. Other users

---

[23] John S. Bell, Physics World **3** (8), 3340 (1990).

[24] As John Bell may have heard the bell of the tea trolley. Hearing something, of course, is a personal experience.

can also observe her action and, under appropriate conditions, experience aspects of the world's reaction closely related to those experienced by the original user. But in QBism the immediate outcome of a measurement is a private experience of the person taking the measurement action, quite unlike the public, objective, classical outcome of a Copenhagen[25] measurement.

Because outcomes of Copenhagen measurements are "classical", they are *ipso facto* real and objective. Because in QBism an outcome is a personal experience of a user, it is real only for that user, since that user's immediate experience is private, not directly accessible to any other user. Because the private measurement outcome of a user is not a part of the experience of any other user, it is not as such real for other users. Some version of the outcome can enter the experience of other users and become real for them as well, only if the other users have also experienced aspects of the world's response to the user who took the measurement-action, or if that user has sent them reliable verbal or written reports of her own experience.

This is, of course, nothing but the famous story of Wigner and his friend, but in QBism Wigner's Friend is transformed from a paradox to a fundamental parable. Until Wigner manages to share in his friend's experience, it makes sense for him to assign her and her apparatus an entangled state in which her possible reports of her experiences (outcomes) are strictly correlated with the corresponding pointer readings (digital displays, etc.) of the apparatus.

Even versions of Copenhagen that do not prohibit mentioning users, would draw the line at allowing a user to apply quantum mechanics to another user's reports of her own internal experience. Other users are either ignored entirely (along with *the* user), or they are implicitly regarded as part of "the classical world". But in QBism each user may assign quantum states in superposition to all of her still unrealized potential experiences, including possible future communications from users she has yet to hear from. Asher Peres' famous Copenhagen mantra, "Unperformed experiments have no results", becomes the QBist user's tautology: "Unexperienced experiences are not experienced."

Copenhagen, as expounded by Heisenberg and Peierls, holds that quantum states encapsulate "our knowledge". This has a QBist flavor to it. But it is subject to John Bell's famous objection: Whose knowledge? Knowledge about what?[26] QBism replaces "knowledge" with "belief". Unlike "knowledge", which implies something underlying it that is known, "belief" emphasizes a believer, in this case the user of quantum mechanics. Bell's questions now have simple answers. Whose belief does the quantum state encapsulate? The belief of the person who has made that state assignment. What is the belief about? Her belief is about the implications of her past experience for her subsequent experience.

No version of Copenhagen takes the view that "knowledge" is the state of belief of the particular person who is making use of quantum mechanics to organize her

---

[25]I shall stop adding the phrase "or any other interpretation", but in many cases the reader should supply it.

[26]Bell used the word "information", not "knowledge", but his objection has the same force with either term.

experience. Peierls may come closest in a little-known 1980 letter to John Bell:[27] "In my view, a description of the laws of physics consists in giving us a set of correlations between successive observations. By observations I mean…what our senses can experience. That we have senses and can experience such sensations is an empirical fact, which has not been deduced (and in my opinion cannot be deduced) from current physics." Had Peierls taken care to specify that when he said "we", "us", and "our" he meant each of us, acting and responding as a user of quantum mechanics, this would have been an early statement of QBism. But it seems to me more likely that he was using the first person plural collectively, to mean all of us together, thereby promulgating the Copenhagen confusion that Bell so vividly condemned.

Copenhagen also comes near QBism in the emphasis Bohr always placed on the outcomes of measurements being stated in "ordinary language". I believe he meant by this that measurement outcomes were necessarily "classical". In QBism the outcome of a measurement is the experience the world induces back in the user who acts on the world. "Classical" for any user is limited to her experience.[28] So measurement outcomes in QBism are necessarily classical, in a way that has nothing to do with language. Ordinary language comes into the QBist story in a more crucial way than it comes into the story told by Bohr. Language is the only means by which different users of quantum mechanics can attempt to compare their own private experiences. Though I cannot myself experience your own experience, I can experience your verbal attempts to represent to me what you experience. It is only in this way that we can arrive at a shared understanding of what is common to all our own experiences of our own external worlds. It is this shared understanding that constitutes the content of science.

A very important difference of QBism, not only from Copenhagen, but from virtually all other ways of looking at science, is the meaning of probability 1 (or 0).[29] In Copenhagen quantum mechanics, an outcome that has probability 1 is enforced by an objective mechanism. This was most succinctly put by Einstein, Podolsky and Rosen,[30] though they were, notoriously, no fans of Copenhagen. Probability-1 judgments, they held, were backed up by "elements of physical reality".

Bohr[31] held that the mistake of EPR lay in an "essential ambiguity" in their phrase "without in any way disturbing". For a QBist, their mistake is much simpler than that: probability-1 assignments, like more general probability-p assignments are personal expressions of a willingness to place or accept bets, constrained only by the

[27] Peierls to Bell, 13/11/1980, *Selected Correspondence of Rudolf Peierls*, vol. 2, Sabine Lee [ed], World Scientific, 2009, p. 807.

[28] Indeed, the term "classical" has no fundamental role to play in the QBist understanding of quantum mechanics. It can be replaced by "experience".

[29] A good example to keep in mind is my above mentioned assignment of probability 1 to my belief that you have personal experiences of your own that have for you the same immediate character that my experiences have for me.

[30] A. Einstein, B. Podolsky, and N. Rosen, Phys. Rev. **47**, 777–780 (1935).

[31] N. Bohr, Phys. Rev. **48**, 696–702 (1935).

requirement[32] that they should not lead to certain loss in any single event. It is wrong to assert that probability assignments must be backed up by objective facts on the ground, even when $p = 1$. An expectation is assigned probability 1 if it is held as strongly as possible. Probability-1 measures the intensity of a belief: supreme confidence. It does not imply the existence of a deterministic mechanism.

We are all used to the fact that with the advent of quantum mechanics, determinism disappeared from physics. Does it make sense for us to qualify this in a footnote: "Except when quantum mechanics assigns probability 1 to an outcome"? Indeed, the point was made over 250 years ago by David Hume in his famous critique of induction.[33] Induction is the principle that if something happens over and over and over again, we can take its occurrence to be a deterministic law of nature. What basis do we have for believing in induction? Only that it has worked over and over and over again.

That probability-1 assignments are personal judgments, like any other probability assignments, is essential to the coherence of QBism. It has the virtue of undermining the temptation to infer any kind of "nonlocality" in quantum mechanics from the violation of Bell inequalities.[34] Though it is alien to the normal scientific view of probability, it is no stranger or unacceptable than Hume's views of induction.[35] What is indisputable is that the QBist position on probability 1 bears no relation to any version of Copenhagen. Even Peierls, who gets closer to QBism than any of the other Copenhagenists, takes probability 1 to be backed up by underlying indisputable objective facts.

Since this is a meeting in celebration of John Bell, I conclude with a few more comments on the quotations from Bell's little-known[36] correspondence with Peierls at the head of my text.

The first quotation suggests a riddle: Why is quantum mechanics like a bicycle? *Answer*: Because while it is possible to learn how to use either without knowing what you are doing, it is impossible to make sense of either without taking account of what people actually do with them.

The second quotation indicates Bell's willingness to consider concepts, as fundamental as "particle" or the person to whom he is writing his letter, as "inventions" that help him to make better sense of the data that constitute his experience.

The third reveals a willingness to regard measurements as particular responses of particular people to particular experiences induced in them by their external world.

These are all QBist views. Does this mean that John Bell was a QBist? No, of course not—no more than Niels Bohr or Erwin Schrödinger or Rudolf Peierls or Sig-

[32]Known as Dutch-book coherence. See the Fuchs-Schack Revs. Mod. Phys. article cited above.

[33]David Hume, *An Enquiry concerning Human Understanding* (1748).

[34]C. A. Fuchs, N. D. Mermin, and R. Schack, Am. J. Phys. **82**, 749–754 (2014).

[35]I would have expected philosophers of science, with an interest in quantum mechanics, to have had some instructive things to say about this connection, but I'm still waiting.

[36]I have had no success finding any of them with Google. For example, there is no point in googling "Bell bicycle." " 'John S. Bell' bicycle" does no better. Even " 'John S. Bell' bicycle quantum" fails to produce anything useful, because there is a brand of bicycle called "Quantum", and Quantum bicycles have bells.

mund Freud were QBists. Nobody before Fuchs and Schack has pursued this point of view to its superficially shocking,[37] but logically unavoidable and, ultimately, entirely reasonable conclusions. On the other hand, what Bell wrote to Peierls, and the way in which he criticized Copenhagen, lead me to doubt that Bell would have rejected QBism as glibly and superficially as most of his contemporary admirers have done.

John Bell and Rudolf Peierls are two of my scientific heroes, both for their remarkable, often iconoclastic ideas, and for the exceptional elegance and precision with which they put them forth. Yet in their earlier correspondence, and in their two short papers in *Physics World* at the end of Bell's life, they disagree about almost everything in quantum foundations. Peierls disliked the term "Copenhagen interpretation" because it wrongly suggested that there were other viable ways of understanding quantum mechanics. Bell clearly felt that Copenhagen was inadequate and downright incoherent. I like to think that they too, like Bohr and Schrödinger, might have found common ground in QBism.

Acknowledgments I am grateful to Chris Fuchs and Rüdiger Schack for their patient willingness to continue our arguments about QBism, in spite of my inability to get their point for many years. And I thank them both for their comments on earlier versions of this text.

[37]Ninety years after the formulation of quantum mechanics, a resolution of the endless disagreements on the meaning of the theory has to be shocking, to account for why it was not discovered long, long ago.

Chapter 5
On the Quantum Measurement Problem

Časlav Brukner

Abstract In this paper, I attempt a personal account of my understanding of the measurement problem in quantum mechanics, which has been largely in the tradition of the Copenhagen interpretation. I assume that (i) the quantum state is a representation of knowledge of a (real or hypothetical) observer relative to her experimental capabilities; (ii) measurements have definite outcomes in the sense that only one outcome occurs; (iii) quantum theory is universal and the irreversibility of the measurement process is only "for all practical purposes". These assumptions are analyzed within quantum theory and their consistency is tested in Deutsch's version of the Wigner's friend gedanken experiment, where the friend reveals to Wigner whether she observes a definite outcome without revealing which outcome she observes. The view that holds the coexistence of the "facts of the world" common both for Wigner and his friend runs into the problem of the hidden variable program. The solution lies in understanding that "facts" can only exist relative to the observer.

Two Measurement Problems

There are at least two measurement problems in quantum mechanics.[1] The less prominent of the two (the "small" problem) is that of *explaining why a certain outcome – as opposed to its alternatives – occurs in a particular run of an experiment.* The bigger problem of the two (the "big" problem) is that of explaining the ways in

[1] Two problems are assumed in Refs. [1, 2] and three problems are assumed in Ref. [3].

Č. Brukner(✉)
Institute for Quantum Optics and Quantum Information (IQOQI), Austrian Academy of Sciences, 1090 Vienna, Austria
e-mail: caslav.brukner@univie.ac.at

Č. Brukner
Faculty of Physics, University of Vienna, 1090 Vienna, Austria

R. Bertlmann and A. Zeilinger (eds.), *Quantum [Un]Speakables II*,
The Frontiers Collection, DOI 10.1007/978-3-319-38987-5_5

which an experiment arrives at an outcome. It addresses the question of *what makes a measurement a measurement*.[2]

In the following, I would like to present a personal account of my understanding of the measurement problems in quantum mechanics. My intention is not to argue that the approach I chose is the "best" way in any particular sense, but rather to demonstrate its logical consistency and to investigate what consequences the requirement for its consistency have for our understanding of physical reality. I will first present a probabilistic argument that explains why the measurement process is irreversible "for all practical purposes". Furthermore, by analyzing Deutsch's version of the Wigner's friend gedanken experiment, I will show that any attempt to assume that the measurement records (or "facts" or experiences) coexist for both Wigner and his friend will run into the problems of the hidden variable program, for which I propose a Bell-type experiment. The conclusion is that these records can have meaning only relative to the observers; there are no "facts of the world per se".

Although I see my view of the quantum measurement problem broadly in the tradition of the Copenhagen interpretation, particularly within the information-theoretical approach to quantum mechanics [4], it contains elements from Qbism [5], the relative interpretation of Rovelli [6] and even the many-worlds interpretation. This indicates that the various interpretations might have much more in common than their supporters are willing to accept.

The solutions to the small measurement problem which have been offered to date basically present two underlying premises. They either introduce "hidden" causes that determine which outcome will occur in a given experimental run (as in Bohm's hidden-variables theory), or they refute the basic notion of measurements resulting in definite outcomes (as in the Everett interpretation). None of that is really necessary. My position is that measurements have definite outcomes in the sense that only one outcome can be the result of a single experimental run. This is rather obvious. If it were otherwise, the notion of measurement would become ambiguous. If the outcome is not definitive, then no observation has occurred. This, however, does not exclude the possibility that the conditions that define a measurement are fulfilled for one observer but not for another. As far as discussions of the small measurement problem are concerned, I fail to see the reality of that problem. If one accepts the possibility of quantum probabilities being fundamentally irreducible, this problem vanishes.

Let me explain that in more detail. Within quantum theory, a description of the quantum state of a system and of the measurement apparatus allows us to calculate the probability $p(a|x)$ to observe outcome a, for a measurement choice x. The probabilities are "irreducible" if there are no additional variables λ in the theory, which potentially are yet to be discovered, such that when one conditions predictions on them, one has either $P(a|x, \lambda) = 1$ or 0, i.e. they allow the outcome to be

---

[2]The proposed formulation of the two problems is inspired but not equivalent to the one of Refs. [1, 2] where the two categories of measurement problems were first introduced with the designations "small" and "big".

predicted perfectly.[3] Not only quantum mechanics, but every probabilistic theory in which probabilities are taken to be irreducible "must have" the small measurement problem. (The "hypothetical collapse" models [8–10] that predict the breakdown of quantum-mechanical laws on a macroscopic scale, if not supplemented by non-local hidden variables, also fall into this category.) The lack of the small measurement problem in the probabilistic theories would contradict the very idea of having irreducible probabilities.[4]

The big measurement problem is more subtle. It can be illustrated by the following situation. As students, we are taught that there are two processes a quantum state can undergo. First, the deterministic, unitary, and continuous time evolution of the state (of a system, possibly together with its environment) that obeys the Schroedinger equation or its relativistic counterpart. Second, the probabilistic, non-unitary, and discontinuous change of the state, called "projection" or "collapse", brought about by measurement. Equipped with this knowledge, we attend a practical training in a quantum optics laboratory, where we see various pieces of equipment, such as photonic sources, beam-splitters, optical fibers, mirrors, photodiodes, phase shifters etc., for the first time. The instructor sets us the task of computing the evolution of the photonic quantum state in the set-up prepared on the optical table. We soon realize that we are in trouble. There is nothing in the theory to tell us which device in the laboratory corresponds to a unitary transformation and which to a projection! We start to ask questions. What makes a photodiode a good detector for photons? And why is a beam splitter a bad detector? At least manufacturers of photon detectors should know the answers to these questions, shouldn't they? Or perhaps the measurement is not completed in the detector, but only when the result is finally recorded in a computer, or even in the observer's mind? Bell sardonically commented [11]: "What exactly qualifies some physical systems to play the role of 'measurer'? Was the wave function of the world waiting to jump for thousands of millions of years until a single-celled living creature appeared? Or did it have to wait a little longer, for some better qualified system … with a Ph.D.?".

---

[3]The notion of irreducibility can be weakened to the requirement that the predictions conditioned on the variables are not more informative about the outcomes of future measurements than the predictions of quantum theory [7]. Formally, for every measurement, the probability distribution conditioned on the variable cannot have lower (Shannon) entropy than the quantum probability distribution.

[4]The so-called "non-local" features of quantum theory are not a subject of the present article. I should, however, like to mention that once one accepts the notion that probabilities can be irreducible, there is no reason to restrict them to be *locally causal* [11], i.e. to be decomposable as: $p(a, b|x, y) = \int d\lambda \rho(\lambda) P(a|x, \lambda) P(b|y, \lambda)$, where x and y are choices of measurement settings in two separated laboratories, a and b are respective outcomes and $\rho(\lambda)$ is a probability distribution. It appears that the main misunderstanding associated with Bell's theorem stems from a failure to acknowledge the irreducibility of quantum probabilities *irrespectively* of the relative experimental space-time arrangements [12]. Bell's local causality accepts that probabilities for local outcomes can be irreducibly probabilistic, but requires those for correlations to be factorized into (a convex mixture of) probabilities for local outcomes. There is no need for imposing such a constraint on a probabilistic theory, where probabilities are considered to be fundamental. Rather, the notion of locality should be based on a operationally well-defined no-signaling condition, and it is this condition whose violation is at odds with special relativity.

One possibility to address these questions would be to dismiss the big measurement problem as a pseudo-issue, just as we did for the small one. If quantum theory is understood as a fundamental theory of observations and observers' actions upon these observations, then measurement can be introduced as a primitive notion, which cannot be subject to a complete analysis, not even in principle. At most, one could motivate it informally, through an appeal to intuition and everyday experience. It seems to me that this path is taken by some proponents of the Copenhagen interpretation and Quantum Bayesianism (QBists), for example when Fuchs and Scheck write [13] "a measurement is an action an agent takes to elicit an experience. The measurement outcome is the experience so elicited". Such a view is consistent and self-contained, but in my opinion, it is *not* the whole story. It is silent about the question: what makes a photon counter a better device for detecting photons than a beam splitter? Yet the question is scientifically well posed and has an unambiguous answer (which manufacturers of photodetectors do know!).

I would like to express clearly that I do agree with the Qbists and the Copenhagenists on the necessity of a *functional* distinction between the object and the subject of observation. This distinction is at the heart of Bohr's epistemological argument that measurement instruments lie outside the domain of the theory, insofar as they serve their purpose of acquiring empirical knowledge. Regretfully, this argument has repeatedly been misinterpreted in textbooks and articles and "replaced by the crude physical assumption that macroscopic systems behave classically, which would introduce an artificial split of the physical world into a quantum microcosms and a classical macrocosms." [14]. The "cut" is not between the macro and micro worlds but between the measuring apparatus and the observed quantum system. It is of epistemic, not of ontic origin.

Bohr and Heisenberg seem to have disagreed about the movability of the cut [15]. As Heisenberg recalls in his letter to Heelan [16] (quoted in Ref. [15]): "I argued that a cut could be moved around to some extent while Bohr preferred to think that the position is uniquely defined in every experiment". In my understanding, the two views are not conflicting and can be brought into accordance. Heisenberg acknowledges the universality of the laws of quantum mechanics in the sense that every system, including the measuring instrument, is *in principle* subject to these laws. Of course, in moving the cut, the measurement instrument loses its function as a means for acquiring knowledge about a quantum system and becomes itself a quantum system—an object that can be observed by a further set of measurement instruments. Bohr, however, believes "that for a *given* (my italics) experimental setting the cut is determined by the nature of the problem ...", as he writes in a 1935 letter to Heisenberg [17] (quoted in Ref. [15]). The cut is hence movable in principle, but is fixed in any concrete experimental set-up. Still, we might wonder what fixes the position of the object–instrument cut in a concrete experimental set-up? Here, Bohr encounters the big measurement problem.

The question of the meaning of the quantum state is closely related to the measurement problem(s). Which approach one takes in addressing the later depends on the specific view one has on the former. The next section is devoted to this question.

What Is the Quantum State?

The discussion over the meaning of quantum states is often presented as a conflict between two fundamentally opposed approaches. The first approach speaks of "states of reality" that are independent of any empirical access, and implicitly assumes the existence of such states. The second approach refers to observations, and what we can know about them and deduce from them. This approach requires differentiation with regard to the question "Knowledge about what?". Insofar as the quantum state is seen as representing the observer's incomplete knowledge about an assumed "state of reality", it is not fundamentally dissimilar to the first approach. This is why, to use modern terminology, the distinction between a realist interpretation of a quantum state that is "psi-ontic" and one that is "psi-epistemic" [18]—which actually is a distinction between two kinds of hidden variable theory—is only relevant to supporters of the first approach.

An alternative exists. The quantum state can be seen as a mathematical representation of what the observer has to know in order to calculate probabilities for outcomes of measurements following a specific preparation. However, one could also object to this "operationalist's view". Malin phrased it nicely [19]: "What if the knower is a physicist who had a martini before trying to 'know'? What if a person who knows just a little physics learns of the result? What if he had a martini? Somehow we feel that such questions are irrelevant." He continues: "To avoid difficulties of this kind regarding the epistemic interpretation, we can consider a quantum state as representing not actual knowledge (which requires a knower), but the available or potential knowledge about a system."

Of course, the argument that quantum theory does not apply in the absence of observers has not been made. Yet when calculating a quantum state, it might help to think of a *hypothetical* observer for whom the quantum state stands for her knowledge.[5] For example, when quantum cosmologists talk about the pressure of a primordial state of the universe, we can make sense of it if we imagine a well-defined experimental procedure a hypothetical observer could apply on the state to provide an operational meaning to the term "pressure". The ultimate meaning is given by current cosmological observations, based on which we reconstruct the idea of the early universe's pressure. (The observer here is always considered to be external to the universe. The "wave function of the universe" that would include the observer is a problematic concept, as it negates the necessity of the object–subject cut.). This is compatible with Malin's view [19] that "quantum states represent the available knowledge about the potentialities of a quantum system, knowledge from the perspective of a particular location in space", not of any actual observer.

I share Malin's view on the meaning of the quantum state, which is essentially the one supported by Copenhagenists and Qbists. I would like to add just one, but an

[5]Peres correctly notes that considering hypothetical observers is not a prerogative of quantum theory [20]. They are also used in thermodynamics, when we say that a perpetual-motion machine of the second kind cannot be built, or in the theory of special relativity, when we say that no signal can be transferred faster than the speed of light.

important, aspect to this view: *The quantum state is a representation of knowledge necessary for a hypothetical observer – respecting her experimental capabilities – to compute probabilities of outcomes of all possible future experiments*. An explicit reference to the observer's experimental capabilities is crucial to address the big measurement problem. The "knowledge" here refers to Wigner's definition of the quantum state [21]: "... the state vector is only a shorthand expression of that part of our information concerning the past of the system which is relevant for predicting (as far as possible) the future behaviour thereof."

The available experimental precision will in every particular arrangement determine to which objects the observer can meaningfully assign quantum states. This agrees with Bohr's view "that for a given experimental setting the cut is determined by the nature of the problem ..." That there is nothing in the theory that would prohibit reaching the necessary experimental precision to allow a meaningful state assignment to objects of increasingly large sizes—eventually as large as our measurement devices—reflects Heisenberg' view that the "cut can be shifted arbitrarily far in the direction of the observer", as he wrote in an unpublished paper [22] from 1935, in which he outlined his response to the Einstein, Podolsky, and Rosen paper from the same year.[6] The measurement instrument and the observer can be included in the quantum mechanical description, and then observed by someone else, a "superobserver", for whom the original measurement instrument loses its previous status as a means for acquiring knowledge. For this purpose, she needs another set of "superinstruments" that are superior to the original instruments regarding their precision.

FAPP Irreversibility

The distinct outcomes of a measurement apparatus are associated with "macroscopically distinct states". (Only in that aspect does "macroscopicity" play a role in the measurement process.) These are defined as states that can still be differentiated even in those cases where the observations are imprecise and coarse-grained. If, for example, just a few spins of a large magnet are flipped, then the entire quantum state of the magnet will change into an orthogonal one. Yet, at our macroscopic level, we will still perceive it as the very same magnet. In order for the change to become noticeable even in a coarse-grained measurement, the quantum states of a sufficiently large number of spins need to be changed.

[6]In the same paper, Heisenberg concludes: "... the quantum mechanical predictions about the outcome of an arbitrary experiment are independent of the location of the cut ..." This can be seen as a consequence of "purification" in quantum theory, which states that every mixed state of system A can always be seen as a state belonging to a part of a composite system AB that itself is in a pure state. This state is unique up to a reversible transformation on B. The assumption of purification is one of the central features of quantum theory, which, taken as an axiom together with a few other axioms, makes it possible to explain why the theory has the very mathematical structure it does [23].

In order to quantify the distinguishability of such macroscopic states, consider a spin-j system, with $j \gg 1$, and the coarse-grained measurements with the POVM elements

$$P_{\Omega_0} = \frac{2j+1}{4\pi} \iint_{\Omega_0} d\Omega |\Omega\rangle\langle\Omega| \tag{1}$$

as a model example, where $|\Omega\rangle = \sum_{m=-j}^{+j} \binom{2j}{j+m}^{1/2} \cos^{j+m}\frac{\theta}{2} \sin^{j-m}\frac{\theta}{2} e^{-im\phi_0}|m\rangle$ is the spin coherent state and θ and ϕ are the polar and azimuthal angles respectively, corresponding to the solid angle Ω. The size of the integration region around the solid angle $\Omega_0 = (\theta_0, \phi_0)$ is taken to be such that its projection Δm along the z axis is much larger than the intrinsic uncertainty of the coherent states,[7] $\Delta m \gg \sqrt{j}$. Under the coarse grained measurement with elements (1), any state $\hat{\rho}$ can effectively be described in terms of a positive probability distribution (the well-known Q-function) [26].

$$Q(\Omega) = \frac{2j+1}{4\pi} \langle\Omega|\hat{\rho}|\Omega\rangle. \tag{2}$$

Specifically, the probabilities for the POVM outcomes can be obtained by averaging the Q-function: $P_{\Omega_0} = \iint_{\Omega_0} Q(\Omega)$. Hence, the description in terms of $Q(\Omega)$ is effectively classical and it leads to the classical limit of quantum mechanics[8] [26]. Since $Q(\Omega)$ represents a complete description of the system under coarse-grained mea-

---

[7]When we introduce coarse-grained observables, we need to define the states that are "close" to each other to conflate them into coarse-grained outcomes. However, the terms "close" or "distant" make sense in a classical context only. There, "close" states correspond to neighboring outcomes in the real configuration space. For example, the coherent states conflated in the single outcome Ω_0 of the POVM all correspond to approximately the same direction Ω_0 in real space. Therefore, certain features of classicality need to be presumed before macroscopic states can be defined. An alternative would be the attempt to reconstruct the notions of closeness, distance, and space—and consequently, also the theories referring to these notions, such as quantum field theory—from within the formalism of the Hilbert space only. Useful tools for this attempt might be preferred tensor factorizations, coarse-grained observables, and symmetries. The results of Refs. [24, 25] present the first progress towards this goal. The most elementary quantum system, the qubit, resides in an abstract state space with SU(2) symmetry. This is locally isomorphic to the group SO(3) of rotations in three-dimensional space. Considering directional degrees of freedom (spin), this symmetry is found to be operationally justified in the symmetry of the configuration of macroscopic instruments used for transforming the spin state. Hereby one assumes that quantum theory is "closed": the macroscopic instruments do not lie outside of the theory, but are described from within it in the limit of a large number of its constituents (as coherent states or "classical fields") [25].

[8]The classical world arises from within quantum theory when neighboring outcomes are not distinguished but bunched together into slots in the measurements of limited precision. What would the classical world look like if *non-neighboring* outcomes were conflated to slots? To address this question, one could imagine an experiment on a person whose nerve fibers behind the retina are disconnected and again reconnected at different, *randomly* chosen, nerve extensions connecting to the brain. It seems reasonable to assume that the neighboring points of the object that is illuminated with light and observed by the person's eye will no more be perceived by the person as neighboring points. One may wonder if, in the course of further interaction with the environment, the person's brain will start to make sense out of the seen "disordered classical world", or if it will post-process the signals to search for more "ordered" structures as a prerequisite for making sense out of them.

surements, I will call it the "macroscopic state". This approach to classicality differs conceptually from and is complementary to the decoherence program that is dynamical and describes correlations of the system with other degrees of freedom which are integrated out [27].

A measure of the distinguishability between two probability distributions $P(\Omega)$ and $Q(\Omega)$ is the Euclidean scalar product $(\mathbf{P}, \mathbf{Q}) := \iint d\Omega \sqrt{P(\Omega)Q(\Omega)}$. If two probability distributions are perfectly distinguishable $(\mathbf{P}, \mathbf{Q}) = 0$, while if they are identical $(\mathbf{P}, \mathbf{Q}) = 1$. Consider two pure quantum states $|\psi_1\rangle$ and $|\psi_2\rangle$ with the Q-functions $|\langle\psi_1|\Omega\rangle|^2$ and $|\langle\psi_2|\Omega\rangle|^2$, respectively. Then $\frac{2j+1}{4\pi} \iint d\Omega\, |\langle\psi_1|\Omega\rangle||\langle\Omega|\,\psi_2\rangle| \geq \frac{2j+1}{4\pi} | \iint d\Omega \langle\psi_1|\Omega\rangle\langle\Omega|\psi_2\rangle| = |\langle\psi_1|\psi_2\rangle|$, where $\frac{2j+1}{4\pi} \iint d\Omega|\Omega\rangle\langle\Omega| = \mathbb{1}$. This shows that distinguishability between the quantum states and distinguishability between macroscopic states are two different notions. The latter implies the former, but the opposite is not true. Say, spin j is composed of N spins-1/2. In this case, a number of spins of the order of $\sqrt{N}$ need to be flipped in order to arrive at a macroscopically distinct state. Only then do we perceive it as a new state of magnetization. It is now a new "fact".

The macroscopic states are robust. This means that they are stable against perturbations, which may for example be caused by repeated coarse-grained observations. In other words, the Q-function before and after a coarse-grained measurement is approximately the same [28]. It therefore becomes possible for different observers to repeatedly observe the same macroscopic state. The result is a certain level of intersubjectivity among them. If we assume, however, that quantum mechanics is universally valid, then it is in principle possible to undo the entire measurement process. Imagine a superobserver who has full control over the degrees of freedom of the measuring apparatus. Such a superobserver would be able to decorrelate the apparatus from the measured system. In this process, the information about the measurement result would be erased. Seen from this perspective, "irreversibility" in the quantum measurement process merely stands for the fact that it is extremely difficult—but not impossible!—to reverse the process. It is irreversible "for all practical purposes" (or "FAPP," to use Bell's acronym).

I have often heard the following objection to FAPP: No matter how low the probability is to reverse the evolution in the measurement process, it is still there. How is it possible to settle the question of what actually exists by an approximation? In my eyes, such questions do not take into consideration the simple fact that quantum theory cannot be both, universal and not irreversible merely FAPP. While on the one hand, measurements have to result in irreversible facts (otherwise, the notion of measurement itself would become meaningless, as no measurement would ever be conclusive), this irreversibility on the other hand must be merely FAPP if quantum theory is in principle applicable to any system. "Any system" here means that the measuring apparatus and even the observer herself can also be subject to the laws of quantum theory. My main point is the following. While it is obviously possible to

---

(Footnote 8 continued)
The latter may eventually nullify the effect of the random reconnection of nerves, and the person will again perceive the ordinary classical world.

describe the subject as an object, it then has to be the object for another subject. So, one can make sense of a quantum state of the measurement apparatus as well as the observer only with respect to another set of measurement instruments and another observers. In my eyes, not enough thought has gone into the fundamental nature of FAPP. More research on the philosophy of FAPP, if you like, should be done by philosophers of physics. This, in my eyes, would contribute to the resolution of the problem in a much deeper way than the perpetual attempts to expel this term from the foundations of physics based on presupposed philosophical doctrine.

Detection devices, such as photographic plates or photo-diodes, consist of a large number of constituents in a certain "metastable state". Their interaction with the observed quantum systems brings them into a "stable state" that can be distinguished from the initial one even under coarse-grained observations. This transition is signified by the "click" in the detector or a new position of the pointer label. In both the metastable and the stable state, the constituents of the instrument can be in any of a large number of quantum states that correspond to the respective macroscopic states. In order to understand how irreversibility FAPP is possible, it is crucial to realize that not only the initial and final quantum states of the instrument are imprecisely known, but also the full details of the interactions (i.e. Hamiltonian) among its constituents and with the environment. Even if it were possible to know the initial and final states precisely, the lack of precise knowledge of these interactions prevents us from reversing the measurement process. Moreover, a photodetector also does not spontaneously "de-click". It does not turn itself back into the initial metastable state and and it does not emit the photon into its initial state.

The irreversibility of the measurement process might be explained in quantum mechanical terms, but as metastable and stable states of the detector are macroscopic states, a classical explanation of irreversibility is sufficient. In fact, nothing "quantum" is indispensable for "solving" the big measurement problem. The "problem" is essentially present in classical measurements as well.

For classical (chaotic) systems, the physical state after a certain time can become unpredictable if the solutions of dynamical equations are highly sensitive, either to initial states or to uncontrollable external perturbations of the Hamiltonian. The situation is different in quantum mechanics because of the unitarity of the dynamical evolution: the scalar product between the unperturbed and the perturbed state is constant such that an uncertainty in initial states will not grow in time. However, an uncontrollable external perturbation to the Hamiltonian can explain FAPP irreversibility for both the classical and quantum case. Below I consider one such model.

Consider the detection device to be a classical dynamical system for which the state $\mathbf{x}_t$ at an arbitrary time t is given in terms of a flow $\mathbf{x}_t = f_t(\mathbf{x}_0)$ on the initial state $\mathbf{x}_0 = (\mathbf{q}, \mathbf{p})$, where $\mathbf{q}$ are the positions and $\mathbf{p}$ the momenta of all the system's constituents. The flow is assumed to be reversible, i.e. there exists the involution $\pi(\mathbf{q}, \mathbf{p}) = (\mathbf{q}, -\mathbf{p})$ with $\pi^2 = \mathbb{1}$ for which for all times $\pi f_t \pi = f_t^{-1}$. We now choose two regions A and B of the phase space. We assume a uniform probability distribution of the state over A to exist at the outset. The probability of finding the state in the set B at the time t when, at the time $t = 0$, the state was in the set A is given by the volume fraction of the states from A that evolves in B.

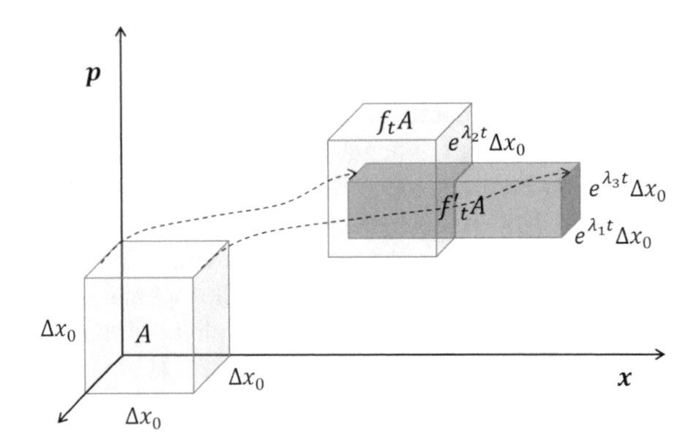

Fig. 5.1 Schematic illustration of the phase space evolution in both the regular and the chaotic case. An infinitesimal element A of volume $(\Delta x_0)^n$, where n is the dimension of the phase space, evolves in f_tA under a regular flow or in $f_t'A$ under a chaotic flow. The volume of the overlap between the two evolved elements has the linear size of Δx_0 along all directions of divergence and $e^{\lambda_i t}\Delta x_0$ ($\lambda_i < 0$) for every direction i of contraction.

$$\mathrm{Prob}[\mathbf{x}_t \in B|\mathbf{x}_0 \in A] = \frac{|A \cap f_t^{-1}B|}{|A|}. \tag{3}$$

Here, $|X|$ is the Lebesgue measure of set X. In the remaining argument I will assume that $B = f_tA$, for which the probability (3) is 1.

Suppose now that we want to reverse the evolution, but do not have precise control of the flow f_t, for example due to uncontrollable influences from the environment. Hence, the inverse flow $f_t' \neq f_t$ is perturbed, where we assume $\pi f_t' \pi = f_t'^{-1}$. At the time t, we inverse all the momenta and set the time again to 0 for simplicity. (Note that inverting momenta does not require measuring them, which would be impossible due to the limited precision of the instruments, nor does it require to know them precisely. An arrangement with elastic bounce of the molecules would be sufficient [29]. If $f_t' = f_t$, it would be sufficient to inverse the momenta to perfectly reverse the evolution.) Then, the probability to find the state in the set πA at the time t when, at time $t = 0$, the state was in the set $\pi B = \pi f_tA$ is given by

$$\mathrm{Prob}[\mathbf{x}_t \in \pi A|\mathbf{x}_0 \in \pi f_tA] = \frac{|\pi f_tA \cap f_t'^{-1}\pi A|}{|\pi f_tA|} = \frac{|f_tA \cap f_t'A|}{|A|}, \tag{4}$$

where we used $|\pi X| = |X|$ and the Liouville theorem $|f_tX| = |X|$. (Once one arrives at the states from the set πA, one can obtain those from the initial set A by simply inverting the momenta. This might induce additional imprecisions, neglected here for simplicity.) The expression (4) has an operational meaning, namely that of the "probability for reversing the evolution". It is the probability to find the system in the initial state under first the forward, and then the reverse, perturbed, flow (Fig. 5.1).

The classical explanation of irreversibility is based on the notions of mixing and coarse graining. Mixing is a property of chaotic systems for which at least one of the Lyapunov exponents is positive. (If the system is a Hamiltonian system, the sum of all Lyapunov exponents is zero. If the system is dissipative, the sum is negative.) Two trajectories in phase space with an initial separation Δx_0 along dimension i diverge at a rate given by $\Delta x_t \approx e^{\lambda_i t}\Delta x_0$, where $\lambda_i > 0$ is a Lyapunov exponent. Suppose that $|A| = (\Delta x_0)^n$ corresponds to the small volume that can still be distinguished from other such volumes in a coarse-grained observation, and n is the dimension of the phase space. Furthermore, suppose that f_t is regular and does not significantly change the form of A, while f'_t is chaotic. One has for the phase space volume $|f'_t A| \approx e^{\sum_i^n \lambda_i}(\Delta x_0)^n$. Then the probability (4) is specified by the volume $|f_t A \cap f'_t A|$ of the overlap between the volume elements $f_t A$ and $f'_t A$. This volume has the linear size Δx_0 along all directions of stretching and $e^{\lambda_i t}\Delta x_0$ along every direction i of contraction. Hence, the probability is bounded as

$$\text{Prob}[\mathbf{x}_t \in \pi A | \mathbf{x}_0 \in \pi f_t A] \leq e^{-\sum'_i \lambda_i t}, \tag{5}$$

where the sum $\sum'_i$ is over positive Lyapunov exponents. The probability of reversing the evolution and arriving at the initial state is negligibly small after several multiples of the characteristic time $t \sim 1/\sum'_i \lambda_i$. The above argument can also explain the classical irreversibility of “macroscopic states” in detection instruments in quantum experiments, but for completeness, I will below present a quantum version for it.

One can define a quantum mechanical measure of the state revival when an imperfect time-reversal evolution is applied to a quantum system. We will illustrate this with an example of a spin system. Suppose that an initial quantum state $|\psi\rangle$ evolves during a time t under a Hamiltonian H_0 into the final state $|\psi(t)\rangle$. The two states define macroscopic states (Q-functions) $P(\Omega, 0) = |\langle\Omega|\psi\rangle|^2$ and $P(\Omega, t) = |\langle\Omega|\psi(t)\rangle|^2$, respectively. Any attempt to reverse the evolution and arrive back at the initial macroscopic state will result in an application of a perturbed, slightly different Hamiltonian $\hat{H} = \hat{H}_0 + \hat{V}$ with perturbation $\hat{V}$. Perfect recovery of macroscopic state could be achieved only if one could have a perfect control over $\hat{H}$. FAAP, however, such a control is impossible.

As a measure of reversibility of macroscopic states we use the scalar product

$$(\mathbf{P}, \mathbf{Q}) = \iint d\Omega \sqrt{P(\Omega, 0)Q(\Omega, t)} \tag{6}$$

between the probability distribution $P(\Omega, 0) = |\langle\Omega|\psi\rangle|^2$ of finding initially the system in a macroscopic “phase point” Ω and the probability distribution

$$Q(\Omega, t) = |\langle\Omega| e^{\frac{i\hat{H}t}{\hbar}} e^{\frac{-i\hat{H}_0 t}{\hbar}} |\psi\rangle|^2 \tag{7}$$

of finding it there after a combined evolution: forward evolution in duration of t under the Hamiltonian $\hat{H}_0$ and then backward evolution in duration of t under $-\hat{H}$.

The combined evolution embodies the notion of time-reversal. If for some $t > 0$, one has $(\mathbf{P}, \mathbf{Q}) \approx 1$, the evolution is reversed at the macroscopic level. (Note that the reversed quantum state $e^{\frac{i\hat{H}t}{\hbar}} e^{\frac{-i\hat{H}_0 t}{\hbar}} |\psi\rangle$ does not need to be identical to the initial one $|\psi\rangle$ to have reversibility at the macroscopic level. It is only important that they approximately correspond to the same macroscopic state.)

Consider for simplicity a non-degenerative Hamiltonian $\hat{H}_0$ with eigenstates $|\alpha_0\rangle$ for eigenvalues E^0_α, and the perturbed Hamiltonian $\hat{H}$ with eigenstates $|\alpha\rangle$ for eigenvalues E_α. For simplicity, I assume an extremely weak perturbation $\hat{V}$ for which $E_\alpha = E^0_\alpha + \langle\alpha_0|\hat{V}|\alpha_0\rangle$ and $\langle\alpha|\beta_0\rangle = \delta_{\alpha,\beta_0}$. Expanding $|\psi\rangle = \sum_\alpha \psi_\alpha |\alpha\rangle$, one has

$$Q(\Omega, t) = \sum_{\alpha,\beta} \psi_\alpha \psi^*_\beta \langle\Omega|\alpha\rangle\langle\beta|\Omega\rangle e^{\frac{i(V_\alpha - V_\beta)t}{\hbar}}, \tag{8}$$

where $V_\alpha := \langle\alpha|\hat{V}|\alpha\rangle$.

The value of (8) depends on the statistical distribution of $V_\alpha - V_\beta$ over different perturbations [30]. This means that in every repetition of our procedure the system might be differently perturbed during its backwards evolution. For chaotic systems, one assumes that this distribution is governed by a random matrix theory [31]. According to this theory, V_α are independent random numbers, and for a large number of eigenstates, the distribution can be approximated by a Gaussian one $g(V_\alpha) = \frac{1}{\sqrt{\pi}\sigma} e^{-\frac{(V_\alpha - W_\alpha)^2}{\sigma^2}}$ around the mean value W_α. Taking an ensemble average over different perturbations, one obtains $\langle e^{\frac{iV_\alpha t}{\hbar}}\rangle_{pert} = \int_{-\infty}^{\infty} dx g(x) e^{\frac{ixt}{\hbar}} e^{\frac{iW_\alpha t}{\hbar}} = e^{\frac{iW_\alpha t}{\hbar}} e^{-\frac{\sigma^2 t^2}{4\hbar^2}}$. In the model the distribution spread σ is taken to be much smaller than the level spacing to ensure no correlations in the distribution, $\langle e^{\frac{i(V_\alpha - V_\beta)t}{\hbar}}\rangle_{pert} = \langle e^{\frac{iV_\alpha t}{\hbar}}\rangle_{pert} \langle e^{\frac{-iV_\beta t}{\hbar}}\rangle_{pert}$. Finally, we obtain

$$\langle Q(\Omega, t)\rangle_{pert} = \sum_{\alpha,\beta} \psi_\alpha \psi^*_\beta \langle\Omega|\alpha\rangle\langle\beta|\Omega\rangle e^{\frac{i(W_\alpha - W_\beta)t}{\hbar}} e^{-\frac{\sigma^2 t^2}{2\hbar^2}} = |\langle\Omega|\phi(t)|^2 e^{-\frac{\sigma^2 t^2}{2\hbar^2}}, \tag{9}$$

where in the final step we introduce $\phi_\alpha(t) := \psi_\alpha e^{\frac{iW_\alpha t}{\hbar}}$ and $|\phi(t)\rangle := \sum_\alpha \phi_\alpha(t)|\alpha\rangle$.

Using $\langle\sqrt{Q(\Omega, t)}\rangle_{pert} \leq \sqrt{\langle Q(\Omega, t)\rangle_{pert}}$ one obtains for the measure of reversibility

$$\langle(\mathbf{P}, \mathbf{Q})\rangle_{pert} \leq \iint d\Omega |\langle\Omega|\psi\rangle||\langle\Omega|\phi(t)\rangle| e^{-\frac{\sigma^2 t^2}{4\hbar^2}} \leq e^{-\frac{\sigma^2 t^2}{4\hbar^2}}. \tag{10}$$

We see that for random perturbations on average the macroscopic state will significantly change after first forward evolution in duration of time t, and then reverse evolution in duration of time t, if $t > \hbar/\sigma$, indicating FAPP irreversibility. Specifically, for $t \rightarrow \infty$, one has $\langle(\mathbf{P}, \mathbf{Q})\rangle_{pert} \rightarrow 0$. The regime beyond the validity of weak perturbation can be treated using the results from the field of quantum chaos [32].

For the present distribution over perturbations both the evolution of the macroscopic state and of the quantum state is FAPP irreversible. One can find, however, such distributions for which the evolution of the macroscopic state is reversible and the evolution of the quantum states is irreversible FAPP, but we will not analyse them here further.

We conclude that the lack of the complete knowledge of the Hamiltonian circumvents the time reversal objection, also known as Loschmidt's paradox, which states that it should not be possible to deduce time irreversibility from an underlying time reversal theory. A similar argument could be applied to address the recurrence objection, which is based on the Poincaré recurrence theorem, that all finite systems are recurrent, i.e. return arbitrarily close to their initial state after a possibly very long time. Results show that recurrence times in the dynamics of quantum states could be extremely large [33]. Hence, quantum measurements are irreversible FAPP. A comprehensive study of various models of quantum measurement can be found in Ref. [34].

Deutsch's Thought Experiment

After arguing in favour of understanding measurement irreversibility to be FAPP, we now turn back to conceptual issues. In Ref. [35], Deutsch proposed an experiment which he claims can distinguish experimentally between the Copenhagen and the Everett interpretations of quantum mechanics. While I do not in any way see the necessity of assuming that the two interpretations might have distinct predictions in the experiment, I acknowledge that the experiment most strikingly demonstrates the necessity of a radical revision of our attitude to physical reality in quantum physics.

The thought experiment involves measurements on the observer by another, superobserver, and is a variant of the Wigner's-friend thought experiment [36]. Four systems are involved in the experiment as illustrated in Fig. 5.2. System 1 is a spin-1/2 atom which passes through a Stern-Gerlach apparatus in such a way that the two trajectories, corresponding to outcomes "spin up" and "spin down", pass over systems 2 and 3. These two systems, also spin-1/2 atoms, represent part of the observer's "sense organ". Their receptive states at the outset are "spin down". They are coupled with atom 1 in such away that when atom 1 follows the "spin up" trajectory, it passes over atom 2. The spin of atom 2 now flips to "spin up". Meanwhile, the spin of atom 3 remains the same. If in a similar way atom 1 follows the "spin-down" trajectory, the spin of atom 3 will flip while the spin of atom 2 will remain unchanged. System 4 is the observer, and it couples only to sense organs 2 and 3. Potentially, there are further systems that constitute an environment of the four systems. They all are isolated from the rest in a sealed laboratory. The experiment is then performed a sufficient number of times to collect statistics.

Initially, the state of the four systems is factorized with atom 1 in state $|x+\rangle_1 = \frac{1}{\sqrt{2}}(|z+\rangle_1 + |z-\rangle_1)$ and the observer in some definite state, $|0\rangle_4$, whose exact proper-

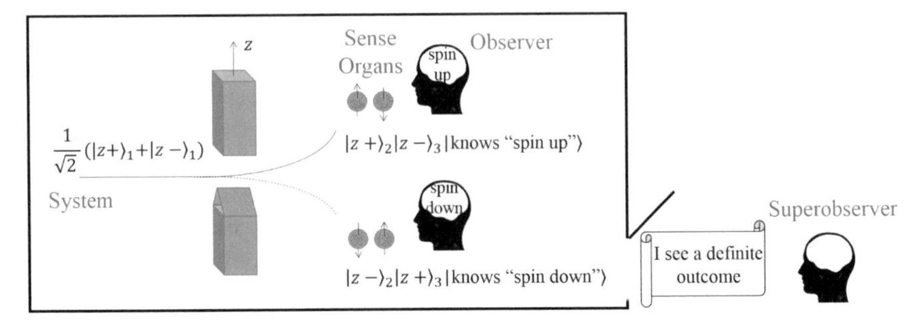

Fig. 5.2 Deutsch's version of the Wigner's friend gedanken experiment. An observer performs the Stern-Gerlach experiment on a spin-1/2 atom (system 1) in a closed laboratory. The outcome "spin up" or "spin-down" is recorded in sense organs, which are also spin-1/2 atoms (systems 2 and 3) and finally in the observer's brain (system 4). The outside observer, the superobserver, describes the experiment as a coherent evolution of a large entangled state. The observer communicates a message to the superobserver outside, which contains information about whether she sees a definite outcome or not, without revealing which outcome she sees. What will the observer then experience? Will the superobserver in principle be able to perform an interference experiment on the systems and the observer and confirm the appropriateness of his state assignment?

ties do not need to be specified, except that she is capable of completing a measurement:

$$|\psi(0)\rangle = \frac{1}{\sqrt{2}}(|z+\rangle_1 + |z-\rangle_1)|z-\rangle_2|z-\rangle_3|0\rangle_4. \tag{11}$$

One can also consider mixed states, but this assumption would complicate the situation unnecessarily. The Stern-Gerlach magnet is assumed to be oriented along the z-axis. After atom 1 has passed through the Stern-Gerlach apparatus and has interacted with the sense organs at time t, the state is

$$|\psi(t)\rangle = \frac{1}{\sqrt{2}}(|z+\rangle_1|z+\rangle_2|z-\rangle_3 + |z-\rangle_1|z-\rangle_2|z+\rangle_3)|0\rangle_4. \tag{12}$$

Finally, after the interaction between the observer and the sense organs at time t', the state becomes

$$|\psi(t')\rangle = \frac{1}{\sqrt{2}}(|z+\rangle_1|z+\rangle_2|z-\rangle_3|\text{knows "up"}\rangle_4 + |z-\rangle_1|z-\rangle_2|z+\rangle_3|\text{knows "down"}\rangle_4), \tag{13}$$

where $|\text{knows "up"}\rangle_4$ and $|\text{knows "down"}\rangle_4$ denote the observer's states after recording the result. If there are further systems in the laboratory, their states can eventually also get correlated with the two amplitudes in Eq. (13) in a huge entangled state. Unless stated otherwise, I will assume that there are no further systems in the laboratory.

Strictly speaking, the quantum state (13) can have an operational meaning only for the superobserver, who is stationed outside the sealed laboratory. To him on the outside, on the basis of all the information that is in principle available to him and conditioned on having sufficient experimental capabilities, the physical description of the state in the laboratory will be the superposition (13). For example, he can test the validity of the state assignment by performing an interference experiment with the output states:

$$|\psi+\rangle = \frac{1}{\sqrt{2}}(|z+\rangle_1|z+\rangle_2|z-\rangle_3|\text{knows "up"}\rangle_4 + |z-\rangle_1|z-\rangle_2|z+\rangle_3|\text{knows "down"}\rangle_4)$$

$$|\psi-\rangle = \frac{1}{\sqrt{2}}(|z+\rangle_1|z+\rangle_2|z-\rangle_3|\text{knows "up"}\rangle_4 - |z-\rangle_1|z-\rangle_2|z+\rangle_3|\text{knows "down"}\rangle_4).$$

This requires a special experimental arrangement and instruments of high measurement precision, which allow measuring the systems' and the observer's brain states in coherent superpositions.

What will the observer in the laboratory perceive in state (13) after completing of her measurement? Will she definitively know if she has observed one single outcome or not? It is tempting to answer such questions within the standard quantum framework: Within the laboratory, the actual observation projects the quantum state into one of the two possibilities. The observer will therefore either observe outcome "spin up" or outcome "spin-down". We know that for the projection to occur, it is not necessary for the observer to actually read out the information from the measurement device; it is sufficient that the information is available in principle [37]. Then, if the superobserver were to project his state onto the basis of "all information that is in principle available" to him, would that not include information that is available to the observer? Should the mere availability of the information about the outcome somewhere—specifically, in the observer's brain—not collapse the quantum state that the superobserver assigns? Or does the observer observe some kind of "blurred reality", while the superobserver keeps describing the situation in terms of the superposition state? Deutsch's ingeniously contrived experiment could answer these questions at least in principle, albeit its execution is impractical.

The idea is that the superobserver could learn whether the observer has observed a definite outcome, without himself learning which outcome she has observed. It is enough for the observer to communicate "I observe a definite outcome" or "I observe no definite outcome" to the superobserver. (For this purpose, the laboratory may be opened to pass only this message, keeping all other degrees of freedom still fully isolated. While being practically demanding, this is possible in principle.) The message could, for example, be written on a piece of paper and passed on to the superobserver. The key element of the experiment is that the message contains no information about which outcome has occurred and thus should not lead to a collapse of the quantum state assigned by the superobserver. Imagine that the observer encodes her message in state $|\text{message}\rangle_5$ of system 5. This state is factorized out from the total state, $|\psi(t')\rangle = \frac{1}{\sqrt{2}}(|z+\rangle_1|z+\rangle_2|z-\rangle_3|\text{knows "up"}\rangle_4 + |z-\rangle_1|z-\rangle_2|z+\rangle_3|\text{knows "down"}\rangle_4)\ |\text{message}\rangle_5$, and thus the communication of the message does not destroy the superposition.

What will be written in the message? Will the superobserver see the interference? Three different results of the experiment are possible[9]:

1. The quantum state collapses due to a breakdown of the quantum-mechanical laws when applied to states of brain or to systems of sufficiently large size, mass, complexity, and the like. The collapse models Ghirardi-Rimini-Weber [8] or Diosi-Penrose [9, 10] fall into this category. One could also argue in favor of the collapse within the view according to which a quantum state is a representation of the observer's knowledge. Every measurement yields new information, and the representation of this knowledge update is the state projection. Since the new information about the outcome is available somewhere—specifically in the observer's brain—the state has to collapse for *all* observers, including the superobserver.[10] Independently of the specific rationale behind the state collapse, the observer sends the message that she observers a definite outcome. The superobserver concludes that although he could exclude all known effects caused by conventional decoherence, the state is not in the superposition. This he can confirm in the interference experiment by observing that both outputs in the interference experiments occur with equal probability.
2. The superobserver's state assignment is the superposition state, and the observer perceives a "blurred reality" that she associates with not seeing a definite outcome. She sends a message: "I observe no definite outcome". The superobserver confirms the superposition state in the interference experiment by observing a single output state in the interference experiment. I personally have trouble to make sense of this option. If quantum theory describes an observer's probability assignments in well-defined experimental procedures, where, to quote Bohr [39] "... by the word 'experiment' we refer to a situation where we can tell others what we have done and what we have learned ...", then experience of "blurred reality" seems to be outside of the standard quantum framework. Moreover, such a situation would install a fundamental asymmetry between the observers, those who see and those who do not see "blurred reality".[11]

---

[9]In a quantum mechanical experiment, the "observer" could be simulated by a qutrit with the following encoding [38]: $|0\rangle$ for "knowing spin up", $|1\rangle$ for "knowing spin down" and $|2\rangle$ for "I see no definite outcome". The message is then encoded either in $|2\rangle$ or in a state with the two-dimensional support spanned by vectors $|0\rangle$ and $|1\rangle$ (for example $\frac{1}{2}(|0\rangle\langle 0| + |1\rangle\langle 1|)$. The superobserver applies the measurement with the projectors $\hat{P}_1 = |0\rangle\langle 0| + |1\rangle\langle 1|$ and $\hat{P}_2 = |2\rangle\langle 2|$.

[10]It seems to me that Deutsch had this particular view in mind when he claimed that the Copenhagen interpretation predicts the occurrence of the collapse. I see this view at most as a variant of the interpretation and (to my knowledge) not widely spread.

[11]A stronger argument against the possibility of seeing "blurred reality" has been brought to my attention by Jacques Pienaar. Consider a different experimental scenario where a referee prepares a state of the system either in a definite state $|z+\rangle$ or $|x+\rangle$ and the observer still performs the spin measurement along z-axis. In the first case the observer sees spin up and informs the superobserver about his outcome (for example by writing a message: "I see spin up along z-axis"). If the observer in the second case instead sees no definite outcome, e.g. a "blurred reality", he informs the superobserver about this too, e.g. through the message "I observe no definite outcome". The protocol

3. The quantum laws are unmodified. The superobserver's state assignment is the superposition state. And yet, the observer observes a definite outcome. The assigned superposition state can be confirmed in the interference experiment.

In my eyes, outcomes 1 and 2 would indicate fundamentally new physics. I will not consider these cases further and regard quantum theory to be a *universal* physical theory. This leaves us with situation 3 as the only possible outcome of Deutsch's thought experiment. The outcome is compatible with the Everett interpretation: each copy of the observer observes a definite but different outcome in different branches of the (multi)universe. The outcome is compatible with the Copenhagen interpretation too, but it is rarely discussed what the implications of this claim are for our understanding of physical reality within the interpretation. The rest of the current manuscript is devoted to this problem.

Note that in situation 3 of the thought experiment, the two observers have complementary pieces of information. *Taken together*, they would violate the complementarity principle of quantum physics. The observer has complete knowledge about the value of observable A_1 with eigenstates $|z+\rangle_1|z+\rangle_2|z-\rangle_3|\text{knows "up"}\rangle_4$ and $|z-\rangle_1|z-\rangle_2|z+\rangle_3\,|\text{knows "down"}\rangle_4$, whereas the superobserver has complete knowledge about the value of observable A_2 with eigenstates $|\psi+\rangle$ and $|\psi-\rangle$. The two observables are non-commuting. One might be tempted to interpret outcome 3 of Deutsch's experiment as implying that the two pieces of information coexist. After all, the superobserver has evidence—in form of the message—that the observer had perfect knowledge about A_1. And yet, on the very same state (13), he can learn the value of A_2. It appears that even the observer herself, retrospectively, after completion of the interference experiment, can be convinced that there is a discrepancy between her message and the fact that she always ends up in one output state in the interference experiment (thereby forgetting which outcomes she had observed). This is because, if she previously were in a state observing a definite outcome, then by applying standard quantum mechanical predictions on the systems and relevant degrees of freedom of herself (which in itself might be seen as a problematic step because without specifying which degrees of freedom belong to the "system" and which to the "observer" one ignores the necessity of the object-subject cut), she should have equal probability to end up in either of the two output states.

The trouble with the assumption that values for A_1 and A_2 *coexist* in a single logical framework is that it introduces "hidden variables", for which a Bell's theorem can be formulated with its known consequences. To this end, consider a pair of superobservers, Alice and Bob, who reside in their local laboratories and conduct an experiment involving observers Anton and Bertlmann, respectively, who in turn each perform a Stern-Gerlach experiment of the type explained above. More specifically, Alice has four systems in her laboratory: atom A1, sense organs A2 and A3, and observer Anton A4. Similarly, Bob has four systems: atom B1, sense organs B2

(Footnote 11 continued)
would allow the superobserver to distinguish between nonorthogonal states perfectly. This is in disagreement with the laws of quantum mechanics.

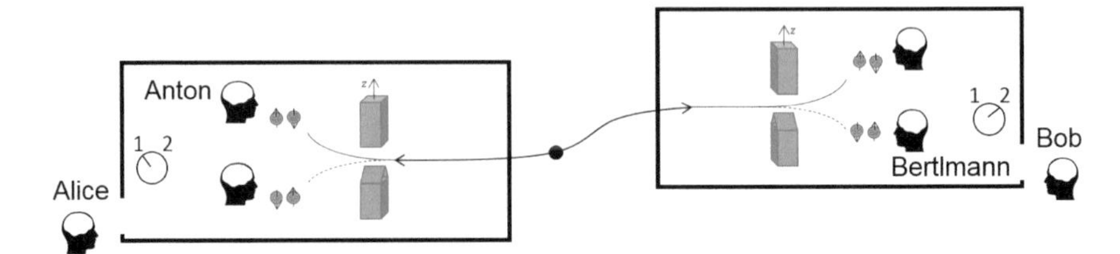

Fig. 5.3 Bell's experiment to exclude the coexistence of "facts" (i.e. measurement outcomes or records) for both the observer and the superobserver. Alice and Bob (both of them superobservers) reside in their space-like separated laboratories in which two further observers, Anton and Bertlmann respectively, perform a Stern-Gerlach type of measurement. By choice of local measurement setting (*1* or *2*), each of the superobservers can either interrogate which outcome the respective observer in his laboratory has observed or perform the interference experiment jointly on the observer and the spin. With a suitable entangled state (14), the superobservers can violate Bell's inequality.

and B3, and observer Bertlmann B4. Suppose that the two superobservers share an entangled state:

$$|\psi\rangle_{AB} = \frac{1}{\sqrt{2}}(|A_{up}\rangle|B_{down}\rangle - |A_{down}\rangle|B_{up}\rangle), \tag{14}$$

where

$$\begin{aligned} |A_{up}\rangle &= |z+\rangle_{A1}|z+\rangle_{A2}|z-\rangle_{A3}|\text{Anton knows "up"}\rangle_{A4}, \\ |A_{down}\rangle &= |z-\rangle_{A1}|z-\rangle_{A2}|z+\rangle_{A3}|\text{Anton knows "down"}\rangle_{A4}, \\ |B_{up}\rangle &= |z+\rangle_{B1}|z+\rangle_{B2}|z-\rangle_{B3}|\text{Bertlmann knows "up"}\rangle_{B4}, \\ |B_{down}\rangle &= |z-\rangle_{B1}|z-\rangle_{B2}|z+\rangle_{B3}|\text{Bertlmann knows "down"}\rangle_{B4}. \end{aligned}$$

Using these states, one can define observables that are analogues to spin projections along the z and x axes of a spin-1/2 particle, respectively: $A_z = |A_{up}\rangle\langle A_{up}| - |A_{down}\rangle\langle A_{down}|$, $A_x = |A_{up}\rangle\langle A_{down}| + |A_{down}\rangle\langle A_{up}|$ for Alice, and similarly for Bob. Note that eigenstates of A_z correspond to the observer's states "knowing the spin-z to be up" and "knowing the spin-z to be down", and those of A_x to the possible outcomes of the superobserver's interference experiment (Fig. 5.3).

Let us assume that in the Bell experiment, Alice chooses between two measurement settings $A_1 = A_z$ and $A_2 = A_x$ and Bob between $B_1 = \frac{1}{\sqrt{2}}(B_z + B_x)$ and $B_2 = \frac{1}{\sqrt{2}}(B_z - B_x)$. In a local (deterministic) hidden variable theory, one assumes that there jointly exist predetermined values for A_1, A_2, B_1 and B_2 which are +1 or -1. It is a well-known fact that state (14) with the chosen settings leads to a violation of the Bell-Clauser-Horne-Shimony-Holt inequality $|\langle A_1B_1\rangle + \langle A_1B_2\rangle + \langle A_2B_1\rangle - \langle A_2B_2\rangle| \leq 2$, where $\langle A_iB_j\rangle, i, j = 1, 2$, is the correlation function. The maximal quantum value for the Bell expression is $2\sqrt{2}$. Just like in every other Bell test, we conclude that the definite values for the observables cannot coexist if one keeps the assumption of locality.[12] We conclude that the two pieces of information, one of the observer and another of the superobserver, cannot be taken to coexist.

What consequences does outcome 3 of Deutsch's thought experiment have for our understanding of physical reality? Let us assume that the observers' and superobservers' laboratories contain a large number of degrees of freedom which allow the information about respective measurement records to be FAPP redundantly imprinted in their respective "environments". I will call these records "facts". This could be a click in a photodetector, a certain position of a pointer device, a printout of a computer or a written page in the lab-book, or a definite human brain state of a colleague who read the lab-book. If we assume that all these records in the observer's laboratory get correlated with the spin atoms and her brain state, *and* the superobserver can still perform the interference experiment, the result of which is also recorded in his laboratory, one has to accept that the two pieces of information can redundantly be imprinted in *two* environments: the sealed laboratory and the outside, respectively. As long as there is no communication on the relevant information (the actual measurement outcome) between the two laboratories, they will remain separate.

If we respect that there should be no preferred observers, then there is no reason to assume that the "facts" of one of them are more fundamental than those of the other.[13] But then, the observers' records cannot be comprised as "facts of the world", independent of the "environment" in which they have occurred. Any attempt to introduce "facts of the world per se" would run into problems of the hidden variable program.

The implications of the present Bell experiment are stronger than those of the standard Bell test. In the latter, we can exclude the view according to which the outcomes for measurements are (locally) predetermined, no matter if any measurement—and no matter which measurement—is actually performed. Still, between the partners there is no ambiguity with respect to whether measurements take place and about the coexistence of their records. The records can be accomplished as "facts of the world", which they share and even need to communicate in order to evaluate the experimental bound of the Bell expression. This is no longer the case in the present Bell experiment. What the Bell experiment excludes is the coexistence of the "facts" themselves. Everettians solve this by assuming that mutually complementary facts never coexist in between two branchings of the (multi)universe. Copenhagenists (can) take the position that *there are no facts of the world per se, but*

[12]Here "locality" means that, for example, value A_1 depends only on the local setting of Alice and not on the distant one of Bob. In a non-local hidden variable theory, we would need to distinguish between A_{11} and A_{12}, depending on whether Bob's setting is 1 or 2, respectively. It is not necessary to assume local deterministic values to derive Bell' inequalities. Bell's local causality is sufficient [11]. This however does not change the conclusions [12].

[13]One might be tempted to assume that the "facts" of the superobserver are the "real" ones, as he definitely has more reliable measurement instruments than the observer. This view cannot withstand the objection that the superobserver himself might be an object observed by yet another observer, the supersuperobserver, who describes the interference experiment of the superobserver quantum mechanically. The regression of increasingly more powerful observers might eventually find its end in a universe with a finite amount of resources.

only relative to observers. This is similar to Quantum Bayesianism, which treats the state of a quantum system as being observer-dependent, and to Rovelli's relational quantum mechanics [6], according to which "quantum mechanics is a theory about the physical description of physical systems relative to other systems ..." There are, however, important differences.

In Rovelli's relational interpretation, the "observer" does not "make any reference to a conscious, animate, or computing, or in any other manner special, system" [6]—each system provides its own frame of reference relative to which states of other systems can be assigned. Taking this position and outcome 3 of the Deutsch's experiment and applying them to, for example, two entangled electrons, one would conclude that, although the observer has no information about the state of a single electron, one electron "knows" in which state is the other. Relative to one electron, other one has a definite state. Obviously, we are here encountering the limits of meaningful language when we associate the terms "knowledge" or "taken" to single electrons. In this respect, quantum theory (in my eyes) remains a fundamental theory of observations in which a (hypothetical) observer, measurement and probabilities play a central role.

The two dominant approaches to the probability interpretation are the frequentist approach and the Bayesian approach. Qbism views the quantum state to be a user's manual—a mathematical tool that an observer uses to make decisions and take actions on the surrounding world upon observations. Central to this position is a Bayesian or personalist probabilistic approach to the probabilities that appear in quantum theory. To me, however, the problem of probability interpretation is prior to quantum theory, the solution of which alone will not be able to answer the question: What are the invariant features that characterize quantum theory in ways that are not relative to observers? By taking the subject matter of quantum theory to be restricted to an individual agent's decisions and experiences, Qbism runs into the danger of denying any objective elements in the notion of the quantum state. I agree with the Qbist's notion of subjective quantum states as representatives of an agent's beliefs, but only to the extent where a fundamental limit on maximal possible degree of belief of any agent is respected. This limit is represented FAPP by a pure quantum state. The fact that the predictions of agents cannot be "improved" over and above this limit in my eyes indicates that probabilities are not just personal and subjective, but also formed by the aforementioned invariant features of the theory. The role of reconstructions of quantum theory is to identify these invariants.[14]

The difference to the Everett interpretation is more evident. In the view adopted here, no meaning is given to "the universal wave function", nor is there an attempt to arrive at the probabilities from within such a concept alone. Here, the probabilities are always given by the Born rule, which is part of the formalism. This applies also to superobservers of any order: probabilities acquire meaning only when the measurement arrangement is specified, in which these probabilities are observed.

[14]In recent years, there have been several attempts to account for the origin of the basic principles from which the structure of quantum theory can be derived without invoking mathematical terms such as "rays in Hilbert space" or "self-adjoint operators." [23, 40–42].

Finally, I comment on the view [43] that the cut cannot be moved to include measurement instruments, observers etc. as objects under observation, since an object can never grow up to the point that it includes measurement contexts that, in turn, are unavoidably given in terms of classical concepts in accordance to Bohr's doctrine [44]: "However far the (quantum) phenomena transcend the scope of classical physical explanation, the account of all evidence must be expressed in classical terms." According to this view [43], the necessity of unambiguous usage of classical concepts fixes the object-subject cut whose position is therefore fundamental and equal for *all* observers. Consequently, one can retain the objectivity of the "facts of the world". I do not think that this view stands up to closer scrutiny. The description of any quantum mechanical experiment is expressed "in common language supplemented with the terminology of classical physics" [45]. Although this observation has played an important role in clarifying misconceptions in debates over the interpretation of quantum theory, it is in retrospective rather self-evident. For example, the description of a double-slit experiment with atoms, includes the depiction of the source of atoms directed towards the diaphragm normal to the beam, where the diaphragm contains two slits and a photographic plate with a characteristic interference pattern on the plate where the atoms are deposited. By extending the experiment to larger and larger systems, eventually as large as measurement instruments, nothing should change in the epistemic basis of the theory: we will still give an unambiguous account of the phenomenon in terms of classical language including a suitable "source", "beam" and "observation screen". This should not be confused with the impossibility of giving a classical explanation of the phenomenon, e.g. in terms of well-defined classical trajectories, which is present both for atoms and for macroscopic objects. To conclude, the cut can be shifted with no change in the epistemic foundation of the theory. Negating this would either mean negating Wigner-type experiments as legitimate quantum mechanical experiments or predicting outcome 1 in Deutsch's experiment. In my view, both choices indicate an acceptance that quantum theory is not universal, though the authors of [43] have a different opinion.

The above-mentioned Bell's theorem for "facts" implies a striking departure from naive realism. This brings us to the question of the role of our physical theories. If physical theories do not describe "physical reality per se", what do they describe then? A possible answer is given by Bohr as communicated by Petersen [46]: "It is wrong to think that the task of physics is to find out how nature is. Physics concerns what we can say about nature".

Acknowledgments This work has been supported by the Austrian Science Fund (FWF) through CoQuS, SFB FoQuS, and Individual Project 2462. I would like to acknowledge discussions with Mateus Araujo, Borivoje Dakić, Philippe Grangier, Richard Healey, Johannes Kofler, Luis Masanes, Jaques Pienaar and Anton Zeilinger.

References

1. J. Bub, I. Pitowsky, Two dogmas about quantum mechanics, in *Many Worlds? Everett, Quantum Theory, and Reality*, ed. by S. Saunders, J. Barrett, A. Kent, D. Wallace (Oxford University Press, 2010), pp. 431–456
2. I. Pitowsky, Quantum mechanics as a theory of probability, in *Festschrift in honor of Jeffrey Bub*, ed. by W. Demopoulos, I. Pitowsky (Springer, Western Ontario Series in Philosophy of Science, New York, 2007)
3. T. Maudlin, Three measurement problems. Topoi **14**(1), 7–15 (1995)
4. C. Brukner, A. Zeilinger, Information and fundamental elements of the structure of quantum theory, in *Time, Quantum, Information*, ed. by L. Castell, O. Ischebeck (Springer, 2003)
5. C.A. Fuch, R. Schack, Quantum-Bayesian coherence. Rev. Mod. Phys. **85**, 1693 (2007)
6. C. Rovelli, Relational quantum mechanics. Int. J. Theor. Phys. **35**, 1637–1678 (1996)
7. R. Colbeck, R. Renner, No extension of quantum theory can have improved predictive power. Nat. Commun. **2**, 411 (2011)
8. G.C. Ghirardi, A. Rimini, T. Weber, Unified dynamics for microscopic and macroscopic systems. Phys. Rev. D **34**, 470 (1986)
9. L. Diosi, Models for universal reduction of macroscopic quantum fluctuations. Phys. Rev. A **40**, 1165–1174 (1989)
10. R. Penrose, On gravity's role in quantum state reduction. Gen. Relat. Gravit. **28**, 581–600 (1996)
11. J.S. Bell, *Speakable and Unspeakable in Quantum Mechanics, Collected Papers on Quantum Philosophy* (Cambridge University Press, 2004)
12. M. Zukowski, Č. Brukner, Quantum non-locality—it ain't necessarily so.., Special issue on 50 years of Bell's theorem. J. Phys. A: Math. Theor. **47**, 424009 (2014)
13. C.A. Fuchs, R. Schack, *QBism and the Greeks: why a quantum state does not represent an element of physical reality*. arXiv:1412.4211 (2014)
14. S. Osnaghi, F. Freitas, O. Freire Jr., The origin of the Everettian heresy. Stud. Hist. Philos. Mod. Phys. **40**(2), 97–123 (2009)
15. K. Camilleri, M. Schlosshauer, Niels Bohr as philosopher of experiment: does decoherence theory challenge Bohr's doctrine of classical concepts? Stud. Hist. Philos. Mod. Phys. **49**, 73–83 (2015)
16. P. Heelan, Heisenberg and radical theoretical change. Z. Allgemeine Wissenschaftstheorie **6**, 113–138 (1975)
17. AHQP, Archives for the History of Quantum Physics—Bohr's Scientific Correspondence, 301 microfilm reels (American Philosophical Society, Philadelphia, 1986)
18. N. Harrigan, R.W. Spekkens, Einstein, incompleteness, and the epistemic view of quantum states. Found. Phys. **40**, 125 (2010)
19. S. Malin, What are quantum states? Quantum Inf. Process. **5**, 233–237 (2006)
20. A. Peres, When is a quantum measurement?. Ann. New York Acad. Sci. **480**, New Tech. Ideas Quantum Meas. Theory **438** (1986)
21. E. Wigner, Symmetries and Reflections (Indiana University Press, 1967), p. 164
22. W. Pauli, Wissenschaftlicher Briefwechsel mit Bohr, Einstein, Heisenberg, vol. 2, ed. by K. von Meyenn, A. Hermann, V. F. Weisskopf (Springer, Berlin, 1985), pp. 1930–1939. For the English translation of Heisenberg's manuscript with an introduction and bibliography see E. Crull, G. Bacciagaluppi. http://philsci-archive.pitt.edu/8590/
23. G. Chiribella, G. D'Ariano, P. Perinotti, Informational derivation of quantum theory. Phys. Rev. A **84**, 012311 (2011)
24. M.P. Mueller, L. Masanes, Three-dimensionality of space and the quantum bit: an information-theoretic approach. New J. Phys. **15**, 053040 (2013)
25. B. Dakic, Č. Brukner, The classical limit of a physical theory and the dimensionality of space, in *Quantum Theory: Informational Foundations and Foils*, ed. by G. Chiribella, R. Spekkens. (Springer, 2016) pp. 249–282. arXiv:1307.3984

26. J. Kofler, Č. Brukner, Classical world arising out of quantum physics under the restriction of coarse-grained measurements. Phys. Rev. Lett. **99**, 180403 (2007)
27. W.H. Zurek, Decoherence, einselection, and the quantum origins of the classical. Rev. Mod. Phys. **75**, 715 (2003)
28. J. Kofler, Č. Brukner, Conditions for quantum violation of macroscopic realism. Phys. Rev. Lett. **101**, 090403 (2008)
29. A. Peres, *Quantum Theory: Concepts and Methods* (Kluwer Academic Publishers, New York, 1995)
30. A. Peres, Stability of quantum motion in chaotic and regular systems. Phys. Rev. A **30**, 1610 (1984)
31. H.-J. Stöckmann, Q. Chaos, *An Introduction* (Cambridge University Press, Cambridge, 1999)
32. P. Jacquod, I. Adagideli, C.W.J. Beenakker, Decay of the Loschmidt echo for quantum states with sub-Planck-scale structures. Phys. Rev. Lett. **89**, 154103 (2012)
33. A. Peres, Recurrence phenomena in quantum dynamics. Phys. Rev. Lett. **49**, 1118 (1982)
34. A.E. Allahverdyan, R. Balian, T.M. Nieuwenhuizen, Understanding quantum measurement from the solution of dynamical models. Phys. Rep. **525**, 1 (2013)
35. D. Deutsch, Quantum theory as a universal physical theory. Int. J. Theor. Phys. **24**, I (1985)
36. E.P. Wigner, Remarks on the mind-body question, in *The Scientist Speculates*, ed. by I.J. Good (London, Heinemann, 1961)
37. X.Y. Zou, T.P. Grayson, L. Mandel, Observation of quantum interference effects in the frequency domain. Phys. Rev. Lett. **69**, 3041 (1992)
38. C. Bennett, Private Communication
39. N. Bohr, Discussion with Einstein on epistemological problems in atomic physics, in *Albert Einstein: Philosopher-Scientist*, ed. by P.A. Schilpp (The Library of Living Philosophers, Evanston, Illinois, 1949)
40. L. Hardy, Quantum theory from five reasonable axioms. arXiv:quant-ph/0101012 (2001)
41. B. Dakic and C. Brukner, Quantum theory and beyond: is entanglement special?, in *Deep Beauty: Understanding the Quantum World through Mathematical Innovation*, ed. by H. Halvorson (Cambridge University Press, 2011), pp. 365–392
42. L. Masanes, M. Müller, A derivation of quantum theory from physical requirements. New J. Phys. **13**, 063001 (2011)
43. A. Auffléves, P. Grangier, Contexts, systems and modalities: a new ontology for quantum mechanics. arXiv:1409.2120 (2014)
44. N. Bohr, *The Philosophical Writings of Niels Bohr 3* (Ox Bow Press, Woodbridge, Conn., 1987)
45. N. Bohr, On the notions of causality and complementarity. Dialectica **2**, 312–319 (1948)
46. As quoted in "The philosophy of Niels Bohr" by Aage Petersen, in the Bull. Atom. Sci. **19**(7) (1963); "The Genius of Science: A Portrait Gallery" (2000) by Abraham Pais, p. 24, and "Niels Bohr: Reflections on Subject and Object" (2001) by Paul. McEvoy, p. 291

Chapter 6
Causarum Investigatio and the Two Bell's Theorems of John Bell

Howard M. Wiseman and Eric G. Cavalcanti

Abstract "Bell's theorem" can refer to two different theorems that John Bell proved, the first in 1964 and the second in 1976. His 1964 theorem is the incompatibility of quantum phenomena with the joint assumptions of LOCALITY and PREDETERMINATION. His 1976 theorem is their incompatibility with the single property of LOCAL CAUSALITY. This is contrary to Bell's own later assertions, that his 1964 theorem began with the assumption of LOCAL CAUSALITY, even if not by that name. Although the two Bell's theorems are logically equivalent, their assumptions are not. Hence, the earlier and later theorems suggest quite different conclusions, embraced by operationalists and realists, respectively. The key issue is whether LOCALITY or LOCAL CAUSALITY is the appropriate notion emanating from RELATIVISTIC CAUSALITY, and this rests on one's basic notion of causation. For operationalists the appropriate notion is what is here called the Principle of AGENT-CAUSATION, while for realists it is REICHENBACH's Principle of common cause. By breaking down the latter into even more basic Postulates, it is possible to obtain a version of Bell's theorem in which each camp could reject one assumption, happy that the remaining assumptions reflect its *weltanschauung*. Formulating Bell's theorem in terms of causation is fruitful not just for attempting to reconcile the two camps, but also for better describing the ontology of different quantum interpretations and for more deeply understanding the implications of Bell's marvellous work.

H.M. Wiseman (✉)
Centre for Quantum Dynamics, Griffith University, Brisbane, QLD 4111, Australia
e-mail: H.Wiseman@griffith.edu.au

E.G. Cavalcanti
Centre for Quantum Dynamics, Griffith University, Gold Coast, QLD 4222, Australia
e-mail: E.Cavalcanti@griffith.edu.au

E.G. Cavalcanti
School of Physics, University of Sydney, Sydney, NSW 2006, Australia

R. Bertlmann and A. Zeilinger (eds.), *Quantum [Un]Speakables II*,
The Frontiers Collection, DOI 10.1007/978-3-319-38987-5_6

Motivation

The work presented here grew from my[1] observation, over the years, but particularly at a quantum foundations conference in 2013 (see Ref. [1]), that different 'camps' of physicists and philosophers have a different understanding of what Bell's theorem actually is, and a different understanding of the words (in particular 'locality') often used in stating it. As a consequence they often talk (or shout) past one another, and come no closer to understanding each other's perspective. I have friends in both camps, and I would like to think that they are all reasonable people who should be able to come to terms that allow the pros and cons of different interpretations of Bell's theorem to be discussed calmly.

More recently, my thinking has evolved beyond this original motivation, as a consequence of the 50th anniversary of Bell's *annus mirabilis*,[2] which led to three invitations to present on the topic: for a special issue of J. Phys. A [4], for an opinion piece in Nature [5], and for the Quantum [Un]speakables conference (and thus to a fourth invitation, for these Proceedings). These challenged me to think more deeply about what lay behind the different positions of the two camps, and how this could lead to a deeper understanding of Bell's theorem. The key to my reconsideration is to be found, conveniently, on the the magnificent ceiling of the *Festsaal* of the Austrian *Akademie der Wißenschaften*, host to Alain Aspect's public lecture at the Quantum [Un]speakables conference. There, one finds the role of natural philosophy defined as *causarum investigatio* (Fig. 6.1). The investigation of causes, or, more particularly, of notions of causation, has proven to be a very fruitful way to analyse Bell's theorem, and the disagreements over it [4–8].

While this chapter, Ref. [4], and Ref. [5], each has independent material, they all share some core material, which can be summarized as follows. Bell actually proved two theorems (Sect. "The Two Bell's Theorem of John Bell"). The 1964 one [3] is that

> [some of] the statistical predictions of quantum mechanics are incompatible with separable [i.e. local] predetermination.

This involves two assumptions: "separability or locality" and the "predetermination [of] the result of an individual measurement" [3]. The 1976 one [9] is that

> quantum mechanics is not embeddable in a locally causal theory

and involves a single assumption, "local causality" [9]. Although each theorem is a corollary of the other (Sect. "A Fine Distinction"), they are embraced by differ-

---

[1] This first section is written in first person by one of us (Wiseman), who spoke at the Quantum [Un]speakables conference. The other of us (Cavalcanti) has been a long-time co-worker and correspondent with Wiseman on Bell's theorem. In particular, since the conference, their discussions have convinced Wiseman of a better way to formulate the causal assumptions in Bell's theorem, and this is reflected in the latter parts of the paper, and in its authorship.

[2] It was the year he wrote his review of hidden variables (HVs) [2], by misfortune not published until 1966, in which he dismissed von Neumann's anti-HV proof, gave the first proof of the necessity of contextuality for deterministic HV theories, and raised the question of the necessity of nonlocality, which he immediately answered in the positive in his 1964 paper [3].

Fig. 6.1 Scene from the ceiling of the *Festsaal* of the Austrian *Akademie der Wissenschaften*.

ent intellectual camps, whom I will call operationalists and realists respectively (Sect. "The Two Camps"). The latter, however, deny that there are really two different theorems, claiming that in his 1964 paper Bell used 'locality' to mean LOCAL CAUSALITY, and that from it he derived PREDETERMINATION[3] rather than assuming it. I have argued in depth [4, 11] that this is a misrepresentation of what Bell proved in 1964.

Whether LOCALITY or LOCAL CAUSALITY is the appropriate notion emanating from Einstein's Principle of relativity rests on one's underlying concept of causation. Operationalists and realists implicitly hold to quite different notions of causation and it is fruitful to make this explicit (Sect. "Back to Basics"). These last points lead on to the notable new material in this chapter: a form of Bell's theorem that could be acceptable to both camps (Sect. "Reactions and Reconciliation"), and discussion of the many other advantages of formulating Bell's theorem in terms of causation (Sect. "Conclusion").

[3] In earlier, albeit recent, publications [4, 5] I used the term 'determinism' for Bell's second assumption in 1964. However, Bell did not actually use this word in 1964, and it is useful to reserve it for a different notion [10]. Note the use here of small-capitals for terms with a precise technical meaning, even when the relevant definition is given only at a later point in the chapter.

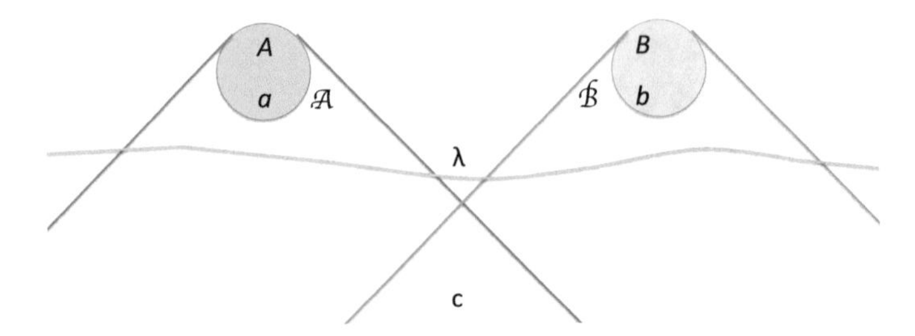

Fig. 6.2 Minkowski diagram for the scenario Bell introduced in 1964. Time is the *vertical axis*, space *horizontal*, and the *diagonal lines* indicate the boundaries of light-cones.

The Situation Bell Considered

The experimental situation Bell considered is shown as a Minkowski (space-time) diagram in Fig. 6.2. This is closely based on such diagrams which Bell used in Refs. [9, 12], for example. He did not use such a diagram in 1964, and only briefly referred to relativistic concepts, but the diagrams he used in his definitive paper on the subject [12] are applicable to, and even use the same notation as, his original 1964 paper. It is convenient to use something close to the form of diagram which Bell finally settled upon [12], as it allows the assumptions from both those theorems (LOCALITY and PREDETERMINATION in 1964; LOCAL CAUSALITY in 1976) to be stated naturally. Considering more general distributions of events in space-time lead to different conclusions about which assumptions can be stated naturally; see Sect. "Conclusion" and Ref. [10].

The experiment involves two observers, Alice and Bob, with spatially separated laboratories, where they perform experiments that are independent, but may have correlated outcomes. This allows them to perform experiments in space-like-separated regions of space-time, labelled $\mathscr{A}$ and $\mathscr{B}$ respectively. In her region, Alice makes a free choice a of setting for her measurement, which yields an outcome A, and Bob likewise, *mutatis mutandis*. In the overlap of the past light-cones of $\mathscr{A}$ and $\mathscr{B}$ is a preparation event c which is necessary to produce the correlations between the outcomes. Even for a fixed preparation c, the ability of Alice and Bob to make free choices, and the existence of multiple possible outcomes, gives rise to a PHENOMENON described by the relative frequencies[4]

$$\{f(A,B|a,b,c)\ :\ A,B,a,b\}. \tag{1}$$

The existence of correlations in the outcomes in a physics experiment is typically 'explained' by stepping away from operationalist language, in a manner such as this: the event c prepares a correlated pair of particles, one of which goes to $\mathscr{A}$ and the other to $\mathscr{B}$, each at subluminal speed.

[4]Note that here we are not making a commitment to frequentism, but rather simply recognising that real experimental data are relative frequencies.

Stepping even further away from operationalism, we follow Bell in allowing for the possibility of non-observable (or "hidden") variables in the formalism, introducing variables λ, defined at a time—i.e. on a space-like hypersurface (SLH)—before a and b but after c, that could describe these particles, or anything else of relevance.[5] Note in particular that λ could represent a pure *quantum state* $|\lambda\rangle$, since this is a mathematical object defined on a SLH. If the source c reliably produces this particular pure state then this $|\lambda\rangle$ would contain no more information than is already given by specifying c. But this is an idealization, and in general $|\lambda\rangle$ would have some distribution given c, and so is just like any other hidden variable. In general, we say that a MODEL (which Bell [3] called a 'theory') θ for the above PHENOMENON, comprises a probability distribution for hidden variables, $P_\theta(\lambda|c)$, and further probabilities $P_\theta(A, B|a, b, c, \lambda)$, such that[6]

$$\sum_{\lambda} P_\theta(A, B|a, b, c, \lambda)P_\theta(\lambda|c) = f(A, B|a, b, c). \tag{2}$$

Note that by considering a MODEL with non-trivial dependence on λ, one is *not* presuming PREDETERMINATION of outcomes:

Definition 1 (PREDETERMINATION) $\forall\, A, a, B, b, \lambda,\ P_\theta(A, B|a, b, c, \lambda) \in \{0, 1\}$.

(See the next section for the source of this terminology.) An example of a hidden variable MODEL that is stochastic (i.e. that violates PREDETERMINATION) is the one mentioned above, where c prepares a mixed quantum state ρ_c and λ is taken to define a pure quantum state $|\lambda\rangle$ such that $\rho_c = \sum_\lambda P_\theta(\lambda|c)\,|\lambda\rangle\,\langle\lambda|$.

Note also that PREDETERMINATION should not be confused with the stronger and purely operational notion of PREDICTABILITY [13]:

Definition 2 (PREDICTABILITY) $\forall\, A, a, B, b,\ f(A, B|a, b, c) \in \{0, 1\}$.

While we have defined PREDICTABILITY in terms of the experimental frequencies $f(A, B|a, b, c)$, for a given MODEL it is of course possible to determine whether the PHENOMENON it is supposed to describe has this property, via Eq. (2). The point is that a MODEL may satisfy PREDETERMINATION even though the PHENOMENON does not satisfy PREDICTABILITY.

The Two Bell's Theorems of John Bell

In this section we present the two theorems, in chronological order, and then discuss the relationship between them.

[5]More generally [12], one could sandwich these variables between two suitable SLHs, but the above formulation will suffice.

[6]Given footnote 4, this equation is not to be read as a strict equality, but as carrying the same meaning as that of any probabilistic prediction.

Bell's 1964 Theorem

In Bell's 1964 paper he states what he has proven most clearly in his Conclusion (Sect. VI):

> *In a theory in which parameters* are added to quantum mechanics to *determine the results of individual measurements*, without changing the statistical predictions, there must be a mechanism whereby *the setting of one measuring device can influence the reading of another instrument, however remote.*

Here the italics, added by us, emphasize the two assumptions that lead to a contradiction with the statistical predictions of quantum mechanics; the second assumption is stated in the negative, since its negation follows if one holds to the first assumption. These two assumptions are stated positively, with equal status, in the immediate preceding sentence, at the end of his Sect. V:

> for at least one ... state ... the statistical predictions of quantum mechanics are incompatible with separable predetermination

as quoted in Sect. "Motivation" above. As was noted there, Bell did not distinguish separability from locality, and he is explicit that PREDETERMINATION means "predetermination [of] the result of an individual measurement". Thus his theorem is:

Theorem 1 (Bell-1964) *There exist quantum* PHENOMENA *for which there is no* MODEL *satisfying* PREDETERMINATION *and* LOCALITY.

Although this was not the case in 1964, the quantum phenomena relevant to Bell's theorem have long since been verified experimentally, albeit with a few challenging loopholes [5, 14].

It is presumably uncontroversial to understand PREDETERMINATION as per Definition 1 already given above. The meaning of LOCALITY is more controversial, to say the least (compare Refs. [15, 16] with Refs. [4, 11]). However, by our reading, Bell is quite clear:

> the requirement of locality [is] more precisely that the result of a measurement on one system be unaffected by operations on a distant system

This is of course the positive form of the final notion in the first quote in this section, and Bell states the same assumption (the irrelevance of Alice's measurement choice to Bob's outcome) twice more in the paper. Although he successfully applies the notion of locality only to theories with predetermined outcomes, he introduces it prior to making the assumption of predetermination.[7] Thus it seems that we should adopt a definition that accords with his words and applies to probabilistic theories:

---

[7]He introduces it in the first paragraph of the paper proper, which serves as motivation for the formulation of the assumptions he will use. There, Bell unfortunately misapplies his notion, in attempting to derive predetermination via EPR-Bohm-correlations and locality. See Refs. [4, 11] for discussions of the irrelevance of this flawed paragraph to Bell's 1964 theorem (i.e. the "result to be proved", as he calls it, which he does indeed prove).

Definition 3 (LOCALITY)

$$\forall\, a, B, b, \lambda,\ P_\theta(B|a, b, c, \lambda) = P_\theta(B|b, c, \lambda),$$

and likewise for Alice's result, which will remain unstated in similar definitions below. (The existence of the function $P_\theta(B|b, c, \lambda)$ is also implicit here, and in similar definitions below.) This definition of LOCALITY was also the one adopted by Jarrett in 1984 [17]. In the same year Shimony [18] coined the phrase "parameter independence" for the same concept, to emphasize that it required only that Bob's outcome be statistically independent of Alice's setting, a controllable 'parameter'. We prefer to follow the terminology of Jarrett and (in our reading) Bell.

Note that LOCALITY is not the same [13] as the purely operational notion of

Definition 4 (SIGNAL-LOCALITY)

$$\forall\, a, B, b,\ f(B|a, b, c) = f(B|b, c),$$

the violation of which has never been observed, and, most physicists think, never will be observed. However, for a strictly operational MODEL (that is, one that makes no use of hidden variables, be they pure quantum states or other objects) there is no distinction between LOCALITY and SIGNAL-LOCALITY. Thus, operational quantum mechanics, involving only preparations, settings, and outcomes, satisfies LOCALITY. In fact, LOCALITY is satisfied even in the non-operational quantum theory discussed above, where c prepares ρ_c but one assumes that, conditional on the hidden variable λ, the probabilities of outcomes are determined by $|\lambda\rangle$. In both cases, the reduced quantum state on Bob's side, which is mixed in general, defines probabilities for Bob's outcomes, for any measurement he makes, which are independent of Alice's choice of measurement. Moreover, introducing a realist narrative, involving instantaneous wave-function collapse, makes no difference to the fact that LOCALITY is satisfied, as a mathematical statement about the probability distributions in the theory. Even spontaneous collapse theories such as the celebrated GRW model [19] respect LOCALITY (to the extent that they respect SIGNAL-LOCALITY).

Bell's 1976 Theorem

More than a decade after his 1964 paper, Bell reformulated his theorem in a way that he would cleave to, in essence, for the rest of his life. Building on his own work from 1971 [20] and that of Clauser and Horne from 1974 [21], he introduced a new notion:

Definition 5 (LOCAL CAUSALITY)

$$\forall\, A, a, B, b, \lambda,\ P_\theta(B|A, a, b, c, \lambda) = P_\theta(B|b, c, \lambda).$$

Strictly speaking (and Bell was strict about this [9]), the above is a consequence of the broader Principle of LOCAL CAUSALITY (Principle 2 in Sect. "Realist Principles" below) when applied to the specific set up of Fig. 6.2. Note the crucial difference from LOCALITY in that LOCAL CAUSALITY requires Bob's outcome to be statistically independent of Alice's outcome as well as her setting. This, Bell argued, is a reasonable 'localistic' notion for a theory in which λ provides a complete description of the relevant state of affairs prior to the measurements being performed.

As quoted in Sect. "Motivation", Bell proved in 1976 that

> quantum mechanics is *not* embeddable in a locally causal theory as formulated [above].

In other words, he proved a theorem involving only a single assumption:

Theorem 2 (Bell-1976) *There exist quantum* PHENOMENA *for which there is no* MODEL *satisfying* LOCAL CAUSALITY.

Unfortunately for our purposes, having invented a new concept with a new name, Bell immediately became indiscriminate once more, using "local causality" and 'locality' interchangeably.[8] However, in his most mature treatment of the subject [12], Bell unequivocally showed his preference for the term "local causality" [4], and in following suit we respect Bell's final will.

A Fine Distinction

Bell's 1976 theorem implies Bell's 1964 theorem. This is because, as is easy to see, any MODEL θ satisfying LOCALITY and PREDETERMINATION satisfies LOCAL CAUSALITY. The converse of that last clause is not true; there are theories satisfying LOCAL CAUSALITY that do not satisfy LOCALITY and PREDETERMINATION. Orthodox quantum mechanics with separable states is one example. Nevertheless, the converse of the first sentence of this subsection *is* true. That is, Bell's 1964 theorem implies Bell's 1976 theorem. This is because of the following:

Theorem 3 (Fine-1982) *For any* PHENOMENON, *there exists a* MODEL θ *satisfying* LOCAL CAUSALITY *if and only if there exists a* MODEL θ' *satisfying* PREDETERMINATION *and* LOCALITY.

Although we have given the credit here to Fine [23], this result was known, in essence, by Bell even in 1971 [20]; see Ref. [4] for details. For this reason, it is useful shorthand to introduce the well-known term 'Bell-local' for describing the type of MODEL that satisfies the broader assumptions of Bell's 1976 theorem. That is,

[8] Indeed, by 1981 [22] he was implying that by 'locality' he had always meant LOCAL CAUSALITY. This historical revisionism is perfectly understandable, and probably unconscious—a plausible unfolding of the localistic intuition Bell had in 1964 is LOCAL CAUSALITY, since this would have worked, where LOCALITY failed, in Bell's attempt to reproduce the EPR argument (see footnote 7).

Definition 6 (BELL-LOCAL) A MODEL θ is BELL-LOCAL if and only if

$$\forall\, A, a, B, b,\ P_\theta(A, B|a, b, c) = \sum_\lambda P_\theta(A|a, c, \lambda) P_\theta(B|b, c, \lambda) P_\theta(\lambda|c). \tag{3}$$

If the two Bell's theorems are logically equivalent, why should we bother to distinguish them? The answer is that the two different forms appeal to two different camps of scholars, and indeed these two camps often recognise only the one form that they favour. The broad term 'scholars' here is deliberately chosen to cover the increasing range of disciplines—including (at least) physics, philosophy, and information science—interested in Bell's theorem. But it is important to note that the division into two different camps does not sharply follow these disciplinary divisions. Of course there are many more than two attitudes towards Bell's theorem. Nevertheless, the most common attitudes can be broadly grouped within the two camps, called here operationalist and realist.

The Two Camps

The realist camp [15, 27–29] has the following credo:

> Bell's theorem uses only one assumption: local causality (or 'locality' as we usually call it for short). This is the only reasonable way to apply the principle of relativity for statistical theories. It is essentially what EPR assumed in 1935. They showed that operational quantum mechanics is nonlocal, and Bell showed in 1964 that adding hidden variables cannot solve the problem. Experiments violating a Bell inequality thus leave us with no option: the principle of relativity is false. The world is nonlocal.

The operationalist camp [6, 24–26], on the other hand, could be caricatured by the following:

> Bell's theorem uses two assumptions. The first assumption is locality. This is essentially the same as signal locality, which is all the principle of relativity implies, but also applies to hidden variable theories. Orthodox quantum mechanics respects locality. The second assumption is something else which has been variously called realism, predetermination, determinism, objectivity, classicality, counter-factual-definiteness, and causality (perhaps with slightly different formulations). Clearly it is the second assumption that we should abandon, whatever we call it. Locality is here to stay.

Why do the two camps come to such contrary conclusions? Partly it is just a difference in terminology: realists use 'locality' to mean LOCAL CAUSALITY, while operationalists use it to mean LOCALITY. But the deeper question is why they disagree about which is the 'correct' way to apply the principle of relativity. Bell well explains the motivations of the realist camp [12]:

> The obvious definition of "local causality" does not work in quantum mechanics, and this cannot be attributed to the "incompleteness" of that theory. ...
>
> Do we then have to fall back on "no signalling faster than light" as the expression of the fundamental causal structure of contemporary theoretical physics? That is hard for me to

> accept. For one thing we have *lost the idea that correlations can be explained* .... More importantly, the *"no signalling" notion rests on concepts which are desperately vague* .... The assertion that "we cannot signal faster than light" immediately provokes the question: "Who do we think *we* are?"

Here the italics, added by us (except for the '*we*' in the final question), emphasize the two key tenets of the realist camp: that correlations should be explained, and that anthropocentric notions such as 'signalling' should play no fundamental role.

An operationalist, however, would claim to know well enough who '*we*' are, and point out that statements about what we may, or may not, be able to do are very useful, for example in informational security proofs. From an operationalist perspective, moreover, explanations, in the sense Bell means, might be regarded as superfluous. These differences between realists and operationalists hark back to Einstein's 1919 distinction between constructive theories and principle theories [30]. But a more precise understanding of the origin of the disagreement is possible by breaking down the assumptions used in Bell's theorem to a more basic level (Sect. "Back to Basics"). As we will see, this is also the way towards enabling the two camps to discuss Bell's theorem using the same terms, agreeing on what combinations of assumptions it implies to be impossible,[9] even while disagreeing on which assumption is most implausible (Sect. "Reactions and Reconciliation").

Back to Basics

Here we make a fresh start, aiming to base Bell's theorem on notions more comprehensive and more fundamental than those defined earlier in this chapter with reference to the particular scenario of Sect. "The situation Bell considered". Those earlier notions are temporarily abandoned, but they will gradually be reintroduced and their relation to the deeper concepts indicated.

To begin, we introduce some axioms. In calling them 'axioms' we are not implying that they are unquestionable, only that we will not question their necessity in the formulations of Bell's theorem below. Without further ado:

Axiom 1 (MACROREALITY) An event observed by any observer is a real single event, and not 'relative' to anything or anyone.

This rules out consideration of the "relative state" [31] or "many worlds" [32] interpretation, as well as the extreme subjectivism of the 'QBist' interpretation [33].

[9]This is always assuming that a loophole-free test, expected soon [5, 14] will not turn up any surprises.

Axiom 2 (MINKOWSKI SPACE-TIME) Concepts such as space-like separation, light-cones, and foliations of SLHs are well defined in ordinary laboratory situations.

This rules out short-cuts through space ('wormholes') between distant locations [34]. There is actually no need to restrict to flat space-time; any time-orientable Lorentzian space-time manifold will do, but the above terminology is more straight-forward.

Axiom 3 (TEMPORAL ORDER) For any event *A*, there is a SLH containing *A* that separates events in the PAST of *A* from events that have *A* in their PAST.

Note the term PAST is not, in this axiom, to be understood as having definite meaning; the font used is meant to alert the reader to this fact. In particular, there is no implication that all events on one side of the SLH are in the PAST of *A* and all events on the other side in the future of *A* (i.e. having *A* in their PAST). For example, TEMPORAL ORDER is satisfied if we take PAST to mean the past light-cone, which defines only a partial ordering of events (that is, for some pairs of events, neither is in the PAST of the other). To define an almost-total ordering of events (that is, such that for almost every pair of events, one of them is in the PAST of the other) one would need to define PAST via a fixed foliation of SLHs.

Axiom 4 (CAUSAL ARROW) Any CAUSE of an event is in its PAST.

This axiom, together with Axioms 2 and 3, implies a causal structure describable by a directed acyclic graph, as is standard in modelling of causation [7]. It rules out retrocausal approaches to Bell's theorem such as in Refs. [35, 36]. Note that, like PAST above, CAUSE here does not yet have any precise meaning. The meaning of these concepts will become more defined as more assumptions are added.

In moving now from axioms to postulates, we list assumptions that are more likely to be questioned, or at least that were questioned, in some form, relatively early in the literature on Bell's theorem. Indeed, if one accepts the Axioms then one must (modulo the remaining experimental loopholes) reject one of the postulates below, as we will show in Sect. "Reactions and Reconciliation". The first postulate (which was listed as an Axiom in Ref [4]) begins the process of adding meaning to 'CAUSE':

Postulate 1 (FREE CHOICE) *A freely chosen action has no relevant* CAUSES.

Here, 'relevant' means "in common with, or among, the other events under study." This postulate is not meant to indicate a philosophical commitment to Cartesian dualism, or a religious commitment to Pelagianism, although it is possibly incompatible with Augustinian predestination.[10] Here, it serves to rule out (when combined with other principles) what Bell called 'superdeterminism' [12]. The 'super' in 'superdeterminism' indicates that free choices would not just have causes as a matter of principle (as believed by those who hold to the 'determinist' philosophy of free will). Rather, they would of necessity have causes that are correlated in a very particular way with external physical variables that affect the outcomes of measurements. Although this viewpoint has at least one prominent scientific proponent [37], our personal view is that Postulate 1 is as unquestionable as any of the Axioms above; see footnote 14 below.

Postulate 2 (RELATIVISTIC CAUSALITY) *The* PAST *is the past light-cone.*

Note that a term like "past light-cone" *is* to be understood as having definite meaning, from the Axiom of MINKOWSKI SPACE-TIME.

Postulate 3 (COMMON CAUSES) *If two sets of events $\mathscr{A}$ and $\mathscr{B}$ are correlated, and no event in either is a* CAUSE *of any event in the other, then they have a set of common* CAUSES *$\mathscr{C}$ that* EXPLAINS *the correlation.*

Here, *common* CAUSES means events that are CAUSES of at least one event in $\mathscr{A}$ and at least one event in $\mathscr{B}$. It is important to note that this postulate is not the same as REICHENBACH's Principle of common cause [38]. Rather, following Ref. [8] (although with some differences in details) we have deliberately split Reichenbach's celebrated principle into the above Postulate of COMMON CAUSES, and the below

[10] *Les passions de l'âme* (1649), *De natura* (415), and *De natura et gratia* (415).

Postulate 4 (DECORRELATING EXPLANATION) *A set of* CAUSES $\mathscr{C}$, *common to two set of events* $\mathscr{A}$ *and* $\mathscr{B}$, EXPLAINS *a correlation between* $\mathscr{A}$ *and* $\mathscr{B}$ *only if conditioning on* $\mathscr{C}$ *eliminates the correlation.*

As Reichenbach said [38], "When we say that the common cause explains the [correlation], we refer ... to the fact that relative to the cause the events A and B are mutually independent."

In the above principles we referred always to events, but for statistical concepts such as correlation it is more common for physicists to think in terms of variables. In such cases we will be loose with notation and terminology, and allow, for instance, A to stand for the outcome that Alice gets (a variable) as well as the event that Alice's outcome takes the value A.

Realist Principles

The axioms and postulates above form (as we will see in Sect. "Reactions and Reconciliation") a nice set in that they are sufficient to enforce BELL-LOCALITY, and, with one exception, all necessary. (The one exception is TEMPORAL ORDER, which is not needed if one assumes RELATIVISTIC CAUSALITY.) However they do not correspond to the principles stated in Bell's two theorems, and hence do not obviously connect to the two camps. Thus we will develop principles that do relate to the two camps (starting with realists) and show their relation to the postulates above.

First, since the explanation of correlations is a realist tenet, realists hold to REICHENBACH's Principle, which we state here explicitly,

Principle 1 (REICHENBACH) If two sets of events $\mathscr{A}$ and $\mathscr{B}$ are correlated, and no event in either is a CAUSE of any event in the other, then they have a set of common CAUSES $\mathscr{C}$, such that conditioning on $\mathscr{C}$ eliminates the correlation.

For realists, this is the point of causation—to explain correlations. However we made that a separate assumption, as discussed above, and as captured by this:

Lemma 1 *The postulates of* COMMON CAUSES *and* DECORRELATING EXPLANATION *imply* REICHENBACH*'s Principle.*

For realists, the role of REICHENBACH's principle is in this:

Lemma 2 REICHENBACH*'s Principle and the Postulate of* RELATIVISTIC CAUSALITY *imply the Principle of* LOCAL CAUSALITY,

if we define the last generally, basically as Bell did in 1976, as

Principle 2 (LOCAL CAUSALITY) If two space-like separated sets of events $\mathscr{A}$ and $\mathscr{B}$ are correlated, then there is a set of events $\mathscr{C}$ in their common Minkowski past such that conditioning on $\mathscr{C}$ eliminates the correlation.

Here the *common* Minkowski past means the intersection of the union of past light-cones of events in $\mathscr{A}$ with the union of past light-cones of events in $\mathscr{B}$.

The reader may well ask how this Principle of LOCAL CAUSALITY relates to the Definition 5. Take the two sets of events in Principle 2 to be $\mathscr{A} = (A, a)$ and $\mathscr{B} = (B, b)$. Thus if LOCAL CAUSALITY is satisfied there must exist a set of CAUSES $\mathscr{C}$ in their common Minkowski past such that $P(A, a, B, b|\mathscr{C}) = P(A, a|\mathscr{C})P(B, b|\mathscr{C})$. Thus, $P(B, b|A, a, \mathscr{C}) = P(B, b|\mathscr{C})$, and $P(B|A, a, b, \mathscr{C}) = P(B|b, \mathscr{C})$. Replacing $\mathscr{C}$ by Bell's variables c and λ,[11] a sufficient specification of the causes by assumption, gives Definition 5.

This is not, however, sufficient to derive a contradiction with quantum phenomena. It was sufficient in Sect. "The Two Bell's Theorem of John Bell" because there we were working within the framework of our definition of a MODEL in Eq. (2). Here we want a more principled derivation of the condition of BELL-LOCALITY, in Eq. (3). For this it is necessary to use an additional assumption related to the Postulate of FREE CHOICE. Specifically, we combine this postulate with other postulates:

Lemma 3 *The Postulates of* RELATIVISTIC CAUSALITY *and* COMMON CAUSES *and* FREE CHOICE *imply the Principle of* LOCAL AGENCY,

where we have defined (for terminological reasons to be explained elsewhere [10])

Principle 3 (LOCAL AGENCY) The only relevant events correlated with a free choice are in its future light cone.

This assumption was not stated precisely by Bell in 1976, who said only "It has been assumed that the settings of instruments are in some sense free variables ....".[12]

---

[11]The alert reader will have noticed some sleight of hand. In Bell's 1976 paper, where he introduced LOCAL CAUSALITY, λ denoted all events in the intersection of the past light-cones. But in Ref. [12] Bell took the λs to be defined between two SLHs, and the limit when these become one corresponds to the situation he considered in 1964, where the λs were "initial values of the [relevant] variables at some suitable instant." As in Fig. 6.2, that suitable *instant* (SLH) may not even cross the intersection of the past light-cones. The resolution is that since, by assumption, the variables $\{\lambda\}$ are the only relevant ones, they must carry the information that was present in the common causes $\mathscr{C}$ in the common Minkowski past of A and B. This they can without falling foul of LOCAL CAUSALITY (in the sense of Principle 2) because there is a part of the SLH that is in the future light-cone of the events in $\mathscr{C}$, but in the past light-cone of A, and likewise for B.

[12]Bell was immediately criticised for the vagueness of this statement (and for what followed, some of which was not sufficient for his purpose) by Shimony, Horne and Clauser [39]. The immediacy was, according to Clauser [40], because the latter two authors had originally drafted their 1974 paper [21] using the above Principle of LOCAL CAUSALITY (or something like it), but Shimony pointed out to them that this was not sufficient to derive BELL-LOCALITY without an extra assumption relating to free choice. As a consequence they retreated from such a principled formulation of LOCAL CAUSALITY to the more specific Definition 5, which they said characterized "objective local theories" [21], enabling a "less general and more plausible" [39] assumption (than LOCAL AGENCY,

However, according to Bell in 1977 [41] what he meant by this, in the context of theories satisfying LOCAL CAUSALITY, was that "the values of such variables have implications only in their future light cones," in other words, the above Principle of LOCAL AGENCY.

It is worth noting, however, that Bell does not actually need a Principle as strong as this, but rather only (as he says two sentences later) that "In particular they have no implications for the ... variables in ... [their] backward light cones." This weaker assumption follows (using Axioms 3 and 4 without RELATIVISTIC CAUSALITY) from the following.

Principle 4 (AGENT-CAUSATION) If a set of relevant events $\mathscr{A}$ is correlated with a freely chosen action, then that action is a CAUSE of at least one event in $\mathscr{A}$.

This can be seen as follows. Assume that a is correlated with some event A. From this Principle of AGENT-CAUSATION, a must be a cause of A. From Axiom 4, a must be in the PAST of A. From Axiom 3, this means there must be a SLH separating a and A. Thus, as claimed, A cannot be in the past light-cone of a. If one is making the assumption of LOCAL CAUSALITY, then the Axioms need not be used, and one can directly conclude from Principle 4 that A must be in the future light-cone of a. That is, the Principles of AGENT-CAUSATION and LOCAL AGENCY become equivalent.

Now, to obtain Eq. (3) we must consider $P(A, B|a, b, c)$, which can be written as $\sum_\lambda P(A, B|a, b, \lambda, c)P(\lambda|a, b, c)$. With the location of events in Fig. 6.2, if the events in $\mathscr{C}$ are in the common Minkowski past of $\mathscr{A}$ and $\mathscr{B}$, they are not in the future light-cone of either a or b. Thus from the principle of LOCAL AGENCY, $P(\lambda|a, b, c) = P(\lambda|c)$. Using LOCAL CAUSALITY as above then gives Eq. (3). Thus we have a more principled version of Bell's 1976 theorem:

Theorem 4 (Bell-1976, in principles) *Quantum* PHENOMENA *violate the conjunction of Axioms 1–4, the Principle of* LOCAL CAUSALITY, *and the Principle of* LOCAL AGENCY *(or* AGENT-CAUSATION*).*

From the lemmata in this section (noting that REICHENBACH's Principle implies the Principle of COMMON CAUSES, and applying this in Lemma 3), we can also formulate this theorem using the more fundamental postulates as follows:

Theorem 5 (Bell-1976, in deeper principles) *Quantum* PHENOMENA *violate the conjunction of Axioms 1–4, the Postulate of* FREE CHOICE, *the Postulate of* RELATIVISTIC CAUSALITY, *and* REICHENBACH*'s Principle.*

These two theorems, and the relationship between them are illustrated in Fig. 6.3.

(Footnote 12 continued)
for example) relating to free choice. Clauser and Horne [21] deserve credit for first (as far as we are aware) discussing, in their footnote 13, the need for independence of the hidden variables λ from the free choices a and b which is implicit in Eq. (2). They note that to justify that assumption one has to rule out the "possibility" that "Systems originate within the intersection of the backward light cones of both analyzers and the source ... [which] effect [*sic.*] both the experimenters' selections of analyzer orientations and the emissions from the source."

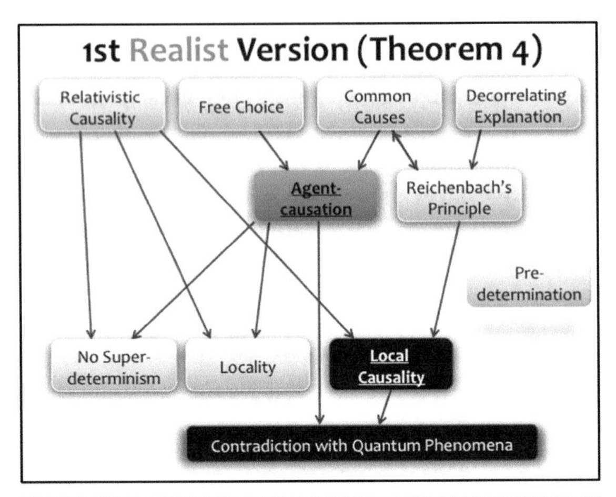

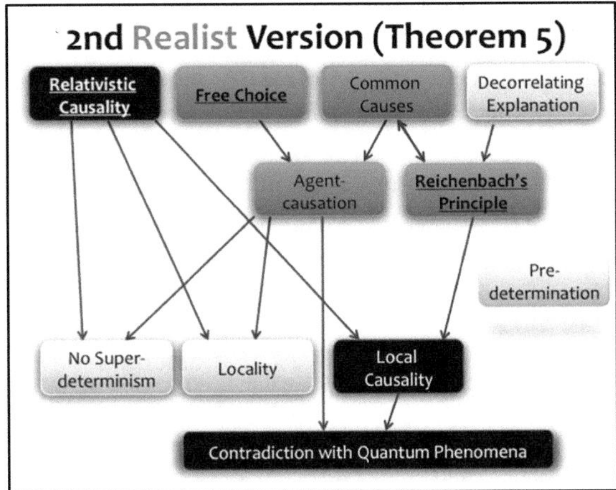

Fig. 6.3 Two realist versions of Bell's Theorems 4 and 5. The *coloured* (*shaded*) boxes are concepts used in the formulation in question, or upheld in the philosophy of the camp (here, realists) in question. The *black boxes* are concepts used in the formulation in question, but rejected in the philosophy of the camp in question. The *white* (*pale*) boxes are concepts not used in the formulation in question and ignored in the philosophy of the camp in question. **Underlined bold font** is used for the fundamental assumptions (2, 3, or 4 in number) of the formulation in question. The arrows indicate that a particular concept holds if all concepts pointing to it (plus the Axioms) hold.

Operationalist Principles

As discussed in Sect. "The Two Camps", a key difference between realists and operationalists is that the former seek to *explain* correlations while the latter do not (or, at least, not in the same sense). We have shown how realists can enshrine this goal in causal terms by REICHENBACH's Principle. Another key difference is that operationalists are happy to put the actors in centre-stage. Thus, operationalists should be happy to accept a notion of causation which is not about explaining all correlations, and which is agent-centric: the Principle of AGENT-CAUSATION (Principle 4 above).

Just as combining REICHENBACH's Principle with the Postulate of RELATIVISTIC CAUSALITY gives LOCAL CAUSALITY (the realists' favoured localist notion), here we have:

Lemma 4 *The Principle of* AGENT-CAUSATION *plus the Postulate of* RELATIVISTIC CAUSALITY *implies the Principle of* LOCALITY*,*

where we can provide the following principled version of LOCALITY:

Principle 5 (LOCALITY) The probability of an observable event A is unchanged by conditioning on a space-like-separated free choice b, even if it is already conditioned on other events not in the future light-cone of b.

This Lemma can be demonstrated as follows. Define $\mathscr{A}$ to be A and other events not in the future light-cone of b. By Postulate 2, none of the events in $\mathscr{A}$ can have b as a CAUSE. If b is a freely chosen action, then by Principle 4, $\mathscr{A}$ cannot be correlated with b. Hence the probability of A, even conditional on other events in $\mathscr{A}$, is independent of b.

Note that LOCALITY is by no means the strongest principle that can be derived from AGENT-CAUSATION and RELATIVISTIC CAUSALITY; one can also show

Lemma 5 *The Principle of* AGENT-CAUSATION *plus the Postulate of* RELATIVISTIC CAUSALITY *implies the Principle of* LOCAL AGENCY*,*

as defined in the preceding section (Principle 3). We use LOCALITY rather than LOCAL AGENCY to remain faithful to Bell's original concept. However, we do also use these principles to derive another principle that was implicit for Bell in 1964:

Lemma 6 *The Principle of* AGENT-CAUSATION*, and the Postulate of* RELATIVISTIC CAUSALITY *imply the Principle of* NO SUPERDETERMINISM*,*

where this last is:

Principle 6 (NO SUPERDETERMINISM) Any set of events on a SLH is uncorrelated with any set of freely chosen actions subsequent to that SLH.

The name of this Principle is taken from Bell [12], and its form chosen in keeping with the assumption of PREDETERMINATION below.

Unlike LOCAL CAUSALITY, LOCALITY is not sufficient to make a theory BELL-LOCAL, even with the Principle of NO SUPERDETERMINISM. We require an additional principle, which, based on Bell's 1964 paper, we formulate as:

Principle 7 (PREDETERMINATION) For any observable event A, and any SLH S prior to it, A has CAUSES on S, which, possibly in conjunction with free choices subsequent to S, determine A.

Applying this to Bell's scenario, choose the SLH to be prior to both a and b, so that Bob's outcome is a function $B(a, b, c, \lambda)$. By the Principle of LOCALITY, the dependence on a must be trivial, because it is space-like-separated from B. Finally, by the Principle of NO SUPERDETERMINISM, the probability of λ cannot depend on a and b. Doing the same for A, and following the same argument as in Sect. "Realist Principles" we see that BELL-LOCALITY is obeyed. Thus,

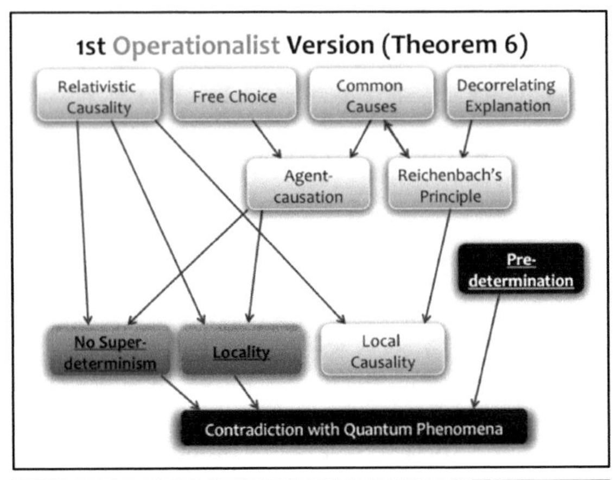

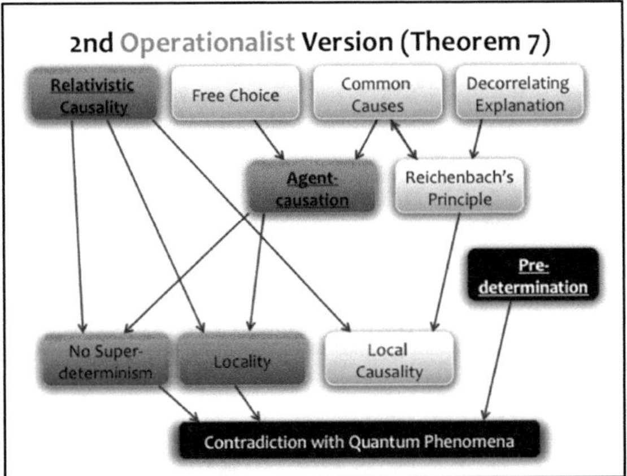

Fig. 6.4 Two operationalist versions of Bell's Theorems 6 and 7. For explanatory details, see Fig. 6.3.

Theorem 6 (Bell-1964, in principles) *Quantum* PHENOMENA *violate the conjunction of Axioms 1–4 and the Principles of* NO SUPERDETERMINISM, LOCALITY, *and* PREDETERMINATION.

From the lemmata in this section, we can also formulate this theorem using more fundamental, and not more numerous, principles, as follows:

Theorem 7 (Bell-1964, in deeper principles) *Quantum* PHENOMENA *violate the conjunction of Axioms 1–4, the Postulate of* RELATIVISTIC CAUSALITY, *and the Principles of* AGENT-CAUSATION *and* PREDETERMINATION.

This has the advantage of not using the term 'locality', which realists generally use with a different meaning, as discussed above. These two theorems, and the relationship between them are illustrated in Fig. 6.4.

Reactions and Reconciliation

Let us now review the forms of Bell's theorem favoured by the two camps, as formulated in terms of causal principles deeper than those used by Bell in either 1964 or 1976. In all of the below (as in the related figures), we assume the usual

1. Axiom of MACROREALITY,
2. Axiom of MINKOWSKI SPACE-TIME,
3. Axiom of TEMPORAL ORDER,
4. Axiom of CAUSAL ARROW.

First, the realist's version makes the additional assumptions of

- The Postulate of FREE CHOICE,
- The Postulate of RELATIVISTIC CAUSALITY,
- REICHENBACH's Principle.

Of these three, realists reject RELATIVISTIC CAUSALITY, leaving the question of what defines PAST in the Axiom of CAUSAL ARROW open to further physical investigation. Operationalists, on the other hand, would reject (or indeed have rejected [6]) REICHENBACH's Principle. But the latter have reason to be unhappy with the remaining list, because the remaining Axioms and Postulates have no empirical consequences. They do not imply even SIGNAL-LOCALITY, for example. Thus the operationalists would be left saying that they privilege the Postulate of RELATIVISTIC CAUSALITY over REICHENBACH's Principle even though they cannot point to any consequences of believing in the former. For this reason the operationalists would, presumably, object to the list of options offered by this form of Bell's theorem.

Next we consider the operationalist's version, which assumes instead

- The Postulate of RELATIVISTIC CAUSALITY,
- The Principle of AGENT-CAUSATION,
- The Principle of PREDETERMINATION.

Of these three, operationalists reject PREDETERMINATION, embracing the idea of intrinsic randomness. Realists, on the other hand, would presumably still reject the Postulate of RELATIVISTIC CAUSALITY. But the latter have reason not to be happy with the list of options offered here. The reason is that PREDETERMINATION is just too easy to reject. Even though many of the theories realists take seriously, such as de Broglie–Bohm mechanics [42, 43], satisfy PREDETERMINATION, it is not a necessary feature of realist theories, and realists would not want to be characterized as basing their rejection of RELATIVISTIC CAUSALITY on a belief in PREDETERMINATION. Realists do not reject the Principle of AGENT-CAUSATION, because it is implied by REICHENBACH's Principle and FREE CHOICE (this is just a weaker version of Lemma 7 below). But it is the fact that AGENT-CAUSATION is weaker than REICHENBACH that is the problem, they might say, because it entails the addition of the third assumption (PREDETERMINATION) which is clearly too strong, and so too easy to reject. Thus realists do, in fact [28], object to a form of Bell's theorem listing PREDETERMINATION as a fundamental assumption.

It would seem that the two camps are still at an impasse. But by returning to the Postulates of Sect. "Back to Basics" we can bridge the gap. From results already discussed, it is apparent that BELL-LOCALITY can be derived from the Axioms plus

- The Postulate of FREE CHOICE,
- The Postulate of RELATIVISTIC CAUSALITY,
- The Postulate of COMMON CAUSES,
- The Postulate of DECORRELATING EXPLANATION.

For realists, little has changed from their preferred formulation, and they will as before, reject RELATIVISTIC CAUSALITY. But operationalists now should be happy to reject DECORRELATING EXPLANATION, and keep the other postulates. The reason is that

Lemma 7 *The Postulates of* FREE CHOICE *and* COMMON CAUSES *imply the Principle of* AGENT-CAUSATION.

This can be seen as follows. Say a set of events $\mathscr{A}$ is correlated with a free choice a. Now by Postulate 1, a has no relevant CAUSES. Thus by Postulate 3, the only option is that a is a CAUSE of at least one event in $\mathscr{A}$, as in Principle 4. Thus operationalists still have their agent-centric notion of causation, which operational quantum mechanics respects, even with the addition of the Postulate of RELATIVISTIC CAUSALITY. Some following this direction believe that a replacement for the Principle of DECORRELATING EXPLANATION is open to further physical investigation [44]. The attitude each camp would be expected to take to this theorem is illustrated in Fig. 6.5.

Conclusion

After thorough investigation, we suggest that the cause of the disagreement between operationalists and realists over Bell's theorem is a disagreement over causes. This leads the two camps to favour Bell's 1964 theorem and Bell's 1976 theorem respectively, because the localistic notions they employ relate to notions of agent-causation and explanatory causation respectively. However, by breaking down notions into more fundamental postulates, we could formulate what we believe is the best version of Bell's theorem, for the purposes of reconciling the two camps:

Theorem 8 (Bell-reconciliation) *Quantum* PHENOMENA *violate the conjunction of Axioms 1–4 and the Postulates of* FREE CHOICE, RELATIVISTIC CAUSALITY, COMMON CAUSES, *and* DECORRELATING EXPLANATION.

This formulation avoids contentious words like 'locality', and allows each camp to reject one assumption (DECORRELATING EXPLANATION and RELATIVISTIC CAUSALITY, respectively), knowing that the remaining assumptions reflect its philosophical

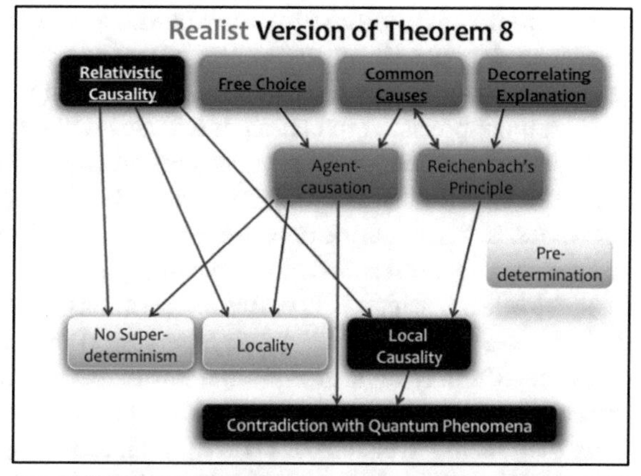

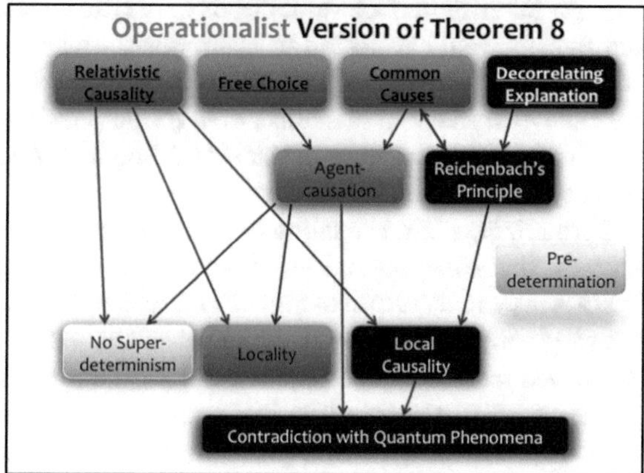

Fig. 6.5 The realist and operationalist interpretations of our 'reconcilation' version of Bell's Theorem 8. For explanatory details, see Fig. 6.3.

position. Of course even if the two camps do agree upon a single form of Bell's theorem, their disagreement about which assumption to reject is still a substantive disagreement. But at least they could discuss the merits of their *weltanschauungen* using a common vocabulary, and so avoid talking past one another, or (the next stage) interminable arguments about what terms like 'locality' should mean. If, in the end, they just agree to disagree, that would still be a great improvement over the present state of affairs [1, 29, 45–47].

Another advantage of the above formulation is that RELATIVISTIC CAUSALITY better reflects the ontology of different quantum interpretations than does a notion like LOCALITY. As has been stated a few times above, orthodox quantum mechanics respects LOCALITY, even when it is understood as a realistic theory, an understanding held by most physicists who don't think long and hard about foundations (and by some who do). That is, even the process whereby, when Alice and Bob share a singlet state, a measurement by Alice in a certain basis causes the quantum state of Bob's system to collapse instantaneously into one of the basis states, does not violate LOCALITY. Yet the very wording of the preceding sentence implies that the described process *does* violate RELATIVISTIC CAUSALITY.[13] By contrast, operational quantum mechanics does not violate RELATIVISTIC CAUSALITY, because it does not entail any causal narrative involving quantum states, but simply uses them as computational tools. A more precise formulation of this idea will be given elsewhere [10].

A similar example of the advantage of talking about causes is the ability to formulate FREE CHOICE in an obvious way: that a freely chosen action has no relevant CAUSES.[14] This can be opposed to Colbeck and Renner's assumption *FR*, which also supposedly corresponds to "the assumption that measurements can be chosen freely" [49], but which is actually the assumption of LOCAL AGENCY as defined in Sect. "Realist Principles". As we have seen (Lemma 3), this can be derived from FREE CHOICE only by also using the Postulates of RELATIVISTIC CAUSALITY and COMMON CAUSES (or something similar). The assumption *FR* clearly embodies far more than merely freedom of choice, and interpreting it as if it does not (i.e. taking their language at face value) leads to bizarre conclusions. For example, Colbeck and Renner would be forced to claim that a typical physicist (see preceding paragraph), adhering to the realistic interpretation of orthodox quantum mechanics, *does not* believe in freedom of choice, *even if* said physicist were to believe that humans are not wholly governed by physical laws and have free will in the strongest possible philosophical sense.

Formalizing notions relating to Bell's theorem as causal principles also makes it apparent that some notions are more natural than others. In particular, the notions of LOCALITY and PREDETERMINATION, which Bell introduced in his first paper, are not very natural. The first is weaker than the notion of LOCAL AGENCY that can be derived from the more fundamental postulates of AGENT-CAUSATION and RELATIVISTIC CAUSALITY, while the second is even more contrived. The latter criticism we would also level, even more strongly, at another notion which has been suggested as a replacement for PREDETERMINATION, namely 'completeness' [17] (or 'Jarrett-completeness' as one of us has called it [4]), also known as 'outcome-independence' [18]. In this context, more natural concepts for formulating a theorem in the style of Bell's 1964 theorem will be considered elsewhere [10].

Finally, approaching Bell's theorem using ideas of causation can lead in new directions. For instance, *if* one assumes the predictions of relativistic quantum

---

[13]Notably, the faster-than-light effect of Alice's choice on Bob's conditioned state has now been verified experimentally with no detector efficiency loophole [48], unlike BELL-NONLOCALITY [14].

[14]This may sound like a strong statement, and the reader may feel tempted to follow neither the operationalist nor the realist camp, but rather to reject Postulate 1 from the list of assumptions in Theorem 8. This temptation should vanish if the reader thinks through what it would actually mean to explain away Bell-correlations through the real (not just in-principle) failure of FREE CHOICE. There is no general theory that does this. If such a theory did exist, it would require a grand conspiracy of causal relationships leading to results in precise agreement with quantum mechanics, even though the theory itself would bear no resemblance to quantum mechanics. Moreover, it is hard to imagine why it should only be in Bell experiments that free choices would be significantly influenced by causes relevant also to the observed outcomes; rather, every conclusion based upon observed correlations, scientific or casual, would be meaningless because the observers's method would always be suspect. It seems to us that any such theory would be about as plausible, and appealing, as, belief in ubiquitous alien mind-control.

mechanics to be correct, then it seems that one can, in Theorem 8, replace the two Postulates of FREE CHOICE and RELATIVISTIC CAUSALITY by the single Postulate of NO FINE-TUNING [7]. This is a postulate that the conditional independence relations between observable events are a consequence only of the causal structure (i.e. which events are CAUSES for which other events), and are not to be explained by fine-tuning of the probabilities of events. Whether this formulation can really be regarded as having fewer postulates than the above, whether it is truly possible to dispense with FREE CHOICE, and how the two camps could be expected to react to it, are interesting questions for exploration.

Acknowledgments This research was supported by the Australian Research Council (ARC) Discovery Project DP140100648, the ARC DECRA project DE120100559, and a grant (FQXi-RFP-1504) from the Foundational Questions Institute Fund (fqxi.org) at the Silicon Valley Community Foundation. We thank Eleanor Rieffel for helpful comments.

References

1. R.F. Werner and respondents, Guest post on Bohmian mechanics. http://tjoresearchnotes.wordpress.com/2013/05/13/guest-post-on-bohmian-mechanics-by-reinhard-f-werner/
2. J.S. Bell, Rev. Mod. Phys. **38** 447–452 (1966). (Reproduced in Ref. [50].)
3. J.S. Bell, Physics **1** 195 1964. (Reproduced in Ref. [50].)
4. H.M. Wiseman, J. Phys. A **47**, 424001 (2014)
5. H.M. Wiseman, Nature **510**, 467469 (2014)
6. B.C. van Fraassen, Synthese **52**, 2538 (1982)
7. C.J. Wood, R.W. Spekkens, New J. Phys. **17**, 033002 (2015)
8. E.G. Cavalcanti, R. Lal, J. Phys. A **47**, 424018 (2014)
9. J.S. Bell, Epistemol. Lett. **9** 11–24 (1976). (Reproduced in Ref. [50].)
10. H.M. Wiseman, E.G. Cavalcanti. (in preparation)
11. H.M. Wiseman, E.G. Rieffel, Int. J. Quantum Found. **1**, 85–99 (2015). http://www.ijqf.org/archives/2093
12. J.S. Bell, La nouvelle cuisine, in *Between Science and Technology* ed. by A. Sarmeljin, P. Kroes (Elsevier/North-Holland, 1990) pp. 97–115. (Reproduced in Ref. [50].)
13. E.G. Cavalcanti, H.M. Wiseman, Found. Phys. **42**, 1329–1338 (2012)
14. J.A. Larsson, J. Phys. A **47**, 424003 (2014)
15. T. Norsen, AIP Conf. Proc. **844**, 281–293 (2006)
16. T. Norsen, Int. J. Quantum Found. **1**, 65–84 (2015). http://www.ijqf.org/archives/2075
17. J. Jarrett, Noûs **18**, 569–89 (1984)
18. A. Shimony, Controllable and uncontrollable non-locality, in *Foundations of Quantum Mechanics in the Light of New Technology* eds. S.K. et al. (Physical Society of Japan, Tokyo, 1984) pp. 225–230
19. G.C. Ghirardi, A. Rimini, T. Weber, Phys. Rev. A **34**, 470 (1986)
20. J.S. Bell, Introduction to the hidden-variable question, in *Foundations of Quantum Mechanics* (Academic, New York, 1971) pp. 171–181. (Reproduced in Ref. [50].)
21. J.F. Clauser, M.A. Horne, Phys. Rev. D **10**, 526–35 (1974)
22. J.S. Bell, J. de Phys. Colloq. **42** 41–62 (1981). (Reproduced in Ref. [50].)
23. A. Fine, Phys. Rev. Lett. **48**, 291–295 (1982)
24. N.D. Mermin, Rev. Mod. Phys. **65**, 803–815 (1993)
25. M.A. Nielsen, I.L. Chuang, *Quantum Computation and Quantum Information* (Cambridge University Press, England, 2000)

26. M. Żukowski, Časlav Brukner, J. Phys. A **47** 424009 (2014)
27. D. Dürr, S. Goldstein, N. Zanghì, J. Stat. Phys. **67**, 843–907 (1992)
28. T. Maudlin, *Quantum Non-Locality and Relativity* (Blackwell, Oxford, 1994)
29. T. Maudlin, J. Phys. A **47**, 424010 (2014)
30. A. Einstein, The Times Nov. **28**, 41–43 (1919)
31. H. Everett, Rev. Mod. Phys. **29**, 454–462 (1957)
32. B. DeWitt, R.N. Graham (eds.), *The Many-Worlds Interpretation of Quantum Mechanics Princeton Series in Physics* (Princeton University Press, Princeton, 1973)
33. C.A. Fuchs, N.D. Mermin, R. Schack, Am. J. Phys. 749–754 (2014)
34. P. Holland, *Quantum Theory of Motion: Account of the de Broglie-Bohm Causal Interpretation of Quantum Mechanics* (Cambridge University Press, 1995)
35. D.T. Pegg, Phys. Lett. A **78**, 233–234 (1980)
36. H. Price, Studies In History and Philosophy of Modern Physics **39**, 752–761 (2008)
37. G. 't Hooft, How does God play dice? (pre-)determinism at the Planck scale, in *Quantum [Un]speakables: from Bell to Quantum Information*, ed. by R.A. Bertlmann, A. Zeilinger (Springer, Berlin, 2002) pp. 307–316
38. H. Reichenbach, *The Direction of Time* (University of California Press, Berkeley, 1956)
39. A. Shimony, M.A. Horne, J.F. Clauser, Epistemol. Lett. **13** 1–8 (1976). (Reproduced in Dialectica **39** (1985).)
40. J.F. Clauser, Early history of Bell's theorem, in *Quantum [Un]speakables: from Bell to Quantum Information* ed. by R.A. Bertlmann, A. Zeilinger (Springer, Berlin, 2002) pp. 61–98
41. J.S. Bell, Epistemol. Lett. **15** 79–84 (1977) (Reproduced in Ref. [50].)
42. L. de Broglie, *Une tentative d'interprétation causale et non linéaire de la mécanique ondulatoire: la théorie de la double solution* (Gauthier-Villars, Paris, 1956)
43. D. Bohm, Phys. Rev. **85**, 166–193 (1952)
44. M. Leifer, R. Spekkens, Phys. Rev. A **88**, 052130 (2013)
45. R.F. Werner, J. Phys. A **47**, 424011 (2014)
46. T. Maudlin, J. Phys. A **47**, 424012 (2014)
47. R.F. Werner (2014). What Maudlin replied to arXiv:1411.2120v1
48. B. Wittmann, S. Ramelow, F. Steinlechner, N.K. Langford, N. Brunner, H.M. Wiseman, R. Ursin, A. Zeilinger, New J. Phys. **14**, 053030 (2012)
49. R. Colbeck, R. Renner, Nat. Commun. **2**, 411 (2011)
50. M. Bell, K. Gottfried, M. Veltman (eds.), *John S. Bell on the Foundations of Quantum Mechanics* (World Scientific, Singapore, 2001)

Chapter 7
Whose Information? Information About *What?*

Jeffrey Bub

My title is from an article by John Bell [1], in which he argues that terms like measurement or information have no place in the formulation of fundamental theories of physics. I want to argue here, against Bell, that the conceptual revolution in the transition from classical to quantum mechanics should be understood as resting on the recognition that there is an *information-theoretic structure* to the mosaic of events, and this structure is not what Shannon thought it was—just as the theory of relativity rests on the recognition that events have a spatio-temporal structure, and this structure is not what Newton thought it was.

Firstly, let me say what I don't mean by this thesis. I don't mean that information is primary and that particles and fields—what physics is usually understood to be about—are in some sense derived from elementary units of information as the basic building blocks of reality. Wheeler's slogan 'it from bit' suggests a view like this. Vlatko Vedral endorses something similar in his book *Decoding Reality*. As Vedral puts it, 'our reality is ultimately made up of information,' and 'the laws of Nature are information about information.'

The classical theory of information was initially developed by Claude Shannon to deal with certain problems in the communication of messages as electromagnetic signals along a channel such as a telephone wire. A communication set-up involves a transmitter or source of information, a communication channel, and a receiver. An information source produces messages composed of sequences of symbols from an alphabet, with certain probabilities for the different symbols. The fundamental question for Shannon was how to quantify the minimal physical resources required to represent messages produced by a source, so that they could be communicated via

J. Bub (✉)
Philosophy Department, Institute for Physical Science and Technology,
Joint Center for Quantum Information and Computer Science, University of Maryland,
College Park, MD 20742, USA
e-mail: jbub@umd.edu

R. Bertlmann and A. Zeilinger (eds.), *Quantum [Un]Speakables II*,
The Frontiers Collection, DOI 10.1007/978-3-319-38987-5_7

a channel and reconstructed by a receiver. For this problem, and related communication problems, the meaning of the message is irrelevant.

As Shannon put it in his seminal paper 'A Mathematical Theory of Communication' [2, p. 379]:

> The fundamental problem of communication is that of reproducing at one point either exactly or approximately a message selected at another point. Frequently the messages have *meaning*; that is they refer to or are correlated according to some system with certain physical or conceptual entities. These semantic aspects of communication are irrelevant to the engineering problem. The significant aspect is that the actual message is one *selected from a set* of possible messages. The system must be designed to operate for each possible selection, not just the one which will actually be chosen since this is unknown at the time of design.

So a theory of information in the sense relevant for physics is about the 'engineering problem' of communicating messages over a channel efficiently. In this sense, the concept of information has nothing to with anyone's knowledge and everything to do with the stochastic or probabilistic process that generates the messages. Shannon showed that it is possible to compress the information required to communicate a message—to reduce the average number of bits per symbol—up to a certain optimal compression, if the probabilities of the different symbols produced by an information source are not all equal.

Quantum mechanics began with Heisenberg's seminal 'Umdeutung' paper of 1925 [3], which he developed shortly afterwards into the first version of quantum mechanics in collaboration with Max Born and Pascual Jordan [4, 5]. The problem, for Heisenberg, was to explain the discrete spectral lines in the emission and absorption spectra of gases, without appealing to the ad hoc quantum rules for electron orbits in Bohr's atomic theory. Bohr's theory stipulates that the energy of an orbiting electron is quantized: there is a discrete set of allowed orbits associated with different energies that an electron can occupy, with lower energy orbits closer to the nucleus. The theory also stipulates that an atom radiates or absorbs energy only when an electron jumps from one of these quantized orbits to another orbit, with the frequency of the radiation depending on the energy gap between the two orbits. These rules for electrons conflict with classical electrodynamics and also with classical mechanics.

The full title of Heisenberg's paper in English is 'On the quantum-theoretical re-interpretation [Umdeutung] of kinematical and mechanical relations.' Heisenberg showed that the discrete frequencies of light emitted by atoms could be explained without referring to electron orbits by *re-interpreting* classical quantities, like position, momentum, energy, angular momentum, as operations. In later versions of the theory, these operations are represented by operators that act on and transform the states of quantum systems. In a letter to Wolfgang Pauli written July 9, 1925, Heisenberg wrote:

> All of my meagre efforts go toward killing off and suitably replacing the concept of the orbital path that one cannot observe.

The algebra of operators contains commuting and noncommuting operators that are 'intertwined,' to use Gleason's term [6, p. 886]: an operator can belong to different mutually commuting sets of operators that don't commute with each other.

Gleason proved, under very weak assumptions, that the quantum probabilities specified by the Born rule are unique as probabilities that can be defined on the operator algebra [6]. Kochen and Specker [7] proved that quantum systems can't have definite values simultaneously for certain finite sets of quantities represented by these operators. The result follows from the way quantities with more than two possible values are intertwined in the operator algebra (there is no intertwinement for quantities with only two possible values), if the value of a physical quantity Q represented by an operator $\boldsymbol{Q}$ is 'noncontextual,' in the sense that the value doesn't depend on the contexts defined by the different commuting sets of operators to which $\boldsymbol{Q}$ belongs. In particular, an electron in atom can't have definite position and momentum values, and so can't have a well-defined orbit in an atom.

If a quantity has a definite value in a quantum state, then certain other quantities are indefinite, and the result of measuring a quantity with an indefinite value is an intrinsically random event, assuming no action at a distance or something similar. (See, e.g., the argument by Colbeck and Renner [8]), who derive intrinsic randomness from a related condition of 'free choice' for the observables measured.) This intrinsic randomness allows *new sorts of nonlocal probabilistic correlations* for entangled quantum states of separated systems, where the probabilities are, as von Neumann put it, 'perfectly new and *sui generis* properties of physical reality' [9]. Schrödinger, who coined the term, referred to entanglement as '*the* characteristic trait of quantum mechanics, the one that enforces its entire departure from classical lines of thought' [10, p. 555].

Schrödinger published a wave mechanical version of quantum mechanics in 1926 [11, 12] that kept the orbits and explained their quantization as a wave phenomenon. Shortly afterwards, he proved the formal equivalence of Heisenberg's noncommutative mechanics and his own wave mechanics. Not surprisingly, physicists found wave mechanics more intuitively appealing as a picture of reality at the subatomic level than the abstract notion of a noncommutative mechanics, but the intuitive appeal is misleading. As Schrödinger pointed out in a lecture to the Royal Institution in London in March, 1928 [13, p. 160], the wave associated with a quantum system evolves in an abstract, multi-dimensional representation space, not real physical space:

> The statement that what *really* happens is correctly described by describing a wave-motion does not necessarily mean exactly the same thing as: what *really* exists is a wave-motion. ...It is merely an adequate mathematical description of what happens.

The idea of a wave as a representation of quantum reality continues to shape contemporary discussions of conceptual issues in the foundations of quantum mechanics. From the perspective adopted here, the later formalization of quantum mechanics by Dirac in 1930 [14] and von Neumann in 1932 [15] as a theory of 'observables' represented by operators on a space of quantum states is fundamentally an elaboration of Heisenberg's 'Umdeutung' rather than a wave theory. What is revolutionary about quantum mechanics is analogous to what is revolutionary about the theory of relativity: a fundamental structural change in the way we represent how events fit together, where the change involves spatio-temporal structure in the case of relativity, and the structure of information in the case of quantum mechanics.

In proposing that quantum mechanics is about the structure of information, I mean that the theory deals with new sorts of probabilistic correlations that are structurally different from correlations that arise in classical theories (and, of course, the theory is also able to handle standard classical correlations). The deep significance of Heisenberg's 'Umdeutung' is that quantum mechanics, as a noncommutative modification of classical mechanics, is a theory about a *structurally different sort of information* than classical information, insofar as information in the physical sense is about probabilistic correlations. What we have discovered is that the 'engineering' possibilities for representing, manipulating, and communicating information in a quantum world are different than we thought, irrespective of what the information is about.

Of course, this is not a historical claim about what the founders of quantum mechanics had in mind. Shannon published 'A Mathematical Theory of Communication' in 1948 and Heisenberg's 'Umdeutung' paper appeared in 1925. Rather, I'm proposing that the conceptual revolution involved in the transition from classical to quantum mechanics should be seen as implicit in the 'Umdeutung,' and that this is about the novel 'engineering' possibilities of information in a quantum world.

The idea is illustrated in the correlation diagram in Fig. 7.1 for correlations between Alice and Bob, who are separated and can each choose to measure one of two binary-valued observables on systems they hold, say A and A' for Alice and B and B' for Bob. Each point in the diagram represents a correlation array of sixteen probabilities for Alice's two possible measurements, each with two possible outcomes, and Bob's two possible measurements, each with two possible outcomes. In a representation proposed by Pitowsky [16], a correlation array is reduced to a correlation four-vector.

Correlations that can be simulated with local resources available to Alice or Bob separately are represented by the points in the innermost square, $\mathcal{L}$, which is a schematic representation of a polytope, the local correlation polytope. The vertices represent all the local deterministic correlations, with 0, 1 probabilities for the out-

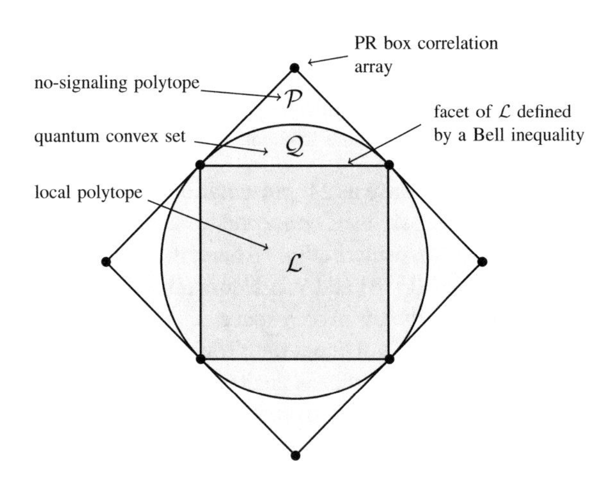

Fig. 7.1 A schematic representation of local and nonlocal correlations.

comes of measurements. In Pitowsky's representation, the local correlation polytope $\mathcal{L}$ is a four-dimensional hyperoctahedron, with eight vertices.

The outermost square $\mathcal{P}$ represents all the probabilistic correlations that satisfy a no-signaling principle: no information should be available to Alice about Bob's choices of measurement, or about whether Bob made any measurement at all, and similarly no information should be available to Bob about Alice's measurements. In Pitowsky's representation, the no-signaling polytope $\mathcal{P}$ is a four-dimensional hypercube, with sixteen vertices, the eight vertices of the local correlation polytope and an additional eight vertices representing extremal nonlocal no-signaling correlations, the Popescu-Rohrlich (PR) correlations [17]. One of these eight no-signaling vertices represents the standard PR correlation: the outcomes are the same, except when Alice measures A' and Bob measures B', in which case the outcomes are different. The remaining seven no-signaling vertices represent correlations defined by replacing A' by A and/or B' by B in the standard PR correlation, and by switching 'same' and 'different' in these correlations.

Quantum correlations are represented by the points inside a convex set $\mathcal{Q}$ with a continuous boundary between the local polytope $\mathcal{L}$ and the no-signaling polytope $\mathcal{P}$. Points between the quantum convex set $\mathcal{Q}$ and the boundary of $\mathcal{P}$ represent superquantum no-signaling correlations. The representation in Fig. 7.1 is only schematic. In particular, the boundary of the quantum set $\mathcal{Q}$ is a complicated three-dimensional manifold, not nicely spherical, as the diagram suggests.

The correlations represented by points in the local correlation polytope $\mathcal{L}$ can be simulated by Alice and Bob with local resources. Local quantum resources provide no advantage over local classical resources, and classical resources (excluding action at a distance) are local, so $\mathcal{L}$ could equivalently be characterized as the polytope of classical correlations.

There are two sorts of explanations for correlations in $\mathcal{L}$: either there is a direct causal connection in which information is transmitted from one event to the other by some physical system moving continuously at finite speed between the correlated events, or there is a common cause that is the source of the same information conveyed to the correlated events, like a flash of lightning that is the common cause of the experience of thunder at two separate locations in the vicinity of an electrical storm. The correlations represented by points outside $\mathcal{L}$ are inconsistent with any explanation by a direct causal connection between the Alice-events and the Bob-events [18, 19], or by a common cause, as Bell showed [20] for the correlations of maximally entangled states represented by points on the boundary of the quantum region $\mathcal{Q}$.

What we've discovered is that we live in a world in which there are correlations outside the local polytope $\mathcal{L}$. Astonishingly, the correlated events outside $\mathcal{L}$ are, separately, intrinsically random. For correlations in $\mathcal{L}$, a deterministic explanation is possible, and Alice and Bob can simulate these correlations with local instruction sets for Alice and Bob separately that determine how they respond to measurements. For correlations outside $\mathcal{L}$, the probabilistic description is complete. Nothing is left out of the story—the probabilities don't reflect our ignorance of variables that would restore a deterministic picture. The indefiniteness of quantum observables and the intrinsic randomness of measurement outcomes is a structural feature, a consequence of the way observables are intertwined in a noncommutative algebra. The content of the Kochen-Specker theorem is that the subalgebra of binary-valued observables, corresponding to the properties of a quantum system, can't be embedded into a Boolean algebra. A Boolean algebra is the structure characterizing the ways in which the properties of a classical system combine under the operations 'and,' 'or', and 'not,' where, for every property, the system either has the property or not in a definite sense (no property is indefinite).

Correlations cry out for explanation, as Bell remarked [21, p. 140]. This raises the problem of what would count as an explanation for correlations outside $\mathcal{L}$, if no common cause explanation or explanation by direct causes is possible. There's a prejudice for deterministic explanations as the gold standard for explanation in physics—or rather, a prejudice for a presupposition of deterministic theories: that probabilities are defined on an event space where the events, which correspond to a system having certain properties, are definite in the sense associated with the Boolean property structure of classical physics.

For example, in Bohm's hidden variable theory [22, 23], the quantum wave function acts as a guiding field for the positions of particles. The particle positions, which are the hidden variables of the theory, are always definite. It's an ingenious feature of the deterministic Bohmian dynamics that if the particles are distributed according to the quantum probabilities defined by the wave function via the Born rule at any time, they remain so distributed as the wave function evolves. So the hidden variables can't be pinned down more precisely than the Born distribution by any possible measurements, which means that they remain hidden. The price to pay for a common cause explanation of nonlocal quantum correlations in Bohm's theory is a violation of the no-signaling principle at the level of the hidden variables. Provided that the hidden variables remain hidden and can't be exploited to refine the quantum probabilities, this does not lead to observable effects.

In the Everett interpretation of quantum mechanics [24, 25], all possible events that are assigned probabilities in the theory actually occur, but in different 'worlds' associated with different branches of the wave function. There are some well-known conceptual difficulties with the Everett interpretation, having to do with how to explain the specific branching associated with quantum probabilities and how to make sense of probability in a universe in which everything that can happen does happen. Assuming that these difficulties can be resolved (and this is a matter of dispute; see, in particular, the critique by Adrian Kent [26]), each Everettian world is perfectly definite in a classical sense.

In both Bohm's theory (with the Born distribution for particle positions) and the Everett interpretation, the randomness of quantum events is only apparent. In Bohm's theory, randomness is only apparent because of a limitation on what we can measure: we can't, in principle, measure position more precisely than allowed by the Born distribution. In the Everett interpretation, randomness is only apparent because of our limited perspective: we live in a multiverse and only have access to our own world in the multiverse.

Is there a new sort of explanation appropriate for correlations outside $\mathcal{L}$, in which the correlated events are intrinsically random? The theory of special relativity provides a case study that is relevant here. Einstein distinguished between two sorts of theories in physics [27]: *constructive theories*, which explain the behavior of complex systems by showing how they are constructed from elementary systems like particles or fields or the ether as basic ontological building blocks, and *principle theories*, which are formulated in terms of principles that express empirical regularities. Einstein's point was that the theory of relativity should be understood as a principle theory, formulated in terms of the relativity principle and the light postulate. Lorentz's rival ether theory [28] is a constructive theory, where the basic building blocks are elementary systems like electrons and electric and magnetic fields and, of course, the ether.

Lorentz thought that a proper account of electromagnetic phenomena requires assuming the existence of the ether as a physically real medium for the transmission of light and other electromagnetic waves. Then it makes sense to take the ether as the state of rest for space and time measurements, in the Newtonian sense of space and time in classical physics. From Lorentz's perspective, Einstein simply assumes the constancy of the velocity of light in empty space, and the impossibility of detecting motion through the ether, either from mechanical phenomena or electromagnetic phenomena. Minkowski spacetime, which is structurally different from Newtonian space and time, follows from these assumptions.

In special relativity, length contraction and time dilation, and more generally the Lorentz transformation, are *kinematic or pre-dynamic* effects of the relativistic structure of Minkowski spacetime. As Lorentz saw it, showing that certain phenomena arise from purely kinematic aspects of this geometrical structure doesn't count as an explanation—what we should be looking for is a dynamical explanation, in terms of forces that have something to do with motion relative to the ether [28, pp. 229–230]:

> [Einstein's] results concerning electromagnetic and optical phenomena ... agree in the main with those which we have obtained in the preceding pages, the chief difference being that Einstein simply postulates what we have deduced, with some difficulty and not altogether satisfactorily, from the fundamental equations of the electromagnetic field. ...
>
> Yet, I think, something may also be claimed in favour of the form in which I have presented the theory. I cannot but regard the ether, which can be the seat of an electromagnetic field with its energy and its vibrations, as endowed with a certain degree of substantiality, however different it may be from all ordinary matter. In this line of thought, it seems natural not to assume at starting that it can never make any difference whether a body moves through the ether or not, and to measure distances and lengths of time by means of rods and clocks having a fixed position relative to the ether.

Bell suggested [29] that to really understand phenomena like Lorentz contraction, it ought to be possible to explain the contraction *dynamically*, with an explicit calculation of the forces involved, and he provided such a dynamical explanation in a clever example of a contracting thread that breaks because the contraction is resisted by two spaceships connected by the thread, which maintain a distance apart greater than the Lorentz contraction. But Bell's dynamical explanation is quite different from a strictly Lorentzian dynamical explanation. Bell doesn't assume the existence of the ether or the validity of Newtonian space and time. His forces are relativistic forces that are consistent with Minkowski spacetime—they are Lorentz invariant. So what Bell shows is that there is a dynamical account of the thread breaking that is *consistent* with the kinematics of special relativity.

On the information-theoretic interpretation, quantum mechanics is a principle theory of information-theoretic structure that provides the framework for the physics of a genuinely indeterministic universe. Just as Minkowski space-time characterizes the structure of space-time in a relativistic universe, the structure of information in a quantum universe is characterized by the intertwinement of commuting and non-commuting observables in Heisenberg's 're-interpretation,' which imposes objective pre-dynamic probabilistic constraints on correlations between intrinsically random events. The Born probabilities arise as a pre-dynamic feature of this information-theoretic structure.

The state of a quantum system can be specified by the probabilities of the outcomes of measurements on some finite 'informationally complete' or 'fiducial' set of observables. If the transition to a definite outcome in a quantum measurement process could be described by a deterministic reversible dynamics, then it would be possible to reverse the dynamical evolution to recover the original quantum state after a measurement. This could be done an indefinite number of times for an informationally complete set of observables, which would allow the identification of an unknown quantum state to any arbitrary accuracy from the statistics of the measurement outcomes.

If this were possible it would violate the no-signaling principle. Suppose that Alice and Bob share an entangled state of two qubits, say the Bell state $|\phi^+\rangle = |0\rangle|0\rangle + |1\rangle|1\rangle = |+\rangle|+\rangle + |-\rangle|-\rangle$. Suppose Alice chooses to measure her qubit either in the computational basis $|0\rangle, |1\rangle$ or in the basis $|+\rangle, |-\rangle$. Bob's qubit will either end up in an equal weight mixture of the states $|0\rangle, |1\rangle$, or an equal weight mixture of the states $|+\rangle, |-\rangle$. If Bob could identify an unknown quantum state by the above procedure, he could distinguish these two mixtures. Alice could signal to Bob, effectively instantaneously, because Alice and Bob could be any distance apart. So a dynamical account of how an individual measurement outcome is selected from among the possible measurement outcomes in a quantum measurement process would be inconsistent with the no-signaling principle. Since every observable belongs to many informationally complete sets, if information is gained about an observable in an informationally complete set in a quantum measurement, the no-signaling principle requires that the measurement must lead to a loss of information about other observables in informationally complete sets.

Lorentz contraction is a physically real phenomenon explained relativistically as a kinematic or pre-dynamic effect of motion in a non-Newtonian space-time structure. We don't worry about how to understand this once we see that space-time has a Minkowski structure rather than the structure Newton thought it had. A dynamical explanation like Bell's is really a demonstration that the relativistic dynamics is consistent with the kinematics. Similarly, the random selection of a particular outcome in a quantum measurement process is a pre-dynamic effect of *any* mode of gaining information about the relevant observable, irrespective of the specific dynamical process involved in the measurement. It's a generic feature of no-signaling correlations that lie outside the local correlation polytope.

A quantum 'measurement' is not really the same sort of thing as the measurement of a property of a classical system. It involves putting a quantum system, like a photon, in a situation, say a beamsplitter or an analyzing filter, where the system has to make an intrinsically random transition between two or more macroscopically distinct alternatives recorded in a device like a photon detector. Given the intrinsic randomness of events, a dynamical description of this process can only demonstrate the *consistency* of the dynamics with the pre-dynamic probabilistic description.

On the standard Copenhagen interpretation of quantum mechanics proposed by Bohr, Heisenberg, and some of the other founders of the theory, quantum phenomena are assumed to take place in a classical arena with observers who can make definite choices and an infrastructure able to record and store stable records that is left out of the quantum theoretical story. As Bohr put it [30, p. 209]:

> The argument is simply that by the word *experiment* we refer to a situation where we can tell others what we have done and what we have learned and that, therefore, the account of the experimental arrangements and the results of the observations must be expressed in unambiguous language with suitable application of the terminology of classical physics.

This doesn't justify applying the presuppositions of classical physics, where systems have definite properties corresponding to definite, pre-existing values of observables, to the macroscopic experimental arrangements used to probe the quantum world. To some extent, the problem of showing that macrosystems in a quantum world are 'quasi-classical' is still an open problem in physics, but there is a large literature on the topic and various ways of showing that, in an idealized sense, the observables of a macrosystem composed of a very large number of microsystems commute, or equivalently, that the properties of a macrosystem have the structure of a Boolean algebra characteristic of classical systems. (See, e.g., the review by Landsman [31], and Landsman's paper on the Bohr-Einstein debate [32].) The problem is not unlike the problem of showing that a macroscopic classical system can have 'emergent' properties that are not properties of its constituent microsystems, and this involves introducing idealizations in the mathematical analysis, as in classical explanations of phenomena like phase transitions from a gas phase to a liquid phase when water vapor cools, or from a liquid phase to a solid phase when water freezes.

If a photon interacts with a macroscopic measuring instrument like a beamsplitter that itself interacts with the environment, the resulting quantum description of the system-plus-beamsplitter (ignoring the environment) is effectively a quantum mixed state with probabilities that are consistent with the kinematic probabilities of the theory. 'Effectively,' because this would be the case if the environmental states were strictly orthogonal. The environmental states aren't strictly orthogonal, but in a typical 'decoherence' process of this sort they become infinitesimally close to orthogonal states, and the approach to orthogonality is virtually instantaneous.

This dynamical analysis does not explain how something definite actually happens in a quantum measurement process, because the full quantum description including the environment is an entangled state. The analysis is often taken in this way, and Bell rightly ridicules such claims as 'FAPP' ('For All Practical Purposes') explanations [1]. But if quantum mechanics is about the structure of information in a similar sense to which relativity is about the structure of space-time, then the explanation that something definite actually happens in a quantum measurement, and happens randomly with the relevant probability, lies in pre-dynamic structural features of the theory, not the dynamics. The random occurrence of a definite outcome in a quantum measurement is a pre-dynamic feature of the structure of information in a quantum world, just as Lorentz contraction is fundamentally a pre-dynamic or kinematic feature of the structure of space-time in a relativistic world.

To sum up: The conceptual framework of Heisenberg's 'quantum theoretical reinterpretation of kinematical and mechanical relations' is a new conceptual framework for physical explanation, in which quantum mechanics describes a genuinely indeterministic world with nonclassical correlations between intrinsically random events. The commutative structure of classical observables is replaced by a noncommutative structure of intertwined observables—equivalently, the Boolean structure of classical properties is replaced by a non-Boolean structure that can't be embedded into a Boolean structure (so it's not just the classical structure with something left out). The possibility of correlations outside the local correlation polytope is a feature of the non-Boolean structure. If the world is genuinely indeterministic in this way, then it isn't possible to provide a dynamical explanation of how a system produces a particular outcome in a quantum measurement—the outcome is *intrinsically random*. Finally, the pre-dynamic measurement probabilities obtained by applying the Born rule to a measured system match up extremely well with the description of the entangled system resulting from a dynamical interaction between the measured system, the macroscopic measuring instrument, and the environment, taking account of decoherence.

Is this a solution to the notorious 'measurement problem' of quantum mechanics? It's not a solution if the problem is to explain how the events associated with quantum probabilities come about in the physical processes that couple microsystems and macrosystems that we call quantum measurements, if what's meant by an acceptable explanation is an account of the evolution in space-time. If quantum mechanics is a principle theory of information-theoretic structure that provides the framework for a genuinely indeterministic universe, then there is no explanation of this sort within the conceptual framework of the current theory. Rather, what the theory provides are probabilities for a range of possibilities associated with macroscopic measuring and recording devices. Decoherence involves an arbitrary split between system and environment, and the decoherence dynamics, which can be applied at any level of precision however the split is defined, is consistent with these probabilities. If this is unsatisfactory, the complaint lies with quantum mechanics itself. There is no remaining problem associated with a sensible research program in quantum mechanics. Current 'solutions to the measurement problem' have a reactionary, Lorentzian air. For any further explanation of what is thought to be unsatisfactory in the quantum description of events, we should look to future physics.

Acknowledgments The research for this paper was supported in part by a University of Maryland Research and Scholarship (RASA) award during the academic year 2014–2014.

References

1. J.S. Bell, Against measurement. Phys. World, **8**, 33–40 (1990). Reprinted in *Sixty-Two Years of Uncertainty: Historical, Philosophical and Physical Inquiries into the Foundations of Quantum Mechanics*, ed. by A. Miller (Plenum, New York, 1990), pp. 17–31
2. C.E. Shannon, A mathematical theory of communication. Bell Syst. Tech. J. **27**(379–423), 623–656 (1948)
3. W. Heisenberg, 'Über quantentheoretischer umdeutung kinematischer und mechanischer beziehungen. Zeitschrift für Physik **33**, 879–893 (1925)
4. M. Born, W. Heisenberg, P. Jordan, Zur quantenmechanik ii. Zeitschrift für Physik **35**, 557–615 (1926)
5. M. Born, P. Jordan, Zur quantenmechanik. Zeitschrift für Physik **34**, 858–888 (1925)
6. A.N. Gleason, Measures on the closed sub-spaces of Hilbert spaces. J. Math. Mech. **6**, 885–893 (1957)
7. S. Kochen, E.P. Specker. On the problem of hidden variables in quantum mechanics. J. Math. Mech. **17**, 59–87 (1967)
8. R. Colbeck, R. Renner, No extension of quantum theory can have improved predictive power. Nat. Commun. **2**, 411 (2011)
9. J. von Neumann, Quantum logics: strict- and probability-logics, in *Collected Works of John von Neumann IV*, ed. by A.H. Taub (Pergamon Press, New York, 1961), pp. 195–197
10. E. Schrödinger, Discussion of probability relations between separated systems. Proc. Camb. Philos. Soc. **31**, 555–563 (1935)
11. E. Schrödinger, Quantisierung als eigenwertproblem. Annalen der Physik **79**, 361–376 (1926)
12. E. Schrödinger, An undulatory theory of the mechanics of atoms and molecules. Phys. Rev. **28**, 1049–1070 (1926)
13. E. Schrödinger, Four lectures on wave mechanics, in *Collected Papers on Wave Mechanics* (Chelsea Publishing Company, New York, 1982), pp. 1745–1748
14. P.A.M. Dirac, *The Principles of Quantum Mechanics* (Clarendon Press, Oxford, 1958)
15. J. von Neumann, *Mathematical Foundations of Quantum Mechanics* (Princeton University Press, Princeton, 1955)
16. Itamar Pitowsky, Correlation polytopes, their geometry and complexity. Math. Program. A **50**, 395–414 (1991)
17. S. Popescu, D. Rhorlich, Quantum nonlocality as an axiom. Found. Phys. **24**, 379 (1994)
18. J.-D. Bancal, S. Pironio, A. Acin, Y.-C. Liang, Valerio Scarani, Nicolas Gisin, Quantum nonlocality based on finite-speed causal influences leads to superluminal signaling. Nat. Phys. **8**, 867 (2012)
19. V. Scarani, J.-D. Bancal, A. Suarez, N. Gisin, Strong constraints on models that explain the violation of bell inequalities with hidden superluminal influences. Found. Phys. **44**, 523–531 (2014)
20. J.S. Bell, On the Einstein-Podolsky-Rosen paradox. Physics **1**, 195–200 (1964)
21. J.S. Bell, Bertlmann's socks and the nature of reality, in *Speakable and Unspeakable in Quantum Mechanics* (Cambridge University Press, Cambridge, 1987), pp. 139–158
22. D. Bohm, A suggested interpretation of quantum theory in terms of 'hidden' variables. I and II. Phys. Rev. **85**, 166–193 (1952)
23. S. Goldstein, Bohmian mechanics, in *The Stanford Encyclopedia of Philosophy*, ed. by E.N. Zalta (2001). http://plato.stanford.edu/entries/qm-bohm
24. H. Everett III, 'Relative state' formulation of quantum mechanics. Rev. Mod. Phys. **29**, 454–462 (1957)

25. D. Wallace, *The Emergent Multiverse: Quantum Theory according to the Everett Interpretation* (Oxford University Press, Oxford, 1912)
26. A. Kent, One world versus many: the inadequacy of Everettian accounts of evolution, probability, and scientific confirmation, in *Many Worlds? Everett, Quantum Theory, and Reality*, ed. by S. Saunders, J. Barrett, A. Kent, D. Wallace (Oxford University Press, Oxford, 2010), pp. 307–354
27. A. Einstein, What is the Theory of Relativity? in *The London Times*, p. 13. First published 28 Nov 1919. Also in A. Einstein, *Ideas and Opinions* (Bonanza Books, New York, 1954), pp. 227–232
28. H.A. Lorentz, *The Theory of Electrons and its Applications to the Phenomena of Light and Radiant Heat* (B.G Teubner, Leipzig, 1909)
29. J.S. Bell, How to teach special relativity, in *Speakable and Unspeakable in Quantum Mechanics* (Cambridge University Press, Cambridge, 1987), pp. 67–80
30. N. Bohr, Discussion with Einstein on epistemological problems in modern physics, in *Albert Einstein: Philosopher-Scientist*, vol. VII, ed. by P.A. Schilpp, pp. 201–241. *The Library of Living Philosophers* (Open Court, La Salle, IL, 1949)
31. N.P. Landsman, Between classical and quantum, in *Philosophy of Physics Part A*, ed. by J. Butterfield, J. Earman (North-Holland, Amsterdam, 2007), pp. 417–553
32. N.P. Landsman, When champions meet: re-thinking the Bohr-Einstein debate. Stud. His. Philos. Mod. Phys. **37**, 212–241 (2006)

Chapter 8
Quantum Theory: It's Unreal

Terence Rudolph

In this essay I try to explain in layman's terms how quantum theory challenges our ability to give a plausible story of what is "really going on" at microscopic scales. I have also tried to channel the spirit of John Bell, who famously advocated banning certain words from discussions of the foundations of quantum theory, and have forbidden myself from using words such as entanglement, superposition, measurement, interference, wave-particle duality and other words that are essentially just labels for things we do not understand, and which therefore carry no useful explanatory power.

This essay is about a simple question: when we zoom closely in on some small object what is in there and what is up to? You might say "I know, I've heard there are tiny particles—things like atoms, protons, electrons, photons and neutrons interacting with each other". It's a nice story on the surface, but, a bit like looking into the promises of a politician, we find when we look into the details of the story, which we do using a theory of physics called "quantum mechanics" then it all begins to fall apart. I want to explain to you how the simple story falls apart and why it matters. I'm then going to make a fool of myself by hazarding some ill formed thoughts for a potential resolution.

I should emphasize that I think stories are a vital and perhaps the only route to our understanding of the world. Narrative understanding of the world is, I believe, deeply ingrained into us, passed down from our hominid ancestors who followed this process to explain the phenomena that governed their lives.

Much of this essay is based on my inaugural lecture "Quantum Theory: It's Unreal", which can be found on YouTube.

T. Rudolph (✉)
Faculty of Natural Sciences, Imperial College London, London SW7 2AZ, UK
e-mail: t.rudolph@imperial.ac.uk

R. Bertlmann and A. Zeilinger (eds.), *Quantum [Un]Speakables II*,
The Frontiers Collection, DOI 10.1007/978-3-319-38987-5_8

A Story About Muons

Let's begin with the story of some particles called muons. You may never have heard of muons—they are some of the more exotic of the tiny charged particles that make up the universe. The story is that a few hundred of these suckers bore straight through you at very high speed every second. They are produced high in the Earth's atmosphere when extremely energetic particles called cosmic rays, which come from the far reaches of outer space, collide with the air. The muons in turn barrel through the atmosphere, through the walls of the room you are sitting in, and then through you!

However you can't feel or see these muons. So should you believe this story? Whenever we are faced with existentially vexing questions I know many of you would agree with me that it is important to turn to... alcohol. In fact you *can* directly observe these muons using alcohol in a pint glass! You can build a small version of a device called a cloud chamber: Place alcohol soaked cotton in a pint glass, cool the bottom of the glass with extremely cold dry ice, which creates a super cold vapour of alcohol that is just desperate to turn into clouds. It looks a bit like a light alcohol fog (by which I do not mean the fog in your head after you have had a few too many pints!). Watch this fog very carefully and you will see much denser streaks of clouds that look a bit like tiny shooting stars! These are the tracks that the tiny muon particles make as they pass through the pint glass. Just as water droplets in clouds use dust particles as a seed to condense on, the muons passing through knock some air molecules around in such a way they become a seed for little alcohol cloud droplets, and these trails are the result. It's similar to how jet planes leave a cloud behind them, called a contrail—because their exhaust fumes can 'seed' the formation of clouds (Fig. 8.1).

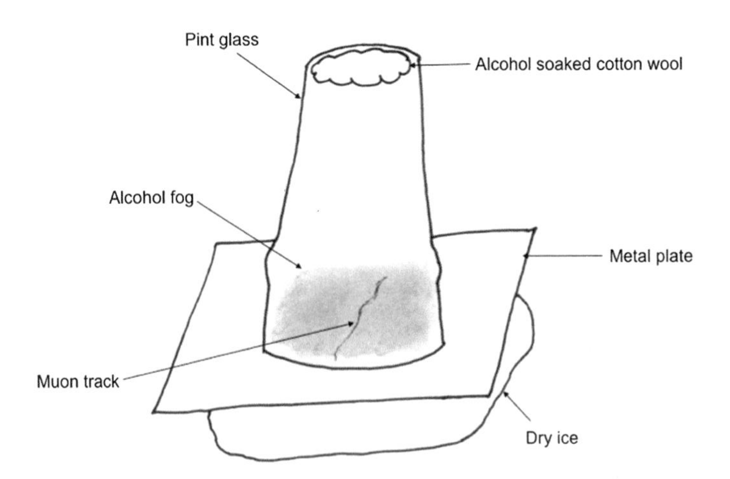

Fig. 8.1 Nic Harrigan's "muon detector in a pint glass" experimental setup. Sketch by Geraldine Cox.

Fig. 8.2 Photo of a muon track in a pint glass, taken by Pete Shadbolt.

This is a very cool experiment to do, and I recommend at least watching a video of someone doing it if you don't want to try it yourself. But, even if you do build a muon detector are you really seeing muons directly? Nope, in fact what you are observing is a vapour trail of tiny alcohol droplets, but, as small as those are, they are huge compared to the actual muons. The observations are consistent with the story I told—that there really are tiny muons flying through the glass—but ultimately you and I are monkeys whose interactions with the world around us are optimized for chasing after bananas and other monkeys whom we think look cute, not for looking at muons. So everything microscopic needs to be amplified up to the scale of bananas before we can claim to have knowledge of it (Fig. 8.2).

The Assumption of Physical Properties

This brings us to our first challenge to understanding what's happening at the small scale. To describe muons, I used words like they are massive, energetic, charged, travelling at high speed—things we would call the physical properties of muons. The assumption that things have physical properties is completely hardwired into us—consider a banana. You would describe the colour as yellow and the shape as curved and the inside as soft and squishy but the outside as firm and the banana as located in the bowl of fruit and so on. What we understand by this banana is in some sense the totality of the large scale physical properties we perceive, many of which are defined only with respect to the other things nearby and by association, the properties of those things as well.

In physics we combine these descriptions of physical properties with causal language—saying things like "eating the banana causes me to be happy", or to talk a bit more like a physicist "the banana flew into the wall, bounced off and came back and hit Terry in the head, causing him to be knocked out". OK, well my experiments never work, it's why I'm a theorist. The point is we extend this type of causal language to the small scale—"the muons knock electrons off some of the air molecules and those molecules now become little seeds for a cloud of alcohol to form upon". Such causal language—that A causes B and B causes C is extremely useful—understanding the causes of things is pretty much what science is about. This whole framework of systems that have physical properties and interact with each other causally is known as realism.

Monkey Realism

Our monkey brains build in some core assumptions about the world—one being that stuff exists and has physical properties. Another is that if I leave a banana here it stays here and doesn't spread out or fall through the table; If it disappears I'm pretty sure another monkey ate it! Another simple assumption is that if I shake a branch in my tree it doesn't instantaneously have an effect on some branch in a tree on the other side of the world, i.e. that influences have to take some time to propagate from the cause to the effect via some tangible thing, which could usually be blocked if we wanted. I am going to call this simple and natural view of the world and how it works "monkey realism". These ideas are so obvious to us, that it sounds strange that I would even point them out, but....

Quantum Realism

There is a conundrum at the heart of quantum theory, because we have discovered that it is extremely difficult to tell such common sense physical stories about what is actually going on and remain consistent with the theory and with the experiments that we do which confirm the theory. That is, if we insist that stuff happens because the little particles really exist, have real physical properties and really cause a sequence of events that culminates in our observations, we find that the descriptions of what is going on are very difficult to construct. It's not impossible, but when we do succeed we always get forced into a corner where the whole story ends up extremely weird and counterintuitive. That is, our story of the world does not end up having the sensible features at all of monkey realism. So I am going to call this weird realism "quantum realism". And unfortunately there happen to be many different options for quantum realism—all very strange (that is to say very different from the kind of monkey realism we are used to taking for granted) but also all radically different from each other.

An Intellectual Mess

Let me give you an example of the intellectual mess we can end up in. I'm going to do this by describing a simple thought experiment where everyone agrees with what would actually be observed on the monkey scale, but for which different stories about what is going on can be proffered.

If it was cold and dark in my office I could open a box and release a single particle. The simplest way of interpreting what quantum theory says goes on is the particle begins to disembody in some sense; it spreads out completely through the room. However; when I turn the lights back on and perhaps use some fancy apparatus to locate the particle, I would only find it as a single lump of stuff at one location—although every time I repeat the experiment the specific location at which I find it will be different.

If that disembodying process is real, when I decided to observe where the particle is by turning the lights on, it must be the case that it immediately springs back to all being concentrated now at one location, the point where I actually observe the particle. This is a little bit crazy—if the particle, really is a tiny lump of stuff as the name suggests, then why when we are not looking at it will it become like some ethereal ghost and spread itself out through the room, but if we do look then the spread out particle springs back instantaneously?

We call the springing back "collapse". It's important to realise that the collapse process is highly unintuitive for a physicist. Normally we would expect the physical stuff which has spread out through the room to take some time to re-morph (consider an elastic band snapping back—we would expect that one piece of band affects the piece beside it and pulls it along and this process happens continuously and takes time even if it is rapid) until all the pieces of particle stuff have somehow regrouped into the complete particle we eventually observe. Which in turn would raise the question why, if the particle really is disembodied through the room, we cannot ever see a small part of it on one side and a small part on the other?

Knowledge Quite Naturally Spreads and Collapses

So what is really going on when we do that experiment with the particle? Is it really spreading out into some disembodied gloop that can re-morph itself instantly? An alternative option we consider seriously is that what spreads out is not the particle and its associated properties, but rather our information, or our knowledge about the particle.

To get a feel for this, let us think about an experiment that does not involve tiny particles, but rather something larger, like a parrot. Imagine you close your eyes, and a friend of yours releases a parrot inside the room you are sitting. The parrot flies around quietly, and after a minute or so you really have no clue where it is. Your knowledge of the parrot's location has become completely spread out—a kind of disembodiment.

When you open your eyes and see where the parrot is, your knowledge of the parrot collapses instantly to something like "aha—the parrot is sitting right there!". Now it is not strange that the collapse process is instantaneous because it is not in some sense physical at all.

Going back to the particle, a natural thing to consider is that when I release the particle it really is somewhere, but due to my lack of information quantum theory insists I talk about it as if it becomes disembodied. Moreover in this picture it is intuitive why every time we repeat the experiment we find the particle/parrot at a different location.

To me this is very compelling. The collapse is not real, the disembodiment and spreadoutness is not real, those are things just to do with my knowledge or information about the world. It wouldn't necessarily get us all the way back to monkey realism, but it would go some of the way.

Sadly, It Cannot Just Be All About Knowledge

Unfortunately, 5 or 6 years ago, some work I did with Nicholas Harrigan opened a small crack in this story. It showed that not quite all of that disembodied spreadoutness of a particle is attributable to a lack of information; Unlike with the parrot—it is not possible to entirely take care of the problem with that approach. This was mildly disconcerting at the time, but I hoped that some sort of subtle trick could be found to avoid the nasty consequences, and that we could still have a sensible account of many weird quantum things like collapse.

Even more unfortunately, a few years later a different student, Matthew Pusey exploded the crack open and caused the dam to flood. The conclusion of that work is that, under some extremely mild assumptions, any assumption of realism whatsoever forces you into accepting that the disembodied spreadoutness is also real. This is something I hate, not least because quantum theory insists that when the collapsing happens, it is completely instantaneous. This seems a very ugly feature to have forced into our causal story of the world and it is certainly not compatible with monkey realism.

Other Ways Quantum Realism Is Weird

So now we have one example—collapse—showing that quantum realism is highly incompatible with monkey realism. There are many other examples: quantum realism allows a quantum sized banana to fall through the table, or disappear from where you left it because it instantaneously spread out and then when you looked it ended up over there, and shaking a quantum branch over here can instantaneously shake a branch over on the other side of the universe, via a completely unblockable influence. The list goes on and on into even weirder stuff.

Things get so weird that people entertain nutty notions to try and avoid these conclusions of quantum realism, things like assuming that there are bazillions of parallel universes and so on! Other serious physicists say that somehow for stuff going on at the microscopic scale that we cannot see directly, any notion of realism must be abandoned; If correct this would be a radical change to our whole view of the world and I'm not sure how we could even really do science, because the whole of scientific endeavour to date has been premised on the notion of stuff existing and having properties that we can understand. For those of you who know a bit about quantum mechanics I emphasize that the weirdness I am talking about is not because of randomness and unpredictability, things that are often thought to be the weird parts of quantum mechanics– those are completely compatible with monkey realism as the parrot story shows—the issues are much deeper.

Faced with this, my way of coping is to turn to consuming lots of alcohol in a pint glass. This consistent program of drinking has helped me consider a whole bunch of whacky ways of trying to get around the conclusion that the disembodied spreadoutness of a particle is real. But, before I outline some crazy thoughts on what might fix the problem, I want to take a bit of time to explain……

Why Thinking About These Pretty Abstract and Philosophical Things Is Sometimes Practically Important

Our brains are very good at thinking and making sense of large-scale things around us, which helps us find bananas and mates, but only under the assumption that monkey realism is true. These days we can build a device that actually helps us to think—that can help us to find those bananas and mates. I mean of course, a computer.

It is a very, very deep feature first understood by Alan Turing, that we can build our computer assuming that the world obeys monkey realism. That is, we build it out of stuff that obeys the simple principles of monkey realism and it also can do the same kind of thinking as us, as long as that thinking doesn't violate the principles of monkey realism. It's very beautiful and deserves a whole essay on its own. Don't be fooled by the smallness of the components in your computer, they are basically just bananas, they behave only in ways that are completely understandable from the perspective of monkey realism.

A really cool twist is that these days we can go and build a whole new type of computer, one built out of quantum sized systems. These microscopic systems that are doing the computing can use all the nutty features of quantum realism. Of course we monkeys have no way of understanding how they are doing what they do, because it's radically incompatible with monkey realism. But, the very cool thing is that—even though we don't know precisely how they do it—we know of a number of ways that they can be highly useful for us—that they can let us solve some problems which we have no hope of solving with our brains or with regular computers, constrained as those are to monkey-realistic principles. For example, one thing these quantum computers will be able to do is to let us design new types of pharmaceuticals, because designing new drugs is about understanding chemistry, understanding chemistry is about understanding how quantum sized particles—atoms and molecules—interact with each other, and of course monkey realism is incompatible with that, it's where the whole problem began!

On the flip side, however, it is extremely difficult for us to understand how we can use such a new type of computer, because it is operating on the different set of principles of quantum realism, whatever the heck they turn out to be. The only understanding we have obtained to date about this practical question has come from making precise and quantitative the differences between monkey realism and quantum realism. This is the remit of the field of quantum foundations and the worryings about realism that we were musing on earlier! So, here at least is one application for this abstract philosophical quest, which is why I am allowed to work in the physics department at Imperial College and haven't been shipped off to a philosophy department somewhere in Oxfordshire.

Back to Crazy Speculations About a Route to a Solution

Perhaps you don't care about the practical applications; you just want to understand what's going on better. Perhaps, like me, you aren't happy with just accepting the bizarre features of all the current proposals for quantum realism, you wonder if there isn't some hidden assumption that has trapped us into an intellectual corner like this. Well, you end up spending your weekends thinking a lot of crazy thoughts, and your colleagues begin to think you are a bit strange, never mind the particles! But I believe such crazy thinking is necessary, that we should show students it's OK to have crazy ideas, and so I will finish by describing one of the craziest I am currently thinking about. I should emphasize—I put very little faith in any of the details I discuss below being "true", whatever that means. I am essentially trying to point out that not everything has been thought about carefully yet.

To set the scene, let me remind you of the story of epicycles—when people believed that the earth was at the centre of the universe, they were forced into thinking that the planets took weird spirally trajectories around earth—because this was the only way that they could fit what they were seeing when they looked up at the planets in the sky, with the earth centred idea that they dogmatically held on to. Once they shifted view and stopped thinking of themselves as so important, they saw that things were greatly simpler if they took the perspective that everything actually revolved around the Sun.

So, what could be the hidden assumption that is leading us to the weirdness of quantum realism? Well I wonder about whether the very concepts of space and time really are more than just something useful for the purposes of chasing bananas. Perhaps space and time themselves are not relevant to the particles of the universe.

Perhaps they are useful to us and so we evolved to see them, but irrelevant on microscopic scales. A glib way of summarising this would be to say "what if space and time are like the taste of a banana—important to monkeys, but not to the basic physical properties we think are essential to obtain a realistic description of the world?".

This is a bit off-the-wall, obviously: Space and time are two of the fundamental physical properties we use to describe everything—where and when something is, is pretty much what we mean by calling a thing a thing at all.

So that you can see how much of a mess we could get into if time and space were wrong assumptions, let me give a brief—tried and tested—example, of how things can get messy if we choose the wrong fundamental properties out of which to construct a theory. Consider the theory called thermodynamics that's been around for some time and is still going strong. This was the theory that gave us steam engines, and the physical properties that this theory uses are things like heat, power, temperature, pressure, entropy and so on. We now understand this theory in terms of a deeper theory of atoms bouncing around in a container. But, in that deeper theory it makes no sense to assign those atoms the same physical properties that were important at the larger scale—we don't say things like "each atom has a pressure, a temperature, a work, an entropy" and so on. If you tried to understand the one theory using the same properties that are relevant in the other you potentially end up with nonsense.

How to Construct a Theory with a Different Space/Time Experience?

I don't really have any clue how to construct a theory in which the space and time experienced by fundamental particles, if they experience it at all, is different to that experienced by us. I have one simple example that I think is cute and which at least indicates how things could be very different—even if it isn't the way that they do actually turn out to be different.

Previously I described an experiment where I could let a particle go and it would spread throughout the room. Let me describe a slightly different experiment which operates similarly: We release a particle at one end of a narrow bench, wait exactly 1 s and then look for it. We arrange things so it can only be found somewhere along the length of the bench.

As before, we don't always find the particle in the same place. Although we prepare the experiment the same every time, we only ever find the particle in certain equally spaced locations along the bench and we never find it in the in-between these locations. We can never be sure exactly, at which of the equally spaced spots we will find it, that is just the randomness of quantum theory, but more strangely if we never find it in-between does it mean it never actually crosses those points?

Fig. 8.3 A sketch by Geraldine Cox of the view from the old man's window.

Let me tell you a story, a parable as it were, which is meant to illustrate how—if our concept of time is not compatible with the time experienced by particles, this behaviour might not be so strange. Things didn't work out so well for the last guy who spent time telling parables about 2,000 years ago, so I'm a bit nervous.

The story goes that a Malawian boy who lives on the shores of Lake Malawi has the job of, every morning at the same time, picking up fish from the fishermen at the lakeshore and running them through the village to his father's store. There is an old man in the village who owns a clock. It is one of those round-dialled clocks, and because this is Africa and pretty much everything there is broken, the clock only has one hand. Every day when he wakes up he looks at the clock and he sees the one hand is pointing at the number one. He believes this means he wakes up at the same instant every day. When he wakes up he looks out the window and he always sees the boy running on the path past his house. He notices after a few months that he only ever sees the boy in one of a few locations along the path and never in-between (Fig. 8.3).

What is a natural explanation of the old man's observation? Is it to say the boy disembodies along the path into some kind of wavy thing that re-morphs only at certain special places? No—it is to say the old man is mistaken about the relationship between the time on his clock and the time experienced by the boy. If it is the case that some days when he sees the number one on the clock it has in fact already done a full revolution before pointing to the number one, and on other days it has done two or three or more revolutions first, then it is quite natural that (assuming the boy really does set out at the same time by his reckoning and runs at constant speed) the old man only ever sees the boy at discretely spaced locations. By understanding that there is a many-to-one relationship between the old man's time and that of the boys, something which seemed to require a quantum-realistic type of explanation becomes more amenable to an monkey-realistic explanation. I emphasize that this proposal is meant to be illustrative; I have only marginal evidence that these types of explanations are tenable more generally. But since this is meant to be a parable, I should say the end of the story is a sad one; That the old man accused the boy of being a witch, and belief in witchcraft still being very strong in Malawi, the boy was burned at the stake. I hope the same fate does not await me when any of my colleagues read these crazy thoughts!

Chapter 9
The Universe Would Not Be Perfect Without Randomness: A Quantum Physicist's Reading of Aquinas

Valerio Scarani

Abstract Randomness is an unavoidable notion in discussing quantum physics, and this may trigger the curiosity to know more of its cultural history. This text is an invitation to explore the position on the matter of Thomas Aquinas, one of the most prominent philosophers and theologians of the European Middle Ages.

Introduction

Physical determinism has been a powerful methodological assumption in science since the dawn of the modern era. It is so engrained in our culture, that the sentence "to find a scientific explanation" usually means "to point to an antecedent situation, which existed (or can reasonably be assumed to have existed) in the past, from which the present situation follows by law of necessity". Since the first half of the 20th century, physical determinism has been famously challenged by quantum physics, with the theorem of Bell that we are celebrating here as one of the milestones. But neither 50 years of a theorem, nor a century of quantum physics, can easily dispose of an intellectual option that dates back at least to the time of Democritus and Lucretius, and has deeply shaped the last few centuries of human thought. The urge for physical determinism explains why many laymen are still associating quantum physics with some esoteric quest rather than with the most successful scientific endeavor of all times, popular journals thriving on the misconception by suggesting that the latest research paper may be the one lifting the

---

This text has been written for the Proceedings of the Conference Quantum [Un]Speakables II: 50 Years of Bell's Theorem (Vienna, 19–22 June 2014).

V. Scarani (✉)
Centre for Quantum Technologies and Department of Physics,
National University of Singapore, Singapore 117543, Singapore
e-mail: valerio.scarani@gmail.com

R. Bertlmann and A. Zeilinger (eds.), *Quantum [Un]Speakables II*,
The Frontiers Collection, DOI 10.1007/978-3-319-38987-5_9

veil. Determinism appeals to the specialists too[1]: for those who support Bohmian mechanics or the many-worlds interpretation, determinism is a cornerstone that should not be removed; one should rather abandon other supposed features of the physical world and of our assessment of it.

Surely, chance and randomness in their various meanings have also had their supporters in the recent scientific and philosophical debates. Nevertheless, I was intrigued when I chanced over the following statement in Thomas Aquinas' *Summa contra gentiles* (CG), most probably written around the years 1260–64:

> **it would be contrary to [...] the perfection of things, if there would be no chance events.**
>
> **(CG, Book 3, Chapter 74)**

Which arguments in favor of chance as a feature of nature could Aquinas bring up, several centuries prior to Darwinian evolution, deterministic chaos and quantum physics—and obviously without any hint thereof?

I decided to write about this text and a few related passages by the same author[2] for these proceedings. I knew it would be a challenge: in the words of a philosopher friend of mine, I opened a window and discovered an ocean, for whose exploration I am not well equipped. The purpose of this text is not to claim a place among the explorers, but to invite others to come at the window.[3]

Aquinas on God's Providence

The third book of the CG is devoted to "Providence", which is *God's governance of creation*. As an orthodox Christian philosopher and theologian, Aquinas had to juggle to accommodate both an all-powerful God and really free human beings. Alongside this extremely important anthropological question, certainly still debated today,[4] comes a more neutral but general observation of nature: all the beings that we perceive are limited in their being. Aquinas integrates both considerations in a response that rings surprisingly modern: not only "human free will", but *the whole*

---

[1]For a recent perspective on determinism by a quantum colleague: L. Vaidman, Quantum theory and determinism, Quantum studies: mathematics and foundations **1**, 5–38 (2014); http://link.springer.com/article/10.1007/s40509-014-0008-4.

[2]Interestingly, a quantum colleague, who contrary to me cannot be suspected of Catholic leaning, has also advocated recently a re-discovery of Aquinas' thought: D.M. Appleby, Mind and Matter, http://arxiv.org/abs/1305.7381. This text sketches a much more ambitious program than mine here; Aquinas' philosophy is proposed as a possible way of avoiding Cartesian dualism.

[3]A very convenient summary of Aquinas' philosophical work is available as: Aquinas, *Selected philosophical writings* (Oxford World's Classics, Oxford University Press, 2008). The Latin text and English translation of most of Aquinas works can be found online in http://www.dhspriory.org/thomas/. When available, I'll give the link to this website.

[4]See e.g. http://plato.stanford.edu/entries/providence-divine/; Aquinas' position is discussed in paragraph 6.

creation, including its material aspect, possesses a relative autonomy from God. This autonomy is going to be the foundation for the discussion of "fortune and chance".

If the reader is yawning at the apparent banality of this observation, they have better wake up quickly because the statement is not trivial at all. In fact, if prompted to define "autonomy from God", we tend to refer to the *cosmology*[5] of the time of Leibniz and Newton. God's role is that of a watchmaker who builds a perfect mechanism, which later evolves "autonomously"—meaning here, according to the deterministic laws of classical physics. The autonomy of the watch from the watchmaker is indeed the fitting setting for physical determinism.

Aquinas lived at a time where the watch had not been invented yet. The cosmology of his time is twice removed from us, transiting as we are to a new cosmology of evolutionary flavor. Aquinas' metaphor for the universe is that of an orderly kingdom[6]:

> Political life offers a parallel: for all the members of the household are ordered to one another by subordination to the master of the house, and then that master and all other masters of households in a city ordered to one another and to the ruler of the city; and he with all his fellows in a kingdom, ordered to the king.
>
> (CG 3, 98)

This cosmology contains two elements that are currently banned from much of the intellectual discourse on nature. These are *finality* and the existence of a *hierarchy of beings*.

Concerning *finality*, or finalism, I feel incapable of providing an even moderately competent discussion[7]; however, there is a point that I want to bring to the attention of the reader. The absence of finalism was a defining feature of the atomism of Democritus and Leucippus, as noticed explicitly by Aristotle. Early in his career, Aquinas writes about this[8]:

> First, we have to know that some stated that there is no providence for anything, that everything happens by chance: this was the position of Democritus and the other ancient authorities who denied agent causes and affirmed only material causes. But this position has been sufficiently refuted in philosophy.
>
> (*Scriptum super Liber Sententiarum* I, d.39)

---

[5]In this text, "cosmology" will be used in the sense of *Weltanschauung*, not in the sense of the discipline of physics that studies the universe at large.

[6]http://www.dhspriory.org/thomas/ContraGentiles3b.htm#98. I am citing from the more readable translation of Timothy McDermott in the Oxford book cited above.

[7]For a text that mentions Aquinas extensively and provides a glimpse of the complexity of the issue, see http://inters.org/finalism, paragraphs I and II.

[8]I am grateful to V. Cordonier for sharing with me her text "La doctrine aristotélicienne de la Providence divine selon Thomas d'Aquin" [in: P. D'Hoine, G. van Riel (ed.), Fate, Providence and Moral Responsibility in Ancient, Medieval and Early Modern Thought. Studies in Honor of Carlos Steel (Peters, Leuven, 2014) pp. 495–515, in which I found this very relevant citation (footnote 14). My translation.

A doctrine that "affirms only material causes": to our ears, this would make of Democritus a precursor of determinism, and as such he is indeed presented in philosophy textbooks. Aquinas' highlights rather the fact that, in such a doctrine, "everything happens by chance", that is "without any providence"—with a more modern twist, we may say "without meaning". Atomism and determinism have gone hand-in-hand for centuries—till quantum mechanics cast its shadow on the idyll. Is it a full-scale betrayal or a test that will further consolidate the relationship? This is what we are still debating.

Let us now discuss the *hierarchy of beings*. In his theory of knowledge, Aquinas considers "being" as the first we grasp[9]: that is, before knowing that it is (say) a tree, or that it is in that place, or that it is good for us, we know that it "is". From the perspective of our knowledge, thus, the existence of various beings is not a truth to be derived, but the starting point. In CG, Aquinas rather takes God's creation as starting point, and argues why such a creation must consist of multiple finite beings[10]:

> Now, created things must fall short of the full goodness of God, so, in order that things may reflect that goodness more perfectly, there had to be variety in things, so that what one thing could not express perfectly could be more perfectly expressed in various ways by a variety of things. [...] And this also draws attention to how great God's perfection is: for the perfect goodness that exists one and unbroken in God can exist in creatures only in a multitude of fragmented ways.
>
> (CG 3, 97)

My last introductory comments will be on the following citation from the same chapter:

> Clearly then, the dispositions of providence have their reasons, but reasons that presuppose God's will. All this allows us to avoid two kinds of mistakes. First, the mistake of those who believe everything comes from the simple will, devoid of reason [...]; and secondly, the mistake of those who say the causal order is a necessary consequence of God's providence.
>
> (CG 3, 97)

The first mistake, that Aquinas is convinced of having disposed of, is the belief that there is no rationality in the world, that everything is pure arbitrariness from God's decisions.[11] The second mistake addresses our concern. It is beyond human

---

[9]It is acknowledged that this is a foundational element of Aquinas' theory of knowledge. If prompted to find a citation, probably the most famous is *primum enim quod in intellectum cadit, est ens* (*De Pot.* 9, 7, ad 15 http://www.dhspriory.org/thomas/QDdePotentia9.htm#9:7).

[10]http://www.dhspriory.org/thomas/ContraGentiles3b.htm#97; I am citing again the translation from McDermott.

[11]Aquinas was aware that some Arabic philosophical schools of Andalusia promoted this doctrine. One century after Aquinas' death, it was going to be championed again, this time in the Christian world by William of Ockham. It still lurks behind many anti-scientific attitudes of our times.

capacity to assess the purpose of God,[12] which is the ultimate final cause; but whatever it is, it does not determine a unique unfolding[13] because created beings can act as causes, including final (i.e. they can act for some purpose that is *their* purpose, not God's). This is the "relative autonomy" that was mentioned. Some beings being more perfect than others, their causality is also of different power: in particular, it may fail to produce the desired effect. This brings us to the main objective: a discussion of the case for "fortune and chance".

The Case for Fortune and Chance

In our main reference, CG 3, 74, Aquinas lists five arguments to defend the thesis that fortune and chance are compatible with God's providence.[14] I rephrase them in my words:

1. If nothing rare would happen, we would conclude to necessity. Thus "fortune and chance" are the manifestation of contingency, which is God's respect of the autonomy of created beings.
2. The second argument combines finality and finiteness: all beings act for an end, but finite beings may fail with regards to the intended end, thus bringing about unintended effects.
3. The third argument is different: it is the classic *concursus causarum*. Since God does not determine everything and each being has its own autonomy, it is possible that initially independent causal chains collide to produce an unexpected effect. The example of Aquinas is more than clear: "For example, the discovery of a debtor, by a man who has gone to market to sell something, happens because the debtor also went to market".
4. In yet another chance of perspective, the finite beings are no longer considered as agents, but as beings, whose properties are not all necessary. The actual text, a scholastic demonstration, sounds very convoluted to us; so let me try my own example. A given woman is a human being, is tall, is dressed in blue, and is a physicist. "While being human" is obviously essential, the other features look accidental and uncorrelated among them— but who knows, maybe there is a deep common cause for all the features of this woman? Aquinas argues that

[12]Aquinas inherited the "negative theology" of Pseudo-Dionysius: we can't know anything of God's plan, besides what He chooses to reveal to us. The Revelation accepted by both deals with the finality set by God for human beings, but says close to nothing about that for the material world (indeed, basically all that Christian belief has to say on the final destiny of matter is that the final destiny of humans does involve a material element, in the following of Jesus, who resurrected with what we could call an "upgraded version" of his own body).

[13]Notice again the discrepancy with the later cosmology: Liebniz argued that the God-watchmaker must have created the best possible world in all the details of its gears.

[14]http://www.dhspriory.org/thomas/ContraGentiles3a.htm#74 .

against such a higher causality: it is proper of finite beings to have indeed many accidental features.

5. The fifth argument has some flavor of the second and the third: the power of a finite cause is necessarily finite and therefore cannot extend to all things that can happen.

A comprehensive commentary is beyond my capacity, since I have not dived into the ocean. But there is another island visible from the window, which is impossible to ignore. Indeed, fortune and chance had already been discussed by Aristotle in his *Physics*, Book II, Chaps. 4–6. After having introduced his classification of four causes, Aristotle discusses the opinion that fortune, chance or "spontaneity" are also causes; he reasons that all these are indeed real, but are not proper causes, thus justifying his previous classification. More or less in the same years[15] as CG, Aquinas wrote a commentary to this work, the *Commentaria* [or *Expositio*] *in Octo Libris Physicorum*. A scholastic commentary was in fact a series of lectures reviewing the text of an authority point by point. It is not easy to follow, insofar as the commentator can be critical of one statement and will nevertheless go ahead with the commentary of the next. I just want to point to a few points that shed some light on the text from CG.

- Even if it comes unexpectedly last, I want to mention first that "chance" and "fortune" are defined in Aristotle's Chap. 6, commented by Aquinas in his 10th lecture.[16] Fortune (misfortune) implies happiness (sadness) and thus is proper of beings that can experience happiness; whereas chance is a neutral word that applies to all beings. The fact that these definitions are elaborations on the common meaning of those words may explain why Aquinas did not define those terms in CG.
- The example of the two men meeting by chance in the market, which we saw in argument #3 above, comes directly from Chap. 4 of Physics. As stressed by Aquinas,[17] the example here is meant to show that "fortune" is certainly not always a cause: indeed, here one would speak of fortune (especially for the creditor), but the cause of each person going to the market was not "fortune", it was "to buy something".
- At the beginning of Chap. 5 of *Physics*, Aristotle makes what we may call nowadays call a "phenomenological study" of the cases in which chance or fortune are invoked as causes. It opens with an observation similar to the one in argument #1 above: one speaks of chance when things happen rarely. In his Lecture 8,[18] Aquinas writes: "it seems that this division [in things that happen

[15]The most probable date is around 1268, which would put it some 5 years after the most probable date for CG—anyway, all that matters here is that the two texts belong to the same period, so that in first approximation we can assume them to be consistent with each other.

[16]*In Physic.* II, l.10 (http://www.dhspriory.org/thomas/Physics2.htm#10).

[17]*In Physic.* II, l.7 (http://www.dhspriory.org/thomas/Physics2.htm#7).

[18]*In Physic.* II, l.8 (http://www.dhspriory.org/thomas/Physics2.htm#8).

always, frequently or rarely] of the Philosopher is insufficient, for there are some happenings which are indeterminate". From what I understand, "indeterminate happenings" (*contingentia ad utrumlibet*) refers to events whose frequency cannot even be defined.

Message in an Old Bottle

In these notes from the window, I tried to grasp Aquinas' effort of rationalization. It's a message in a bottle from a cultural world that is no longer ours: in particular it would be grossly anachronistic to read Aquinas as a precursor of quantum physics.[19] But we are allowed to read the message and derive some inspirations for our times.

Aquinas' study is very far from a naive god-of-the-gaps argument, which would run: "there are things I cannot predict, the only possible explanation is to invoke the intervention of God or some other spirit". Chance and fortune are neither God's doing nor the devil's: they are the manifestation of the finiteness of created beings, and of the autonomy that God's providence gave them. Since this autonomy is a sign of God's respect for his creation, chance and fortune are to be considered positive realities.

I want to stress that the existence of God is not an assumption for the argument. As we said, Aquinas does believe in an all-powerful God, and his challenge was to present a doctrine of providence that does not end up in determinism. One could make a God-free case for randomness along similar lines, as long as one accepts the existence of finite autonomous beings. This is far from universally accepted: many philosophies and mysticisms around the world and across the centuries have upheld the doctrine that behind the appearance of a multitude of beings there is only one Being.[20] Such "holistic" or "pantheistic" doctrines may have their own way to deal

[19]Authors like Heisenberg, Jauch and Piron, have used the wording of "potency and act" in their attempts to appraise quantum physics. Inspired by this, some years ago I browsed extensively Aquinas' works to see if a hint of the quantum could be found there: I can say with high confidence that such is not the case. Let me give an example. For Aquinas (and Aristotle), the statement "I am in potency of being at B" means that I actually am at A, and by motion I could exchange my "being at A" with "being at B". In no way they had thought of "not being actually localized anywhere", which is what Heisenberg, Jauch and Piron were aiming at—as to whether this extension is legitimate and successful, I am skeptical but with no strong feelings.

[20]When speaking of "holism" or "pantheism", Indian-born religions come immediately to mind, but similar hints can be found even in Plato. Very relevant for our story is the fact that Averroes explicitly commented Aristotle in a holistic sense, making him very suspicious in the Christian world: the "redemption" of Aristotle from that interpretation was arguably Aquinas' greatest challenge (see e.g. G.K. Chesterton, *Saint Thomas Aquinas, the "Dumb Ox"*, several editions). In later times, some humanists will promote again the doctrine of an *anima mundi*, and Spinoza will champion a renewed form of pantheism. Presumably some members of the Church of the Larger Hilbert Spaces have a similar doctrine in mind.

with freedom and chance, of which I know little, but certainly they won't follow Aquinas' path.

I want to conclude on a recollection from the conference, whose proceedings you are reading. I happened to give two talks there: the first one was my own, the second one was the one of Nicolas Gisin, who had been retained in Geneva by urgent family matters. This second talk led to a broad discussion, during which Anton Zeilinger dropped the suggestion that a better appreciation of quantum physics may pass through a rediscovery of *finality*. There and then, I thought of Aquinas. Now, after reading CG 3, 74, it's easy to recall that Zeilinger and Brukner have also promoted the explanation of quantum randomness as a consequence of the *finiteness* of the information that can be stored in a quantum system. Surely, if all nature cares is that some statistics are respected, the concrete way to get there may be left to "chance". Happy as I am with this argument, I feel it does not explain why nature wanted the statistics to violate Bell inequalities. Maybe God wanted us humans to be able to *certify* intrinsic randomness?

Acknowledgments This text has greatly benefitted from discussions with and feedback or encouragement from: Antonio Acín, Michael Brooks, Jeff Bub, Jonathan Chua Yi, Valérie Cordonier, Artur Ekert, Berge Englert, Nicolas Gisin, Jenny Hogan, Matthew Leifer, Antoine Suarez, Giuseppe Tanzella-Nitti and Anton Zeilinger. My research on randomness is sponsored by the Ministry of Education, Singapore, under the Academic Research Fund Tier 3 MOE2012-T3-1-009 "Random numbers from quantum processes".

Chapter 10
Bell's Theorem Tells Us *Not* What Quantum Mechanics *Is*, but What Quantum Mechanics *Is Not*

Marek Żukowski

Abstract Non-locality, or quantum-non-locality, are buzzwords in the community of quantum foundation and information scientists, which purportedly describe the implications of Bell's theorem. When such phrases are treated seriously, that is it is claimed that Bell's theorem reveals non-locality as an inherent trait of the quantum description of the micro-world, this leads to logical contradictions, which will be discussed here. In fact, Bell's theorem, understood as violation of Bell inequalities by quantum predictions, is consistent with Bohr's notion of complementarity. Thus, if it points to anything, then it is rather the significance of the principle of Bohr, but even this is not a clear implication. Non-locality is a necessary consequence of Bell's theorem only if we reject complementarity by adopting some form of realism, be it additional hidden variables, additional hidden causes, etc., or counterfactual definiteness. The essay contains two largely independent parts. The first one is addressed to any reader interested in the topic. The second, discussing the notion of local causality, is addressed to people working in the field.

PACS numbers: 03.65.Ta · 03.65.Ud

Introduction

In the first part I shall present a simple version of the Bell's theorem based on the works of GHZ [1] and Mermin [2], and its relation to the EPR paradox [3]. Simple but not oversimplified, I hope. The aim of the presentation will be to show that the EPR work does not contain a complete analysis of the problem, and some crucial assumptions are not overtly expressed therein, but can nevertheless be pinpointed.

M. Żukowski (✉)
Institute of Theoretical Physics and Astrophysics, University of Gdánsk,
PL-80-308 Gdánsk, Poland
e-mail: marek.zukowski@univie.ac.at

R. Bertlmann and A. Zeilinger (eds.), *Quantum [Un]Speakables II*,
The Frontiers Collection, DOI 10.1007/978-3-319-38987-5_10

The whole section is formulated in as a counter-factual story about EPR attempting to write their paper, but learning about GHZ results before finishing it. I have chosen GHZ paradox, instead of Bell's original work "On the Einstein-Podolsky-Rosen" paradox [4], only because GHZ-type reasoning is very accessible, and does not require any knowledge about statistical methods, or probability. The presentation of the situation is for laypersons. It will require only elementary algebra, and logical intuition. The second section is also elementary, however it is addressed to scientists interested in Bell's theorem.

All what I present here is not new. It is a concise presentation of the basic facts about quantum theory, Einstein-Podolsky-Rosen Paradox (EPR) and Bell's Theorem. The aim is to reveal logical inconsistency in attempts to claim that the sole assumptions of Bell's theorem are locality, freedom to choose the setting of experimental devices, and quantum predictions. Some authors following such views claim that no other assumption is needed at all (currently some claim that it is enough to assume local causality, which is correct, but they seemingly treat it as the opposite notion to non-local causality, which not the case). Some claim the missing assumption of hidden variables or realism, or counterfactual definiteness [5] is irrelevant because it can be replaced in the derivations of Bell's inequalities by determinism, which in turn is derivable via an EPR type reasoning [6]. This will be shown to be impossible in the first part of this essay.

The second section concentrates on showing that the assumption of *local causality* is effectively a version of local hidden variable theories. What is important, its negation is *not* non-local causality. As local causality is a compound condition, it may not hold because of several reasons. This may be non-locality or non-existence of causes of elementary quantum events which are outside of the quantum description (that is, that the only predictive tool describing given situation is the quantum state, and quantum formalism allowing, via some kind of form of Born rule, to calculate the probabilities). Thus the violation of Bell inequalities, which can be derived using freedom assumption and local causality, does not indicate whether quantum mechanics it is local or non-local. The property of non-signalling indicates that quantum mechanics is local. Some claim that the collapse of the wave packet postulate (projection postulate) points to non-locality. However, the wave function has real status only in realistic interpretations. In (neo?) Copenhagen Interpretation, which is, admittedly, different for every apostle of it, this is just symbolic mathematical tool allowing predictions for a system which is a member of an ensemble defined by the preparation procedure. At least Bohr himself says: "actual calculations are most conveniently carried out with the help of a Schrodinger state function, from which the statistical laws governing observations obtainable under specified conditions can be deduced by definite mathematical operations. It must be recognized, however, that we are dealing here with a purely symbolic procedure, the unambiguous physical interpretation of which in the last resort requires reference to the complete experimental arrangement." [7] Note also, that the projection postulate within Qbist Interpretation is just the quantum Bayesian update [8], thus cannot be treated as a non-local real phenomenon.

Paradoxes of Interpretations of the EPR Paradox: Elementary Presentation

Let me now present a reasoning which is based on some quantum predictions, and the assumptions of EPR which led Bell in 1964 to formulate his original inequalities, which in turn led to his Theorem. In a way this will be an intellectual game. Let start just like EPR did in 1935, but let us consider the fascinating properties of three photons which have maximally entangled polarizations, discovered 54 years later by Greenberger, Horne and Zeilinger (GHZ). Let us see whether EPR can reach the same conclusions as in their 1935 paper, where they considered a completely different state. Let us also see whether Bohr was right in his 1935 criticism of the EPR paper [9].

Consider first, the physical situation described by GHZ. Imagine that we have a source, which in a single run of the experiment, when excited, emits three particles, each in a different direction, such they reach three observers (for a history of efforts to actually build such source with quantum optical techniques see [10]). The three observes Danny, Mike and Anton are very very far away from each other. The particles reach their labs at the same moment of time. Spin is a kind of internal angular momentum of a particle. Angular momentum is a vector, thus spin must be a vector too. But this is a very strange vector. In the case of photons if we measure any of its components the results are always $+1$ or -1 (times Planck Constant dividend by 2π, but this we shall treat as an irrelevant detail). By the way, polarization of light is a consequence of the fact that photons are massless "spin-1" particles, and measurements of the spin components of photons are in fact specific measurements of their polarization.

After very many repetitions of their experiments, on a big enough statistical sample of the emissions of the particle-triples, the three observes, if they inform each other about the results of their measurements and the settings of their measuring local apparatuses (which is decided randomly by the whim of each local observer separately, in each run of the experiment), notice the following phenomena.

- If two of them decide to measure the vertical component of the spin of the arriving particle (its velocity is assumed to define the third direction in space for the local observer), and the remaining third observer (whoever he is) chooses to measure the horizontal component (at the right angle with the previous one, and perpendicular to the velocity of the local particle), then the product of their local results is always -1. That is either all of them get -1 or for two of them the results are 1 and for one of them -1.
- However in such situations. as well as in the other one considered below, each observer, before he exchanges the information with his partners, sees all his ± 1 local results as fully random, following the statistics of fair coin toss. This is so irrespectively of the local setting of the measuring device (that is, of which component is measured).

- If by chance all three decide to measure the horizontal components of local spins, the product of their local results is always 1: either all of them get 1, or for two of them results are -1, and for one of them is 1.

We have considered four experimental situations. EPR would notice here that in each such a situation any pair of observers would be able to predict with certainty what would be the result of the third one, if he also chooses a setting which allows the correlations described above.

- For example Anton and Mike measure both vertical components, and each receives the value -1. It they exchange the information about their results, they would know that if the choice of Danny was to find out the value of the horizontal component, his results must be -1 (as the product of the three local results must always be -1).
- However, if Anton and Mike measure vertical and horizontal components, respectively, and they receive, say, -1 and 1, they would know (after information exchange) that if the choice of Danny is to measure the vertical component, he would definitely receive -1.

Elements of reality. This is the moment at which EPR could enter with their definition. They did say: "If, without in any way disturbing a system, we can predict with certainty (i.e., with probability equal to unity) the value of a physical quantity, then there exists an element of physical reality corresponding lo this physical quantity." Additionally they stressed that "every element of the physical reality must have a counter part in the physical theory". Notice that all four situations discussed above are cases in which each result of each observer can be predicted with certainty by other two observers. E.g., in the first displayed example Mike and Anton can fix the value of Danny's element of reality pertaining to his possible measurement result for horizontal component. Thus this possible result must be "part of the physical theory". However, the second example fixes the value of the vertical component of spin of Danny's particle. It is also potentially an element of reality.

EPR realized that "one would not arrive at our conclusion if one insisted that two or more physical quantities can be regarded as simultaneous elements of reality only when they can be simultaneously measured or predicted. On this point of view, since either one or the other, but not both simultaneously, of the quantities [horizontal spin component] and [vertical component] can be predicted, they are not simultaneously real. This makes the reality of [horizontal component] and [vertical component] depend upon the process of measurement carried out on the [two other] system[s], which does, not disturb the [third] system in any way. No reasonable definition of reality could be expected to permit this."[1]

Thus the claim is that any "reasonable definition of reality" would treat the values of horizontal and vertical components of the spin as two elements of reality, which

[1]The phrases in brackets indicate the text of EPR transformed in such a way so that it fits the considered three-particle example. Their Q (position) is now horizontal component of the spin, and P (momentum) is the vertical one.

are missing in the quantum theory. Thus quantum theory in incomplete. This is where the argument of EPR ends.

Bohr in his 1935 reply said "... there is essentially the question of an influence on the very conditions which define the possible types of predictions regarding the future behavior of the system... In fact, it is the mutual exclusion of any two experimental procedures, permitting unambiguous definition of complementary physical quantities, which provides room for new physical laws the coexistence of which at first sight appear irreconcilable with the basic principles of science."

Thus, the trouble is that in the considered examples, leading to elements or reality, Mike is to measure either vertical or horizontal components of the spin of his particle. However no experimental device can measure the two components simultaneously (as such a device must must have the property that it singles our a specific direction, linked with the measured spin component; there is no way to *single* out *two* directions). The two components are complementary, they are non-commensurable.[2]

So, who is right EPR of Bohr? The statement of EPR is equivalent to treating two situations, the actual one (say, Mike in a given run of the experiment, on a specific particle, measured vertical component) and the potential one (Mike could have measured instead the horizontal component), on equal footing. Replacing the final "this" in the quoted paragraph of EPR by what is actually meant by it we get: "No reasonable definition of reality could be expected to permit [...]" "that two or more physical quantities can be regarded as simultaneous elements of reality only when they can be simultaneously measured or predicted." Thus EPR effectively treat the actual and potential different situations [measurements] on equal footing.[3] Bohr definitely does not.

Who is right? Let us return to the GHZ predictions for three spins. As all possible combinations of the results which are consistent with the quantum predictions are equiprobable (recall that values obtained by each observer have the statistics of a fair coin toss), the EPR elements of reality for a specific run of an experiment and the first three potential situations (in which only one of the observers measures the horizontal component, other two measure vertical) can be all -1. Why not? Their product is -1, everything is OK. However, this means that if any one of them measures the horizontal component, and if the notion of elements of reality is internally consistent, and

[2]The mathematical formalism of quantum mechanics reflects complementarity of pairs "observables". If this is the case say for observables A and B, e.g. describing two different components of spin, then they "do not commute". This is turn means that in the formal quantum description "operators" associated which the two observables we have the following property: $AB \neq BA$. Of course complementarity can occur in various degrees. We have perfect complementarity when experiments measuring B give completely random results for quantum systems prepared in any state, which was prepared by measuring A and selecting only systems which gave the same result of this measurement. For example, photons which are selected by a polarization analyzer which allows only linearly polarized photons to pass through it, would upon subsequent measurement of circular polarization give fully random results. Either clockwise or anti-clockwise polarized photons would appear, with equal probabilities. like in a coin toss.

[3]Such an approach accepts so called "counterfactual" statements or conditionals. Such statements contain an "if" clause which describes a situation which in fact did not occur: e.g., "If EPR knew the results of the GHZ paper, they would not have written their 1935 work".

counter-factual reasonings allowed, then the result must be -1 (as *all* results mentioned above were assumed to be such, and one of them in all three situations pertains to measurement of the horizontal component by one of the observers, in each case a different observer). This implies that if by their own whim in the considered run of the experiment accidentally all choose to measure horizontal components (the fourth situation), the product of their results must be -1. However quantum mechanics predicts the product to be in such a case *always* $+1$, see above. The reader may check all other combination of possible results which agree with the three first situations considered in our example. This invariably leads to the the same prediction for the fourth situation: if elements of reality are to be a consistent notion then the product must be -1.

Thus if the thesis of EPR holds then $1 = -1$. As the complementary nature of horizontal and vertical spin components prohibits the reasoning of the previous paragraph: if Bohr is right one still has $1 = 1$. If for a given photon Mike measures the horizontal component, he is not allowed to even speak about the possible value for the complementary measurement.

This is the end. Where here is non-locality? One could try to argue that non-locality is the solution of this conundrum, provided one insists that counterfactual situations can be treated on equal footing with the actual ones. Then the fact that Mike and Anton choose to measure horizontal component could, by "a spooky action at a distance", flip the value of Danny's element of reality for the horizontal component to $+1$. Of course, such a non-locality would clearly contradict EPR argument as: "This makes the reality of [horizontal component] and [vertical component] depend upon the process of measurement carried out on the [two other] system[s], which does, not disturb the [third] system in any way. No reasonable definition of reality could be expected to permit this."

Thus non-locality cannot be derived via an EPR-type reasoning, just as the elements of reality (and thus determinism) are not derivable. The bad luck of EPR was to consider in their work the specific quantum state and specific "observables", for which elements of reality seem to be a consistent notion (the original EPR state was a state of two particles with total momentum equal to zero). In 1964 Bell took the simplest possible two spin entangled state, the so-called singlet, and showed that if we accept EPR reasoning and thus elements of reality, then the bound for his original inequality for elements of reality does not hold for quantum predictions (i.e., the inequality is violated). However his reasoning was not as simple as the one for GHZ states, which allows to reveal directly the fallacy of EPR ideas.

The moral of this story is that with EPR-type reasoning it is *not* possible to derive determinism, elements of reality, etc. This is so despite the fact that the tacit assumption of EPR is counterfactual definiteness, which is effectively a rejection of complementarity, as in counterfactual reasonings we can talk about values of measurements which were actually *not* done. So the situation is really tragic. The EPR reasoning, which originally already assumed a form of realism (counterfactual definiteness) when confronted with Bell and GHZ reasonings cannot lead us to a consistent notion of elements of reality, as they imply $1 = -1$. All this invalidates one of their assumptions. If we carefully enumerate the assumptions they are: "free will", counterfactual definiteness, locality and quantum predictions. Their conjunction must be false. As quantum mechanics has not been as yet falsified in any experiment, and "free will" is rather indisputable, it is locality and/or counterfactual definiteness which must be abandoned. But which—there is no answer.

Additionally, we see that EPR aim to falsify complementarity cannot be reached once we consider GHZ correlations. They wrote "Starting then with the assumption that the wave function does give a complete description of the physical reality, we arrived at the conclusion that two physical quantities, with noncommuting operators, can have simultaneous reality." However, their effective assumption of counterfactual definiteness directly implies rejection of complementarity. Thus, the reasoning does not lead to any progress in this question (effectively they show that counterfactual definiteness prohibits complementarity, which is a tautology). Still, with their example, of two particles with vanishing total momentum, everything seems to be internally consistent, although... circular. However, had they considered the GHZ correlations, they would have not been able to reach the consistency, as in such a case we are led to $1 = -1$, which is a pretty inconsistent statement in elementary algebra.

Local Causality

The above fairy tale shows that any attempt to derive determinism, or realism via EPR correlations, is futile (as just one counterexample is needed to disproof such a claim).[4] Thus realism of a form must be separately assumed in any derivation of Bell inequalities. This means that violation of such inequalities is a falsification of a compound assumption which is a conjunction realism of sorts (at least counterfactual definiteness), locality (no action at a distance, constraints of relativistic causality working), and free will (settings decided at the whim of the local observer, or existence of stochastic processes which decide the local settings, which can be statistically independent of any other factor in the experiment).[5]

Still, there is a current fashion to claim that what is sufficient to derive Bell's inequalities is just freedom to choose settings and *local causality*, which is treated as unrelated with any form of realism or hidden variables.

Let me therefore present a standard introduction of local causality, which mainly follows the work of Bell 'La nouvelle cuisine' [11].

We have two space-like separated parties, Alice and Bob. They can choose freely between a number of local measurement settings. Let us denote, by x and y Alice's and Bob's measurement settings and by A and B their results. The predictions for a Bell type experiment are given probabilities $p(A, B|x, y)$.

---

[4]EPR forgot that if a new notion is to be introduced to a theory, then it must checked whether it is consistent with *all* predictions of the theory...

[5]"Free will" is usually not a challenged assumption, thus we assume it to hold throughout the discussion.

It is argued, that the probabilities $p(A, B|x, y)$ are a statistical mixture of different situations, labeled by λ, and called by Bell 'causes'.[6] The probabilities acquire the following form

$$p(A, B|x, y) = \int d\lambda \rho(\lambda) p(A, B|x, y, \lambda),$$

where $\rho(\lambda)$ is a probability distribution of the causes. Standard formulas for conditional probabilities, and the fact that conditional probabilities for the same condition have all properties of unconditioned probabilities, allow one to put

$$p(A, B|x, y, \lambda) = p(A|B, x, y, \lambda) p(B|x, y, \lambda).$$

Local causality assumption as stated by Bell reads 'The direct causes (and effects) of [the] event are near by, and even the indirect causes (and effects) are no further away than permitted by the velocity of light'. This allows one to state that 'what happens on Alice's side does not depend on what happens on Bob's side' and *vice versa* [12]. Thus the following must hold $p(A|B, x, y, \lambda) = p(A|x, \lambda)$ and $p(B|x, y, \lambda) = p(B|y, \lambda)$. By symmetry, which must be assumed in any reasonable approach $p(B|A, x, y, \lambda) = p(B|y, \lambda)$. Thus, we obtain the general mathematical structure of probabilities which allows to derive all two-particle Bell inequalities:

$$p(A, B|x, y) = \int d\lambda \rho(\lambda) p(A|x, \lambda) p(B|y, \lambda), \tag{1}$$

provided one additionally assumes "free will" to choose measurement settings.

Sometimes, local causality is thought to be synonymous to locality. There are claims that introduction of this notion by Bell in 1976 is effectively his second theorem about entanglement.[7] This I cannot understand because stochastic hidden variable theories, giving probabilities of the structure of Eq. (1), were introduced by Clauser and Horne in 1974 [13]. The formula $p(A|B, x, y, \lambda) = p(A|x, \lambda)$ is already implied by in the Bell 1964 condition: local result depends on the local setting and the local hidden variables λ (or "more complete specification"), however this was formulated by Bell in the deterministic context (just replace here 'local result' by 'probability of a local result', and local causality emerges). I shall argue below the local causality is a form of local realism or local hidden variables.[8]

---

[6] Note already here, that λ's do not appear in quantum mechanics, thus they are *hidden variables. Basically this could already end the discussion, as hidden variables are a program of completing quantum mechanics, just like the aim of EPR. As a matter of fact elements of reality are indeed hidden variables.*

[7] Some authors reserve the phrase Bell's second theorem to his independent derivation of the impossibility of non-contextual hidden variables.

[8] Of course there is a full mathematical equivalence between local causal theories and stochastic local hidden variable theories of Clauser and Horne. I shall argue that additionally there is no conceptual difference.

The danger in thinking that local causality is equivalent to locality is the fact the opposite notion to locality is non-locality. However violation of local causality implies either non-local causality, or that we have spontaneous events of a local or non-local nature. Local causality assumes locality and existence of causes λ, which are not present in the quantum formalism.

In the case of *mixed* separable states, one may think that λ specifies the "actual" quantum state in the probabilistic mixture. All this would agree with the formula (1). However when the two particle quantum state is a pure entangled one, denoted here by ψ, there is a trouble. There is just one joint quantum mechanical state describing the two separated systems. No other specification of the situation is allowed in quantum mechanics, all predictions are derivable using the state and the quantum formalism which additionally gives us methods of calculation probabilities, once the state is known. The probabilities are calculable using projectors which depend on the (local) settings. Thus the quantum state ψ is the sole "cause" in quantum theory, except the settings. Applying local causality principle would mean that $p(A|B, x, y, \psi) = p(A|x, \psi)$ and equivalently $p(B|A, x, y, \psi) = p(B|x, \psi)$, and the formula (1) would read

$$p(A, B|x, y) = p(A|x, \psi)p(B|y, \psi), \tag{2}$$

implying no correlations whatsoever! To get correlations, one must introduce at least one *two valued* 'cause', say $\lambda = \lambda_1$ or λ_2, other than ψ. In this way we can get

$$p(A, B|x, y) = \sum_{i=1}^{2} p(A|x, \psi, \lambda_i)p(B|y, \psi, \lambda_i). \tag{3}$$

Such a formula allows for correlations. As such additional λ's do not appear in quantum formalism, they are hidden variables *per se*. They are an attempt to complete the quantum formalism by some additional factors.

The λ's, which enter Eq. (1) get various names: e.g. 'the physical state of the systems as described by any possible future theory' [12], 'local beables' [14], 'the real state of affairs', 'complete description of the state', etc. Bell himself writes 'λ denote any number of hypothetical additional variables needed to complete quantum mechanics in the way envisaged by EPR' ([11], pp. 242). This sentence of Bell's is often forgotten by those who think that local causality differs from local hidden variable theories. As a matter of fact even Bell himself had a tendency to ignore it [15].

The other way of looking at this is to notice that local causality implies existence of an underlying joint probability distribution for results of all possible measurements (commensurable or non-commensurable), which is normalizable to unity and non-negative. Once we have probabilities $p(A|x, \lambda)$ and $p(B|y, \lambda)$ to define such an object is trivial, while in general in quantum mechanics this in general is impossible, and there is no quantum mechanical method to formulate such a distribution. Denote by A_{x_i} the possible values (spectrum) of an observable $\hat{O}^A_{x_i}$ for Alice's system, and also by B_{y_j} the possible values (spectrum) of an observable $\hat{O}^B_{y_i}$ for Bob's system. Indices i, j numerate the observables in whatever way. Then the underlying distribution reads

$$P(A_{x_1}, \ldots, A_{x_n}, B_{y_1}, \ldots, B_{y_m}) = \int d\lambda \rho(\lambda) \prod_{i=1}^{n} p(A_{x_i}|x_i, \lambda) \prod_{j=1}^{m} p(B_{y_j}|y_j, \lambda)$$

and all other probabilities in a local causal theory are marginals of such distributions. If in a probability theory such distributions exist then then all properties of the axioms of Kolmogorov are satisfied, and probabilities can have the classical lack of knowledge interpretation. One can model them by a normalized measure on some sample space Ω. Elements of Ω have all properties of hidden variables. Thus conversely any theory with all $P(A_{x_1}, \ldots, A_{x_n}, B_{y_1}, \ldots, B_{y_m})$ existing and non-negative can be modelled with some hidden variable on a sample space, for details see [16].

Final Remarks

In Sect. "Paradoxes of Interpretations of the EPR Paradox: Elementary Presentation" it was shown that the EPR reasoning cannot be used to derive determinism or realism of sorts, as it has as a tacit assumption acceptance of courterfactual statements as valid. Once one accepts such statements, then for a given quantum system unmeasured quantities have the same status as the one actually measured, even if they are mutually non-commensurable. This is of course prohibited by complementarity principle. Thus, EPR reasoning cannot be used do deny complementarity, because EPR deny complementarity in their initial assumptions.

Still the situation is even worse, as in the counterfactual situation of EPR considering the GHZ correlations, they would have run into a $1 = -1$ contradiction in their denial of complementarity. Thus, EPR reasoning, as it leads in this case to an outright contradiction cannot be useful in any scientific reasoning, and especially it is useless as a tool to derive determinism or realism of any sort (as this leads in such a case to a contradiction). Still, this work is seminal and extremely important—but these are not terms which describe its internal consistency, especially if one considers all quantum predictions, and not only for their state (in the case of which unluckily for the history of physics the contradiction was difficult to spot).[9]

Thus, the consequences of the inconsistency of EPR argument are either that unperformed experiments have no results [19], or non-locality, or both (if we insist to retain "freedom"). With counterfactual reasonings the algebraic identity leading to the CHSH [20] inequality reads:

$$A_{x_1}B_{y_1} + A_{x_1}B_{y_2} + A_{x_2}B_{y_1} - A_{x_2}B_{y_2} = \pm 2. \tag{4}$$

---

[9]For a version Bell's theorem for EPR states see [17], or for a more recent development see [18].

Note that it pertains to all possible choices of local measurements in two-setting per observer Bell experiment, hypothetically applied in the case of a given single pair of particles. With complementarity in mind, and for the case in which the actual setting for the given pair of particles is on both sides the first one, x_1, y_1, as values for the complementary settings are then *unspeakable*, we have

$$A_{x_1}B_{y_1} + A_{x_1}(?) + (?)B_{y_1} - (?)(?) = ?. \tag{5}$$

In the considerations of Sect. "Local Causality", it is shown that local causality is a compound notion. Thus, its opposite is not 'non-locality'. Local causality is equivalent to stochastic local hidden variable theories introduced in [13]. We have locality and causes λ which are not present in quantum description, and therefore are hidden variables. Thus an antonym of local causality may be non-locality or spontaneous events, or both.

The author is supported by FNP (project TEAM), ERC AdG grant QOLAPS, and COPERNICUS award-grant of DFG and FNP.

References

1. D.M. Greenberger, M.A. Horne, A. Zeilinger, in *Bell's Theorem, Quantum Theory, and Conceptions of the Universe*, ed. by M. Kafatos (Kluwer Academic, Dordrecht, 1989); D.M. Greenberger, M.A. Horne, A. Shimony, A. Zeilinger, Am. J. Phys. **58**, 1131 (1990)
2. N.D. Mermin, Phys. Today **43**(6), 9 (1990)
3. A. Einstein, B. Podolsky, N. Rosen, Phys. Rev. **47**, 777 (1935)
4. J.S. Bell, Physics **1**, 195 (1964)
5. See e.g., W.M. de Muynck, W. De Baere, Found. Phys. Lett., **3**, 325 (1990), W.M. de Muynck, W. De Baere, H. Martens, Found. Phys. **24**, 1589–1664 (1994), A. Stairs, unpublished, http://www.terpconnect.umd.edu/~stairs/papers/EPR_Illusion.pdf
6. See e.g., T. Norsen, Found. Phys. Lett. **19**, 633 (2006), T. Norsen, Against Realism, Found. Phys. **37**(3), 311–340 (2007), see also R. Tumulka, Found. Phys. **37**, 186 (2007) for a similar approach
7. N. Bohr, in *Essays 1958–1962 on Atomic Physics and Human Knowledge* (Wiley, New York, 1963), http://www-physics.lbl.gov/~stapp/Complementarity.doc
8. C.A. Fuchs, N.D. Mermin, R. Schack, Am. J. Phys. **82**(8), 749 (2014)
9. N. Bohr, Phys. Rev. **48**, 696 (1935)
10. J.-W. Pan, Z.-B. Chen, C.-Y. Lu, H. Weinfurter, A. Zeilinger, M. Zukowski, Rev. Mod. Phys. **84**, 777 (2012)
11. J.S. Bell, La nouvelle cuisine, in *Speakable and Unspeakable in Quantum Mechanics*, 2nd ed. (Cambridge University Press, 2004)
12. N. Gisin, Found. Phys. **42**, 80 (2012)
13. J. Clauser, M. Horne, Phys. Rev. D **10**, 526 (1974)
14. J.S. Bell, The theory of beables, TH-2053-CERN (1975)
15. M. Zukowski, Stud. Hist. Phil. Mod. Phys. **36B**, 566–575 (2005)
16. M. Zukowski, C. Brukner, J. Phys. A: Math. Theor. **47**, 424009 (2014)
17. K. Banaszek, K. Wodkiewicz, Phys. Rev. A **58**, 4345 (1998)
18. K. Rosolek, M. Stobinska, M. Wiesniak, M. Zukowski, Phys. Rev. Lett. **114**, 100402 (2015)
19. A. Peres, Am. J. Phys. **46**, 747 (1978)
20. J.F. Clauser, M.A. Horne, A. Shimony, R.A. Holt, Phys. Rev. Lett. **23**, 880 (1969)

Part III
Contextuality

Chapter 11
The Unspeakable Why

Adán Cabello

Abstract For years, the biggest unspeakable in quantum theory has been why quantum theory and what is quantum theory telling us about the world. Recent efforts are unveiling a surprisingly simple answer. Here we show that two characteristic limits of quantum theory, the maximum violations of Clauser-Horne-Shimony-Holt and Klyachko-Can-Binicioğlu-Shumovsky inequalities, are enforced by a simple principle. The effectiveness of this principle suggests that non-realism is the key that explains why quantum theory.

Introduction

There is a photograph of John Bell taken in 1982 in front of a blackboard in his office at CERN. The famous Clauser-Horne-Shimony-Holt (CHSH) Bell inequality [1, 2] is written in the blackboard. At the right hand side of the maximum bound for local hidden variable theories it is written "Einstein." Below that, it is written the maximum for quantum theory (QT): "$2\sqrt{2}$." Already in 1969, CHSH noticed that this was the maximum for two-qubit systems [2]. In 1980, Tsirelson proved that it is also the maximum in QT, no matter the dimensionality of the state space [3]. It took a lot of time until somebody asked the obvious question: why? [4]. It took a surprising amount of time until somebody came with a compelling answer [5]. However, it was soon proved that this answer cannot explain the maximum quantum values for some tripartite Bell inequalities [6]. This leads us back to square one: Why the quantum maxima of all Bell inequalities? What is the fundamental reason that limits quantum probabilities?

In the summer of 1964, before submitting the Bell inequality paper, Bell submitted other paper which, for several reasons [7, 8], was not published until 1966 [9]. There, Bell discusses the implications for the hidden variables problem of Gleason's theorem [10], which was directed to reducing the axiomatic basis of QT. The relevant corollary of Gleason's theorem is that, if the dimensionality d of the state space

A. Cabello (✉)
Departamento de Física Aplicada II, Universidad de Sevilla, 41012 Sevilla, Spain
e-mail: adan@us.es

R. Bertlmann and A. Zeilinger (eds.), *Quantum [Un]Speakables II*,
The Frontiers Collection, DOI 10.1007/978-3-319-38987-5_11

is grater than two, then there exists a set of elementary tests such that values t(rue) or f(alse) cannot be assigned to them respecting that: (i) t cannot be assigned to two mutually exclusive tests, and (ii) t must be assigned to exactly one of d mutually exclusive tests. Bell proved this corollary by constructing an explicit infinite set of elementary tests in $d = 3$ for which such an assignment is impossible. A finite set was found by Kochen and Specker in 1962 [11], but not published until 1967 [12], making explicit a result anticipated by Specker in 1960 [13]. These sets prove the impossibility of reproducing QT with theories satisfying the assumption of non-contextuality of results, namely, the assumption that the "measurement of an observable must yield the same value independently of what other measurements may be made simultaneously" [9]. Bell considered that this assumption was not reasonable and finished his paper suggesting that it would be interesting to pursue some proof of impossibility of hidden variables replacing non-contextuality by some assumption of locality. One month later, Bell submitted the Bell inequality paper.

However, the problem of hidden variables in QT can be mathematically formulated in a way which goes beyond whether or not non-contextuality is reasonable. The problem of hidden variables in QT is simply whether or not it is possible to recover the quantum probabilities from a joint probability distribution over a single probability space. What is proven by Kochen-Specker and Bell's examples is that this is not possible in any scenario in which the dimensionality of the state space is three or higher, irrespective of whether or not locality can be invoked (Fig. 11.1).

Fig. 11.1 John Bell at CERN. © 1982 CERN.

The KCBS and the CHSH Inequalities

What if Bell would have derived a Bell-like inequality violated by quantum systems of dimension three? Such an inequality was introduced in 2008 and is called the Klyachko-Can-Binicioğlu-Shumovsky (KCBS) non-contextuality (NC) inequality [14]. The KCBS inequality is the analog for quantum systems of dimension three of the CHSH inequality. The KCBS inequality is the simplest NC inequality violated by quantum systems of dimension three; the CHSH inequality is the simplest NC inequality violated by quantum systems of dimension four. Simplicity is here measured by the number of dichotomic observables used. The quantum violation of NC inequalities shows the impossibility to recover the quantum probabilities from a joint probability distribution over a single probability space.

The KCBS inequality says that, for any non-contextual hidden variable (NCHV) theory,

$$\kappa \overset{\text{NCHV}}{\leq} 3, \tag{1}$$

with

$$\kappa = \langle A_1 A_2 \rangle + \langle A_2 A_3 \rangle + \langle A_3 A_4 \rangle + \langle A_4 A_5 \rangle - \langle A_5 A_1 \rangle, \tag{2}$$

where A_i are observables with two possible results -1 and $+1$, and

$$\langle A_j A_{j+1} \rangle = P(A_j+, A_{j+1}+) - P(A_j+, A_{j+1}-) - P(A_j-, A_{j+1}+) + P(A_j-, A_{j+1}-), \tag{3}$$

where, e.g., $P(A_j+, A_{j+1}-)$ denotes the joint probability of obtaining $+1$ when measuring A_j and -1 when measuring A_{j+1}. Probabilities in Eq. (3) are assumed to be well defined no matter in which order A_j and A_{j+1} are measured. However, for A_j and A_{j+2} this may not be the case.

Similarly, the CHSH inequality says that, for any local hidden variables (LHV) theory,

$$\beta \overset{\text{LHV}}{\leq} 2, \tag{4}$$

with

$$\beta = \langle A_1 A_2 \rangle + \langle A_2 A_3 \rangle + \langle A_3 A_4 \rangle - \langle A_4 A_1 \rangle. \tag{5}$$

The difference between κ and β is that, in β, A_1 and A_3 can be measured on a subsystem while A_2 and A_4 are measured on a distant subsystem. Therefore, in β the choice of measurement on one subsystem and the result on the other subsystem can be space-like separated. This allows us to invoke locality to justify the assumption of non-contextuality.

In contrast with (1) and (4), in QT,

$$\kappa \overset{\text{QT}}{\leq} 4\sqrt{5} - 5 \approx 3.944 \tag{6}$$

and

$$\beta \overset{\text{QT}}{\leq} 2\sqrt{2} \approx 2.828. \tag{7}$$

The big question is not just why QT violates the inequalities for hidden variable theories, but rather why QT violates them *exactly up to these limits*.

Introducing the Exclusivity Principle

Consider non-demolition measurements that are repeatable (i.e., that give the same result when repeated) and cause no disturbance on other measurements (i.e., when combined with these other measurements, all are repeatable). These measurements are called "sharp" [15, 16] and, in QT, are represented by projection operators. These are the measurements that von Neumann called "quantum observables" [17]. Let us define an *event* as the state of the system after some sharp measurements (with certain results) on some initial state. Two events are equivalent when they correspond to indistinguishable states. Two events are exclusive when there is a measurement that distinguishes between them.

A theory satisfies the *exclusivity (E) principle* [18] when any set of n pairwise exclusive events is n-wise exclusive. Therefore, if we assume the E principle, Kolmogorov's axioms of probability lead us to the conclusion that the sum of the probabilities of any set of pairwise exclusive events cannot be higher than 1.

However, the E principle cannot be derived from Kolmogorov's axioms. To see it, consider the maximum value of the following sum of probabilities of events:

$$S = P(A_1+, A_2+) + P(A_2-, A_3-) + P(A_3+, A_1-), \tag{8}$$

where the notation is the same used above. The three events (A_1+, A_2+), (A_2-, A_3-) and (A_3+, A_1-) are pairwise exclusive. Therefore, the only restrictions from Kolmogorov's axioms are that the probabilities are non-negative and that

$$P(A_1+, A_2+) + P(A_2-, A_3-) \leq 1, \tag{9a}$$

$$P(A_2-, A_3-) + P(A_3+, A_1-) \leq 1, \tag{9b}$$

$$P(A_3+, A_1-) + P(A_1+, A_2+) \leq 1. \tag{9c}$$

Therefore, for theories satisfying Kolmogorov's axioms the maximum is $S = 3/2$, since each of the three probabilities in (8) can be 1/2. However, for theories satisfying the E principle, the maximum is $S = 1$, since the E principles forces that

$$P(A_1+, A_2+) + P(A_2-, A_3-) + P(A_3+, A_1-) \overset{\text{E}}{\leq} 1. \tag{10}$$

The E principle can be derived from a variety of axioms. For example, from the axiom that pairwise co-measurability implies joint co-measurability [13], from the principle of fundamental sharpness of measurements [16], from axioms 1 and 2 in Ref. [19], and from the principle of lack of irreducible third order interference [20].

The E principle imposes limits to the sum of probabilities of pairwise exclusive events. Therefore, in order to study the implications of the E principle for the limits of the KCBS and CHSH inequalities, it is convenient to rewrite both inequalities in terms of sums of probabilities of events. For that, it is useful to notice that the condition of normalization of probabilities allows us to write

$$\langle A_j A_{j+1} \rangle = 2P(A_j+, A_{j+1}+) + 2P(A_j-, A_{j+1}-) - 1, \tag{11a}$$

$$-\langle A_j A_{j+1} \rangle = 2P(A_j+, A_{j+1}-) + 2P(A_j-, A_{j+1}+) - 1. \tag{11b}$$

Therefore, we can write

$$\kappa = 2S_{\text{KCBS}} + 2S'_{\text{KCBS}} - 5, \tag{12a}$$

$$\beta = 2S_{\text{CHSH}} - 4, \tag{12b}$$

where

$$S_{\text{KCBS}} = P(A_1+, A_2+) + P(A_2-, A_3-) + P(A_3+, A_4+) + P(A_4-, A_5-) + P(A_5+, A_1-), \tag{13a}$$

$$S_{\text{CHSH}} = P(A_1+, A_2+) + P(A_1-, A_2-) + P(A_2+, A_3+) + P(A_2-, A_3-) + P(A_3+, A_4+) + P(A_3-, A_4-) + P(A_4+, A_1-) + P(A_4-, A_1+) \tag{13b}$$

and S'_{KCBS} is obtained from S_{KCBS} by changing the signs of all the results. Then, we can write the KCBS and CHSH inequalities and their quantum limits as follows:

$$S_{\text{KCBS}} \overset{\text{NCHV}}{\leq} 2 \overset{\text{QT}}{\leq} \sqrt{5} \approx 2.236, \tag{14a}$$

$$S_{\text{CHSH}} \overset{\text{LHV}}{\leq} 3 \overset{\text{QT}}{\leq} 2 + \sqrt{2} \approx 3.414. \tag{14b}$$

The Limit of the KCBS Inequality

Our first target is to explain why S_{KCBS} cannot go beyond $\sqrt{5}$ or, equivalently, why κ cannot go beyond $4\sqrt{5} - 5$. For this purpose, consider two independent experiments both aiming the maximum of S_{KCBS}. Suppose that one of the experiments is performed in Vienna on a certain physical system, while the other experiment is performed in Stockholm on a different physical system. Let us denote by $(A_j+, A_{j+1}+)$ an event of the experiment in Vienna, by S^A_{KCBS} the sum of the corresponding five probabilities, by $(B_k+, B_{k+1}+)$ an event of the experiment in Stockholm and by S^B_{KCBS} the sum of the corresponding five probabilities.

Since the experiments are independent, the probability of an event involving both experiments is the product of the probabilities of the corresponding (single-city) events. For example,

$$P(A_j+, A_{j+1}+, B_k+, B_{k+1}+) = P(A_j+, A_{j+1}+)P(B_k+, B_{k+1}+). \tag{15}$$

Having two copies, we can identify larger sets of pairwise exclusive events. For example, the set with the following events: (A_1+, A_2+, B_1+, B_2+), (A_2-, A_3-, B_4-, B_5-), (A_3+, A_4+, B_2-, B_3-), (A_4-, A_5-, B_5+, B_1-) and (A_5+, A_1-, B_3+, B_4+). The E principle and assumption (15) applied to this set imply that

$$P(A_1+, A_2+)P(B_1+, B_2+) + P(A_2-, A_3-)P(B_4-, B_5-) + P(A_3+, A_4+)P(B_2-, B_3-)$$
$$+P(A_4-, A_5-)P(B_5+, B_1-) + P(A_5+, A_1-)P(B_3+, B_4+) \overset{\mathrm{E}}{\leq} 1. \tag{16a}$$

Similarly, by identifying sets of pairwise exclusive events, we can derive the following inequalities:

$$P(A_1+, A_2+)P(B_3+, B_4+) + P(A_2-, A_3-)P(B_1+, B_2+) + P(A_3+, A_4+)P(B_4-, B_5-)$$
$$+P(A_4-, A_5-)P(B_2-, B_3-) + P(A_5+, A_1-)P(B_5+, B_1-) \overset{\mathrm{E}}{\leq} 1, \tag{16b}$$
$$P(A_1+, A_2+)P(B_5+, B_1-) + P(A_2-, A_3-)P(B_3+, B_4+) + P(A_3+, A_4+)P(B_1+, B_2+)$$
$$+P(A_4-, A_5-)P(B_4-, B_5-) + P(A_5+, A_1-)P(B_2-, B_3-) \overset{\mathrm{E}}{\leq} 1, \tag{16c}$$
$$P(A_1+, A_2+)P(B_2-, B_3-) + P(A_2-, A_3-)P(B_5+, B_1-) + P(A_3+, A_4+)P(B_3+, B_4+)$$
$$+P(A_4-, A_5-)P(B_1+, B_2+) + P(A_5+, A_1-)P(B_4-, B_5-) \overset{\mathrm{E}}{\leq} 1, \tag{16d}$$
$$P(A_1+, A_2+)P(B_4-, B_5-) + P(A_2-, A_3-)P(B_2-, B_3-) + P(A_3+, A_4+)P(B_5+, B_1-)$$
$$+P(A_4-, A_5-)P(B_3+, B_4+) + P(A_5+, A_1-)P(B_1+, B_2+) \overset{\mathrm{E}}{\leq} 1. \tag{16e}$$

The geometry behind these sets is explained in Fig. 11.2. If we sum all five inequalities (16a)–(16e), we obtain

$$S^A_{\mathrm{KCBS}} S^B_{\mathrm{KCBS}} \overset{\mathrm{E}}{\leq} 5. \tag{17}$$

Assuming that the maximum is the same in both experiments, i.e., that $S^A_{\mathrm{KCBS}} = S^B_{\mathrm{KCBS}}$, we can conclude that, for any theory satisfying the E principle,

$$S_{\mathrm{KCBS}} \overset{\mathrm{E}}{\leq} \sqrt{5}. \tag{18}$$

Exactly as in QT. This is an arguably clearer presentation of a result introduced in Ref. [18].

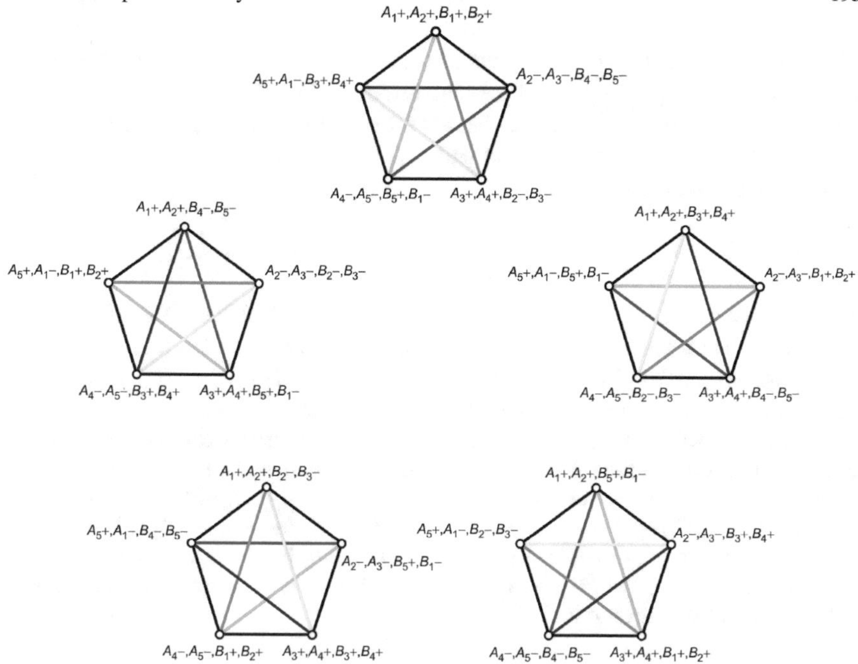

Fig. 11.2 Exclusivity graphs of the five sets of five pairwise exclusive events used in the proof of the limit of the KCBS inequality. Events are represented by nodes and exclusivity relations by edges. The *black* pentagons correspond to the exclusivity relations between the events $(A_j\gamma, A_{j+1}\delta)$. The *coloured* pentagrams correspond to the exclusivity relations between the events $(B_k\epsilon, B_{k+1}\phi)$. Any two graphs differ in a rotation of the pentagram.

The Limit of the CHSH Inequality

Our second target is to explain why S_{CHSH} cannot go beyond $2+\sqrt{2}$ or, equivalently, why β cannot go beyond $2\sqrt{2}$. For this purpose, first notice that the state space on which A_1 and A_3 act is, at least, two-dimensional. Second, notice that the conditions of normalization of probabilities allows us to write,

$$\begin{aligned}4 - S_{\text{CHSH}} = \; & P(A_1+,A_2-) + P(A_1-,A_2+) + P(A_2+,A_3-) + P(A_2-,A_3+) \\ & + P(A_3+,A_4-) + P(A_3-,A_4+) + P(A_4+,A_1+) + P(A_4-,A_1-). \quad (19)\end{aligned}$$

Now consider two independent experiments both testing S_{CHSH}. Suppose that one of the experiments is performed in Vienna and the other experiment in Stockholm. As before, let us denote by $(A_j+,A_{j+1}+)$ one event in Vienna and by $(B_k+,B_{k+1}+)$ one event in Stockholm.

Now notice that A_1 and B_1 are co-measurable and that the state space on which A_1 and B_1 act is, at least, four-dimensional. Therefore, there must exist an observable

C_{11} co-measurable with A_1 and B_1 and such that the result for C_{11} is $+1$ if the results of A_1 and B_1 are equal, and -1 if they are different. C_{11} acts on a four-dimensional state space, but only distinguishes between two subspaces. Therefore, there must be an observable C_{33} co-measurable with C_{11} that distinguishes between other two subspaces and such that

$$P(C_{11}+, C_{33}+) + P(C_{11}+, C_{33}-) + P(C_{11}-, C_{33}+) + P(C_{11}-, C_{33}-) = 1. \quad (20)$$

Since we have made no assumption about A_3 and B_3 (other than each of them acts on a different at least two-dimensional subspace), we can relate C_{33} to A_3 and B_3, the same way we related C_{11} to A_1 and B_1. Similarly, we can start with A_1 and B_3 and define C_{13} and then define a co-measurable C_{31} related to A_3 and B_1.

These observables allow us to identify larger sets of pairwise exclusive events. For example, the set with the following events: $(A_1+, A_2+, B_1+, B_2+, C_{11}+)$, $(A_1+, A_2-, B_1+, B_2-, C_{11}+)$, $(A_3+, A_2+, B_3-, B_2-, C_{33}-)$, $(A_3+, A_2-, B_3-, B_2+, C_{33}-)$, $(A_1-, A_2-, B_1-, B_2-, C_{11}+)$, $(A_1-, A_2+, B_1-, B_2+, C_{11}+)$, $(A_3-, A_2-, B_3+, B_2+, C_{33}-)$, $(A_3-, A_2+, B_3+, B_2-, C_{33}-)$ and $(C_{11}-, C_{33}+)$. Since, by definition of C_{11}, $P(A_1+, A_2+, B_1+, B_2+, C_{11}+) = P(A_1+, A_2+, B_1+, B_2+)$, the E principle and assumption (15) applied to this set imply that

$$\begin{aligned} &P(A_1+, A_2+)P(B_1+, B_2+) + P(A_1+, A_2-)P(B_1+, B_2-) + P(A_3+, A_2+)P(B_3-, B_2-) \\ &+P(A_3+, A_2-)P(B_3-, B_2+) + P(A_1-, A_2-)P(B_1-, B_2-) + P(A_1-, A_2+)P(B_1-, B_2+) \\ &+P(A_3-, A_2-)P(B_3+, B_2+) + P(A_3-, A_2+)P(B_3+, B_2-) + P(C_{11}-, C_{33}+) \overset{\text{E}}{\leq} 1. \end{aligned} \quad (21)$$

As explained in Table 11.1, there are 16 sets like this one. For each of them, there is a inequality like (21). If we sum all of them we obtain,

$$S_{\text{CHSH}}^{A} S_{\text{CHSH}}^{B} + (4 - S_{\text{CHSH}}^{A})(4 - S_{\text{CHSH}}^{B}) + 4 \overset{\text{E}}{\leq} 16. \quad (22)$$

Assuming that the maximum is the same in both experiments, we can conclude that, for any theory satisfying the E principle,

$$S_{\text{CHSH}} \overset{\text{E}}{\leq} 2 + \sqrt{2}. \quad (23)$$

Exactly as in QT. This result is an improved version of an argument introduced in Ref. [21]. A similar argument allows us to derive the quantum limits for n-partite Bell-like inequalities for non-local (but not genuinely n-partite non-local) hidden variable theories [22].

Table 11.1 Eight sets of nine pairwise exclusive events. Each row displays five pairwise exclusive events. For each row, there are other four events (not displayed) which are obtained by changing the results of the first four events. There are other eight sets (not displayed) which are obtained by exchanging $(A_j\gamma, A_k\delta, B_l\epsilon, B_m\phi)$ by $(A_l\epsilon, A_m\phi, B_j\gamma, B_k\delta)$

(A_1+, A_2+, B_1+, B_2+)	(A_1+, A_2-, B_1+, B_2-)	(A_3+, A_2+, B_3-, B_2-)	(A_3+, A_2-, B_3-, B_2+)	$(C_{11}-, C_{33}+)$
(A_1+, A_2+, B_1-, B_2-)	(A_1+, A_2-, B_1-, B_2+)	(A_3+, A_2+, B_3+, B_2+)	(A_3+, A_2-, B_3+, B_2-)	$(C_{11}+, C_{33}-)$
(A_1+, A_2+, B_1-, B_4-)	(A_1+, A_2-, B_1-, B_4+)	(A_3+, A_2+, B_3-, B_4+)	(A_3+, A_2-, B_3-, B_4-)	$(C_{11}+, C_{33}+)$
(A_1+, A_2+, B_1+, B_4+)	(A_1+, A_2-, B_1+, B_4-)	(A_3+, A_2+, B_3+, B_4-)	(A_3+, A_2-, B_3+, B_4+)	$(C_{11}-, C_{33}-)$
(A_1+, A_2+, B_3-, B_4+)	(A_1+, A_2-, B_3-, B_4-)	(A_3+, A_2+, B_1-, B_4-)	(A_3+, A_2-, B_1-, B_4+)	$(C_{13}+, C_{31}+)$
(A_1+, A_2+, B_3+, B_4-)	(A_1+, A_2-, B_3+, B_4+)	(A_3+, A_2+, B_1+, B_4+)	(A_3+, A_2-, B_1+, B_4-)	$(C_{13}-, C_{31}-)$
(A_1+, A_2+, B_3+, B_2+)	(A_1+, A_2-, B_3+, B_2-)	(A_3+, A_2+, B_1-, B_2-)	(A_3+, A_2-, B_1-, B_2+)	$(C_{13}-, C_{31}+)$
(A_1+, A_2+, B_3-, B_2-)	(A_1+, A_2-, B_3-, B_2+)	(A_3+, A_2+, B_1+, B_2+)	(A_3+, A_2-, B_1+, B_2-)	$(C_{13}+, C_{31}-)$

The Unspeakable Why

We have shown that some characteristic limits of QT have a simple explanation. Indeed, we suspect that all the limits of quantum probabilities have the same explanation. If this would be the case, what should we learn about QT?

For some people, a fundamental message of QT is that the world is non-local, i.e., that the results of quantum observables correspond to some reality and change non-locally [24]. However, from this perspective, it is puzzling that the no-signaling principle allows for higher than quantum violations of the CHSH inequality [4]. Why then QT is not more non-local?

The reason why QT is exactly as non-local as it is, apparently, the same reason why QT is exactly as contextual as it is. However, in the contextuality case, there is no Alice and Bob and no communication.

Why nobody paid attention to the E principle before? Arguably, because the E principle is trivial in classical deterministic theories and not well defined in non-local realistic theories.

However, if one takes non-realism (of the results of observables represented in QT by self-adjoint operators) as a fundamental key of the world, then everything makes much more sense. The fundamental non-existence of results makes that not all conceivable combinations of observables allow for joint probability distributions (i.e., makes not all observables to be co-measurable). Indeed, it makes all conceivable relations of co-measurability/non-co-measurability (among sharp measurements) realizable and, as a consequence, makes all conceivable relationships of exclusivity/non-exclusivity (among events) realizable. There is where the E principle makes a profound contribution: The possible sets of probabilities of a given scenario (i.e., a certain structure of co-measurability/non-co-measurability) are restricted by the E principle applied to all conceivable embeddings of the scenario into a larger scenario. In this sense, the E principle acts in an holistic way. In particular, the limits of the probabilities of a given scenario follow from identifying the most (or one of the most) restrictive embedding(s) (details will be presented elsewhere [23]). The resulting picture points out that non-realism is not "a soft option" [24], but rather a fundamental key of the world. QT is a probability theory about things that do not exist and are unpredictable at a fundamental level.

One may argue that the view I have drawn before is just one of the possible options and that the predictions of QT are also compatible with contextual and non-local realistic views of the world. I disagree. Common to all these other views is a certain degree of realism that ranges from hidden variables determining the results of all possible experiments to just taking the quantum state as real. It seems evident that any of these other views, when examined in detail, will lead to predictions that QT does not make. Hopefully, we will soon identify these predictions and test them. Time and experiments will tell.

Acknowledgments This work was supported by the project FIS2011-29400 (MINECO, Spain) with FEDER funds, the FQXi large grant project "The Nature of Information in Sequential Quantum Measurements" and the program Science without Borders (CAPES and CNPq, Brazil).

References

1. J.S. Bell, Physics (Long Island City, N.Y.) **1**, 195 (1964)
2. J.F. Clauser, M.A. Horne, A. Shimony, R.A. Holt, Phys. Rev. Lett. **23**, 880 (1969)
3. B.S. Cirel'son [Tsirelson], Lett. Math. Phys. **4**, 93 (1980)
4. S. Popescu, D. Rohrlich, Found. Phys. **24**, 379 (1994)
5. M. Pawłowski, T. Paterek, D. Kaszlikowski, V. Scarani, A. Winter, M. Żukowski, Nature **461**, 1101 (2009)
6. T.H. Yang, D. Cavalcanti, M.L. Almeida, C. Teo, V. Scarani, New J. Phys. **14**, 013061 (2012)
7. M. Jammer, *The Philosophy of Quantum Mechanics* (Wiley, New York, 1974)
8. M. Jammer, Found. Phys. **20**, 1139 (1990)
9. J.S. Bell, Rev. Mod. Phys. **38**, 447 (1966)
10. A.M. Gleason, J. Math. Mech. **6**, 885 (1957)
11. S. Kochen, in E. Engeler, N. Hungerbühler, and J.A. Makowsky (eds.), Elem. Math. **67**, 1 (2012)
12. S. Kochen, E.P. Specker, J. Math. Mech. **17**, 59 (1967)
13. E.P. Specker, Dialectica **14**, 239 (1960). English translation: arXiv:1103.4537
14. A.A. Klyachko, M.A. Can, S. Binicioğlu, A.S. Shumovsky, Phys. Rev. Lett. **101**, 020403 (2008)
15. M. Kleinmann, J. Phys. A: Math. Theor. **47**, 455304 (2014)
16. G. Chiribella, X. Yuan, arXiv:1404.3348
17. J. von Neumann, Mathematische Grundlagen der Quantenmechanik (Springer-Verlag, Berlin, 1932) *(Mathematical Foundations of Quantum Mechanics* (Princeton University Press, Princeton, New Jersey, 1955)
18. A. Cabello, Phys. Rev. Lett. **110**, 060402 (2013)
19. H. Barnum, M. P. Müller, C. Ududec, New J. Phys. **16**, 123029 (2014)
20. J. Henson, Phys. Rev. Lett. **114**, 220403 (2015)
21. A. Cabello, Phys. Rev. A **90**, 062125 (2014)
22. A. Cabello, Phys. Rev. Lett. **114**, 220402 (2015)
23. A. Cabello (in preparation)
24. N. Gisin, Found. Phys. **42**, 80 (2012)

Chapter 12
A Reconstruction of Quantum Mechanics

Simon B. Kochen

Introduction

Almost a century after the mathematical formulation of quantum mechanics, there is still no consensus on the interpretation of the theory. This may be because quantum mechanics is full of predictions which contradict our everyday experiences, but then so is another, older theory, namely special relativity.

Although the Lorentz transformations initially gave rise to different interpretations, when Einstein's 1905 paper appeared it soon led to a nearly universal acceptance of Einstein's interpretation. Why was this? Einstein began with the new conceptual principle that time and simultaneity are relative to the inertial frame, dropping the classical assumption that they are absolute. By then using the linearity of transformations due to the local nature of special relativity and the experimental fact that the speed of light is constant, Einstein was able to derive the Lorentz transformations. Furthermore, by introducing the natural classical notions of state, observable, and symmetry in the new setting, Einstein derived the new dynamical equations to replace the Newtonian equations. This manifestly consistent derivation allowed for a resolution of the apparent paradoxes which confounded the older ether theory, and led to the adoption of Einstein's interpretation by physicists.

In this paper, we shall endeavor to use Einstein's approach as a model for deriving and interpreting quantum mechanics. We also start with a new conceptual precept which replaces a classical premise. It is a basic assumption of classical physics that experiments measure pre-existing inherent observables and properties of systems, and any disturbance due to the interaction with the apparatus can be minimized or incorporated into its effect on the observables. By contrast, when we measure a particle's component of spin in a particular direction in a Stern-Gerlach experiment, it is the general belief that we are not measuring a pre-existing property. Rather, it is

S.B. Kochen (✉)
Department of Mathematics, Princeton University, Princeton, NJ 08544-1000, USA
e-mail: kochen@math.princeton.edu

R. Bertlmann and A. Zeilinger (eds.), *Quantum [Un]Speakables II*,
The Frontiers Collection, DOI 10.1007/978-3-319-38987-5_12

the interaction of the particle with the magnetic field, which is inhomogeneous in that direction, that creates the value of the spin. We shall say that such properties are *relational* or *extrinsic*, as opposed to the intrinsic properties of classical physics.

That quantum observables and properties take values only upon suitable interactions is, of course, not new to physicists. Bohr, the founder of the Copenhagen interpretation, wrote in [1]: "The whole situation in atomic physics deprives of all meaning such inherent attributes as the idealization of classical physics would ascribe to such objects." This is a radically new consequence of quantum physics that controverts one of the main conceptual assumptions of classical physics, that properties of a physical system are intrinsic.

The aim of this paper is to show that a mathematical formulation of this principle allows us to reconstruct the formalism of quantum mechanics. Let us give the basic idea in defining the structure of extrinsic properties, given in Sect. "Properties". Every experiment yields a σ-algebra of measured properties. For instance, in measuring an quantum observable with spectral decomposition $\sum a_i P_i$, the σ-algebra is generated by the projections P_i. It is shown in Sect. "Properties" that for quantum experiments the different measured σ-algebras cannot all be imbedded into a single σ-algebra. In the case of classical physics, on the other hand, the measured σ-algebras all sit inside the σ-algebra $B(\Omega)$ of intrinsic properties of the system, consisting of the σ-algebra generated by the open sets of the phase space Ω of the system.

To mathematically treat the extrinsic properties of quantum mechanics, we replace the encompassing σ-algebra $B(\Omega)$ of properties by a σ-*complex* Q, consisting of the union of all the σ-algebras of the system elicited by different decoherent interactions, such as measurements.

This change allows us to define in a uniform and natural manner the concepts of state, observable, symmetry, and dynamics, which reduce to the classical notions when Q is a Boolean σ-algebra, and to the standard quantum notions when Q is the σ-complex $Q(\mathcal{H})$ of projections of Hilbert space $\mathcal{H}$. Moreover, we use this approach to derive both the Schrödinger equation and the von Neumann-Lüders Projection Postulate. We also show on the basis of interferometry experiments why Q has the form $Q(\mathcal{H})$.

The most noteworthy feature of this reconstruction of quantum mechanics is that the classical definitions of the key physical concepts such as state, observable, symmetry, dynamics, and the combining of systems take on precisely the same form in the quantum case when they are applied to extrinsic properties.

In the standard formulation, these concepts take on a strikingly different form from the classical one. In particular, the definition of state as a complex function and the complex form of the Schrödinger equation, as opposed to the intuitive, real definitions of classical physics, led Bohr to speak of this formalism as only a symbolic representation of reality.

One purpose of this approach is to show that once the relational character of properties is accepted, the definitions of the basic concepts of quantum mechanics are as real and intuitive as is the case for classical mechanics. Of course, it is not our intention to dispense with the linear complex Hilbert space in treating problems in physics. The linearity of the Schrödinger equation is crucial for solving atomic problems. Our purpose in showing that our intuitive definitions of the notions are equivalent to the standard complex ones is rather to reduce the use of the complex Hilbert space to a technical computational tool, similar to the use of complex methods in classical electromagnetism and fluid mechanics.

At first sight the structure of a σ-complex Q is unusual. Operations between elements of Q are not defined unless they lie in the same Boolean σ-algebra within Q. That however is the whole point of this structure. Operations are only defined when they make physical sense. This points to the main difference of this approach to that initiated by Birkhoff and von Neumann [2], and carried forward by Mackey [3], and Piron [4], among others. They define the logic of quantum mechanics to be a certain kind of lattice, consisting of the set of projection operators of Hilbert space. However, Birkhoff and von Neumann [2] already raised the question:

> What experimental meaning can one attach to the meet and join of two given experimental propositions?

That question has never been adequately answered. Varadarajan, in his book [5] on the lattice approach to quantum mechanics, written some thirty years after the Birkhoff and von Neumann paper, writes:

> The only thing that may be open to serious question in this is [the] assumption ... which forces any two elements of $\mathcal{L}$ to have a lattice sum, ... We can offer no really convincing phenomenological argument to support this.

Replacing the structure of a complex Hilbert space by an equally mysterious structure of a lattice does not achieve the goal of a transparent foundation for quantum mechanics. What is perhaps surprising is that the far weaker structure of a σ-complex suffices to reconstruct the formalism of quantum mechanics. Our approach has nevertheless benefited from the lattice approach, especially as delineated in Varadarajan [5], since theorems using lattices turned out often to have proofs using the weaker σ-complex structure.

One of the aims of a consistent, logical reconstruction of quantum mechanics is to resolve problematic questions and inconsistencies in the orthodox interpretation, such as the Measurement Problem, the Einstein-Podolsky-Rosen paradox, the Kochen-Specker paradox, the problem of reduction and the von Neumann-Lüders Projection Rule, and wave-particle duality. We discuss a resolution of these questions in the context of this reconstruction as they arise in this paper.

At various points in the paper we consider properties of systems as they are measured by experiments. We are not however espousing an operational view of quantum mechanics. We believe quantum mechanics describes general interactions in the world, independently of a classical macroscopic apparatus and observer. We do not subscribe to the Bohrian view that classical physics is needed to give meaning to quantum phenomena. The interactions we describe using a macroscopic apparatus could apply equally well to appropriate decoherent interactions between two systems in general. (See the discussion in Sect. "Properties"). Nevertheless, we refer for the most part to experiments rather than general interactions in order to emphasize that the postulates have operational content and meaning. This has the merit of allowing those who prefer the operational approach to make sense of this reconstruction.

Another point is that since the properties that constitute a σ-complex correspond to the results of possible measurements, they refer to what in the orthodox interpretation are the properties that may hold as a result of reduction. We do not attempt to discuss the conditions under which reduction or decoherence occurs. There are discussions in the literature on the conditions under which reduction can occur. For instance, Bohm [6] analyzes the strength of the inhomogeneity of the magnetic field for a successful reduction to occur in the Stern-Gerlach experiment. We consider these as interesting pragmatic questions which lie outside the purview of this paper.

We have not given a new axiomatization of quantum physics. In fact, there are no axioms in this paper, only definitions of the basic concepts, definitions which are identical with the classical ones. Rather, we have presented a framework that is common to all physical theories. It is the aim of every theory to predict the probabilities of the outcomes of interactions between systems. Experiments are particular instances of such interactions. An experiment gives rise to a Boolean σ-algebra of events which reflects an isomorphic σ-algebra of properties of the system. The different possible experiments yield a family of σ-algebras, reflecting a family of σ-algebras properties of the system, whose union we call a σ-complex. This σ-complex helps determine the underlying theory, and conversely, a given theory determines the kind of σ-complex of perperties that arises, but the general structure of a σ-complex as a union of σ-algebras is independent of any particular theory.

The main aim of the paper is to derive elementary quantum mechanics by applying the natural classical definitions of the physical concepts to extrinsic properties, and then use this derivation to resolve the standard paradoxes and problematic questions. We shall accordingly give only outlines of the proofs of the requisite theorems. To show that we have accomplished the goal of reconstructing the formalism, we shall use the textbook by Arno Bohm [6]. This book has the advantage of explicitly introducing five postulates which suffice to treat the standard topics in quantum theory. We shall specify each of the Bohm postulates as we derive them in the paper.

To avoid repetition, we shall make the blanket assumption that the Hilbert space $\mathcal{H}$ that we deal with is a separable complex Hilbert space. The Appendix has a table which summarizes the reconstruction.

Properties

Scientific theories predict the probabilities of outcomes of experiments. We recall from probability theory that the individual outcomes of an experiment on a system form the *sample space* S. For instance, a Stern-Gerlach experiment which measures the z-components of spin for a spin 1 system has the sample space $S = \{s_{-1}, s_0, s_1\}$ corresponding to the three possible spots labeled s_{-1}, s_0, s_1 on the screen. An experiment to measure the temperature of water by a thermometer has (an interval of) the real line as sample space.

Out of the elementary outcomes, one forms an algebra of more complex outcomes, called *events*, consisting of a Boolean algebra B of subsets of the space S. The operations of B consist of union $a \vee b$, and complementation $a^{\perp}$, and all other Boolean operations, such as intersection $a \wedge b$, which are definable from them. If S is finite, then B consists of all subsets of S. If S is infinite, then the operation of countable union $\bigvee a_i$ of elements a_i of B, is added, and B is called a (Boolean) σ-algebra. (For the definition of and details about Boolean algebras see Koppellberg [7]).

The algebra B of events, i.e. sets of outcomes, reflects the corresponding structure of properties of the system. For instance, in the above Stern-Gerlach experiment, the sets $\{s_{-1}\}$, $\{s_0\}$, and $\{s_1\}$ correspond to the properties $S_z = -1$, $S_z = 0$, and $S_z = 1$; the set $\{s_{-1}, s_1\}$ corresponds to the property $S_z = -1 \vee S_z = 1$ (where $\vee$ denotes 'or'), or equivalently, the property $\neg(S_z = 0)$ (where $\neg$ denotes 'not'), and so on. In this case, the Boolean algebra is clearly the eight element algebra. In the case of the above temperature measurement of the water, the elementary outcomes are open intervals of the real line, and the algebra of events is the σ-algebra of (*Borel*) sets generated by the intervals by complement and countable intersection.

Thus, for both classical and quantum physics, every experiment on a given system S elicits a σ-algebra of properties of S, which are true or false, i.e. have a truth value, for the system.

We come now to a crucial difference between the two theories. In classical physics, we assume that the measured properties of the system already exist prior to the measurement. It may be true that the interaction of the system with the apparatus disturbs the system, but this disturbance can be discounted or minimized. For instance, the thermometer may change the temperature of the water being measured, but this change can be accounted for, and there is no doubt that the water had a temperature prior to the measurement which is approximated by the measured value. The basic assumption is that systems have intrinsic properties, and the experiment measures the values of some them.

The family of intrinsic properties of a system form a Boolean algebra, and in the infinite case a σ-algebra. For classical physics, one introduces the phase (or state) space, with a canonical structure. The open sets of Ω generate a σ-algebra $B(\Omega)$ of *Borel* sets by complement and countable intersection. The algebra $B(\Omega)$ constitutes the σ-algebra of intrinsic properties of the system. Since the σ-algebras of measured properties are aspects of all the intrinsic properties of the system, these different σ-algebras must all be part of the σ-algebra $B(\Omega)$. Hence, the union $\cup B$ of all the σ-algebras arising from possible measurements is embeddable in $B(\Omega)$. In fact, if we assume that every property of the system is, in principle, experimentally measurable then the union $\cup B$ itself forms a σ-algebra.

In quantum mechanics, for measurements such as the Stern-Gerlach experiment, physicists do not believe that the value of the spin component S_z exists prior to the measurement. On the contrary, it is the interaction with the magnetic field, inhomogeneous in the z-direction, that results in a definite spot, say s_1, on the screen, reflecting the value, $S_z = 1$ of the spin of the particle.

This general conviction is, in fact, supported by a theorem, called the Kochen-Specker Paradox. This result showed that the spin component S_z cannot be an intrinsic property of a spin 1 particle. We recall that this result shows that there exist a small number of directions in space (33 suffice) such that any prior assignment of values to the squares of the components of spin in these directions contradicts the condition that $S_x^2 + S_y^2 + S_z^2 = 2$, for an orthogonal triple (x, y, z). Since the squares of the components of spin in orthogonal directions commute for a spin 1 system, we may measure them simultaneously for the triple (x, y, z). For instance, the measurement of the observable $S_x^2 - S_y^2$, with eigenvalues 1,-1, 0 gives us the value 0 for S_x^2, S_y^2, or S_z^2, respectively, and 1 for the other two. We shall call such an experiment a *triple experiment on the frame* (x, y, z).

The operators S_x^2, S_y^2, S_z^2 generate an eight element Boolean algebra:

$$B_{xyz} = \{0, 1, S_x^2, S_y^2, S_z^2, 1 - S_x^2, 1 - S_y^2, 1 - S_z^2\}.$$

The 33 directions give rise to 40 orthogonal triples, and hence 40 Boolean algebras. It is important to note that the Boolean algebras have common sub-algebras. For instance, the algebra $B_{x'y'z}$ of the triple experiment on (x', y', z) has the Boolean algebra $B_z = (0, 1, S_z^2, 1 - S_z^2)$ in common with B_{xyz}.

The 40 Boolean algebras, and hence their union $\cup B_{xyz}$, cannot be embedded into a single Boolean algebra. We may see this directly from the fact that every Boolean algebra has truth values, i.e. a homomorphism onto the Boolean algebra $\{0, 1\}$, so that such an embedding would assign values to all the 40 Boolean algebras simultaneously, and hence to the 40 triples S_x^2, S_y^2, S_z^2, contradicting the Kochen-Specker theorem. (For a proof of this theorem, with the 40 triples, see Conway and Kochen [8]).

The conclusion is that, in general, quantum mechanical properties are not intrinsic to the system, but have truth values created by interactions with other systems. We shall call such interactive or relational properties *extrinsic*. The question now is: what mathematical structure captures the concept of extrinsic properties, to replace the Boolean σ-algebras that characterize intrinsic properties?

Such a structure must contain all the σ-algebras that are elicited by experiments. The minimal structure is then clearly the union $\cup B$, where B ranges over all the σ-algebras that arise in experiments. Intuitively, we may obtain such a structure by gluing together the σ-algebras at the "faces," i.e. the common sub-σ-algebras. This structure is the minimal one which contains all the σ-algebras arising from different experiments. We shall adopt it as embodying the idea of extrinsic properties. We now give the formal definition of this notion.

Definition[1] Let F be a family of σ-algebras. The σ-complex Q_F based on F is the union $\cup B$ of all σ-algebras B lying in F.

We shall generally leave the family F implicit, and simply refer to a σ-complex Q. We shall usually deal with σ-complexes that are closed under the formation of sub-σ-algebras. We can, in any case, always close a σ-complex by adding all its sub-σ-algebras.

---

[1] A Boolean σ-complex is a closely connected generalization of a *partial Boolean algebra* (introduced in Kochen and Specker [9], and further studied in [10] and [11]).

The term σ-complex is based on the notion of a *simplicial complex* in topology. A simplicial complex is obtained by taking a family of simplices, which is closed under sub-simplices, and gluing together common simplicial faces. σ-complexes are not just analogous to simplicial complexes, but have a close correspondence, as we now outline. First recall that an *atom* of a Boolean algebra is an element x such that $y \leq x$ (i.e. $x \wedge y = y$) implies $y = 0$ or $y = x$. The atoms of a Boolean algebra in a closed Boolean complex define the vertices of a simplex, and the union of these simplices yield a simplicial complex. We may conversely define a Boolean complex from a simplicial complex. The graphs called K-S diagrams in the literature define simplicial complexes of the corresponding Boolean complexes. Strictly speaking, a simplicial complex is the family of simplices, and their union is called the carrier, so we should really call F the σ-complex. However, we shall find it convenient and harmless to conflate the two notions of σ-complex and its carrier.

Let $\mathcal{H}$ be a Hilbert space. Every set of pair-wise commuting projection operators closed under the operation of orthogonal complement $P^{\perp}(= 1 - P)$ and countable intersection $\bigwedge P_i$ forms a σ-algebra. We form the family of all such σ-algebras, and name their union, the σ-complex based on this family, $Q(\mathcal{H})$. The σ-complex $Q(\mathcal{H})$ is the structure in quantum mechanics that replaces the σ-algebra $B(\Omega)$ of Borel sets of the phase space Ω in classical mechanics.

We now summarize this discussion of properties in a form that will serve as a template for each of the other concepts we introduce in the later sections. We first give the classical form of the concept in terms of the σ-algebra $B(\Omega)$; then we generalize the concept by simply replacing the σ-algebra by a σ-complex Q; finally, we specialize to quantum mechanics by taking Q to be the σ-complex $Q(\mathcal{H})$. It then requires a theorem to show that the resulting concept is equivalent to the standard quantum definition on $\mathcal{H}$. Some of the classical concepts are defined in terms of the phase space Ω, rather than the σ-algebra $B(\Omega)$. We must then give an equivalent definition of the concept in terms of $B(\Omega)$.

Classical Mechanics

The properties of a system form the σ-algebra $B(\Omega)$ of Borel sets of the phase space Ω of the system.

General Theory

The properties of a system form a σ-complex Q.

Quantum Mechanics

The properties of a system form the σ-complex $Q(\mathcal{H})$ of projections of the Hilbert space $\mathcal{H}$ of the system.

For a system S with a σ-complex Q, an appropriate interaction with another system, such as a measurement, or, more generally, a decoherent interaction, will elicit a σ-algebra B in Q of properties that have truth values. We shall call B *the (current) interactive algebra* for the system S in the interaction.

For instance, B_{xyz} is the interactive algebra in the triple experiment with the frame (x, y, z). Thus, a measurement of the observable $S_x^2 - S_y^2$ has the interactive algebra B_{xyz}. We may also consider an experiment for which the interactive algebra is $B_z =$

$\{0, 1, S_z^2, 1 - S_z^2\}$. For instance, a variant of the Stern-Gerlach experiment with the magnetic field replaced by an inhomogeneous electric field measures the absolute value $|S_z|$ of S_z, since the electric field vector is a polar vector. For a spin 1 system this amounts to measuring S_z^2. Such an experiment is described in Wrede [12].

In general, a measurement of the observable with discrete spectral decomposition $\sum a_i P_i$ has as interaction algebra the σ-algebra generated by the P_i's. The general case, where the observable contains a continuous spectrum, is described in Sect. "Observables".

In the triple experiment, the interaction algebra B_{xyz} of the measured system is reflected in the isomorphic eight element algebra of events consisting of the subsets of the three spots on the detection screen.

This isomorphism is, as we have seen, a general feature of a measurement, but it is also true for any appropriate decoherent interaction. If the state of the combined two interacting systems is $\sum a_i \phi_i \otimes \psi_i$ at the end of the interaction, then the interaction algebras of the systems are the two σ-algebras generated by the P_{ϕ_i} and the P_{ψ_i}, which are isomorphic. It is important to note that the macroscopic nature of the apparatus plays no role in the classical nature of the interaction algebras as Boolean σ-algebras. It simply follows from the nature that we attributed to extrinsic properties, that in every appropriate interaction they have the classical structure of a σ-algebra. As a consequence, we have no need to (and do not) subscribe to the Copenhagen interpretation, especially espoused by Bohr, that it is necessary to presuppose a classical physical description of the world in order explicate the quantum world. Quantum properties are not intrinsic, but the appropriate interaction elicits an interaction algebra with the classical structure of a σ-algebra.

States

Probability Measures

The theory of probability (following Kolmogorov) is based on a *probability measure*, a countably additive, [0, 1]-valued measure, i.e. a function

$$p : B \to [0, 1]$$

with domain B a σ-algebra, such that $p(1) = 1$, and

$$p(\bigvee a_i) = \sum p(a_i) \text{ for pair-wise disjoint elements } a_1, a_2, \ldots \text{ in } B.$$

In the case of a measurement on a system S, the probability function p gives the probabilities of the σ-algebra of events, or equally of the measured properties of S. A physical theory predicts the probabilities of outcomes of any possible experiment, given the present state. This leads to the following concept of a state.

Classical Mechanics: σ-algebra $B(\Omega)$

A *state* of a system with phase space Ω is a probability measure on the σ-algebra $B(\Omega)$.

General Theory: σ-complex Q

A *state* of a system with a σ-complex of properties Q is a map $p : Q \to [0, 1]$ such that the restriction $p|B$ of p to any σ-algebra B in Q is a probability measure on B.

Quantum Mechanics: σ-complex $Q = Q(\mathcal{H})$

Assume that $\mathcal{H}$ has dimension greater than two. There is a one-one correspondence between states p on $Q(\mathcal{H})$ and density operators (i.e. positive Hermitean operators of trace 1) w on $\mathcal{H}$ such that

$$p(x) = \mathrm{tr}(wx) \text{ for all } x \in Q(\mathcal{H}).$$

That a density operator w defines a probability measure p on $Q(\mathcal{H})$ is an easy computation. The converse, that a state p defines a unique density operator w on $\mathcal{H}$, follows from a theorem of Gleason [13]. Gleason's theorem is the affirmative answere to a question of Mackey [3], in which Mackey asked whether a state on the lattice of projections on $\mathcal{H}$ defines a unique density operator. A careful check of Gleason's proof of the theorem shows that, in fact, the stronger theorem stated above is true, and that the lattice operations on non-commuting projections are not needed for the proof.

As this result shows, the intuitive and plausible definition of classical states leads, with the change from intrinsic to extrinsic properties, to a similar characterization of quantum states.

Pure and Mixed States

The set of states on a σ-complex is closed under the formation of convex linear combinations: if $p_1, p_2, \ldots$ are states then so is $\sum c_i p_i$, for positive c_i, with $\sum c_i = 1$. The above one-one correspondence between states of $Q(\mathcal{H})$ and density operators is convexity-preserving. The extreme points of the convex set of states of a system are those that cannot be written as a non-trivial convex combination of states of the system.

Classical Mechanics: σ-algebra $B(\Omega)$

A *pure state* of a system is an extreme point of the convex set of all states of the system.

For $B(\Omega)$, a pure state p has the form $p(s) = \begin{cases} 1 \text{ if } \omega \in s \\ 0 \text{ if } \omega \notin s \end{cases}$.

In other words, the classical pure states correspond to the points in Ω. Thus, the phase space Ω consists of the pure states, and so is also called the state space.

Thus, in the classical case all the properties of the system in a pure state are either true or false. As we would expect for intrinsic properties, measurements simply find out which measured properties are the case. The general states as mixtures of the pure states can then be interpreted as giving the probabilities of the properties which are true. These may be termed epistemic probabilities, based on the knowledge of the actual pure state that subsists.

General Theory: σ-complex Q

A pure state of a system is an extreme point of the convex set of states of the system.

Quantum Mechanics: σ-complex $Q = Q(\mathcal{H})$

There is a one-one correspondence between the pure states of a system and rays $[\psi]$ of unit vectors ψ in $\mathcal{H}$, such that $p(x) = \langle \psi, x\psi \rangle$.

For it is easily seen that the pure states correspond to one-dimensional projections P_ψ (with ψ in the image of P_ψ) and $p(x) = \mathrm{tr}(P_\psi x) = \langle \psi, x\psi \rangle$. As in the classical case, the state space of the system consists of the pure states, and in this case corresponds to the projective Hilbert space of the rays of $\mathcal{H}$.

In the quantum case, even the pure states predict probabilities that are not 0 or 1, and so these are not the probabilities of properties that already subsist. This is, of course, what we should expect of extrinsic properties. A pure state simply predicts the probabilities of properties in possible future interactions, such as measurements. Mixed states are, as in the classical case, mixtures of the pure states. However, in this case there is no unique decomposition of a mixed case into pure states. This has led to a traditional difficulty in interpreting quantum mixed states. We shall postpone a discussion of our interpretation of mixed states until we have treated conditional probabilities in Sect." Reduction and Conditional Probability".

Observables

Some classical concepts such as observables are defined using the phase space Ω rather than the σ-algebra $B(\Omega)$. We can, in general, restate these definitions in terms of $B(\Omega)$. The reason for this is that the Stone Duality Theorem between Boolean algebras and spaces (and its extension by Loomis to σ-algebras) assures us that constructions on the algebras have their counterparts on the spaces and vice versa.

A classical observable is defined as a real-valued function $f : \Omega \to \mathbb{R}$ on the phase space Ω of the system. To avoid pathological, non-measurable functions, f is assumed to be a Borel function, i.e. a function such that $f^{-1}(s) \in B(\Omega)$, for every set s in the σ-algebra $B(\mathbb{R})$ of Borel sets generated by the open intervals of $\mathbb{R}$.

The inverse function $f^{-1} : B(\mathbb{R}) \to B(\Omega)$ is easily seen to preserve the Boolean σ operations, i.e. to be a homomorphism. Moreover, as we see below, any such homomorphism allows us to recover the function f.

For our purposes, the advantage of using the inverse function is that it involves only the σ-algebra $B(\Omega)$ instead of the phase space Ω, allowing us to generalize the definition to a σ-complex.

Classical Mechanics: σ-algebra $B(\Omega)$

An *observable* of a system with phase space Ω is a homomorphism $u : B(\mathbb{R}) \to B(\Omega)$, i.e. a map u satisfying

$$u(s^{\perp}) = u(s)^{\perp},$$
$$u(\bigvee s_i) = \bigvee u(s_i),$$

for all $s, s_1, s_2, \ldots$ in $B(\mathbb{R})$.

There is a one-to-one correspondence between observables u and Borel functions $f : \Omega \to \mathbb{R}$ such that $u = f^{-1}$.

For given the map u we may define the Borel function f by the equation

$$f(x) = \inf\{y \mid y \in \mathbb{Q},\ x \in u((-\infty, y])\}.$$

The proof that f has the requisite properties is direct, using the denumerability of the rationals $\mathbb{Q}$ to apply the countable additivity of u. (See Varadarajan [5, Theorem 14]).

General Theory: σ-complex Q

An *observable* of a system with σ-complex Q is a homomorphism

$$u : B(\mathbb{R}) \to Q.$$

Note that the image of u lies in a single σ-algebra in Q.

Quantum Mechanics: σ-complex $Q = Q(\mathcal{H})$

There is a one-one correspondence between observables $u : B(\mathbb{R}) \to Q(\mathcal{H})$ and Hermitean operators A on $\mathcal{H}$, such that, given $u, A = \int \lambda dP_{\lambda}$, where $P_{\lambda} = u((-\infty, \lambda])$.

Conversely, given a Hermitean operator A on $\mathcal{H}$, the spectral decomposition $A = \int \lambda dP_{\lambda}$ defines the observable u as the spectral measure $u(s) = \int_s dP_{\lambda}$, for $s \in B(\mathbb{R})$. This establishes the one-one correspondence.

It follows easily that if $u : B(\mathbb{R}) \to Q(\Omega)$ is an observable with corresponding Hermitean operator A, then, for the state p with corresponding density operator w, the expectation of u

$$\mathrm{Exp}_p(u) = \mathrm{tr}(Aw).$$

(See Postulates I and II of Bohm [6]).

The theorem shows the close connection between the measurement of an observable and the interaction algebra of measured properties. For instance, for the case of a discrete operator A, the spectral decomposition $A = \sum a_i P_i$ defines the interaction algebra of measured properties generated by the P_i. Conversely, given the interaction algebra of measured properties, its atoms P_i allow us to define, for each sequence of real numbers a_i, the Hermitean operator $\sum a_i P_i$ which is thereby measured. In particular, we may in this way associate an observable with values 0 and 1 with every property in $Q(\mathcal{H})$. If A is a non-degenerate observable with eigenvalue λ belonging to eigenstate ϕ, we shall often speak of the property $A = \lambda$ to mean the projection P_{ϕ} which has image the ray of ϕ.

Combined Systems

An essential part of the formalism of physics is the mathematical description of the physical union of two systems. In this section we answer the question: what is the σ-complex of the union $S_1 + S_2$ of two systems with given σ-complexes Q_1 and Q_2?

In classical physics, given two systems S_1 and S_2 with the phase spaces Ω_1 and Ω_2, the phase space of the combined system $S_1 + S_2$ is the direct product space $\Omega_1 \times \Omega_2$, whereas for quantum systems with Hilbert spaces $\mathcal{H}_1$ and $\mathcal{H}_2$, the Hilbert space of $S_1 + S_2$ is the tensor product $\mathcal{H}_1 \otimes \mathcal{H}_2$. The direct and tensor products are very different constructions. The dimension of the direct product space is the sum of the dimensions of the two factor spaces, whereas the dimension of the tensor product is the product of the dimensions of the factor spaces. It is this difference that lies behind the promise of quantum computers.

We have nevertheless to combine these two operations via a single construction on the σ-complex Q. When $Q = B(\Omega)$, we may get a clue to the construction by means of Stone duality for Boolean algebras and Boolean spaces. The dual of the direct product of two Boolean spaces is the direct sum $B_1 \oplus B_2$ (also called the free product or co-product) of Boolean algebras. (See Koppelberg [7, Chap. 4]). A similar duality extends to σ-algebras. (See [7, Chap. 5]). We now use our general principle of defining a concept on a σ-complex by reducing it to the corresponding concept on its σ-algebras.

Classical Mechanics: σ-algebra $B(\Omega)$

Given two systems S_1 and S_2 with σ-algebras $B(\Omega_1)$ and $B(\Omega_2)$, the combined system $S_1 + S_2$ has the σ-algebra $B(\Omega_1) \oplus B(\Omega_2)$. There is a unique space $\Omega_1 \times \Omega_2$ such that $B(\Omega_1) \oplus B(\Omega_2) \cong B(\Omega_1 \times \Omega_2)$.

The isomorphism is a well-known part of Stone Duality. For a proof see Koppelberg [7, Chaps. 4 and 5].

General Theory: σ-complex Q

Given two systems S_1 and S_2 with σ-complexes Q_1 and Q_2, the combined system $S_1 + S_2$ has the σ-complex $Q_1 \oplus Q_2$, consisting of the closure (i.e. all the sub-σ-algebras) of the direct sums $B_1 \oplus B_2$ of all pairs of σ-algebras B_1 and B_2 in Q_1 and Q_2.

Quantum Mechanics: σ-complex $Q = Q(\mathcal{H})$

Given the combined system $S_1 + S_2$ with the σ-complex $Q(\mathcal{H}_1) \oplus Q(\mathcal{H}_2)$, there is a unique Hilbert space $\mathcal{H}_1 \otimes \mathcal{H}_2$ such that $Q(\mathcal{H}_1) \oplus Q(\mathcal{H}_2) \cong Q(\mathcal{H}_1 \otimes \mathcal{H}_2)$. (See Postulate IVa of Bohm [6]).

We give an outline of the proof when $\mathcal{H}_1$ and $\mathcal{H}_2$ have finite dimensions. It suffices to show that every element of $Q(\mathcal{H}_1 \otimes \mathcal{H}_2)$ lies in $Q(\mathcal{H}_1) \oplus Q(\mathcal{H}_2)$. The elements of $Q(\mathcal{H}_1) \oplus Q(\mathcal{H}_2)$ are generated by the one-dimensional projections $P_{\phi \otimes \psi}$, where $\phi \in \mathcal{H}_1$ and $\psi \in \mathcal{H}_2$. We must show that if Γ is an arbitrary unit vector in $\mathcal{H}_1 \otimes \mathcal{H}_2$, then P_Γ lies in $Q(\mathcal{H}_1) \oplus Q(\mathcal{H}_2)$. One definition of the tensor product allows us to think of Γ as a conjugate-linear map from $\mathcal{H}_2$ to $\mathcal{H}_1$. (See Jauch [14], for example). The proof proceeds by induction on the rank of Γ as such a map. The maps of rank 1 are of the form $P_{\phi \otimes \psi}$, so the basis of the induction is true.

Now suppose Γ has rank n.

The proof is greatly simplified by choosing suitable orthonormal bases in $\mathcal{H}_1$ and $\mathcal{H}_2$ in which to expand Γ. We can construct bases $\{\phi_i\}$ and $\{\psi_i\}$ in $\mathcal{H}_1$ and $\mathcal{H}_2$ such that $\Gamma = \sum c_i \phi_i \otimes \psi_i$, with the c_i real. (Briefly, $\Gamma\Gamma^*$ and $\Gamma^*\Gamma$ have common strictly positive eigenvalues, say a_i, and respective eigenvectors ϕ_i and ψ_i; it follows that $\Gamma = \sum \sqrt{a_i}\phi_i \otimes \psi_i$. See Jauch [14], for example).

Let

$$\Theta = \begin{cases} -c_2\phi_1 \otimes \psi_1 + c_1\phi_2 \otimes \psi_2, & \text{for } n = 2 \\ c_1\phi_3 \otimes \psi_1 + c_2\phi_2 \otimes \psi_3 + c_3\phi_1 \otimes \psi_3 + \sum_{i>3} c_i\phi_i \otimes \psi_1, & \text{for } n > 2 \end{cases}$$

$$\Delta = \; c_1\phi_2 \otimes \psi_1 + c_2\phi_1 \otimes \psi_2 + \sum_{i\geq 3} c_i\phi_i \otimes \psi_2$$

Then Γ, Θ, and Δ are pairwise orthogonal unit vectors. Hence, P_Γ, P_Δ, and P_Θ mutually commute, and $P_\Gamma = (P_\Gamma \vee P_\Theta) \wedge (P_\Gamma \vee P_\Delta)$.

For $n = 2$, let $x_+ = c_2\Gamma + c_1\Theta$ and $x_- = c_1\Gamma - c_2\Theta$. For $n > 2$, let $x_\pm = \Gamma \pm \Theta$. Also, let $y_\pm = \Gamma \pm \Delta$. Then it is easily checked that the four vectors $x_\pm$ and $y_\pm$ are of rank $n - 1$, and x_+ and x_- are orthogonal, as are y_+ and y_-. It follows that $[P_{x_+}, P_{x_-}] = [P_{y_+}, P_{y_-}] = 0$. Moreover, $P_\Gamma \vee P_\Theta = P_{x_+} \vee P_{x_-}$ and $P_\Gamma \vee P_\Delta = P_{y_+} \vee P_{y_-}$. Hence, $P_\Gamma = (P_{x_+} \vee P_{x_-}) \wedge (P_{y_+} \vee P_{y_-})$. Since $P_{x_+}, P_{x_-}, P_{y_+}$, and P_{y_-} inductively lie in $Q(\mathcal{H}_1) \oplus Q(\mathcal{H}_2)$ and each of the pairs (P_{x_+}, P_{x_-}), (P_{y_+}, P_{y_-}), and $(P_{x_+} \vee P_{x_-}, P_{y_+} \vee P_{y_-})$ lie in a common σ-algebra, it follows that P_Γ lies in $Q(\mathcal{H}_1) \oplus Q(\mathcal{H}_2)$. The proof provides an algorithm for constructing $x_\pm$ and $y_\pm$.

The uniqueness (up to isomorphism) is a routine consequence of the fact that $Q_1 \oplus Q_2$ is categorically a co-product (see Koppellberg [7] for a proof in the σ-algebra case).

The infinite dimensional case is discussed in Sect. "Reconstructing the σ-Complex $Q(\mathcal{H})$".

As an illustration we consider the simplest case of the tensor product $\mathcal{H}_1 \otimes \mathcal{H}_2$ of two-dimensional Hilbert spaces, which we may take to represent two spin $\frac{1}{2}$ particles. Each element of $Q(\mathcal{H}_1)$ (resp. $Q(\mathcal{H}_2)$) corresponds to the property $s_z \otimes I = \frac{1}{2}$ (resp. $I \otimes s_z = \frac{1}{2}$) for some direction z. For Γ in $\mathcal{H}_1 \otimes \mathcal{H}_2$ we shall identify $P_{x_+}, P_{x_-}, P_{y_+}$, and P_{y_-}.

We write the vector Γ in the diagonal form $c_1\phi_1 \otimes \psi_1 + c_2\phi_2 \otimes \psi_2$. Hence,

$$x_- = \phi_1 \otimes \psi_1, \; x_+ = \phi_2 \otimes \psi_2$$

$$y_+ = (\phi_1 + \phi_2) \otimes (c_1\psi_1 + c_2\psi_2), \; y_- = (\phi_1 - \phi_2) \otimes (c_1\psi_1 - c_2\psi_2)$$

Now ϕ_1 defines $s_z \otimes I = \frac{1}{2}$ for a direction z, and ψ_1 defines $1 \otimes s_w = -\frac{1}{2}$ in a direction w. Thus, $\phi_1 \pm \phi_2$ defines $s_x \otimes I = \pm\frac{1}{2}$ for a direction x orthogonal to z. Also, if we write $c_1 = \cos(\mu/2)$, then $c_1\psi_1 + c_2\psi_2$ defines $I \otimes s_u = \frac{1}{2}$ in a direction u at an angle μ from the w direction, and $c_1\psi_1 - c_2\psi_2$ defines $I \otimes s_v = \frac{1}{2}$ in a direction v at angle $-\mu$ from the w direction. It follows that

$$\begin{aligned} P_\Gamma &= (P_{x_+} \vee P_{x_-}) \wedge (P_{y_+} \vee P_{y_-}) \\ &= \left(s_z \otimes I = \frac{1}{2} \leftrightarrow I \otimes s_w = -\frac{1}{2}\right) \wedge \left[\left(I \otimes s_u = \frac{1}{2} \to s_x \otimes I = \frac{1}{2}\right) \wedge \left(s_x \otimes I = \frac{1}{2} \to I \otimes s_v = \frac{1}{2}\right)\right] \end{aligned}$$

In this manner every state in a combined system can be interpreted as a compound proposition about the factors.

A particularly interesting case is the singleton state $\Gamma = \sqrt{\frac{1}{2}}(\phi_z^+ \otimes \psi_z^- - \phi_z^- \otimes \psi_z^+)$, (with $s_z\phi_z^\pm = \pm\frac{1}{2}\phi_z^\pm$ and $s_z\psi_z^\pm = \pm\frac{1}{2}\psi_z^\pm$) where

$$\begin{aligned} P_\Gamma &= (P_\Gamma \vee P_\Theta) \wedge (P_\Gamma \vee P_\Delta) \\ &= (S_z = 0) \wedge (S_x = 0) \\ &= (P_{x_+} \vee P_{x_-}) \wedge (P_{y_+} \vee P_{y_-}) \\ &= (s_z \otimes I = \frac{1}{2} \leftrightarrow I \otimes s_z = -\frac{1}{2}) \wedge (s_x \otimes I = \frac{1}{2} \leftrightarrow I \otimes s_x = -\frac{1}{2}). \end{aligned}$$

In Sect. "The Einstein-Podolsky-Rosen Experiment" we shall apply this result to the EPR experiment.

This construction of the direct sum generalizes in an obvious way to the direct sum of an arbitrary number of σ-complexes, representing the union of several systems. The above theorems then generalize to:

$$\begin{aligned} B(\Omega_1) \oplus B(\Omega_2) \oplus \cdots &\cong B(\Omega_1 \times \Omega_2 \times \cdots) \\ Q(\mathcal{H}_1) \oplus Q(\mathcal{H}_2) \oplus \cdots &\cong Q(\mathcal{H}_1 \otimes \mathcal{H}_2 \otimes \cdots). \end{aligned}$$

These general sums are needed in discussing statistical mechanics. It is now routine to define symmetric and anti-symmetric direct sums of σ-complexes, yielding the corresponding symmetric and anti-symmetric tensor products of Hilbert spaces, needed to deal with identical particles. (See Postulate IVb of Bohm [6]. The spin-statistics connection that Bohm adds can also be added here).

Symmetries

As Noether, Weyl, and Wigner showed, observables such as position, momentum, angular momentum, and energy arise from global symmetries of space and time, and the conservation laws for them arise from the corresponding symmetries of interactions. Other observables arise from local symmetries. In classical physics the symmetries appear as canonical transformations of phase space, and in quantum physics they appear as unitary or anti-unitary transformations of Hilbert space. For us they naturally appear as symmetries of a σ-complex.

Definition An automorphism of a σ-complex Q is a one-one transformation $\sigma : Q \to Q$ of Q onto Q such that for every σ-algebra B in Q and all $a, a_1, a_2, \ldots$ in B

$$\sigma(a^\perp) = \sigma(a)^\perp \text{ and } \sigma(\bigvee a_i) = \bigvee \sigma(a_i).$$

General Theory: σ-complex Q

A *symmetry* of a system with σ-complex Q is given by an automorphism of Q.

A symmetry σ defines a natural convexity-preserving map $p \to p_\sigma$ on the states of Q by letting $p_\sigma = p \circ \sigma^{-1}$, i.e. $p_\sigma(x) = p(\sigma^{-1}(x))$, for all $x \in Q$.

Quantum Mechanics: σ-complex $Q = Q(\mathcal{H})$

There is a one-one correspondence between symmetries $\sigma : Q(\mathcal{H}) \to Q(\mathcal{H})$ and unitary or anti-unitary operators u on $\mathcal{H}$ such that $\sigma(x) = uxu^{-1}$, for all $x \in Q(\mathcal{H})$.

If a state p corresponds to the density operator w, then

$$p_\sigma(x) = p(\sigma^{-1}(x)) = \mathrm{tr}(wu^{-1}xu) = \mathrm{tr}(uwu^{-1}x),$$

so that the state p_σ corresponds to the density operator uwu^{-1}.

It is easy to check that unitary and anti-unitary operators define a symmetry on $Q(\mathcal{H})$. For the converse we use a well-known theorem of Wigner. (See Bargmann [15]). The original theorem of Wigner posits a one-one map of the set of rays of $\mathcal{H}$ onto itself which preserves the inner product. Uhlhorn [16] was able to weaken this to preserving the orthogonality of rays. As Bargmann states in [15], the proof he gives of Wigner's theorem may be easily modified to prove Uhlhorn's result. (For a proof see Varadarjan [5]).

Now assume that σ is a symmetry of $Q(\mathcal{H})$. Then σ is a one-one map of the set of atoms, i.e. one-dimensional projections P_ψ, of $Q(\mathcal{H})$ onto atoms of $Q(\mathcal{H})$. In other words, rays $[\psi]$ of $\mathcal{H}$ are one-to-one mapped onto rays of $\mathcal{H}$. Moreover, since σ-algebras are mapped by σ to σ-algebras, the orthogonality of rays is preserved. The Uhlhorn version of Wigner's theorem then shows there is a unique (up to a multiplicative constant) unitary or anti-unitary map u on $\mathcal{H}$ such that $\sigma(x) = uxu^{-1}$.

In the case of classical physics, with $Q = B(\Omega)$, a symmetry is defined by a canonical transformation of the manifold. Every such transformation defines an automorphism of the σ-algebra $B(\Omega)$. However, the converse is not true. Although the automorphism still defines a continuous map from Ω to itself, the structure of a σ-algebra is too weak to recover the canonical structure. It is remarkable that the σ-complex structure is sufficient to allow one to define the symmetries of the Hilbert space. In that sense, quantum physics allows a more satisfactory reconstruction than classical physics. As Sect. "From Quantum Physics to Classical Physics" suggests, we may recover the classical canonical structure from the quantum structure in the limit of an increasing number of particles.

Dynamics

Now that we have shown that the symmetries of $Q(\mathcal{H})$ are implemented by symmetries of $\mathcal{H}$, we may use time symmetry to introduce a dynamics for systems.

To define dynamical evolution, we consider systems that are invariant under time translation. For such systems, there is no absolute time, only time differences. The change from time 0 to time t is given by a symmetry $\sigma_t : Q \to Q$, since the structure of the system of properties is indistinguishable at two values of time. We assume that if the state evolves first for a time t and then the resulting state for a time t', then this yields the same result as the original state evolving for a time $t + t'$. Moreover, we assume that evolution over a small time period results in small changes in the probability of properties occurring.

The passage of time is thus given by a continuous representation of the additive group $\mathbb{R}$ of real numbers into the group $\mathrm{Aut}(Q)$ of automorphisms of Q under composition:
i.e. a map $\sigma : \mathbb{R} \to \mathrm{Aut}(Q)$, such that

$$\sigma_{t+t'} = \sigma_t \circ \sigma_{t'}$$

and $p_{\sigma_t}(x)$ is a continuous function of t.

The image of σ is then a continuous one-parameter group of automorphisms on Q.[2]

We have seen that an automorphism σ corresponds to a unitary or anti-unitary operator. Anti-unitary operators actually occur as symmetries, for instance in time reversal. However, for the above representation only unitary operators u_t corresponding to the symmetry σ_t can occur, since $u_t = u_{t/2}^2$, which is unitary.[3]

It follows that the evolving state p_{σ_t} corresponds to the density operator $w_t = u_t w u_t^{-1}$. By Stone's Theorem,

$$u_t = e^{-\frac{i}{\hbar}Ht},$$

where $\hbar$ is a constant to be determined by experiment; so

$$w_t = e^{-\frac{i}{\hbar}Ht} w \, e^{\frac{i}{\hbar}Ht}.$$

---

[2]The group $\mathrm{Aut}(Q)$ may, in fact, be construed as a topological group by defining, for each $\epsilon > 0$, an ϵ-neighborhood of the identity to be $\{\sigma \mid |p_\sigma(x) - p(x)| < \epsilon$ for all x and $p\}$. We may then directly speak of the continuity of the map σ, in place of the condition that $p_{\sigma_t}(x)$ is continuous in t.

[3]More precisely, we have a projective unitary representation of $\mathbb{R}$, but such a representation of $\mathbb{R}$ is equivalent to a vector representation. (See, e.g., Varadarajan [5]).

Differentiating,

$$\partial_t w_t = -\frac{i}{\hbar}[H, w_t].$$

This is the Liouville-von Neumann Equation.

Conversely, this equation yields a continuous representation of $\mathbb{R}$ into $\mathrm{Aut}(\mathcal{Q}(\mathcal{H}))$. For $w = P_\psi$, a pure state, $w_t = P_{\psi(t)}$ and this equation reduces to the Schrödinger Equation:

$$\partial_t \psi(t) = -\frac{i}{\hbar} H \psi(t).$$

(See Postulate Va of Bohm [6]. Postulate Vb is the Heisenberg form of the equation, and follows similarly).

We stop here without specifying any further the form of the Hamiltonian H. This form depends upon calculating the linear and angular momentum observables as operators from the homogeneity and isotropy of space, using the corresponding unitary representations that we have used for time homogeneity. This a well-known part of quantum mechanics and need not be explored further here. (See Jauch [14], for example). We have treated the non-relativistic dynamical equation. The connection between automorphisms of $\mathcal{Q}(\mathcal{H})$ and unitary operators given above allows to us to treat the relativistic dynamical equations in a similar manner, following Wigner's work. (See Varadrajan [5]).

Reduction and Conditional Probability

Conditional States

With these results, which cover four of Bohm's five postulates, we can now recover much of quantum theory. So far however, we will never predict interference. The states we introduced are probability measures on $\mathcal{Q}$, which for any experiment is a classical probability measure on the σ-algebra of properties being measured. In fact, the probability must be classical, since it is mirrored in the probability measure on the experiment's σ-algebra of events, which are generated by macroscopic spots on a screen.

How then does interference enter the picture? In dealing with experiments, we have omitted a key ingredient that is usually referred to as "the preparation of state." To calculate the probability $p(x)$ of a property holding at the end of an experiment, we need to know both the property x and the state p. In general, when we are presented with a particle to be measured, we do not know its state. One way to know the state is to prepare it by means of a prior interaction.

For instance, the book [17] by Feynman, Leighton, Sands introduces quantum mechanics via a spin 1 system by discussing the probability of, for instance, going to state $S_x = 1$, given that it is in state $S_z = 0$. The particle is prepared in state $S_z = 0$

by sending it through a Stern-Gerlach field in the z direction, and then filtering it through a one-slit screen to allow only the central beam through. If the system is not detected as hitting the filtering screen, then it is reduced to the state $S_z = 0$. If allowed to hit a final detection screen it is certain to register the central spot. But we are free to send it through another Stern-Gerlach field in the x direction to measure $S_x = 1$, say. This is a reduction by preparation of the original, possibly unknown, state to the state $S_z = 0$.

Some physicists think that reduction is a phenomenon unique to quantum mechanics that has no counterpart in classical mechanics, but this not the case. Consider a one slit experiment with bullets. If we shoot at a target, we get a probability distribution on the target that defines a mixed state for the bullet. Since the target screen can be placed anywhere from the gun to any distant point, the probability distribution is a function of time that gives a time evolution of this state, satisfying the classical Liouville equation for mixed states. If we now interpose a one-slit screen between the gun and the target screen, we find that after the evolution of the state p up to the one-slit screen, the bullet either has hit this screen, or if not, has passed through with a new state $p(\cdot \mid y)$, where y is the property that it has not hit the screen. This is classically called conditionalizing the state p to y. The new state $p(\cdot \mid y)$ is defined by $p(x \mid y) = p(x \wedge y)/p(y)$, as the frequency definition of probability can verify. This filtering to a new state is entirely similar to the filtering of a spin 1 system described earlier, and is the classical equivalent of reduction.

Now that we have the classical form of reduction as conditionalization, we can follow our prescription by generalizing from a σ-algebra to a σ-complex.

Classical Mechanics: σ-algebra $B(\Omega)$

Let p be a state on the σ-algebra $B(\Omega)$ and $y \in B(\Omega)$ such that $p(y) \neq 0$. By a *state conditionalized on* y we mean a state $p(\cdot \mid y)$ such that for every x in $B(\Omega)$,

$$p(x \mid y) = p(x \wedge y)/p(y).$$

General Theory: σ-complex Q

Let p be state on a σ-complex Q and $y \in Q$ such that $p(y) \neq 0$. By a *state conditionalized on* y we mean a state $p(\cdot \mid y)$ such that for every σ-algebra B in Q containing y and every $x \in B$,

$$p(x \mid y) = p(x \wedge y)/p(y).$$

In the literature, there exist generalizations of probability measures and conditional probability to non-commutative algebras, and, in particular, to lattices of projections. (See Beltrametti and Cassinelli [18]). In general, it is by no means clear that such a state $p(\cdot \mid y)$ either exists or is unique, as is obviously the case for classical mechanics. However, for the quantum σ-complex $Q(\mathcal{H})$ this can be proved:

Quantum Mechanics: σ-complex $Q = Q(\mathcal{H})$

If p is a state on $Q(\mathcal{H})$ and $y \in Q(\mathcal{H})$ such that $p(y) \neq 0$, then there exists a unique state $p(\cdot \mid y)$ conditionalized on y. If w is the density operator corresponding to p, then

$ywy/\operatorname{tr}(ywy)$ is the density operator corresponding to the state $p(\cdot \mid y)$.

To see that the operator $ywy/\operatorname{tr}(ywy)$ corresponds to the state $p(\cdot \mid y)$, note that if x lies in the same σ-algebra as y, then x and y commute, so

$$\operatorname{tr}(ywyx)/\operatorname{tr}(ywy) = \operatorname{tr}(wxy)/\operatorname{tr}(wy) = p(x \wedge y)/p(y) = p(x \mid y).$$

For uniqueness, it suffices to consider the case when $x \in Q(\mathcal{H})$ is a one-dimensional projection. Let $p(\cdot \mid y)$ be a state conditionalized on y, and let v be the corresponding density operator. Let ϕ be a unit vector in the image of x. We can write $\phi = y\phi + y^{\perp}\phi$. Then

$$\begin{aligned} p(x \mid y) &= \operatorname{tr}(vx) = \langle \phi, v\phi \rangle \\ &= \langle y\phi, vy\phi \rangle + \left\langle y\phi, vy^{\perp}\phi \right\rangle + \left\langle y^{\perp}\phi, vy\phi \right\rangle + \left\langle y^{\perp}\phi, vy^{\perp}\phi \right\rangle . \end{aligned}$$

Now, $\operatorname{tr}(vy^{\perp}) = p(y^{\perp} \mid y) = p(y^{\perp} \wedge y)/p(y) = 0$, so $vy^{\perp}\phi = 0$. Hence,

$$p(x \mid y) = \langle y\phi, vy\phi \rangle = \|y\phi\|^2 \operatorname{tr}(vP_{y\phi}) = \|y\phi\|^2 p(P_{y\phi})/p(y),$$

since $P_{y\phi} \leq y$. If $p'(\cdot \mid y\}$ is another state conditionalized on y, then

$$p'(x \mid y) = \|y\phi\|^2 p(P_{y\phi})/p(y) = p(x \mid y),$$

proving uniqueness.

The change from w to $ywy/\operatorname{tr}(wy)$ in state preparation or measurement is the general formula for the reduction of state given by the von Neumann-Lüders Projection Rule. In the orthodox interpretation this rule is an additional principle that is appended to quantum mechanics. Here it appears as the unique answer to conditionalizing a state to a given property. (See Postulate IIIa of Bohm [6]).

The natural definition of applying a symmetry σ to a conditionalized state $p(\cdot \mid y)$ is given by

$$p_\sigma(x \mid y) = p(\sigma^{-1}(x) \mid \sigma^{-1}(y)).$$

Classical and Quantum Conditional Probability

In the well-known paper [19], Feynman writes that the basic change from classical to quantum mechanics lies in the revision in the probability rule called the Law of Alternatives,
$p(a \mid c) = \sum_i p(a \mid b_i)p(b_i \mid c)$ for disjoint b_i, to the quantum law that $\langle \alpha \mid \beta \rangle = \sum_i \langle \alpha \mid \beta_i \rangle \langle \beta_i \mid \gamma \rangle$, giving an additional interference term.

We agree that this is an important difference in the two theories. However, we shall derive it from what we consider the more basic difference, that between intrinsic and extrinsic properties.

Let $y_1, y_2, \ldots$ lie in a σ-algebra with $y_i \wedge y_j = 0$ for $i \neq j$, and let $y = \bigvee y_i$. Then

Classical Mechanics:

$$\begin{aligned} p(x \mid y) &= p(\bigvee(x \wedge y_i))/p(y) \\ &= \sum(p(x \wedge y_i)/p(y_i)) \cdot (p(y_i)/p(y)) \\ &= \sum p(x \mid y_i)p(y_i \mid y), \end{aligned}$$

The Law of Alternatives in classical probability theory.

On the other hand, by Sect. "Conditional States", we have

Quantum Mechanics:

$$\begin{aligned} p(x \mid y) &= \mathrm{tr}(ywyx)/\,\mathrm{tr}(wy) \\ &= \mathrm{tr}(\bigvee_{i,j} y_i w y_j x)/\,\mathrm{tr}(wy) \\ &= \sum \mathrm{tr}(y_i w y_i x)/\,\mathrm{tr}(wy) + \sum_{i \neq j} \mathrm{tr}(y_i w y_j x)/\,\mathrm{tr}(wy) \\ &= \sum p(x \mid y_i)p(y_i \mid y) + \sum_{i \neq j} \mathrm{tr}(y_i w y_j x)/\,\mathrm{tr}(wy). \end{aligned}$$

This shows that in conditionalizing for the extrinsic properties of quantum mechanics an interference term must be added to the classical law of alternatives.

Conditionalizing on Several Properties

There is a different kind of preparation of state, one which leads to a mixed state. This occurs when, instead of all but one of the beams being blocked, as in Sect. "Conditional States", the beams are allowed to pass through the filter, while being registered. For instance, [17] describes a version of the two-slit experiment in which the particle scatters high frequency photons that register which slit the particle passed through. In this case, the property y_1 of passing through slit 1 is true or the property y_2 of passing through slit 2 is true, so that the state of the particle is either the conditional state $p(\cdot \mid y_1)$ or the state $p(\cdot \mid y_2)$.

If we consider an ensemble of particles, then each of the particles in the ensemble will be in the state $p(\cdot \mid y_i)$ with probability $p(y_i)$, for $i = 1, 2$, so that the ensemble is in the mixed state $p(y_1)p(\cdot \mid y_1) + p(y_2)p(\cdot \mid y_2)$. Thus, by registering the results of passage through each of the two slits, we restore the classical Law of Alternatives.

For a single particle, the same mixed state describes its predicted state upon passage through the registering two-slit screen. However, upon actual passage through the registered slits, the state is either $p(\cdot \mid y_1)$ or $p(\cdot \mid y_2)$. We may say that even after the passage, the state of the particle for an experimenter who is not aware of the registered result the state remains the mixed state. In this regard, the mixture has a similar interpretation as in the classical case, viz., the ignorance interpretation of mixtures.

A measurement of an observable is the most familiar example of conditionalizing with respect to several properties. If the observable has a spectral decomposition $\sum a_i P_i$, then measuring the observable amounts to registering the values of the properties given by the P_i. The interaction algebra B is the σ-algebra generated by the P_i.

We now formulate this notion of conditioning with respect to several conditions. Given a system with σ-complex Q and disjoint elements $y_1, y_2, \ldots$ in a common σ-algebra in Q, and a state p, we define the state conditionalized on $y_1, y_2, \ldots$ to be $p(\cdot \mid y_1, y_2, \ldots) = \sum p(y_i) p(\cdot \mid y_i)$. We shall also write this more succinctly as $p(\cdot \mid B)$, the state conditionalized on the interaction algebra B, the σ-algebra generated by the y_i.

For quantum mechanics, with $Q = Q(\mathcal{H})$, if w is the density operator corresponding to the
state p:

$$p(\cdot \mid B) = \sum \mathrm{tr}(w y_i)(y_i w y_i / \mathrm{tr}(w y_i)) = \sum y_i w y_i,$$

so that for each x the probability $p(x \mid B) = \sum \mathrm{tr}(y_i w y_i x)$. This gives the state of an ensemble without selection. (See Postulate IIIb of Bohm [6]).

The natural definition for applying a symmetry to the conditioned state is given by

$$p_\sigma(x \mid B) = p(\sigma^{-1}(x) \mid \sigma^{-1} B).$$

Note that the non-uniqueness of the decomposition of a degenerate density operator into pure states causes no problems in this interpretation. This is because mixed states arise as mixtures of given pure states in the conditionalization from an experiment or the evolution of the mixture. The σ-algebra B generated by the $y_1, y_2, \ldots$ is simply the current interaction algebra of the σ-complex, and is always given to us as part of the interaction.

The fact that degenerate density operators do not have a unique decomposition into pure states has led some to put mixed and pure states on an equal footing, and to deny them the role as mixtures. This puts the cart before the horse, and ignores the historical development of the concept of mixed states. Mixtures of pure states were in long use in quantum mechanics (as well as in classical statistical mechanics) when von Neumann introduced the invariant formulation of a mixed state as a density operator. The use of the density operator has the advantage of allowing the introduction of the abstract notion of mixed state, without requiring the explicit mention of any basis of pure states, which could be recovered in the non-degenerate case. For us, however, in any interaction (and subsequent evolution) the interaction algebra is always given, which yields a unique decomposition of the mixed state as a mixture of pure states even in the degenerate case.

Reconstructing the σ-Complex $Q(\mathcal{H})$

We saw in Sect. "Properties" that if we restrict ourselves to classical experiments, then the σ-complex of interaction algebras can be imbedded into a σ-algebra. On the other hand, the 40 quantum triple experiments yield a σ-complex that cannot be so imbedded. Thus, increasing the set of experiments has changed the structure of the σ-complexes of systems. It may then be possible that a sufficiently comprehensive family of experiments may force the structure of the σ-complex Q to be isomorphic to $Q(\mathcal{H})$. In this section we shall see that this is indeed the case.

The result is based on the paper Reck, Zeilinger, Bernstein, Bertani [20]. The interactions arise from a composition of interferometers. First, Mach-Zender interferometers together with beam splitters allow one to construct $Q(\mathcal{H}_2)$, where $\mathcal{H}_2$ is a two-dimensional Hilbert space. A standard theorem, which allows one to decompose n-dimensional unitary operators as a product of two-dimensional ones, is then used to treat the σ-complex of higher dimensional Hilbert spaces.

We outline the construction in [20] (from which the diagrams below are copied). The experimental realization of a general two-dimensional unitary matrix is obtained by a Mach-Zender interferometer consisting of two mirrors, two 50-50 beam splitters, an ω-phase shifter, and a ϕ-phase shifter at one output port:

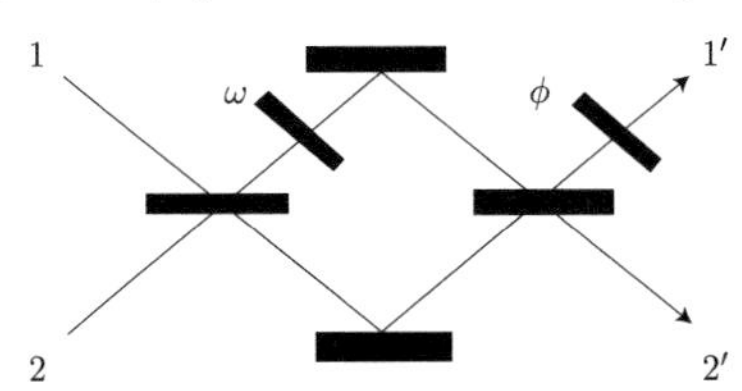

This device transforms the input state with modes (k_1, k_2) into the output state with modes (k_1', k_2'), which are related by the unitary matrix:

$$\begin{pmatrix} k_1' \\ k_2' \end{pmatrix} = \begin{pmatrix} e^{i\phi} \sin\omega & e^{i\phi} \cos\omega \\ \cos\omega & -\sin\omega \end{pmatrix} \begin{pmatrix} k_1 \\ k_2 \end{pmatrix}.$$

We can then realize all 2-dimensional unitary matrices by varying the phase shifters.

To treat $n \times n$ unitary matrices, the authors in [20] show how to eliminate the off-diagonal element u_{jk} of a unitary matrix U by multiplying U by the matrix T_{jk} which is obtained from the $n \times n$ identity matrix I by replacing the (jj), (jk), (kj), (kk) entries by the entries of a matrix of the above 2-dimensional unitary form. This inductively results in the product

$$UT_{nn-1}T_{nn-2} \cdots T_{32}T_{31}T_{21} = D$$

where D is a diagonal unitary matrix with diagonal entries of modulus 1. Hence

$$U = DT^{\dagger}_{21}T^{\dagger}_{31}T^{\dagger}_{32}\cdots T^{\dagger}_{n1}T^{\dagger}_{n2}\cdots T^{\dagger}_{nn-1}.$$

We now combine copies of the above interferometers so that the outputs of one are the inputs of the succeeding one, corresponding to the above product of the $T^{\dagger}_{jk}$ matrices, followed by n phase shifters to account for the matrix D. The result is a device which realizes the matrix U. For instance, for $n = 3$, we have:

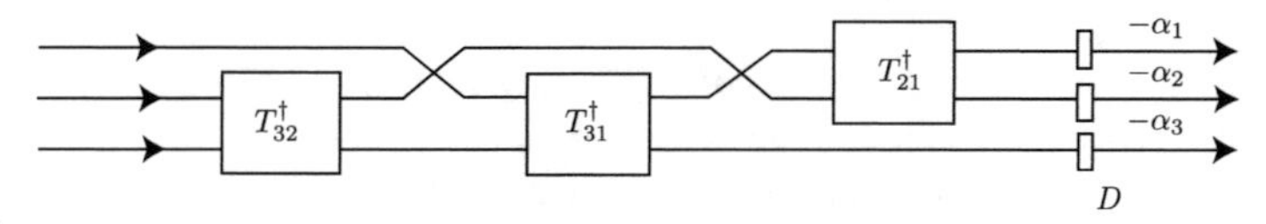

(Each box represents an interferometer of the above type.)

To realize an n-dimensional Hermitean matrix A, we use additional beam splitters to superpose those beams that correspond to the same eigenspace of A, and then add detectors for the resulting beams. The use of beam splitters to superpose beams is well-known. (See e.g. Zukoowski, Zeilinger, and Horne [21]).

This is a précis of the construction in [20]. It allows us to realize every element of $Q(\mathcal{H})$, where $\mathcal{H}$ is an n-dimensional complex Hilbert space. What is significant is that we can also realize the σ-complex structure of $Q(\mathcal{H})$. To see this it suffices to consider the two Boolean operations of complementation $x^{\perp}$ and join $x \vee y$. The output for a projection x consists of two beams, labeled the 1-beam and the 0-beam according to the eigenvalues of x. The operation of complementation $x^{\perp}$ requires only a transposition of the 1 and 0 labels. The join $x \vee y$ of two projections corresponds to superposing the two 1-beams of x and y. These two operations suffice to define all the Boolean operations, and therefore the σ-complex structure of $Q(\mathcal{H})$. Note that this realization of the σ-complex of properties via the different σ-algebras generated by the outcomes of interferometer experiments follows the general prescription given in Sect. "Properties" for defining the σ-complex of properties of a system by means of the different σ-algebras of events defined by the experimental outcomes.

It is instructive to contrast the simple experimental counterparts to the σ-complex structure with the lattice structure of the set of projections. We know of no corresponding experimental realization to the lattice join (or meet) of two non-commuting projections. This is due to the difficulty of relating the eigenspaces of two non-commuting operators to the eigenspaces of their sum (or, for projections, to their union), while for commuting operators there is a simple relation. It is this difficulty that is alluded to in our earlier quotations from Varadarajan [5] and Birkhoff and von Neumann [2] in the introduction.

We have seen that if we can in principle form arbitrarily large networks of interferometers, then we can realize the σ-complex $Q(\mathcal{H})$ for Hilbert spaces of all finite dimensions. The single minimal space $\mathcal{H}$ for which $Q(\mathcal{H})$ realizes *all* the interferometer experiments, and hence contains all finite dimensional Hilbert spaces, is an infinite dimensional separable pre-Hilbert space, i.e. an inner product space $\mathcal{H}$, whose completion forms a separable Hilbert space $\mathcal{H}_{\omega}$. To see this note that $\mathcal{H}$ may be construed as the space of all complex sequences $\{a_i\}$ that are non-zero for only finite many i, with inner product $\langle\{a_i\},\{b_i\}\rangle = \sum a_i b_i$.

Thus in the infinite dimensional case we must add ideal elements which are limits of sequences of realized elements. We cannot expect to realize $Q(\mathcal{H}_\omega)$ via experiments without adding limits since the world itself may be finite. This is similar to the use of probability in physica as an ideal limit of relative frequency for longer and longer sequences of experiments. Of course, even the above realization of $Q(\mathcal{H})$ in the finite dimensional case is an idealization, since it requires ω-phase shifters for arbitrary real ω, in $[0, 2\pi]$.

We may now extend the result $Q(\mathcal{H}_1) \oplus Q(\mathcal{H}_2) \simeq Q(\mathcal{H}_1 \otimes \mathcal{H}_2)$ of Sect. "Combined Systems" to the infinite dimensional case.

The fact that $\mathcal{H}$ is the *minimal* space such that $Q(\mathcal{H})$ is realized by the above interferometry experiments highlights the open-ended nature of our reconstruction. If we restrict ourselves to experiments of classical physics, then the σ-complex reduces to a σ-algebra, and the concepts lead to classical physics. If we add the forty triple experiments, the resulting σ-complex cannot be imbedded into a σ-algebra. If we allow for the interferometry experiments of this section, then Q must take the form $Q(\mathcal{H})$. It thus suffices to consider these interferometry experiments to realize the structure of quantum physics. We may then apply the resulting theory to general interactions.[4] As we have emphasized throughout the paper, the special nature of experiments, with the macroscopic apparatus, plays no role in the theory. Any appropriate decoherent interaction gives rise to isomorphic σ-algebras for the two systems. Experiments do play the pragmatic role of allowing us to become cognizant of a sufficient number of interactions to help deteremine the theory.

It is possible that other experiments may require a different realization of the σ-complexes. For instance, if we consider systems which satisfy *superselection rules* (see e.g. [18]), then the σ-complex Q has a non-trivial σ-algebra which is common to all the σ-algebras B in Q. In this case Q is not of the form $Q(\mathcal{H})$, but is a sub-σ-complex of $Q(\mathcal{H})$. $\mathcal{H}$ takes the form of a direct sum $\oplus\mathcal{H}_i$ of Hilbert spaces with the pure states forced to lie in a factor $\mathcal{H}_i$.

From Quantum Physics to Classical Physics

With the description in Sect. "Combined Systems" of the σ-complex of combined systems, it is possible to treat the statistics of a large number of particles such as macroscopic bodies. This is, of course, a major subject in quantum statistics, and we shall not venture there. However, we wish to say a few words on how the σ-complex of quantum mechanics tends to a classical σ-algebra with an increasing number of particles, so that the quantum system becomes effectively classical.

[4]Historically, of course, it was not such interferometry experiments, but rather spectroscopic experiments that lead Schrödinger to his equation.

We shall adapt a remark in Finkelstein [22] for this purpose. Let S be an ensemble of n non-interacting copies of a system S_i, $i = 1, 2, \ldots, n$, with σ-complex $Q(\mathcal{H}_i)$. Then S has the σ-complex

$$Q(\mathcal{H}_i) \oplus Q(\mathcal{H}_2) \oplus \cdots \oplus Q(\mathcal{H}_n) \simeq Q(\mathcal{H}_1 \otimes \mathcal{H}_2 \otimes \cdots \otimes \mathcal{H}_n).$$

Suppose each S_i is in the pure state ϕ. Then S is in the state $\Phi = \phi \otimes \phi \otimes \cdots \otimes \phi$. Consider the observable $\mathbf{A}$ of S which is the average of the same observable A of each S_i:

$$\mathbf{A} = (A \otimes I \otimes \cdots \otimes I + I \otimes A \otimes \cdots \otimes I + \cdots + I \otimes I \otimes \cdots \otimes A)/n.$$

We recall that the uncertainty ΔR of an operator R is the square root of the variance: $(\Delta R)^2 = \mathrm{Exp}((R - \mathrm{Exp}\, R)^2)$. Hence,

$$\begin{aligned}(\Delta \mathbf{A})^2 &= \langle \Phi, (\mathbf{A} - \mathrm{Exp}\, \mathbf{A})\Phi \rangle \\ &= (1 - 1/n) \langle \phi, (A - \mathrm{Exp} A)\phi \rangle^2 + (1/n) \left\langle \phi, (A - \mathrm{Exp} A)^2 \phi \right\rangle \\ &= (\Delta A)^2/n.\end{aligned}$$

Hence, if

$$\mathbf{B} = (B \otimes I \otimes \cdots \otimes I + I \otimes B \otimes \cdots I + \cdots + I \otimes I \otimes \cdots \otimes B)/n$$

is another such averaged observable, then for the commutator $[A, B]$ we have

$$\Delta[\mathbf{A}, \mathbf{B}] = \Delta[A, B]/n.$$

Thus, $\lim_{n\to\infty} \Delta[\mathbf{A}, \mathbf{B}] = 0$. It follows that the averaged observables of S all commute in the limit, and so the σ-complex of S becomes essentially a σ-algebra for very large n, as in a macroscopic body.

This calculation was made under the assumption that S is an ensemble of non-interacting replicas of one particle. In a real body the states and observables need not be identical. Without going into details, it is possible to give conditions on the allowed variation of the states of the particles and the averaged observables so that $\Delta[\mathbf{A}, \mathbf{B}]$ still tends to zero with increasing n. In any case, the result is at least suggestive that in a real body, the σ-complex of S will be very close to a σ-algebra.

The change in dynamics accompanying the move from the Hilbert space $\mathcal{H}$ to the phase space Ω has been well-studied. In essence, the quantum bracket $\frac{i}{\hbar}[X, Y]$ is replaced by the Poisson bracket $\{X, Y\}$, so that the von Neumann-Liouville equation $\partial_t w_t = -\frac{i}{\hbar}[H, w_t]$ is replaced by the classical Liouville equation $\partial_t f_t = -\{H, f_t\}$. (See Faddeev and Yakubovskii [23], for example). We saw in Sects. "Symmetries" and "Dynamics" that the lack of sufficient structure of a σ-algebra did not allow us to derive the classical dynamics from the automorphisms of $B(\Omega)$, whereas we could do so in the quantum case $Q(\mathcal{H})$. We can see now how it is possible to recover the classical dynamical equation by an excursion into the quantum structure $Q(\mathcal{H})$.

Interpreting and Resolving Quantum Paradoxes

The K-S Paradox and the Projection Rule

We have already applied this reconstruction to treat several issues in the interpretation of the formalism. One of these, the Kochen-Specker Paradox, which showed that the assumption that all properties are intrinsic leads to a contradiction, was the motivation for introducing the σ-complex of extrinsic properties. Conversely, assuming the relational nature of properties resolves this paradox. Another issue, discussed in Sect. "Reduction and Conditional Probability", is the nature of reduction and the von Neumann-Lüders Projection Rule, which here appears as the counterpart to classical conditionalizing, not as an ad hoc addition to quantum theory. We now consider a number of other controversial questions from the literature.

Wave-Particle Duality

We discuss wave-particle duality in the context of the two-slit experiment. Let y_1 and y_2 be the projections of position in the regions of the two slits δ_1 and δ_2. Then $y_1 \vee y_2$ is the projection of position for the union $\delta_1 \cup \delta_2$. Let x be the property of position in a local region Δ on the detection screen.

If passage through each of the two slits is registered, then the Law of Alternatives of Sect. "Classical and Quantum Conditional Probability" tells us that $p(x|y_1 \vee y_2) = p(x|y_1)p(y_1|y_1 \vee y_2) + p(x|y_2)p(y_2|y_1 \vee y_2)$,which, in the case of symmetrical positioned slits, is propotional to the sum $p(x|y_1) + p(x|y_2)$ of the probabilities of passage through the individual slits, just as in the classical case.

In the case where the passage through the two slits by the quantum particle is not registered, we have shown in Sect. "Classical and Quantum Conditional Probability" that there is an additonal interference term

$$[tr(y_1 w y_2)x) + tr(y_2 w y_1 x)]/tr(w(y_1 \vee y_2)).$$

Note that if x and y_1 and y_2 commute, this interference term vanishes. This happens if the detector is right next to the two-slit screen. If the detector is a distance from the two-slit screen, then the particle undergoes free flight evolution σ_t, so $\sigma_t(y_i) = u_t y_i u_t^{-1}$ no longer commutes with x, giving rise to the non-zero interference term.

An explanation of the interference effect that is often given is that the particle is, or acts as, a pair of waves emanating from the slits, which exhibit constructive and destructive interference effects. This was, of course, the explanation for Young's original experiment with the classical electromagnetic field. For individual quantum particles however, it leads to the paradoxical effect that the wave suddenly collapses to a local region at the detection screen.

The explanation given here is a different one. A system forms a localized particle if there is a position operator for the system, so that a measurement of position detects the system at a localized region in space. Until the position is measured the position has no value, since position in a region is an extrinsic property. We may view the two-slit screen as a preparation of state for the particle, for which the position is conditionalized, or reduced, to the region $\delta_1 \cup \delta_2$. This reduction is not a position measurement, since $\delta_1 \cup \delta_2$ is not a localized region (as it would be for a single-slit screen). It is only at the detection screen, where the particle, in interaction with the screen, is reduced to the local region Δ, that its position has a value.

The question of why the particle shows the interference effects of a wave is answered in Sect. "Dynamics", where the evolution of the quantum particle was defined by a trajectory in the space $\mathrm{Aut}(Q)$. This yielded the Schrödinger equation, which is a wave equation. On the other hand, a trajectory in the phase space of a classical particle passing through a two-slit screen is governed by the classical Liouville equation, without any wave properties. Thereby, the wave-like properties of a quantum particle are explained by the extrinsic character of its properties.

The Measurement Problem

The Measurement Problem refers to an inconsistency in the orthodox interpretation of quantum measurement. The interpretation assumes that an isolated system undergoes unitary evolution via Schrödinger's equation. We quote from Bohm [6, Chap. 12]:

> If time evolution is a symmetry transformation, then the mathematical structure (in particular the algebraic relations) of the algebra of observables does not change in time; this means that the physical structure is indistinguishable at two different points in time. Our experience shows that there are physical systems that have this property and in fact it is this property that defines the isolated systems. Thus isolated physical systems do not age, an absolute value of time has no meaning for these systems, and only time differences are accessible to measurement. Irreversible processes do not take place in isolated physical systems defined as above.

Accordingly, in the orthodox interpretation, for a measurement of an observable A of a system S by an apparatus T, the total system $S + T$, which is assumed to be isolated, undergoes unitary evolution.

We outline the standard description of an ideal measurement. Suppose the spectral decomposition of an observable is $A = \sum a_i P_i$, where each P_i is a one-dimensional projection with eigenstate ϕ_i. The apparatus is assumed to be sensitive to the different eigenstates of A. Hence, if the initial state of S is ϕ_k and the apparatus T is in a neutral state ψ_0, so that the state of $S + T$ is $\phi_k \otimes \psi_0$, then the system evolves into the state

$\phi_k \otimes \psi_k$, where the ψ_i are the states of the apparatus co-ordinate corresponding to the states ϕ_i of the system. By linearity, if S is in the initial state $\phi = \sum a_i \phi_i$, then $S + T$ evolves into the state $\Gamma = \sum a_i \phi_i \otimes \psi_i$. The intractable problem for the orthodox interpretation is that the completed measurement gives a particular apparatus state ψ_k, indicating that the state of S is ϕ_k, so that the state of the total system is $\phi_k \otimes \psi_k$, in contradiction to the evolved state $\sum a_i \phi_i \otimes \psi_i$. We may also see the reduction from the viewpoint of the conditionalization of the states. If the state p of $S + T$ just prior to measurement is P_Γ, then after the measurement it is the conditionalized state

$$p(\cdot\,|P_{\phi_k} \otimes I \wedge I \otimes P_{\psi_k}) = (P_{\phi_k} \otimes I \wedge I \otimes P_{\psi_k})P_\Gamma(P_{\phi_k} \otimes I \wedge I \otimes P_{\psi_k})/tr((P_{\phi_k} \otimes I \wedge I \otimes P_{\psi_k})P_\Gamma)) = P_{\phi_k \otimes \psi_k}.$$

Hence, the new conditionalized state of $S + T$ is the reduced state $\phi_k \otimes \psi_k$.

The orthodox interpretation then has to reconcile the unitary evolution of $S + T$ with the measured reduced states of S and T. The present interpretation stands the orthodox interpretation on its head. We do not begin with the unitary development of an isolated system, but rather with the results of a measurement, or, more generally, of a decoherent interaction. In fact, the original motivation for forming a σ-complex of properties was via the set of measured, and hence reduced, properties which form the current interaction algebra. For us, it is the conditions under which dynamical evolution occurs that is to be investigated, rather than the reduced state. We cannot take for granted what is assumed in the orthodox interpretation, as in the above quotation, that an isolated system evolves unitarily. So we must answer the question whether in a measurement the σ-complex structure of $S + T$ undergoes a symmetry transformation at different times of the process. As Sect. "Dynamics" showed, this is formalized as the condition for the existence of a representation $\sigma : \mathbb{R} \to \mathrm{Aut}(Q)$.

It is easy to see, however, that in the process of a completed measurement or a state preparation there are two distinct elements of $Q(\mathcal{H})(= Q(\mathcal{H}_1 \otimes \mathcal{H}_2))$ at initial time 0 which end up being mapped to the same element at a later time t. We have seen that an initial state $\phi \otimes \psi_0$ results in a state $\phi_k \otimes \psi_k$, for some k. However, $\phi_k \otimes \psi_0$ also results in the state $\phi_k \otimes \psi_k$. If we choose the state ϕ to be distinct from ϕ_k, then the two elements $P_{\phi \otimes \psi_0}$ and $P_{\phi_k \otimes \psi_0}$ of $Q(\mathcal{H})$ both map to the same element $P_{\phi_k \otimes \psi_k}$. However, any automorphism σ_t is certainly a one-to-one map on Q, so the measurement process cannot be described by a representation $\sigma : \mathbb{R} \to \mathrm{Aut}(Q)$, and hence a unitary evolution.

In our interpretation, the Measurement Problem is thus resolved in favor of reduction rather than unitary evolution. The point can be made intuitively that points of absolute time do exist in a measurement and also in state preparation, namely the point (or, better, small interval) of time at which reduction takes place. If for instance, we consider a Stern-Gerlach experiment with a state preparation in which a filter registers the passage of a particle through one of several slits, before the particle reaches a detection screen, then the interval of time of passage through the slit, in which the state of the particle is reduced, is such an absolute point of time: the state after passing through the slit is the conditionalized state, whereas before it is not.

Time and its passage is a problematic concept in physics, so to reinforce the point we shall give another example, in which time homogeneity is tied to spatial symmetry. Consider a particle resulting, say, from decay in which its state has spherical symmetry. Assume that the particle is initially at the center of a spherical detector system. During the passage of the particle until it hits the detector, the combined system of particle and detector is spherically symmetric and time homogeneous. At the moment of registering the impact on a local region of the detector, the system loses both its isotropy in space and its time symmetry. If it is difficult to argue against this breaking of space symmetry in favor of a particular direction, it seems to us to be equally hard to gainsay the breaking of time symmetry at the moment this non-isotropy occurs.

For a composite system it is not only outside forces that can break symmetry, but internal interactions. As opposed to the quotation of Bohm [6] above, we believe that symmetry-breaking processes do take place in isolated compound systems with internal decoherent interactions during reduction of state. To argue that nevertheless symmetry has not been broken for the combined system is to favor the theoretical formalism ahead of the facts on the ground. It is notable that with this interpretation the system consisting of the universe as a whole, for which there are no external systems, acquires reduced or, as we say, conditionalized states as a result of the interactions of component systems.

Note that our alternative term *interactive property* is more appropriate here than *extrinsic property*. The reduction of the state to $\phi_k \otimes \psi_k$ happens for the composite system $S_1 + S_2$ because of the interaction of the component systems S_1 and S_2 which are internal to $S_1 + S_2$ rather than an interaction of $S_1 + S_2$ with an external system.

The Einstein-Podolsky-Rosen Experiment

We shall discuss the EPR phenomenon in the Bohm form of two spin $\frac{1}{2}$ particles in the combined singlet state Γ of total spin 0. Suppose that in that state the two particles are separated and the spin component s_z of particle 1 is measured in some direction z. That means that the observable $s_z \otimes I$ of the combined system is being measured.

Let $P_z^{\pm} = \frac{1}{2}I \pm s_z$. We have the spectral decomposition

$$s_z \otimes I = \tfrac{1}{2}P_z^+ \otimes I + (-\tfrac{1}{2})P_z^- \otimes I,$$

so the interaction algebra $B = \{0, 1, P_z^+ \otimes I, P_z^- \otimes I\}$. We expand the singlet state

$$\Gamma = \sqrt{\tfrac{1}{2}}(\phi_z^+ \otimes \psi_z^- - \phi_z^- \otimes \psi_z^+),$$

where $P_z^{\pm}\phi_z^{\pm} = \phi_z^{\pm}$ and $P_z^{\pm}\psi_z^{\pm} = \psi_z^{\pm}$. Thus, if particle 1 has spin up, the state $p(\cdot \mid P_z^+ \otimes I)$ of the system is, by Sect. "Conditional States", given by

$$p(\cdot \mid P_z^+ \otimes I) = (P_z^+ \otimes I)P_\Gamma(P_z^+ \otimes I)/\operatorname{tr}((P_z^+ \otimes I)P_\Gamma) = P_{\phi_z^+ \otimes \psi_z^-}.$$

This is, of course, equivalent to projecting the vector Γ into the image of $P_z^+ \otimes I$:

$$P_z^+ \otimes I(\Gamma) = \sqrt{\tfrac{1}{2}}(\phi_z^+ \otimes \psi_z^-).$$

Similarly, if particle 1 has spin down the state $p(\cdot \mid P_z^- \otimes I)$ is given by the vector

$$P_z^- \otimes I(\Gamma) = \sqrt{\tfrac{1}{2}}(\phi_z^- \otimes \psi_z^+).$$

This shows that if s_z is measured for particle 2, it is certain to have opposite value of s_z for particle 1. It does *not* mean that after s_z is measured for particle 1, then s_z has a value for particle 2. The properties $I \otimes P_z^+$ and $I \otimes P_z^-$ do not lie in the interaction algebra $B = \{P_z^+ \otimes I, P_z^- \otimes I, 0, 1\}$, and so have no value. The spin components are extrinsic properties of each particle, which do not have values until the appropriate interaction. To claim otherwise is to revert to the classical notion of intrinsic properties.

This is a necessary consequence of our interpretation, but it also follows from a careful application of standard quantum mechanical principles. For after the measurement of s_z on particle 1 gives a value of $\frac{1}{2}$, the state of the combined system is $\phi_z^+ \otimes \psi_z^-$, which is an eigenstate of $I \otimes s_z$. Born's Rule implies that an eigenstate of an observable will yield the corresponding eigenvalue as value only if and when that observable is measured.

The situation is entirely similar to the unproblematic triple experiment. A triple experiment on the frame (x, y, z) yields the interaction algebra B_{xyz}. If $S_z^2 = 0$, then $S_x^2 = S_y^2 = 1$. If (x', y', z) is another frame, then it is also the case that $p(S_{x'}^2 = 1 \mid S_z^2 = 0) = 1$, so that $S_{x'}^2$ is certain to have the value 1 if the triple experiment on the frame (x', y', z) is performed. But $S_{x'}^2$ does not have a value unless and until that experiment is carried out since $S_{x'}^2 = 1$ does not lie in the interaction algebra B_{xyz}.

We have not in this discussion mentioned a word about special relativity. Indeed, the spin EPR phenomenon has nothing to do with position or motion and is independent of relativistic questions. However, EPR with space-like separated particles has been used to put in question the full Lorentz invariance of quantum mechanics. This is replaced by a weaker notion that EPR correlations cannot be used for faster than light signaling. We believe that Lorentz invariance is a fundamental symmetry principle, which gives rise to basic observables, and is not simply an artifact of signaling messages between agents.

The relativistically invariant description of the EPR experiment is that if experimenters A_1 and A_2 measure particles 1 and 2, and the directions of spin in which they are measured are the same, then an experimenter B in the common part of the future light cones of A_1 and A_2 will find that the spins are in opposite directions.

Now that we have studied what EPR actually says, we shall treat the question of how correlations can exist between the different directions of spins of two particles when such spins cannot simultaneously have values.

To set the stage for EPR, we again first consider the triple experiment. For a spin 1 particle the proposition $S_z^2 = 1$ defines the same projection in $Q(\mathcal{H})$ as the proposition

$$S_x^2 = 0 \leftrightarrow S_y^2 = 1 \tag{1}$$

If we perform the (x, y, z) triple experiment with interaction algebra B_{xyz} and find that $S_z^2 = 1$, then we can check that either $S_x^2 = 0$ and $S_y^2 = 1$ or $S_x^2 = 1$ and $S_y^2 = 0$, so that (1) is true. However, for the orthogonal triple (x', y', z)

$$S_{x'}^2 = 0 \leftrightarrow S_{y'}^2 = 1 \tag{2}$$

is the same projection as (1) and so is also true. But $S_{x'}^2$ and $S_{y'}^2$ do not lie in the interaction algebra B_{xyz}, and so have no truth value unless and until the (x', y', z) triple experiment is performed. Thus, the correlation (2) is true without its component properties $S_{x'}^2$ and $S_{y'}^2$ having truth values.

Now consider the EPR experiment. We have seen in Sect. "Combined Systems" tha $S = 0$ is the same projection as $(S_z = 0) \wedge (S_x = 0)$, and $S_z = 0$ and $S_x = 0$ are in turn respectively the same projections as

$$s_z \otimes I = \frac{1}{2} \leftrightarrow I \otimes s_z = -\frac{1}{2} \tag{3}$$

and

$$s_x \otimes I = \frac{1}{2} \leftrightarrow I \otimes s_x = -\frac{1}{2}. \tag{4}$$

If the projections $S_z = 0$ and $S_x = 0$ are true, then so are the correlations (3) and (4) since they define the same projections. As in the triple experiments, we see that these correlations subsist simultaneously, even though the spins s_z and s_x for each particle cannot have values simultanously. Thus, the existence of seemingly paradoxical EPR correlations in different directions can be understood via the logic of extrinisic properties.

In summary, the extrinsic properties of a σ-complex may have relations subsisting among its elements because of general laws of physics, such as conservation laws, which are timeless and independent of particular interactions. The σ-complex structure accommodates such relations in the form of compound formulas such as (3) and (4), which are true, even when the constituent parts do not have truth values. This fact allows us to interpret the EPR phenomenon in a fully relativistically invariant way. For extrinsic properties a compound property may have truth values even when the component parts do not.

On the Logic of Quantum Mechanics

As we have stressed thoughout this paper, the major transformation from classical to quantum physics in this approach lies not in modifying the basic classical concepts such as state, observable, symmetry, dynamics, combining systems, or the notion of probability, but rather in the shift from intrinsic to extrinsic properties.

Now properties, whether considered as predicates or propositions, are the domain of logic. Boolean algebras correspond to propositional logic and σ-algebras to predicate logic. Hence the change to a σ-complex of exrinsic properties should entail a new logic of properties. At first sight however, it would appear that the logic of extrinsic properties as elements of a σ-complex Q is no different than classical propositional logic, since these elements can only be compounded when they lie in the same σ-algebra in Q. This is far from the case; in fact, the difference in logic plays an important role in resolving some of the quantum paradoxes. The underlying reason is that a compound property such as $x \vee y$ may be lie in an interaction algebra and so have a truth value, even though neither x nor y lie in the algebra, and have no truth value.

The logic of extrinsic properties has been sysematically studied in Kochen, Specker [10, 11], where a complete axiomatization of the propositional calculus of extrinsic properties is given. Here we shall confine ourselves to pointing out some uses of this logic that appeared in this paper.

1. The simplest such case is $x \vee x^{\perp}$, which equals 1 in Q, and so is always true, even though x may have no truth value.[5] Thus, for a spin $\frac{1}{2}$ particle, $s_z = \frac{1}{2} \vee s_z = -\frac{1}{2}$ is true simultaneously for all directions z, though s_z may have no value.

2. In the two-slit experiment (Sect. "Classical and Quantum Conditional Probability"), we saw that it is this lack of truth value that leads to the interference pattern at the detector screen. The source of the interference pattern is not some non-classical probability, but rather the applications of classical Kolmogorov axioms of probability to the logic of extrinsic propeties. The conditional probability $p(x|y)$ is the probability of x given that y has happened and so has a truth value. Therefore the probability $p(x|y_1 \vee y_2)$ implies that $y_1 \vee y_2$ is true. However, neither y_1 nor y_2 has happened. We should not expect the classical Law of Alternatives connecting

[5]This is reminiscent of Aristotle's famous sea battle in *De Interpretatione*: "A sea battle must either take place tomorrow or not, but it is not necessary that it should take place tomorrow neither is it necessary that it should not take place, yet it is necessary that it either should or should not take place tomorrow."

$p(x|y_1 \vee y_2)$ to $p(x|y_1)$ and $p(x|y_2)$ to be valid unless y_1 and y_2 are events that have happened. In that case the Law of Alternatives is in fact valid in quantum mechanics.

3. In the EPR experiment, the singleton state $S = 0$ implies that $s_z \otimes I = \frac{1}{2} \leftrightarrow I \otimes s_z = -\frac{1}{2}$ is true for any direction z. In fact, as shown in Sect. "Combined Systems" the element $S = 0$ equals

$$(s_z \otimes I = \frac{1}{2} \leftrightarrow I \otimes s_z = -\frac{1}{2}) \wedge (s_x \otimes I = \frac{1}{2} \leftrightarrow I \otimes s_x = -\frac{1}{2}).$$

Thus, the correlation exists in both the z and x directions even though the spins cannot simultaneously have values in these directions. Section "Combined Systems" shows how general superpositions of states of combined systems may be reformulated as compound statements of this quantum logic.

4. The K-S Paradox in Sect. "Properties" can be stated as a proposition that is classically true but false in quantum mechanics. To see this, let $+$ denote exclusive disjunction. Then $x + y + z + x \wedge y \wedge z$ is true if and only if exactly one of x, y, and z is true.

The statement $\bigvee_{i \le 40}(x_i + y_i + z_i + x_i \wedge y_i \wedge z_i)^{\perp}$, where (x_i, y_i, z_i) range over the orthogonal triples of the 40 triple experiments of Sect. "Properties" is classically true, but false under a substitutions $x_i \mapsto S^2_{x_i}, y_i \mapsto S^2_{y_i}, z_i \mapsto S^2_{z_i}$.

For two spin $\frac{1}{2}$ particles there is a K-S Paradox in Mermin [24] which yields a much simpler such proposition in four dimensional Hilbert space:

$$[(x \leftrightarrow y) \leftrightarrow (z \leftrightarrow w)] \leftrightarrow [(x \leftrightarrow z) \leftrightarrow (y \leftrightarrow w)].$$

This classically true proposition is false under the substitution

$$x \mapsto s_z \otimes I = \frac{1}{2},\ y \mapsto I \otimes s_z = \frac{1}{2},\ w \mapsto s_x \otimes I = \frac{1}{2},\ z \mapsto I \otimes s_x = \frac{1}{2}.$$

(see [6] for details.) Kochen, Specker[6] [9] Theorem 4 shows that every K-S Paradox corresponds to a classically true proposition which is false under a substitution of quantum properties.

Acknowledgments Mathematics Department, Princeton University. Dedicated to the memory of Ernst Specker. This work was partially supported by an award from the John Templeton Foundation.

[6]For details, see J. Conway and S. Kochen, *The Geometry of the Quantum Paradoxes, Quantum [Un]speakables*, R.A. Bertlemann, A. Zeilinger (ed.), Springer-Verlay, Berlin, 2002, 257.

Appendix: Summary Table of Concepts

	General mechanics	Classical mechanics	Quantum mechanics
Properties	σ-complex $Q = \cup B$, with B a σ-algebra	σ-algebra $B(\Omega)$	σ-complex $Q(\mathcal{H}\}$
States	$p : Q \to [0, 1]$ $p \mid B$, a probability measure	$p : B(\Omega) \to [0, 1]$ a probability measure	$w : \mathcal{H} \to \mathcal{H}$ Density operator $p(x) = \mathrm{tr}(wx)$
Pure states	Extreme point of convex set	$\omega \in \Omega$	1 dim operator i.e. unit $\phi \in \mathcal{H}$ $p(x) = \langle x, x\phi\rangle$
Observables	$u : B(\mathbb{R}) \to Q$ homomorphism	$f : \Omega \to \mathbb{R}$ Borel function	$A : \mathcal{H} \to \mathcal{H}$ Hermitean operator
Symmetries	$\sigma : Q \to Q$ automorphism	$h : \Omega \to \Omega$ canonical transformation	$u : \mathcal{H} \to \mathcal{H}$ unitary or anti-unitary operator $\sigma(x) = uxu^{-1}$
Dynamics	$\sigma : \mathbb{R} \to \mathrm{Aut}(Q)$ representation	Liouville equation $\partial_t \rho = -[H, \rho]$	von Neumann -Liouville equation $\partial_t w_t = -\frac{i}{\hbar}[H, w_t]$
Conditionalized states	$p(x) \to p(x \mid y)$ for $x, y \in B$ in Q $p(x \mid y) = p(x \mid y)/p(y)$	$p(x) \to p(x \mid y)$ $= p(x \wedge y)/p(y)$	$w \to ywy/\,\mathrm{tr}(wy)$ von Neumann -Lüders Rule
Combined systems	$Q_1 \oplus Q_2$ direct sum of σ-complexes	$\Omega_1 \times \Omega_2$ direct product of phase spaces	$\mathcal{H}_1 \otimes \mathcal{H}_2$ tensor product of Hilbert spaces

References

1. N. Bohr, Causality and complementarity. Phil. Sci. **4**, 289 (1937)
2. G. Birkhoff, J. von Neumann, The logic of quantum mechanics. Ann. Math. **37**, 823 (1936)
3. G.W. Mackey, *Mathematical Foundations of Quantum Mechanics* (Benjamin, Amsterdam, 1963)
4. C. Piron, *Foundations of Quantum Physics* (Benjamin, Reading, MA, 1976)
5. V.S. Varadarajan, *The Geometry of Quantum Theory, Van Nostrand* (Princeton, NJ., 1968)
6. A. Bohm, *Quantum Mechanics: Foundations and Applications* (Springer, New York, 2001)
7. S. Koppelberg, *Handbook of Boolean Algebras*, vol. 1 (North-Holland, Amsterdam, 1989)
8. J. Conway, S. Kochen, The strong free will theorem. Amer. Math. Soc. Not. **56**, 226 (2009)
9. S. Kochen, E.P. Specker, The problem of hidden variables in quantum mechanics. J. Math. Mech. **17**, 59 (1967)
10. S. Kochen, E.P. Specker, *Logical Structures Arising in Quantum Mechanics* (Symposium at Berkeley, The Theory of Models, 1967), p. 177
11. S. Kochen, E.P. Specker, *The Calculus of Partial Propositional Functions* (Congress at Jerusalem, Methodology and Philosophy of Science, 1964), p. 45
12. E. Wrede, Über die Ablenkung von Molekularstrahlen elektrischer Dipolmolekule iminhomogenen elektrischen Feld. Z. Phys. A **44**, 261 (1927)

13. A.M. Gleason, Measures on the closed subspaces of a Hilbert space. J. Math. Mech. **6**, 885 (1957)
14. J.M. Jauch, *Foundations of Quantum Mechanics* (Addison-Wesley, Reading, MA, 1968)
15. V. Bargmann, Note on Wigner's theorem on symmetry operations. J. Math. Phys. **5**, 862 (1964)
16. U. Uhlhorn, Representation of symmetric transformations in quantum mechanics. Arkiv Fysik **23**, 307 (1963)
17. R.P. Feynman, R.B. Leighton, M. Sands, *The Feynman Lectures on Physics*, vol. 3 (Mass, Addison-Wesley, Reading, 1966)
18. E.G. Beltrametti, G. Cassinelli, *The Logic of Quantum Mechanics* (Addison-Wesley, Reading, Mass, 1981)
19. R.P. Feynman, Space-time approach to non-relativistic quantum mechanics. Rev. Mod. Phys. **20**, 36 (1948)
20. M. Reck, A. Zeilinger, H.J. Bernstein, P. Bertani, Experimental realization of any discrete unitary operator. Phys. Rev. Lett. **73**, 58 (1994)
21. M. Zukowski, A. Zeilinger, M.A. Horne, Realizable higher-dimensional two-particle entanglements via multiport beam splitters. Phys. Rev. A **55**, 2564 (1997)
22. D. Finkelstein, The logic of quantum physics. Trans. N. Y. Acad. Sci. **25**, 621 (1963)
23. L.D. Faddeev, O.A. Yakubovskii, *Lectures on Quantum Mechanics for Mathematics Students* (American Mathematical Society, Providence, RI, 2009)
24. N.D. Mermin. Phys. Rev. Lett. **65**, 3373–3376 (1990)

Part IV
Bell Inequalities—Theory

Chapter 13
A Quantum Mechanical Bound for CHSH-Type Bell Inequalities

Michael Epping, Hermann Kampermann and Dagmar Bruß

Abstract Many typical Bell experiments can be described as follows. A source repeatedly distributes particles among two spacelike separated observers. Each of them makes a measurement, using an observable randomly chosen out of several possible ones, leading to one of two possible outcomes. After collecting a sufficient amount of data one calculates the value of a so-called Bell expression. An important question in this context is whether the result is compatible with bounds based on the assumptions of locality, realism and freedom of choice. Here we are interested in bounds on the obtained value derived from quantum theory, so-called Tsirelson bounds. We describe a simple Tsirelson bound, which is based on a singular value decomposition. This mathematical result leads to some physical insights. In particular the optimal observables can be obtained. Furthermore statements about the dimension of the underlying Hilbert space are possible. Finally, Bell inequalities can be modified to match rotated measurement settings, e.g. if the two parties do not share a common reference frame.

Introduction

Since the advent of quantum theory physicists have been struggling for a deeper understanding of its concepts and implications. One approach to this end is to carve out the differences between quantum theory and "classical" theories, i.e. to explicitly point to the conflicts between quantum theory and popular preconceptions, which evolved in each individual and the scientific community from decoherent macroscopic experiences. Plain formulations of such discrepancies and convincing experimental demonstrations are crucial to internalizing quantum theory and replacing

M. Epping · H. Kampermann · D. Bruß (✉)
Heinrich-Heine-University Düsseldorf,
40225 Düsseldorf, Germany
e-mail: dagmar.bruss@uni-duesseldorf.de

R. Bertlmann and A. Zeilinger (eds.), *Quantum [Un]Speakables II*,
The Frontiers Collection, DOI 10.1007/978-3-319-38987-5_13

existing misconceptions. For this reason the double-slit-experiments (and similar experiments with optical gratings) [1–5], which expose the role of state superpositions in quantum theory, are so very fascinating and famous. Other examples of "eye-openers" are demonstrations of tunneling [6, 7, pp. 33–12], the quantum Zeno effect [8] and variations of the Elitzur-Vaidman-scheme [9–11], to pick just a few.

Bell experiments [12–15], which show entanglement in a particularly striking way, belong to this list. Informally, entanglement is the fact that in quantum theory the state of a compound system (e.g. two particles) is not only a collection of the states of the subsystems. This fact can lead to strong correlations between measurements on different subsystems. Before going into more detail here, we would like to note that the described differences between the relatively new quantum theory and our old preconceptions are obvious starting points when to look for innovative technologies which were even unthinkable before. This is in fact a huge motivation for the field of quantum information, where Bell experiments play a central role.

Bell Experiments Bring Three Fundamental Common Sense Assumptions to a Test

The idea of Bell was to show that some common sense assumptions lead to predictions of experimental data which contradict the predictions of quantum theory. In the following we employ a black box approach to emphasize that this idea is completely independent of the physical realization of an experiment. For example the measurement apparatuses get some input (an integer number which will in the following be called "setting") and produce some output (the "measurement outcomes"). We refer readers preferring a more concrete notion to Sect. "The CHSH inequality can be violated in experiments with entangled photons", where physical implementations and concrete measurements are outlined.

In the present paper we consider the following (typical) Bell experiment, see also Fig. 13.1. There are three experimental sites, two of which we call the parties Alice (A) and Bob (B), and the third being a preparation site which we call source (S). Alice and Bob have a spatial separation large enough such that no signal can travel from one party to the other at the speed of light during the execution of our experiment. The source is separated such that no signal can travel from A or B to it at the speed of light before it finishes the state production. The importance of such separations will become clear later.

The source produces a quantum system, and sends one part to Alice and one to Bob. We will exemplify this in Sect. "The CHSH inequality can be violated in experiments with entangled photons". A and B are in possession of measurement apparatuses with a predefined set of different settings. In each run they choose the setting randomly, e.g. they turn a knob located at the outside of the apparatus, measure the system received from the source and list the setting and outcome. In the present paper the measurements are two-valued and the outcomes are denoted by -1 and $+1$. Let

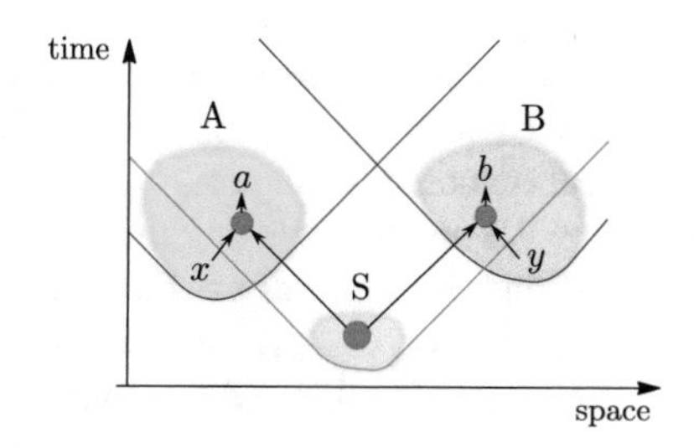

Fig. 13.1 Two parties, Alice (*A*) and Bob (*B*), perform a Bell experiment. Both of them receive parts of a quantum system from the source (*S*). They randomly choose a measurement setting, denoted by $x = 1, 2, \ldots, M_1$ and $y = 1, 2, \ldots, M_2$, and write down their outcomes $a = -1$ or 1 and $b = -1$ or 1, respectively. The experiment is repeated until the accumulated data is analyzed according to the text. Angles of 45° in the space-time-diagram correspond to the speed of light. The future light cones of *A*, *B* and *S* show, that the setting choice and outcome of one party cannot influence the other and that A and B also cannot influence any event inside the source.

M_1 and M_2 be the number of different measurement settings at site A and B, respectively. We label them by $x = 1, 2, 3, \ldots, M_1$ for Alice and $y = 1, 2, 3, \ldots, M_2$ for Bob. This preparation and measurement procedure of a quantum system is repeated until the amount of data suffices to estimate the expectation value of the measured observables, up to the statistical accuracy one aims at. The expectation value of an observable is the average of all possible outcomes, here ± 1, weighted with the corresponding probability to get this outcome.

Let us sketch the preconceptions that are *jointly* in conflict with the quantum theoretical predictions for Bell tests. These are mainly three concepts: Locality, realism and freedom of choice. This forces us to question at least one of these ideas, because any interpretation of quantum theory, as well as any "postquantum" theory, cannot obey all of them. We invite the reader to pick one to abandon while reading the following descriptions. Do not be confused by our comparison with the textbook formalism of quantum theory: so far you are free to choose any of them.

Locality is the assumption, that effects only have nearby direct causes, or the other way around: any action can only affect directly nearby objects. If some action here has an impact there, then something traveled from here to there. And, according to special relativity, the speed of this signal is at most the speed of light. In our setup, this means that whatever Alice does cannot have any observable effect at Bob's site. In particular, the measurement outcome at one side cannot depend on the choice of measurement setting at the other site. While the formalism of quantum theory has some "nonlocal features", e.g. a global state, it is strictly local in the above sense, because any local quantum operation on one subsystem does not change expectation values of local observables for a different subsystem.

Realism is the concept of an objective world that exists independently of subjects ("observers"). A stronger form of realism is the "value-definiteness" assumption meaning that the properties of objects always have definite values, also if they are not measured or even unaccessible for any observer. It seems to be against common sense to assume that objects cease to have definite properties if we do not measure them any longer. In particular the natural sciences were founded on the assumption, that nature and its properties exist independently of the scientist. In our setup realism implies, that the measurement outcomes of unperformed measurements (in unchosen settings) have some value. We do not know them, but we can safely assume that they exist, give them a name and use them as variables. If possible outcomes are -1 and $+1$, for example, we might use that the outcome squared is 1 in any of our calculations. In general the (usual) formalism of quantum theory does not contain definite values for measurement outcomes independent of a measurement.

Freedom of choice, which is also sometimes called the free will assumption, means it is possible to freely choose what experiment to perform and how. Because this idea is elusive, we are content with a decision that is statistically independent of any quantity which is subject of our experiment. The idea of fate seems to be tempting to many people. However, dropping freedom of choice makes science useless. Just imagine you "want" to investigate the question whether a bag contains black balls but your fate is to pick only white balls (and put them back afterwards), even though there are many black balls inside. In our setup, freedom of choice implies, that A's and B's choice of measurement setting does not depend on the other's choice or the outcomes. In quantum theory, there is freedom of choice in the sense that random measurement outcomes of some other process can be used to make decisions.

If you decided that you preferably take leave of locality you are in good company. Many scientists conclude from Bell's theorem, that the locality assumption is not sustainable. This is particularly interesting when you consider the above comparison with the standard textbook formalism of quantum theory, which is apparently not realistic but local in the described sense. The fact that in this context many scientists speak about "quantum nonlocality" thus leads to controversy [16]. We therefore want to stress again, that the experimental contradiction only tells us that at least one of all the assumptions that lead to the predictions needs to be wrong. We cannot decide which assumption is wrong from Bell's theorem alone.

We now focus on a tool to show the contradiction in the described experiment between the above assumptions and quantum theory, the so called Bell inequalities. These are inequalities of measurable quantities which are (mainly) derived from locality, realism and freedom of choice and therefore hold for all theories which obey these principles, while they are violated by the predictions of quantum theory. We consider a special kind of Bell inequalities which are linear combinations of joint expectation values of Alice's and Bob's observables. The joint expectation value of the two observables of Alice and Bob is the expectation value of the product of the measurement outcomes, which again takes values ± 1. It depends on the setting choice x at Alice's site and y at Bob's site and we denote it by $E(x, y)$. If we denote the (real) coefficient in front of the expectation value $E(x, y)$ as $g_{x,y}$, then we can write such Bell inequalities as

$$\sum_{x=1}^{M_1} \sum_{y=1}^{M_2} g_{x,y} E(x, y) \leq B_g, \tag{1}$$

where the bound B_g depends only on the coefficients $g_{x,y}$. These coefficients form a matrix g which has dimension $M_1 \times M_2$. Any real matrix g defines a Bell inequality via Eq. (1). The may be most famous example is the Clauser-Horne-Shimony-Holt (CHSH) [17] inequality, which reads

$$E(1,1) + E(1,2) + E(2,1) - E(2,2) \leq 2. \tag{2}$$

Here the corresponding matrix g is

$$g = \begin{pmatrix} 1 & 1 \\ 1 & -1 \end{pmatrix}. \tag{3}$$

Due to its prominence we call the class of Bell inequalities in the form of Eq. (1) CHSH-type Bell inequalities. For completeness we sketch the derivation of B_g. It turns out that it suffices to consider deterministic outcomes only, as a probabilistic theory, where the outcomes follow some probability distribution, cannot achieve a higher value in Eq. (1): it can be described as a mixture of deterministic theories and the value of Eq. (1) is the sum of the values for the deterministic theories weighted with the corresponding probability in the mixture. For deterministic theories the expectation value is merely the product of the two (possibly unmeasured) outcomes a of Alice and b of Bob, which we are allowed to use when assuming *realism*. Due to *locality* a only depends on the setting x of Alice, which has no further dependence due to *freedom of choice*. Analogously b depends only on the setting y of Bob, which in turn has no further dependence. Thus the expectation value is

$$E(x, y) = a(x)b(y). \tag{4}$$

Now we can calculate B_g by maximizing Eq. (1) over all possible assignments of -1 and $+1$ values to $a(x)$ and $b(y)$. In Eq. (2) the maximal value is $B_g = 2$, which is achieved for $a(1) = a(2) = b(1) = b(2) = 1$, for example. Note that the sign of $E(2, 2)$ cannot be changed independently of the other three terms, because $E(1, 2)$ and $E(2, 1)$ contain $b(2)$ and $a(2)$, respectively.

We point out that any function that maps the probabilities of different measurement outcomes to a real number may be used to derive Bell inequalities, and different types of Bell inequalities can be found in the literature (e.g. [18]). However, here we focus on Bell inequalities of the form of Eq. (1).

The CHSH Inequality Can Be Violated in Experiments with Entangled Photons

We recapitulate some basics of quantum (information) theory. Analogously to a classical bit the quantum bit, or qubit, can be in two states 0 and 1, but additionally in every possible superposition of them. Mathematically this state is a unit vector in the two-dimensional Hilbert space (a vector space with a scalar product) $\mathbb{C}^2$ spanned by the basis vectors

$$\mathbf{0} := \begin{pmatrix} 1 \\ 0 \end{pmatrix} \quad \text{and} \quad \mathbf{1} := \begin{pmatrix} 0 \\ 1 \end{pmatrix}. \tag{5}$$

An example of a superposition of these basis states is $\psi = \frac{1}{\sqrt{2}}(\mathbf{0} + \mathbf{1})$. Any observable on a qubit with outcomes $+1$ and -1 can be written as

$$A = a_x \underbrace{\begin{pmatrix} 0 & 1 \\ 1 & 0 \end{pmatrix}}_{\sigma_x} + a_y \underbrace{\begin{pmatrix} 0 & -i \\ i & 0 \end{pmatrix}}_{\sigma_y} + a_z \underbrace{\begin{pmatrix} 1 & 0 \\ 0 & -1 \end{pmatrix}}_{\sigma_z}, \tag{6}$$

where the vector $\mathbf{a} = (a_x, a_y, a_z)^T$ (here ^{T} denotes transposition) defines the measurement direction and the matrices σ_x, σ_y and σ_z are called Pauli matrices. The expectation value of this observable given any state ψ can be calculated as $E = \psi^\dagger A \psi$ (here † denotes the complex conjugated transpose), which is between -1 and $+1$.

Any quantum mechanical system with (at least) two degrees of freedom can be used as a qubit. In the present context the spin of a spin-$\frac{1}{2}$-particle, two energy levels of an atom and the polarization of a photon are important examples of qubits. The spin measurement can be performed using a Stern-Gerlach-Apparatus [19], the energy level of an atom may be measured using resonant laser light, or the polarization of a photon can be measured using polarization filters or polarizing beam splitters.

The Hilbert space of two qubits is constructed using the tensor product, i.e. $\mathbb{C}^2 \otimes \mathbb{C}^2 = \mathbb{C}^4$. The tensor product of two matrices (of which vectors are a special case) is formed by multiplying each component of the first matrix with the complete second matrix, such that a bigger matrix arises. The state of the composite system of two qubits in states $\phi^A = (\phi_1^A, \phi_2^A)^T$ and $\phi^B = (\phi_1^B, \phi_2^B)^T$ then reads $\phi^{AB} = \phi^A \otimes \phi^B = (\phi_1^A\phi_1^B, \phi_1^A\phi_2^B, \phi_2^A\phi_1^B, \phi_2^A\phi_2^B)^T$. The states of such composite systems might be superposed, which leads to the notion of entanglement.

Out of several physical implementations of the CHSH experiment we sketch the ones with polarization entangled photons (see [20]). We identify $\mathbf{0}$ with the horizontal and $\mathbf{1}$ with the vertical polarization of a photon. Nonlinear processes in special optical elements can be used to create two photons in the state

$$\phi_+ = \frac{1}{\sqrt{2}}(1, 0, 0, 1)^T, \tag{7}$$

i.e. an equal superposition of two horizontally polarized photons and two vertically polarized photons. The measurements of Alice and Bob in setting 1 and 2 are

$$A_1 = \cos(2 \times 22.5°)\sigma_x + \sin(2 \times 22.5°)\sigma_z, \quad (8)$$
$$A_2 = \cos(-2 \times 22.5°)\sigma_x + \sin(-2 \times 22.5°)\sigma_z, \quad (9)$$
$$B_1 = \cos(2 \times 0°)\sigma_x + \sin(2 \times 0°)\sigma_z \quad (10)$$
$$\text{and } B_2 = \cos(2 \times 45°)\sigma_x + \sin(2 \times 45°)\sigma_z, \quad (11)$$

respectively. Here the angles are the angles of the polarizer and the factor 2 is due to the fact that in contrast to the Stern-Gerlach-Apparatus a rotation of the polarizer of 180° corresponds to the same measurement again. One can now calculate the value of Eq. (2):

$$\begin{aligned} E(1,1)+E(1,2)+E(2,1)-E(2,2) = \; & \phi_+^\dagger(A_1 \otimes B_1)\phi_+ + \phi_+^\dagger(A_1 \otimes B_2)\phi_+ \\ & +\phi_+^\dagger(A_2 \otimes B_1)\phi_+ - \phi_+^\dagger(A_2 \otimes B_2)\phi_+ \\ & = \frac{1}{\sqrt{2}} + \frac{1}{\sqrt{2}} + \frac{1}{\sqrt{2}} - \left(-\frac{1}{\sqrt{2}}\right) \\ & = 2\sqrt{2}. \end{aligned} \quad (12)$$

The value $2\sqrt{2} \approx 2.82$ is larger than 2 and therefore the CHSH inequality is violated. One can ask whether it is possible to achieve an even higher value, e.g. when using higher-dimensional systems than qubits, because at the first glance a value of up to four seems to be possible. This question is addressed in the following sections (the answer, which is negative, is given in Sect. "The singular value bound is a simple Tsirelson bound").

The Quantum Analog to Classical Bounds on Bell Inequalities Are Tsirelson Bounds

Analogously to the "classical" bound one can ask for bounds on the maximal value of a Bell inequality obtainable within quantum theory, so-called Tsirelson bounds [21], and the observables that should be measured to achieve this value. In other words: which observables are best suited to show the contradiction between quantum theory and the conjunction of the three discussed common sense assumptions. This question, which is also of some importance for applications of Bell inequalities, is the main subject of the present essay.

The scientific literature contains several approaches to derive Tsirelson bounds, some of which we want to mention. The problem of finding the Tsirelson bound of Eq. (1) can be formulated as a semidefinite program. Semidefinite programming is a method to obtain the global optimum of functions, under the restriction that the variable is a positive semidefinite matrix (i.e. it has no negative eigenvalues). This implies that well developed (mostly numerical) methods can be applied [22, 23]. The interested reader can find a Matlab code snippet to play around with in the appendix “MATLAB snippets”. Furthermore there has been some effort to derive Tsirelson bounds from first principles, amongst them the non-signalling principle [24], information causality [25] and the exclusivity principle [26].

The non-signalling principle is satisfied by all theories, that do not allow for faster-than-light communication. Information causality is a generalization of the non-signalling principle, in which the amount of information one party can gain about data of another is restricted by the amount of (classical) communication between them. The exclusivity principle states, that the probability to see one event out of a set of pairwise exclusive events cannot be larger than one.

The Singular Value Bound

Here we will discuss a simple mathematical bound for the maximal quantum value of a CHSH-type Bell inequality defined via a matrix g, which we derived in [27]. While it is not as widely applicable as the semidefinite programming approach, it is an analytical expression which is easy to calculate and it already enables valuable insights. For “simple” Bell inequalities, like the CHSH inequality given above, it is sufficient to use the method of this paper.

We will make use of singular value decompositions of real matrices, a standard tool of linear algebra, which we now shortly recapitulate.

Any Matrix Can Be Written in a Singular Value Decomposition

A singular value decomposition is very similar to an eigenvalue decomposition, in fact the two concepts are strongly related. Any real matrix g of dimension $M_1 \times M_2$ can be written as the product of three matrices V, S, W^T, i.e.

$$g = VSW^T, \tag{13}$$

where these three matrices have special properties. The matrix V is orthogonal, i.e. its columns, which are called left singular vectors, are orthonormal. It has a dimension of $M_1 \times M_1$. The matrix S is a diagonal matrix of dimension $M_1 \times M_2$, which is not necessarily a square matrix. Its diagonal entries are positive and have non-increasing order (from upper left to lower right). They are called singular values of g. The matrix W is again orthogonal. It has dimension $M_2 \times M_2$ and its columns are called right singular vectors.

The largest singular value can appear several times on the diagonal of S. We call the number of appearances the degeneracy d of the maximal singular value. Due to the ordering of S, these are the first d diagonal elements of S. Here we note the

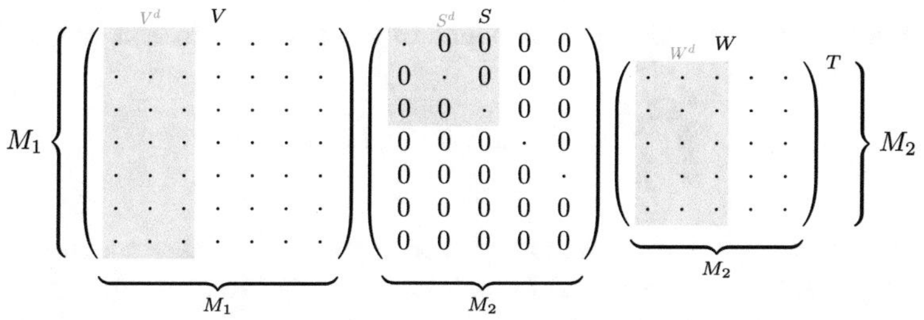

Fig. 13.2 The matrices involved in the singular value decomposition of a general real $M_1 \times M_2$ matrix g: V and W are orthogonal matrices, S is diagonal. V and W contain the *left* and *right* singular vectors, respectively, as columns, and S contains the singular values on its diagonal. The *shaded* parts belong to a truncated singular value decomposition of g. We denote the parts corresponding to the maximal singular value as $V^{(d)}$, $S^{(d)}$ and $W^{(d)}$.

concept of a truncated singular value decomposition: instead of using the full decomposition one can approximate g by using only parts of the matrices corresponding to, e.g., the first d singular values (i.e. only the maximal ones). These are the first d left and right singular vectors, and the first part of S, which is just a $d \times d$ identity matrix multiplied by the largest singular value. Since these matrices play an important role in the following analysis we will give them special names: $V^{(d)}$, $S^{(d)}$ and $W^{(d)}$. All these matrices are depicted in Fig. 13.2.

The matrix g maps a vector $\mathbf{v}$ to a vector $g\mathbf{v}$ which, in general, has a different length than $\mathbf{v}$. Here the length is measured by the (usual) Euclidean norm $||\mathbf{v}||_2 = \sqrt{v_1^2 + v_2^2 + \cdots + v_{M_2}^2}$. The largest possible stretching factor for all vectors $\mathbf{v}$ is a property of the matrix: its matrix norm induced by the Euclidean norm. The value of this matrix norm coincides with the maximal singular value S_{11}. We can therefore express the maximal singular value using

$$S_{11} = \max_{\mathbf{v} \in \mathbb{R}^{M_2}} \frac{||g\mathbf{v}||_2}{||\mathbf{v}||_2} =: ||g||_2. \tag{14}$$

The notation $||g||_2$ for the maximal singular value of g is more convenient than S_{11}, as it contains the matrix as an argument.

The Singular Value Bound Is a Simple Tsirelson Bound

It turns out that the matrix norm of g, i.e. its maximal singular value, leads to an upper bound on the quantum value for a Bell inequality, defined by g via Eq. (1). This is the central insight of this essay. It is remarkable that a *mathematical* property, solely due to the rules of linear algebra, leads to a bound for a *physical* theory, here the theory of quantum mechanics. With the definition of the matrix norm given above, we can now write this singular value bound of g, a simple Tsirelson bound [27]. It reads

$$\sum_{x=1}^{M_1} \sum_{y=1}^{M_2} g_{x_1,x_2} E(x_1, x_2) \leq \sqrt{M_1 M_2} ||g||_2, \tag{15}$$

where E now denotes the expectation value of a quantum measurement in setting x_1 and x_2. Equation (15) is the central formula of this essay. Note that this bound is not always tight, i.e. there exist examples where the right hand side cannot be reached within quantum mechanics. However for many examples it is tight. The proof of this bound is sketched in Appendix "Tsirelson's theorem carries the Tsirelson bound to Linear Algebra".

We now calculate this bound for the CHSH inequality given in Eq. (2). We see, that here the matrix of coefficients is

$$g = \begin{pmatrix} 1 & 1 \\ 1 & -1 \end{pmatrix} = \underbrace{\begin{pmatrix} \frac{1}{\sqrt{2}} & \frac{1}{\sqrt{2}} \\ \frac{1}{\sqrt{2}} & -\frac{1}{\sqrt{2}} \end{pmatrix}}_{V} \underbrace{\begin{pmatrix} \sqrt{2} & 0 \\ 0 & \sqrt{2} \end{pmatrix}}_{S} \underbrace{\begin{pmatrix} 1 & 0 \\ 0 & 1 \end{pmatrix}}_{W^T}. \tag{16}$$

It is easy to check that the given decomposition of g is a singular value decomposition, i.e. V, S and W have the properties described above. From this we read, that the maximal singular value of g is $||g||_2 = \sqrt{2}$. Then Eq. (15) tells us, that the maximal value of the CHSH inequality (Eq. (2)) within quantum theory is not larger than $2\sqrt{2}$, a value which can also be achieved when using appropriate measurements and states (see Sect. "The CHSH inequality can be violated in experiments with entangled photons")

Tightness of the Bound Can Be Checked Efficiently

We already mentioned that the inequality (15) is not always tight, i.e. sometimes it is not possible to find observables and a quantum state such that there is equality. From the derivation of Eq. (15) sketched in Appendix "Tsirelson's theorem carries the Tsirelson bound to Linear Algebra" one understands, why this is the case. The value $\sqrt{M_1 M_2}||g||_2$ is achieved if and only if there exists a right singular vector $\mathbf{v}$ to the maximal singular value and and a corresponding left singular vector $\mathbf{w}$ which fulfill further normalization constraints.

It is common to denote the element in the i-th row and j-th column of a matrix A as A_{ij}. We will extend this notation to denote the whole i-th row by A_{i*} and the whole j-th column by A_{*j}, i.e. the $*$ stands for "all". For example, the l-th $M_1 + M_2$ dimensional canonical basis vector, with a one at position l and 0 everywhere else, can then be written as $\mathbb{1}_{*,l}^{(M_1+M_2)}$.

With this notation at hand we write down the normalization constraint from above as the system of equations

$$\left\|\alpha^T V_{x*}^{(d)}\right\|^2 = 1 \text{ for } x = 1, 2, ..., M_1 \tag{17}$$

$$\text{and } \left\|\sqrt{\frac{M_2}{M_1}}\alpha^T W_{y*}^{(d)}\right\|^2 = 1 \text{ for } y = 1, 2, ..., M_2, \tag{18}$$

where the $d \times d'$ matrix α is the unknown. The bound in Eq. (15) is tight if and only if such matrix α solving this system of equations can be found. Here d is the degeneracy of the maximal singular value of g and d', the dimension of the vectors $\mathbf{v}_x = \alpha^T V_{x*}^{(d)}$ and $\mathbf{w}_y = \alpha^T V_{y*}^{(d)}$, is a natural number. The steps leading to Eqs. (17) and (18) can be found in the supplemental material of [27]. Because Eqs. (17) and (18) are quadratic in α it may not be obvious how to solve it. In [27] we described an algorithm to solve the above system of equations in polynomial time with respect to the size of g. The interested reader may also find a Matlab snippet in the Appendix "MATLAB snippets". Often the solution α is obvious, e.g. when it is proportional to the identity matrix.

Optimal Measurements Are Obtained from the SVD

From the previous considerations we understand that the existence of the unit vectors $\mathbf{v}_x = \alpha^T V_{x*}^{(d)}$ and $\mathbf{w}_y = \alpha^T V_{y*}^{(d)}$, i.e. the existence of the matrix α that allows this normalization, is crucial to the satisfiability of the singular value bound. Furthermore they have a physical meaning, because they are related to the observables in the following way.

Let us again consider the example of Eq. (2), with the singular value decomposition

$$g = \begin{pmatrix} 1 & 1 \\ 1 & -1 \end{pmatrix} = \underbrace{\begin{pmatrix} \frac{1}{\sqrt{2}} & \frac{1}{\sqrt{2}} \\ \frac{1}{\sqrt{2}} & -\frac{1}{\sqrt{2}} \end{pmatrix}}_{V} \underbrace{\begin{pmatrix} \sqrt{2} & 0 \\ 0 & \sqrt{2} \end{pmatrix}}_{S} \underbrace{\begin{pmatrix} 1 & 0 \\ 0 & 1 \end{pmatrix}}_{W^T}. \tag{19}$$

which we repeat from Eq. (16). The multiplicity $d = 2$ of the maximal singular value $\sqrt{2}$ equals the number of measurement settings M_1 and M_2, so each of the rows $V_{x*}^{(d)}$ and $W_{y*}^{(d)}$ are already normalized due to orthogonality of V and W. Therefore we can choose $\alpha = \mathbb{1}^{(2)}$ to solve Eqs. (17) and (18). We then have $\mathbf{v}_1 = (1, 1)^T/\sqrt{2}$, $\mathbf{v}_2 = (1, -1)^T/\sqrt{2}$, $\mathbf{w}_1 = (1, 0)^T$ and $\mathbf{w}_2 = (0, 1)^T$. We are looking for a state and observables such that $E(x, y) = \mathbf{v}_x \cdot \mathbf{w}_y$, which is always possible to find (see Tsirelson's theorem, Appendix "Tsirelson's theorem carries the Tsirelson bound to Linear Algebra").

Consider for example two spin-$\frac{1}{2}$ particles in the state $\phi_+ = \frac{1}{\sqrt{2}}(1, 0, 0, 1)^T$. Alice and Bob can measure their particles' spin with Stern-Gerlach apparatuses along any orientation in the x-z-plane. The observable of Alice corresponding to a measurement along the direction $(a_x, a_z)^T$ is

$$A = a_x \begin{pmatrix} 0 & 1 \\ 1 & 0 \end{pmatrix} + a_z \begin{pmatrix} 1 & 0 \\ 0 & -1 \end{pmatrix}, \tag{20}$$

where the matrices are two of the so-called Pauli matrices. Bob's measurement reads analogously. The reader can easily verify that the expectation value of the joint observable $A \otimes B$ is given by

$$\phi_+^\dagger (A \otimes B) \phi_+ = \mathbf{a} \cdot \mathbf{b}. \tag{21}$$

Therefore optimal measurement directions leading to equality in In Eq. (15) are given by $\mathbf{v}_x$ and $\mathbf{w}_y$. For this reason we will call $\mathbf{v}_x$ and $\mathbf{w}_y$ the measurement directions, even though they can have a dimension greater than three for general g.

We note how this construction of observables generalizes: The state can be taken to be $\phi_+ = \frac{1}{\sqrt{D}} \sum_{i=1}^{D} \mathbf{e}_i \otimes \mathbf{e}_i$ and the observables can be constructed as $A_x = \mathbf{v}_x \cdot \mathbf{X}$ and $B_y = \mathbf{v}_y \cdot \mathbf{X}$, where $\mathbf{X}$ is a vector of matrices X_i generalizing Pauli matrices in some sense (they anticommute, i.e. $X_i X_j + X_j X_i = 0$ for $i \neq j$).

Bell Inequalities Allow to Lower Bound the Hilbert Space Dimension

In the previous example we chose α to be a square matrix, namely $\alpha = \mathbb{1}^{(2)}$. We will now illustrate the role of the dimension of the measurement directions d' with an example of a trivial Bell inequality, where $d' = 1$ suffices to obtain the Tsirelson bound. For this example the coefficients are $g = \mathbb{1}^{(2)}$. An obvious singular value decomposition of this identity matrix is to choose $V = S = W = \mathbb{1}^{(2)}$. Just as before we can say that $\alpha = \mathbb{1}^{(2)}$ is a solution to Eqs. (17) and (18), thus the bound is achievable with $d' = 2$. But we can also choose $\alpha = (1, 1)^T$, which also solves the system of equations. In this case the measurement directions are one-dimensional ($d' = 1$), in fact they are all equal to 1. Then the expectation value given by the scalar product of the measurement directions reduces to the "classical" expectation value of deterministic local and realistic theories given in Eq. (4). Both quantum theory and local realistic theories can achieve the maximal value of two. This inequality is therefore unable to show a contradiction between quantum theory and locality, realism and freedom of choice. You might have expected this, since the matrix of coefficients does not even contain a negative coefficient, which implies that the maximum value is achieved if all outcomes are +1.

Let us discuss a more interesting example. It is a special instance of the family of Bell inequalities discussed by Vertési and Pál [28]. You can also find the following analysis for the whole family in the supplemental material of [27]. The coefficients are

$$g = \begin{pmatrix} 1 & 1 & 1 & 1 \\ -1 & 1 & 1 & 1 \\ 1 & -1 & 1 & 1 \\ -1 & -1 & 1 & 1 \\ 1 & 1 & -1 & 1 \\ -1 & 1 & -1 & 1 \\ 1 & -1 & -1 & 1 \\ -1 & -1 & -1 & 1 \end{pmatrix}. \tag{22}$$

Please note, that the columns of g are orthogonal, thus it is easy to find a truncated singular value decomposition of g: We can choose $V^{(d)} = \frac{1}{2\sqrt{2}}g$, $S^{(d)} = 2\sqrt{2}\mathbb{1}^{(4)}$ and $W = \mathbb{1}^{(4)}$. One can easily check, that $\alpha = \sqrt{2}\mathbb{1}^{(4)}$ is a solution for the $(d \times d')$-matrix α of Eqs. (17) and (18), so the maximal quantum value of 16 (see Eq. (15)) is achievable with $(d' = 4)$-dimensional measurement directions. It turns out, that the system of equations is not solvable if we choose $d' = 3$, i.e. α to be a (4×3)-dimensional matrix. This has some very interesting physical implications. Since $(d' = 3)$-dimensional measurement directions do not suffice to obtain the maximal value of the Bell inequality, we can conclude from a measured value of $Q = 16$, that our measurement directions were at least four-dimensional. Of course one will never measure this value perfectly in experiment, so what one has to do in practice is to calculate the maximum of the Bell inequality over all three-dimensional measurement directions (this is analog to the calculation of the classical bound B_g described above). If we call this value T_3, then any value between T_3 and 16 witnesses the dimension of the measurement directions to be at least four (see Fig. 13.3).

For spin-$\frac{1}{2}$ particles, there are three orthogonal measurement directions (orientations of the Stern-Gerlach-apparatus), i.e. x-, y- and z-direction, corresponding to the three Pauli matrices (see Eq. (6)) and not more. This holds for all quantum systems

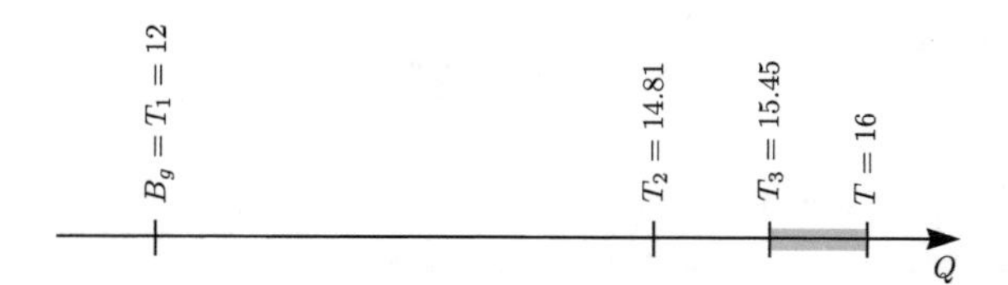

Fig. 13.3 Depending on the dimension d' of the measurement directions different values $T_{d'}$ are maximal for the Bell inequality given by coefficients in Eq. (22). An experimentally obtained value Q of the Bell inequality inside the shaded area witnesses, that the produced quantum system had a greater Hilbert space dimension than qubits (see text). The values are taken from [28].

with two-dimensional Hilbert space (qubits). Thus if in some Bell experiment the value of the Vertési-Pál-inequality given by the coefficients in Eq. (22) is found to be 16 (or larger than T_3), one can conclude that the produced and measured systems were no qubits. In particular they were not single spin-$\frac{1}{2}$ particles. Please note, that this argument is independent of the physical implementation of the source and the measurement apparatuses. For this reason the concept is often called device independent dimension witness.

Satisfiability of the Bound Can Be Understood Geometrically

With $\mathbf{r} = V^{(d)}_{x*}$ Eq. (17) can be written as $\mathbf{r}^T \alpha\alpha^T \mathbf{r} = 1$. This quadratic form defines an ellipsoid with semi-axes $\frac{1}{\sqrt{\lambda_1}}, \frac{1}{\sqrt{\lambda_2}}, ..., \frac{1}{\sqrt{\lambda_d}}$ where $\lambda_1, \lambda_2, ..., \lambda_d$ are the eigenvalues of $\alpha\alpha^T$. Analogously the vectors $\mathbf{r}' = \sqrt{\frac{M_2}{M_1}} W^{(d)}_{y*}$ lie on the same ellipsoid (see Eq. (18)).

We therefore state, that the singular value bound is obtainable if and only if the vectors $V^{(d)}_{x*}$ and $\sqrt{\frac{M_2}{M_1}} W^{(d)}_{y*}$ lie on an ellipsoid. As we mentioned before, in many cases (e.g. from the literature), α can be chosen to be proportional to the identity matrix. Thus in these cases the vectors lie on a d-dimensional sphere, i.e. for $d = 2$ they are on a circle, which is shown for the CHSH inequality [17] in Fig. 13.4.

If α is not square or not full rank (i.e. at least one eigenvalue of α is zero), then at least one of the eigenvalues of $\alpha\alpha^T$ is zero, too. We define the corresponding semi-axis to be infinite.

The measurement directions lie in the image of the linear transformation associated with α. Thus the dimension of the measurement directions cannot be larger than the rank of α. For $g = \mathbb{1}$ we show the degenerate ellipsoid with one infinite semi-axis corresponding to the solution $\alpha = (1, 1)^T$ (see above) in Fig. 13.5.

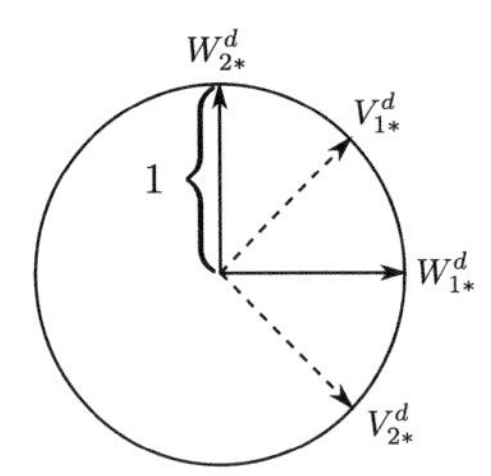

Fig. 13.4 The singular value bound is achievable if and only if the vectors $V^{(d)}_{x*}$ and $\sqrt{\frac{M_2}{M_1}} W^{(d)}_{y*}$ lie on the surface of an ellipsoid. These vectors and the ellipsoid (here a *circle*) are shown for the CHSH inequality.

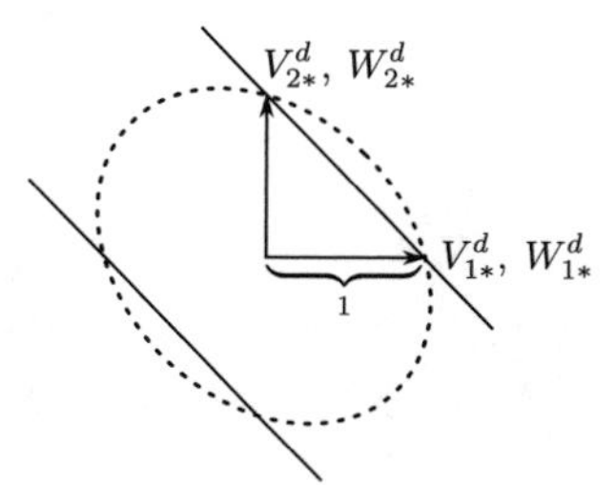

Fig. 13.5 The vectors $V^{(d)}_{x*}$ and $\sqrt{\frac{M_2}{M_1}}W^{(d)}_{y*}$ of $g = \mathbb{1}^{(2)}$ lie on the dotted ellipse. Increasing the larger semi-axis while keeping the vectors on the ellipse leads to the solid (degenerate) ellipse in the limit. Infinite semi-axes of the ellipsoid imply, that lower dimensional measurement directions (here $d' = 1$) suffice to achieve the Tsirelson bound.

Changing g Without Changing the Tsirelson Bound

The parts of the SVD of g which do not correspond to the maximal singular value of (i.e. the non-shaded areas in Fig. 13.2) did not appear in our discussion of the Tsirelson bound. Therefore any changes of these singular vectors in V and W and singular values in S will not affect our analysis. The last is, of course, only true as long as these new singular values do not become bigger than the (previously) maximal singular value. While this changes the matrix g, i.e. leads to a new Bell inequality, the quantum bound remains obtainable and its value remains the same.

From the geometric picture we immediately understand, that rotations of the vectors $V^{(d)}_{x*}$ and $\sqrt{\frac{M_2}{M_1}}W^{(d)}_{y*}$ which keep them on the ellipsoid (see Figs. 13.4 and 13.5) also do not change the value and satisfiability of the singular value bound.

We give an example to illustrate that the measurement directions can be rotated without affecting the singular value bound and its tightness. Consider the CHSH test described above, but now Alice and Bob did not agree on a common coordinate system before performing the experiment, see Fig. 13.6. Let us assume for simplicity that their local coordinate systems are only rotated relative to each other by an angle φ around their common y-axis. This angle φ is unknown to Alice and Bob at the time of collecting the measurement data. The quantum state is still $\psi = \frac{1}{\sqrt{2}}(1, 0, 0, 1)^T$, independent of φ.

Let us analyze the effect of the relative rotation on the violation of the CHSH inequality. The first idea might be to measure the observables of Sect. "Optimal measurements are obtained from the SVD" in the local basis and insert the estimated expectation values into the CHSH inequality. For a relative angle $\varphi = 0°$ these observables are optimal, but for an angle of $\varphi = -45°$ Alice and Bob measure in the same direction and their data will not violate the CHSH inequality. From the previous considerations we know that it is also possible to "rotate" the Bell inequality such that the actually performed measurements are optimal for that inequality. This can be done by applying a rotation matrix to the matrix W. However, twisting the

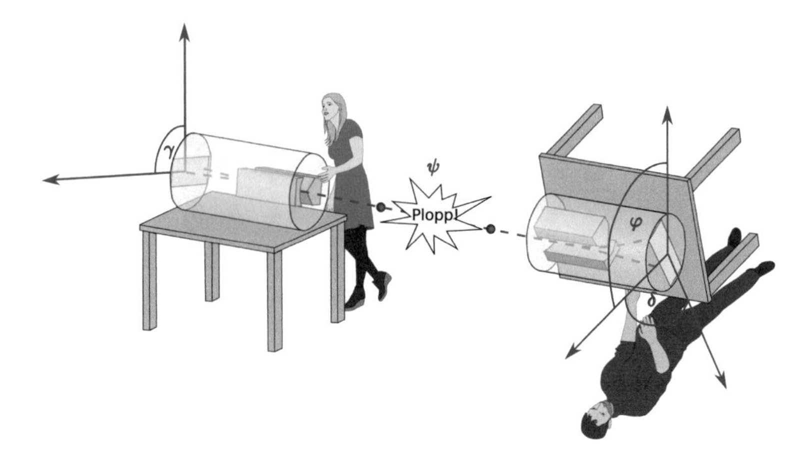

Fig. 13.6 Alice and Bob share pairs of particles in a spin-entangled state ψ and want to violate a Bell inequality. They each can measure the spin of their particle along transversal axes with different angle relative to the table's up. Unfortunately they were not able to agree on what "up" means, yet, and their local coordinate systems are twisted by a relative angle φ. The text explains that one possibility is to measure using (local) angles $\gamma_1 = 45°$, $\gamma_2 = -45°$, $\gamma_3 = 0°$, $\gamma_4 = 90°$ at Alice's site and $\delta_1 = 0°$ and $\delta_2 = 90°$ at Bob's site and "rotate" the Bell inequality.

original CHSH inequality by 45° gives $\sqrt{2}\mathbb{1}$ (up to relabeling of the measurement settings), see Figs. 13.4 and 13.5. And as it is shown in Fig. 13.5 all but one semiaxis of the ellipse associated with α can be chosen to be infinite, which is equivalent to the fact that the classical bound and the quantum bound coincide. This implies that the inequality given by coefficients $g = \sqrt{2}\mathbb{1}$ cannot be violated.

The trick is to include more measurement directions. If the measurement directions of Alice already uniquely define the ellipsoid associated with α, then the rotation of the measurement directions of Bob does not change the fact that the Bell inequality can be violated. One obvious possibility to achieve this is to add all settings of Bob to Alice. We do this for the CHSH inequality (see Eq. (16)) and get

$$g(\varphi) = \begin{pmatrix} \frac{1}{\sqrt{2}} & \frac{1}{\sqrt{2}} \\ \frac{1}{\sqrt{2}} & -\frac{1}{\sqrt{2}} \\ 1 & 0 \\ 0 & 1 \end{pmatrix} \begin{pmatrix} \cos(\varphi) & -\sin(\varphi) \\ \sin(\varphi) & \cos(\varphi) \end{pmatrix}. \tag{23}$$

If we call the different measurement angles $\gamma_1, \gamma_2, \gamma_3, \gamma_4$ at Alice's site and δ_1, δ_2 at Bob's site we have for $\alpha = \sqrt{2}\mathbb{1}$ that $\gamma_1 = 45°$, $\gamma_2 = -45°$, $\gamma_3 = 0°$, $\gamma_4 = 90°$, $\delta_1 = 0°$ and $\delta_2 = 90°$ are optimal measurement settings. The quantum value $T = 4$ of this inequality does not depend on φ, but the classical bound B does. Figure 13.7 shows the violation of the Bell inequality depending on the relative rotation φ. As expected it is always strictly larger than one. The maximal violation of $4 - 2\sqrt{2}$ can

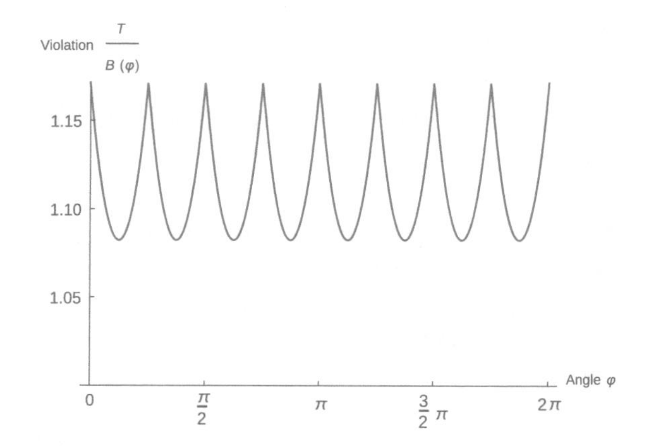

Fig. 13.7 The ratio of the maximal quantum and classical value, the violation, is plotted for the Bell inequality given by the coefficients of Eq. (23) as a function of the relative rotation of the two laboratories φ.

be obtained for $\varphi = k\frac{\pi}{4}$, where k is an integer number. We remark that if Alice and Bob even do not agree on a common coordinate system for the analysis of the data, they still can maximize the violation over the angle φ.

A similar analysis for a general rotation in three dimensions given by three Euler angles was done in [29]. Different approaches to Bell inequalities without a common coordinate system have been described in the literature. We want to mention the following strategy. Each party measures along random but orthogonal measurement directions. Afterwards the violation of the CHSH inequality is calculated for all combinations of pairs of measured settings of Alice and Bob. The result is similar to the one in this section: if the parties measure along more than two directions, then one can find a Bell inequality that is violated with certainty [30].

A deeper understanding of the correlations between measurements on separated systems possible according to quantum theory, including the maximal value of Bell inequalities, is an aim of ongoing research in the field of quantum information theory. In this essay we saw how more measurement settings and higher-dimensional quantum systems can lead to stronger violations of Bell inequalities, e.g. in the context of device-independent dimension witnesses or Bell experiments without a shared reference frame. The insights gained from these simple examples may help to find Bell inequalities well suited for different situations and applications.

Acknowledgments We thank Jochen Szangolies and Michaela Stötzel for feedback which helped to improve this manuscript. ME acknowledges financial support of BMBF, network Q.com-Q.

Appendix

Tsirelson's Theorem Carries the Tsirelson Bound to Linear Algebra

We now sketch the derivation of Eq. (15) following [27]. It is strongly based on a theorem by Boris Tsirelson [31]. It links the expectation values of quantum measurements to scalar products of real vectors. While the full theorem shows equivalence of five different ways of expressing the expectation value, we will repeat two of them here.

Remember that in the formalism of quantum theory observables are hermitean operators, i.e. they equal their complex conjugated transpose. And quantum states can be described by density matrices, which are convex mixtures of projectors onto pure quantum states, with the weights being the probability to find the system in the corresponding pure state. This implies that the density matrix is positive and has trace one.

Consider two fixed sets of observables with eigenvalues in $[-1, 1]$, $\{A_1, A_2, ..., A_{M_1}\}$ and $\{B_1, B_2, ..., B_{M_2}\}$, and a quantum state given in terms of its density matrix ρ. Then the expectation value of the joint measurement of A_x and B_y, $A_x \otimes B_y$, is $E(x, y) = \mathrm{tr}(A_x \otimes B_y \rho)$ according to quantum theory. Tsirelson's theorem states, that there exist real $M_1 + M_2$ dimensional unit vectors $\{\mathbf{v}_1, \mathbf{v}_2, ..., \mathbf{v}_{M_1}\}$ and $\{\mathbf{w}_1, \mathbf{w}_2, ..., \mathbf{w}_{M_2}\}$ such that all expectation values can be expressed as $E(x, y) = \mathbf{v}_x \cdot \mathbf{w}_y$. This is the direction we need, because it allows us to replace the expectation value in Eq. (1) by the scalar product of some real vectors. Tsirelson also proved the converse direction: given the vectors $\mathbf{v}_1, \mathbf{v}_2, ..., \mathbf{v}_{M_1}$ and $\mathbf{w}_1, \mathbf{w}_2, ..., \mathbf{w}_{M_2}$ there exist observables $A_1, A_2, ..., A_{M_1}$ and $B_1, B_2, ..., B_{M_2}$ and a state ρ such that the expectation value $E(x, y) = \mathrm{tr}(A_x \otimes B_y \rho)$ equals the scalar product $\mathbf{v}_x \cdot \mathbf{w}_y$.

After application of Tsirelson's theorem Eq. (1), i.e. $\sum_{x,y} g_{x,y} E(x, y)$, takes the form

$$\sum_{x=1}^{M_1} \sum_{y=1}^{M_2} g_{x,y} \sum_{i=1}^{M_1+M_2} v_{x,i} w_{y,i} = \sum_{x=1}^{M_1} \sum_{y=1}^{M_2} \sum_{i=1}^{M_1+M_2} \sum_{j=1}^{M_1+M_2} v_{x,i} g_{x,y} \delta_{ij} w_{y,j}$$
$$= \mathbf{v}^T (g \otimes \mathbb{1}^{(M_1+M_2)}) \mathbf{w}. \tag{24}$$

Here we expressed the scalar product as a matrix product using the $M_1 + M_2$ dimensional identity matrix $\mathbb{1}^{(M_1+M_2)}$ and defined the vectors $\mathbf{v}$ and $\mathbf{w}$, which arise if one concatenates all $\mathbf{v}_x$ and $\mathbf{w}_y$, respectively. For the decomposition given in Eq. (16) with $\alpha = \mathbb{1}^{(2)}$, for example, $\mathbf{v}_1 = (\frac{1}{\sqrt{2}}, \frac{1}{\sqrt{2}})^T$ and $\mathbf{v}_2 = (\frac{1}{\sqrt{2}}, -\frac{1}{\sqrt{2}})^T$ and thus $\mathbf{v} = (\frac{1}{\sqrt{2}}, \frac{1}{\sqrt{2}}, \frac{1}{\sqrt{2}}, -\frac{1}{\sqrt{2}})^T$. From Eq. (24) we see, that the maximal quantum value of the Bell inequality is given by the maximal singular value (the maximal stretching factor) of $g \otimes \mathbb{1}^{(M_1+M_2)}$ times the length of the vectors $\mathbf{v}$ and $\mathbf{w}$. The matrix $g \otimes \mathbb{1}^{(M_1+M_2)}$ has the same singular values as g, except that each of them appears

$M_1 + M_2$ times. Because the $\mathbf{v}_x$ and $\mathbf{w}_y$ constituting $\mathbf{v}$ and $\mathbf{w}$ are all unit vectors, the length of $\mathbf{v}$ is $\sqrt{M_1}$ and the length of $\mathbf{w}$ is $\sqrt{M_2}$. Putting these factors together we arrive at Eq. (15).

MATLAB Snippets

```
function [ T ] = singularvaluebound( g )
%SINGULARVALUEBOUND Calculates the SV-bound of g
%   the returned value is a Tsirelson bound for
%   the CHSH-type inequality given by g
T=sqrt(numel(g))*norm(g);
end

function [ a ] = alphamatrix( g )
%ALPHAMATRIX Links SVD to measurement directions
%   see PRL 111, 240404 (2013)
[M1 M2]=size(g);
[V S W]=svd(g);
acc=1E-4; % adjust to numerical precision
d=sum(diag(S)>=S(1,1)-acc);
% the vectors to be normalized by alpha:
A=[V(1:M1,1:d); sqrt(M2/M1)*W(1:M2,1:d)];
Q=(A*A').^2;
c=pinv(Q)*ones(M1+M2,1);
if sum(abs(Q*c-ones(M1+M2,1)) > acc)
    error('alphamatrix:nosol','No solution alpha found.');
else
    X=A'*diag(c)*A;
    if eigs(X,1,'sm')<0
        error('alphamatrix:norealsol',
              'No real solution alpha found.');
    end
end
a=X^0.5;
end

function [ T ] = tsirelsonbound( g )
%TSIRELSONBOUND Calculates the Tsirelson bound for g
%   Uses the semidefinite programm described by
%   Stephanie Wehner in PRA 73, 022110 (2006).
[M1 M2]=size(g);
W=[zeros(M1,M1) g; g' zeros(M2,M2)];
G=sdpvar(M1+M2,M1+M2);
obj=trace(G*W)/2;
F=set(G>0);
for i=1:M1+M2
```

```
    F=F+set(G(i,i) == 1);
end
solvesdp(F,obj,sdpsettings('verbose',0));
T=-double(obj);
end
```

References

1. T. Young, *A Course of Lectures on Natural Philosophy and the Mechanical Arts* (Taylor and Walton, 1845)
2. R.P. Feynman, R.B. Leighton, M. Sands, *The Feynman Lectures on Physics*, vol. 3 (Addison Wesley, 1971). http://www.feynmanlectures.caltech.edu/
3. R. Bach, D. Pope, S.H. Liou, H. Batelaan, New J. Phys. **15**(3), 033018 (2013). doi:10.1088/1367-2630/15/3/033018
4. B. Brezger, L. Hackermüller, S. Uttenthaler, J. Petschinka, M. Arndt, A. Zeilinger, Phys. Rev. Lett. **88**, 100404 (2002). doi:10.1103/PhysRevLett.88.100404
5. M. Arndt, K. Hornberger, Nat. Phys. **10**, 271 (2014). doi:10.1038/nphys2863
6. M. Razavy, *Quantum Theory of Tunneling* (World Scientific Pub Co Inc, 2003)
7. R.P. Feynman, R.B. Leighton, M. Sands, *The Feynman Lectures on Physics*, vol. 2 (Addison Wesley, 1977). http://www.feynmanlectures.caltech.edu/
8. W.M. Itano, D.J. Heinzen, J.J. Bollinger, D.J. Wineland, Phys. Rev. A **41**, 2295 (1990). doi:10.1103/PhysRevA.41.2295
9. A.C. Elitzur, L. Vaidman, Found. Phys. **23**(7), 987 (1993). doi:10.1007/BF00736012
10. P.G. Kwiat, A.G. White, J.R. Mitchell, O. Nairz, G. Weihs, H. Weinfurter, A. Zeilinger, Phys. Rev. Lett. **83**, 4725 (1999). doi:10.1103/PhysRevLett.83.4725
11. G. Barreto Lemos, V. Borish, G.D. Cole, S. Ramelow, R. Lapkiewicz, A. Zeilinger, Nature 409–412 (2014)
12. A. Aspect, P. Grangier, G. Roger, Phys. Rev. Lett. **49**, 91 (1982). doi:10.1103/PhysRevLett.49.91
13. G. Weihs, T. Jennewein, C. Simon, H. Weinfurter, A. Zeilinger, Phys. Rev. Lett. **81**, 5039 (1998). doi:10.1103/PhysRevLett.81.5039
14. M.A. Rowel, D. Kielpinski, V. Meyer, C.A. Sackett, W.M. Itano, C. Monroe, D.J. Wineland, Nature **409**, 791 (2001). doi:10.1038/35057215
15. J.M. Martinis et al., Nature **461**, 504 (2009). doi:10.1038/nature08363
16. M. Zukowski, C. Brukner, J. Phys. A: Math. Theor. **47**(42), 424009 (2014). doi:10.1088/1751-8113/47/42/424009
17. J.F. Clauser, M.A. Horne, A. Shimony, R.A. Holt, Phys. Rev. Lett. **23**, 880 (1969). doi:10.1103/PhysRevLett.23.880
18. J.F. Clauser, M.A. Horne, Phys. Rev. D **10**, 526 (1974). doi:10.1103/PhysRevD.10.526
19. C. Gerthsen, *Physik* (D. Meschede, 2001)
20. D. Dehlinger, M.W. Mitchell, Am. J. Phys. **70**, 903 (2002). doi:10.1119/1.1498860
21. B. Tsirelson, Lett. Math. Phys. **4**, 93 (1980)
22. S. Wehner, Phys. Rev. A **73**, 022110 (2006). doi:10.1103/PhysRevA.73.022110
23. M. Navascués, S. Pironio, A. Acín, Phys. Rev. Lett. **98**, 010401 (2007). doi:10.1103/PhysRevLett.98.010401
24. S. Popescu, D. Rohrlich, Found. Phys. **24**(3), 379 (1994). doi:10.1007/BF02058098
25. M. Pawlowski, T. Paterek, D. Kaslikowski, V. Scarani, A. Winter, M. Zukowski, Nature 1101–1104 (2009). doi:10.1038/nature08400
26. A. Cabello, Phys. Rev. Lett. **110**, 060402 (2013). doi:10.1103/PhysRevLett.110.060402
27. M. Epping, H. Kampermann, D. Bruß, Phys. Rev. Lett. **111**, 240404 (2013). doi:10.1103/PhysRevLett.111.240404
28. T. Vértesi, K.F. Pál, Phys. Rev. A **77**, 042106 (2008). doi:10.1103/PhysRevA.77.042106

29. M. Epping, H. Kampermann, D. Bruß, J. Phys. A: Math. Theor. **47**(424015) (2014). doi:10.1088/1751-8113/47/42/424015
30. P. Shadbolt, T. Vértesi, Y.C. Liang, C. Branciard, N. Brunner, J.L. O'Brien, Scientific Reports **2**(470) (2012). doi:10.1038/srep00470
31. B. Tsirelson, *Letters in Mathematical Physics* pp. 93–100 (1980)

Chapter 14
Bell Inequalities with Retarded Settings

Lucien Hardy

Abstract We consider retarded settings in the context of a Bell-type experiment. The retarded setting is defined as the value the setting would have taken were it not for some external intervention (for example, by a human). We derive *retarded Bell inequalities* that explicitly take into account the retarded settings. These inequalities are not violated by Quantum Theory (or any other theory) when the retarded settings are equal to the actual settings. We construct a simple model that reproduces Quantum Theory when the retarded and actual settings are equal, but violates it when they are not. We discuss using humans to choose the settings in this type of experiment and the implications of a violation of Quantum Theory (in agreement with the retarded Bell inequalities) in this context.

Introduction

I first got interested in Bell's theorem [4], many years ago, on account of the following question: if we employed humans to switch the measurement settings at the two ends of the experiment, might we then expect Bell's inequalities to be satisfied and Quantum Theory to be violated? I was particularly interested in whether we might think of this as a test for mind-matter duality. The papers I wrote on this subject did not, of course, get past the referees in 1989. In the meantime, I have come to be much more accepting of Bell style nonlocality in Quantum Theory. By now I more-or-less fully expect that, even if humans were used to switch the measurement settings, we would see a violation of Bell's inequalities in agreement with Quantum Theory. On the other hand, the implications of a violation of Quantum Theory in this context would be so incredibly significant that it is worth discussing how we might go about doing an experiment.

In this contribution I will present modified Bell inequalities that I obtained 1989 (but did not publish) that take account of the possibility that a signal actually passes between the two ends of the experiment at the speed of light carrying information

L. Hardy (✉)
Perimeter Institute for Theoretical Physics, Waterloo, ON N2L 2Y5, Canada
e-mail: lhardy@perimeterinstitute.ca

R. Bertlmann and A. Zeilinger (eds.), *Quantum [Un]Speakables II*,
The Frontiers Collection, DOI 10.1007/978-3-319-38987-5_14

as to the distant setting (this is the *retarded setting*). The inequalities I will present actually include these retarded settings. After all, these are things we can measure and their values would be significant if switching distant settings actually changed the physics.

Although my original motivation for thinking about retarded settings was in the context of having people actually do the switching, we could use the inequalities obtained here in other contexts. For example, we might attempt to collect cosmological signals from regions of space-time that are causally disconnected from our own to implement the switching (see, for example, [13]). We might use them to analyze existing experiments in which the settings are varied in time [2] or in which a random number generator is used to do the switching [14, 16]. Additionally, we might investigate practical applications of such inequalities (in device independent quantum cryptography [1, 3] and communication complexity [7, 11] for example).

Retarded Settings

We will define two notions of retarded settings. We are particularly interested in the second type (whose definition is a little subtle). Consider that the settings a and b at the two ends, 1 and 2 respectively, of a Bell experiment are switched by some means during the course of the experiment. Then we are interested in the retarded settings, a_r (as regarded from side 2), and b_r (as regarded from side 1).

Simple retarded settings. The most obvious interpretation of retarded setting is that it is simply $a_r = a(t_2 - L/c)$ where L is the distance between the two ends and t_2 is the time at which the measurement at side 2 takes place. Likewise, we would have $b_r = b(t_1 - L/c)$.

Predictive retarded settings. A different notion of retarded settings is that a_r is a prediction made at end 2 as to what setting a will take at time t_1 based on information that can be locally communicated to end 2. Thus, if the variation of a were deterministic then a calculation at end 2 would enable us to predict a at time t_1. Now, we can imagine that the variation of a is deterministic except for *interventions*. Then the retarded setting, a_r, is be the value a is predicted to take at time t_1 if there are no interventions on this setting after time $t_2 - L/c$ that alter a at t_1 from the value it would have taken.

If we only allow the measurement setting to be changed by these supposed interventions then the above two definitions of retarded setting coincide. However, this will not be the case in general.

In the introduction we supposed that the interventions are due to a person doing the switching. We will discuss this possibility later and the issues arising. Another possibility, also mentioned above, is that the interventions are due to signals from causally disconnected regions of space. One other possibility is that appropriate random number generators can supply such interventions.

Clauser Horne Shimony Holt Inequalities with Retarded Settings

Consider a Bell type experiment with two ends. Imagine we have a central source of two systems, 1 and 2, described by hidden variables, $\lambda \in \Gamma$, with probability distribution $\rho(\lambda)$ such that

$$\int_{\Gamma} \rho(\lambda) d\lambda = 1 \tag{1}$$

We can obtain Clauser Horne Shimony Holt type Bell inequalities [10] with retarded settings. In this scenario, we have a measurement, A, on system 1 which can take values $+1$ and -1. Similarly, we have a measurement, B, on the right which can take values $+1$ and -1. For simplicity, we will assume that the hidden variable model is deterministic (this assumption could easily be dropped). Let

$$A(a, b_r, \lambda) \tag{2}$$

be the outcome at side 1 when we have setting a, retarded setting b_r, and hidden variable λ. Similarly, we have

$$B(b, a_r, \lambda) \tag{3}$$

at end 2. Note that, at each end, we allow for a dependence on the retarded setting at the other end. Since these settings are retarded, this is a local dependence.

We define the correlation function

$$E(a, b|a_r, b_r) = \int_{\Gamma} A(a, b_r, \lambda) B(b, a_r, \lambda) d\lambda \tag{4}$$

So $E(a, b|a_r, b_r)$ is the expectation value of the product of the outcomes at the two ends.

Clauser, Horne, Shimony, and Holt used the following (easily verified) mathematical identity

$$X'Y' + X'Y + XY' - XY = \pm 2 \tag{5}$$

where $X, X', Y, Y' = \pm 1$. We put

$$X = A(a, b_r, \lambda) \tag{6}$$

$$X' = A(a', b'_r, \lambda) \tag{7}$$

$$Y = B(b, a_r, \lambda) \tag{8}$$

$$Y' = B(b', a'_r, \lambda) \tag{9}$$

Substituting these into (5) and integrating over λ we obtain

$$-2 \leq E(a', b'|a'_r, b'_r) + E(a', b|a_r, b'_r) + E(a, b'|a'_r, b_r) - E(a, b|a_r, b_r) \leq +2 \tag{10}$$

These are the retarded CHSH inequalities.

When Retarded and Actual Settings Are Equal

If the retarded settings are equal to the actual settings for each term in the retarded CHSH inequalities (10) then we have $a'_r = a' = a_r = a$ and $b'_r = b' = b_r = b$ and the inequalities become

$$-2 \leq 2E(a, b|a, b) \leq +2 \tag{11}$$

This inequality is always satisfied (as E is bounded by ± 1) and hence there is no constraint from the retarded CHSH inequalities when the retarded settings are equal to the actual settings.

It is also interesting to consider the case where the retarded setting is equal to the actual setting for one end only. Consider the case when $a'_r = a' = a_r = a$. Then the retarded CHSH inequality reduces to

$$-2 \leq E(a, b'|a, b'_r) + E(a, b|a, b'_r) + E(a, b'|a, b_r) - E(a, b|a, b_r) \leq +2 \tag{12}$$

Now, this inequality is not violated by any theory, T, which has

$$E(a, b|a_r, b_r) = E^T(a, b) \tag{13}$$

i.e. theories in which the retarded settings do not influence the physics (such as Quantum Theory). This is because (12) then reduces to

$$-2 \leq 2E^T(a, b') \leq +2 \tag{14}$$

which cannot be violated. Hence, if we want to test such theories, then we need to be sure the retarded and actual settings are different for each end.

Testing Quantum Theory

The quantum predictions do not depend on the retarded settings so, according to Quantum Theory, we would have

$$E(a, b|a_r, b_r) = E^{\mathrm{QT}}(a, b) \tag{15}$$

and the inequalities would become

$$-2 \le E^{\mathrm{QT}}(a', b') + E^{\mathrm{QT}}(a', b) + E^{\mathrm{QT}}(a, b') - E^{\mathrm{QT}}(a, b) \le +2 \tag{16}$$

As is well known, these inequalities can be violated by the predictions of Quantum Theory.

Hence, we cannot have a local model of the sort used in setting up the retarded CHSH inequalities that reproduces Quantum Theory. However, we have also seen that we can have

$$E(a, b|a, b) = E^{\mathrm{QT}}(a, b) \tag{17}$$

Furthermore, we would expect this to be true since the motivation for considering local models with retarded settings is to reproduce Quantum Theory when the retarded settings are equal to the actual settings. The model we will provide in Sect. "A model" has the property (17) by construction. The retarded CHSH inequalities also allow

$$E(a, b|a, b_r) = E(a, b|a_r, b) = E^{\mathrm{QT}}(a, b) \tag{18}$$

(where the retarded setting equals the actual setting for one end). The model we provide in Sect. "A model" does not have this property. However, it should be possible to build a more sophisticated model that does have this property.

Retarded Versus Standard Bell Inequalities

The standard Bell inequalities do not take account of retarded settings. If we perform an experiment where we actively change the settings during the flight of the systems from the source before they arrive at the measurement apparatuses, then we have to take care to be sure that the retarded and actual settings are different for a sufficiently large proportion of cases. However, in the standard Bell inequalities, we simply ignore the retarded settings and average over all cases. If the probability of any particular retarded setting is independent of the actual settings then we can recover standard Bell inequalities. In this case we can define

$$E_{\mathrm{av}}(a, b) = \sum_{a_r, b_r} p(a_r, b_r) E(a, b|a_r, b_r) \tag{19}$$

where $p(a_r, b_r)$ is the probability that the retarded settings are a_r and b_r. Now we can take the average of the retarded CHSH inequality and obtain

$$-2 \le E_{\mathrm{av}}(a', b') + E_{\mathrm{av}}(a', b) + E_{\mathrm{av}}(a, b') - E_{\mathrm{av}}(a, b) \le +2 \tag{20}$$

These are standard CHSH inequalities (where we ignore the retarded settings). However, this derivation of standard from retarded CHSH inequalities fails when there is a correlation between the retarded and actual settings. Any such correlation could, in principle, lead to a situation where the standard Bell inequalities are violated while the retarded Bell inequalities are satisfied. Hence, if we take seriously the need to actively switch the settings, then we need to use the retarded Bell inequalities.

It is particularly noteworthy that the famous experiment of Aspect, Dalibard and Roger in 1982 [2] used periodic switching. Unfortunately, the switching period was such that the actual and retarded settings were equal. This was pointed out by Zeilinger [17] and formed part of the motivation for the experiment in his group [16] in which ultrafast random switching was used. Under such a scenario, it seems likely that the retarded and actual settings would not be correlated and hence we can obtain standard from retarded Bell inequalities by the above type of averaging (another, even more definitive, experiment was performed by Zeilinger's group in [14]). On the other hand, if we use the retarded Bell inequalities directly, then we do not have to make such an assumption. Retarded Bell inequalities provide a tool for analyzing this kind of experiment. Of course, neither of these experiments used humans or signals from cosmologically disconnected parts of the universe and so it is really models with "simple retarded settings" (as defined in Sect. "Retarded settings") that are being tested (though one might argue that a random number generator forces interventions of the sort we discussed above).

Source Distribution of Hidden Variables

In our model, we supposed that the retarded settings influenced the outcome at the other end (for example, by using a function $A(a, b_r, \lambda)$). Another possibility (considered by Zeilinger [17]) is that the retarded settings influence the distribution of hidden variables at the source. Then we would have $\Gamma_{a_r b_r}$. This would block the derivation of the retarded Bell inequalities above. We can address this concern in the following way. First, rather than associating the hidden variables, λ, with the source alone we associate them with the full situation concerning the experiment at a time, t_0, earlier than both $t_1 - L/c$ and $t_2 - L/c$. Thus, the hidden variables describe the source, measurement apparatuses and every other detail of the physics that might be relevant for the experiment. This means, in particular, that λ also encodes the retarded settings a_r and b_r as long as there is no intervention between t_0 and the relevant retarded time. Let us assume that these retarded settings are equal to a and b. If there are no interventions in the remaining time then the actual settings will be a and b respectively. On the other hand, if there is an intervention at both ends, or just one end then we could have actual settings a' and/or b' accordingly. This is true with the given initial distribution on λ and so we can obtain retarded Bell inequalities as follows

$$-2 \leq E(a', b'|a, b) + E(a', b|a, b) + E(a, b'|a, b) - E(a, b|a, b) \leq +2 \quad (21)$$

Note that every term has the same retarded settings (consistent with assumption above). This inequality is interesting as only the $E(a', b'|a, b)$ term has different retarded and actual settings on both sides. The inequality can be violated by Quantum Theory if we substitute (15) in. Thus, in the unlikely event we saw a violation of Quantum Theory, the term, $E(a', b'|a, b)$, is the most likely to be the place we would see it.

A Model

It is interesting to construct an explicit model reproducing the predictions of Quantum Theory (for a certain state) when the retarded settings are equal to the actual settings for both ends. Consider a singlet state

$$|\psi\rangle = \frac{1}{\sqrt{2}} \left(|+\rangle_1|-\rangle_2 - |-\rangle_1|+\rangle_2\right) \tag{22}$$

We can subject this to a measurement of spin in the xy plane at angle a at end 1 and angle b at end 2. Then a simple calculation shows that the correlation function is

$$E^{\Psi}(a, b) = -\cos(a - b) \tag{23}$$

Now consider a hidden variable model with a hidden variable λ having

$$0 \leq \lambda < 2\pi \qquad \Gamma = \frac{1}{2\pi} \tag{24}$$

We define the result functions

$$A(a, b_r, \lambda) = \left\{ \begin{array}{l} +1 \text{ for } \theta_L \leq \lambda < \theta_L + \pi \\ -1 \text{ for } \theta_L + \pi \leq \lambda < \theta_L + 2\pi \end{array} \right\} \tag{25}$$

and

$$B(b, a_r, \lambda) = \left\{ \begin{array}{l} +1 \text{ for } \theta_R \leq \lambda < \theta_R + \pi \\ -1 \text{ for } \theta_R + \pi \leq \lambda < \theta_R + 2\pi \end{array} \right\} \tag{26}$$

where we understand λ to be an angle (so angles greater than, or equal to 2π are identified with angles in the interval $[0, 2\pi)$ in the usual way) and where θ_L is a function of a and b_r and θ_R is a function of b and a_r. It is easy to prove that

$$E(a, b|a_r, b_r) = 1 - \frac{2|\theta_R - \theta_L|}{\pi} \tag{27}$$

Hence, if we set

$$\theta_L = -\frac{\pi}{4}(1 + \cos(a - b_r)) \quad \theta_R = \frac{\pi}{4}(1 + \cos(a_r - b)) \tag{28}$$

we obtain

$$E(a, b|a_r, b_r) = -\frac{1}{2}(\cos(a - b_r) + \cos(a_r - b)) \tag{29}$$

When the retarded settings are equal to the actual settings we get

$$E(a, b|a, b) = -\cos(a - b) \tag{30}$$

in agreement with Quantum Theory. If the actual and retarded setting differ for one side only then the model does not give quantum theory (although the retarded CHSH inequalities would allow quantum theory to be reproduced).

The retarded Bell inequalities are not violated by this model. To illustrate this consider the special case

$$a = \frac{\pi}{2}, \quad a' = 0, \quad b = -\frac{\pi}{4}, \quad b' = \frac{\pi}{4} \tag{31}$$

If we substitute (29) into (21) with these settings then we obtain

$$E(a', b'|a, b) + E(a', b|a, b) + E(a, b'|a, b) - E(a, b|a, b) = -\sqrt{2} \tag{32}$$

This satisfies the particular retarded CHSH inequalities. It is interesting that we do not saturate the inequalities with this model. A better model may saturate the inequality.

Clauser Horne Inequalities

We can also derive retarded Clauser Horne inequalities based on the Clauser Horne inequalities [9]. These inequalities are especially useful in experiments since they have 0 as the upper bound. Consequently it is sufficient to measure count rates without normalizing the probabilities with a total count rate. These inequalities pertain to the same setting as before, but now we are interested in the probabilities for some particular outcome (we will take this to be the + outcome) at each end. We let

$$p_1(a, b_r|\lambda) \tag{33}$$

be the probability of that we see a outcome +1 to measurement A with setting a at this end and retarded setting b_r at the other end. Similarly we have

$$p_2(b, a_r|\lambda) \tag{34}$$

for the probability that we see outcome +1 for measurement B on particle 2 with setting b and retarded setting a_r at the other end. The joint probability of seeing a +1 at both ends is

$$p_{12}(a, b|a_r, b_r) = \int_\Gamma p_1(a, b_r|\lambda)p_2(b, a_r|\lambda)d\lambda \tag{35}$$

Note that we allow for a dependence of this joint probability on the retarded settings at the other end. We can also construct the local probabilities

$$p_1(a) = \int_\Gamma p_1(a, b_r|\lambda)d\lambda \tag{36}$$

$$p_2(b) = \int_\Gamma p_2(b, a_r|\lambda)d\lambda \tag{37}$$

We could, without violating locality, also allow these probabilities to depend on the retarded settings. However, this seems less likely and so we will stick with the given functional dependence. If we do want to have such functional dependence then this can easily be inserted in the Bell inequalities we derive below.

Quantum theory does not predict any dependence on the values of retarded settings. Thus, according to Quantum Theory, we will have

$$p_{12}(a, b|a_r, b_r) = p_{12}^{\mathrm{QT}}(a, b) \tag{38}$$

However, it will follow (by adapting the usual Bell analysis) that this cannot actually be the case in a local hidden variable model of the type we are considering.

Now we will derive retarded Clauser Horne inequalities. Consider the following easily verified mathematical inequalities (introduced by Clauser and Horne [9])

$$-1 \leq x'y' + x'y + xy' - xy - x - y \leq 0 \tag{39}$$

where $0 \leq x, y, x', y' \leq 1$. Now we put

$$\begin{aligned} x &= p_1(a, b_r|\lambda) \\ x' &= p_1(a', b'_r|\lambda) \\ y &= p_2(b, a_r|\lambda) \\ y' &= p_2(b, a'a_r|\lambda) \end{aligned}$$

Inserting these into (39) we obtain

$$\begin{aligned} -1 \leq p_{12}(a', b'|a'_r, b'_r) + p_{12}(a', b|a_r, b'_r) + p_{12}(a, b'|a'_r, b_r) \\ -p_{12}(a, b|a_r, b_r) - p_1(a) - p_2(b) \leq 0 \end{aligned} \tag{40}$$

Note that we could allow $p_1(a)$ and $p_2(b)$ to depend on the retarded settings (so we would have $p_1(a|b_r)$ and $p_2(b|a_r)$ instead). We can substitute the quantum predictions (38) in to this inequality. This gives

$$-1 \leq p_{12}^{\mathrm{QT}}(a', b') + p_{12}^{\mathrm{QT}}(a', b) + p_{12}(a, b') - p_{12}^{\mathrm{QT}}(a, b) - p_{1}^{\mathrm{QT}}(a) - p_{2}^{\mathrm{QT}}(b) \leq 0 \quad (41)$$

It was shown by Clauser and Horne that this inequality can be violated by choosing the two systems to be in an appropriate entangled state and by choosing appropriate settings. Note that, to get a violation by Quantum Theory, it is a necessary condition that both $a' \neq a$ and $b' \neq b$.

How to Perform an Experiment

To perform an experiment to test the retarded Bell inequalities we need a source of interventions. We could imagine two subjects, let us call them Alice and Bob, sitting at the two ends each switching the settings by hand. The only problem with this is that the hand and the device it switches both operate at mechanical speeds. To have any chance of having a retarded setting different from the actual setting, with such a system, we would need the distance between the two ends to be very big. However, whenever Alice decides to switch the setting, there is some accompanying electrical activity in the brain. We could use this accompanying electrical activity to do the actual switching (where the device that Alice switches is just a retrospective control). Pockel cells can be used to accomplish fast switching (at electrical speeds) on photons. Hence, with this set up we could realistically perform an experiment over a shorter distance.

It is not actually necessary that Alice (and Bob) actually switch a switch. It only necessary that they engage in some activity such that we want to regard the resulting electrical activity as constituting an intervention. A challenge would be identifying the electrical signals originating in the brain that should be regarded as interventions in the sense understood here. However, we could filter for lots of different types of signature and analyze the data accordingly. Another challenge would be to get the count rates high enough that we can have statistically meaningful results.

A possible control on this type of experiment would be to introduce a delay between the source of interventions and the switching. If this delay were longer than L/c then any supposed effect ought to vanish.

Interpretation

Quantum Theory is a coherent whole. It has been tested extensively at low energy in laboratories around the world. Hence, it seems very unlikely that we could expect to see its violation in circumstances such as those we have discussed. On the other hand, as long as we do these kind of experiments to test the theory, it is worth thinking carefully about what we are testing. We have here suggested that we may see a violation of Quantum Theory in accordance with the retarded Bell inequalities derived here when we actually use humans to switch the settings. If such a violation of Quantum Theory was seen, and yet it was impossible to obtain such a violation where the switching was performed by non-animate systems (such as computer programs, physically chaotic systems, or quantum random number generators) then we would have to seek an explanation of this. The Cartesian idea of mind-matter duality provides a model for a kind of external intervention (of mind on matter) of the type that we have discussed. This kind of duality has been much discussed by philosophers in the context of understanding consciousness. A modern proponent of such dualism is Chalmers [8] while Dennett [12] advocates the opposite point of view. While it is difficult to understand consciousness in terms of matter stuff alone, it is not clear that adding mind stuff into the mix helps us particularly. On the other hand, if it turned out that systems we take to be conscious were capable of things (like violating Quantum Theory under the described circumstances) that ordinary systems were not then that would be a profound challenge to our usual way of thinking about the world. We should not shy away from such experiments.

The situation here is reminiscent of the Turing test [15]. In the Turing test, computers and humans compete over a computer screen interface to convince human interviewers that they are human. This test involves the subjective judgement of the interviewers. Here, instead, pairs of humans compete against pairs of computers (or whatever other physical system we want to use) to violate Quantum Theory by providing the inputs to the setting switches of an experimental apparatus which they, otherwise, have no control over. The test is completely objective—passing the test would entail bringing about a violation of Quantum Theory in accord with the retarded Bell inequalities. As such, this test is interesting simply because it provides a scientific way to investigate a particular model of mind-matter duality (even if, as seems overridingly likely, an actual test will not violate Quantum Theory).

Bell's La Nouvelle Cuisine Remarks

I never met John Bell (though I was in the audience for two talks he gave). I did, however, send him a copy of a paper outlining some of the above ideas (my second attempt attempt at such a paper). He responded by sending me a copy of his La Nouvelle Cuisine paper [5] (now available as the penultimate article in the wonderful collection of papers by Bell et al. [6]). This paper is a beautiful discussion of how to understand causality. While he had clearly thought about using humans to do the switching, I suspect he was not sympathetic to the idea that anything would come of it. I end with a quote from this paper—it is classic Bell:

> The assertion that "we cannot signal faster than light"immediately provokes the question:
>
> Who do we think *we* are?"

> *We* who can make "measurements", *we* who can manipulate "external fields", *we* who can "signal" at all, even if not faster than light? Do *we* include chemists, or only physicists, plants, or only animals, pocket calculators, or only mainframe computers?
>
> The unlikelihood of finding a sharp answer to this question reminds me of the relationship of thermodynamics to fundamental theory.

Acknowledgments I am grateful to the late Euan Squires for discussions on this subject while I was doing my PhD under his supervision. Research at Perimeter Institute is supported by the Government of Canada through Industry Canada and by the Province of Ontario through the Ministry of Economic Development and Innovation. This project was made possible in part through the support of a grant from the John Templeton Foundation. The opinions expressed in this publication are those of the author and do not necessarily reflect the views of the John Templeton Foundation.

References

1. A. Acin, N. Gisin, L. Masanes, From bells theorem to secure quantum key distribution. Phys. Rev. Lett. **97**(12), 120405 (2006)
2. A. Aspect, J. Dalibard, G. Roger, Experimental test of bell's inequalities using time-varying analyzers. Phys. Rev. Lett. **49**(25), 1804 (1982)
3. J. Barrett, L. Hardy, A. Kent, No signaling and quantum key distribution. Phys. Rev. Lett. **95**(1), 010503 (2005)
4. S.J. Bell, On the einstein-podolsky-rosen paradox. Physics **1**(3), 195–200 (1964)
5. J.S. Bell, in *La Nouvelle Cuisine*. ed by A. Sarlemijn, P. Kroes. Between Science and Technology. (Elsevier, 1990), p. 97
6. M. Bell, K Gottfried, M. Veltman (eds.), *John S. Bell on the Foundations of Quantum Mechanics* (World Scientific, 2001)
7. Č. Brukner, M. Żukowski, J.-W. Pan, A. Zeilinger, Bells inequalities and quantum communication complexity. Phys. Rev. Lett. **92**(12), 127901 (2004)
8. D.J Chalmers, *The Conscious Mind: In Search of a Fundamental Theory* (Oxford University Press, 1997)
9. F.J. Clauser, M.A. Horne, Experimental Consequences of Objective Local Theories. Phys. Rev. D **10**(2), 526 (1974)
10. F.J. Clauser, M.A. Horne, A. Shimony, R.A. Holt, Proposed experiment to test local hidden-variable theories. Phys. Rev. Lett. **23**(15), 880 (1969)
11. R. Cleve, H. Buhrman, Substituting quantum entanglement for communication. Phys. Rev. A **56**(2), 1201 (1997)
12. D.C. Dennett, *Consciousness Explained* (Penguin, UK, 1993)
13. J. Gallicchio, A.S. Friedman, D.I Kaiser, Testing bells inequality with cosmic photons: closing the setting-independence loophole. Phys. Rev. Lett. **112**(11), 110405 (2014)
14. T. Scheidl, R. Ursin, J. Kofler, S. Ramelow, X.-S. Ma, T. Herbst, L. Ratschbacher, A. Fedrizzi, N.K. Langford, T. Jennewein et al., Violation of local realism with freedom of choice. Proc. Natl. Acad. Sci. **107**(46), 19708–19713 (2010)
15. A.M. Turing, Computing machinery and intelligence. Mind 433–460 (1950)
16. G. Weihs, T. Jennewein, C. Simon, H. Weinfurter, A. Zeilinger, Violation of bell's inequality under strict einstein locality conditions. Phys. Rev. Lett. **81**(23), 5039 (1998)
17. A. Zeilinger, Testing bell's inequalities with periodic switching. Phys. Lett. A **118**(1), 1–2 (1986)

Chapter 15
How to Avoid the Coincidence Loophole

Jan-Åke Larsson

Bell inequality tests of local realism are notoriously difficult to perform. Physicists have attempted these tests for more than 50 years, and for each attempt, gotten closer and closer to a proper test. So far, every test performed has been riddled by one or more loopholes. While I personally am not overly fond of using the word "loophole", it has become the standard term; the word is usually used in connection to laws, most commonly taxation law, where a "loophole" points to some unintended and/or unexpected circumstance where the law does not apply, so that one can avoid the law without technically breaking it. In connection to Bell tests, the word points to some unexpected circumstance in experiments that makes additional assumptions necessary, we will see some examples of this below.

Here, we will encounter one of the more recently found loopholes, the coincidence loophole, and learn how to avoid it. First, the scene is set by reviewing the relation between the Einstein-Podolsky-Rosen (EPR) paradox and the Bell inequality, and also fix the notation and formal background. Then a few (not all) recent and not so recent experiments will be presented, with some of their qualities and shortcomings. And finally, we have a look at the coincidence loophole, why it seems to be more serious than one could have expected given results from the eighties, and then why the seemingly large impact is in fact not so damaging after all. In short, how to avoid the coincidence loophole.

J.-Å. Larsson (✉)
Institutionen för systemteknik, Linköpings Universitet, 581 83 Linköping, Sweden
e-mail: jan-ake.larsson@liu.se

R. Bertlmann and A. Zeilinger (eds.), *Quantum [Un]Speakables II*,
The Frontiers Collection, DOI 10.1007/978-3-319-38987-5_15

Local Realism

The concept of local realism is motivated by the question posed by Einstein, Podolsky, and Rosen (EPR) in [12]: "Can [the] quantum-mechanical description of physical reality be considered complete?" EPR argue that the answer to the question is negative, while Bohr in [7] argues that the question is meaningless. Although many physicists agree with Bohr, one should note that EPR were actually correct in a sense: for the system used by EPR, there does exist a more complete description as Bell [5] noted. Bell simply constructed a more complete description of that system. His construction gives the same predictions as quantum mechanics for position and momentum, and also has these two explicit in the description.

Another example was used by Bohm [6] to simplify the experiment, going from the infinite-dimensional quantum description of position-momentum to the much simpler finite-dimensional quantum description of two spin systems. The choice of measuring position or momentum now translates into a choice of axis, or direction, along which a spin component measurement is performed. This choice of direction is denoted ϕ and ψ at the two sites in Fig. 15.1. In a spin-1/2 system, there are only two outcomes: magnetic dipole moment parallel to the direction ("spin up", $+1$) or anti-parallel to the direction ("spin down", -1).

The system is such that the outcomes are (anti-)correlated; if the same direction is chosen on both sides ($\psi = \phi$), the outcomes are always opposite, as indicated in Fig. 15.1. This is called a *singlet* state, which is a special case of a so-called *entangled* state. The terminology is due to Schrödinger who translated the German *Verschräkung* in one of his letters to Einstein into *entanglement*. Obtaining an outcome at one site at direction ϕ allows prediction (with probability 1) of the outcome at the other at direction $\psi = \phi$. If the systems are separated, the choice of direction can be performed so that it (the choice of direction) cannot disturb the remote system, such disturbances being limited by the speed of light. Prediction without disturbance is the reason to believe that the outcome exists, as a physical property, independent of measurement (EPR [12]).

In this system, there are more choices than in the EPR system, since the direction parameter ϕ is has a continuous range (uncountably infinite to a mathematician) rather than the two choices of position and momentum in EPR. Still, if one only con-

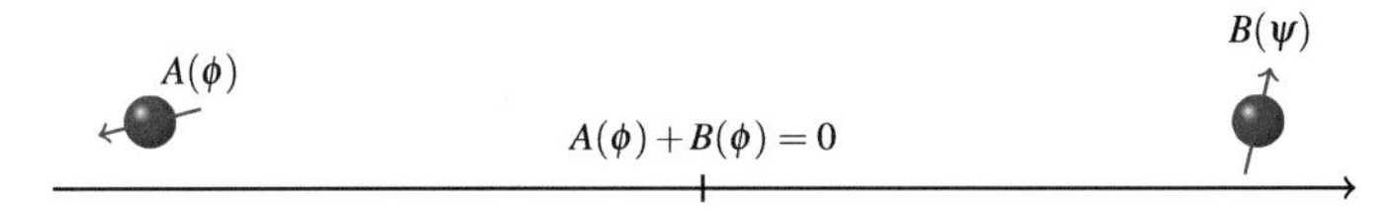

Fig. 15.1 Bohm's modification of the EPR setup. Two spin-1/2 systems are created in a joint state such that measurement of the spin component along the same direction gives opposite results, for all directions. Such a state can be created, it is called a singlet state and has total spin 0. The two systems are then separated, while making sure that the joint state is preserved. Then, a local measurement at one site for one direction can be used to predict, with probability 1, the outcome of a measurement at the remote site, for the same direction.

siders the predictions for equal settings $\psi = \phi$, there exists a more complete description just as for EPR, so that equal settings give opposite outcomes (several authors, but see e.g., [3]). In this sense, EPR are still correct. However, there are many more combinations $\psi \neq \phi$ than in the plain EPR case (position $\neq$ momentum), something John Bell made use of in his famous inequality [4]. The trick is to use three (later four, five, ...) different settings, and use that anticorrelations are large when the two directions are close but not equal to each other.

To write down the inequality we need some concepts and notions from probability theory. In probability theory, measurement outcomes are given by random variables A and B that can take parameters such as measurement directions, and also depend on a random *sample* λ, here often referred to as a *hidden variable*. The hidden variable takes values in a *sample space* Λ, and subsets of the sample space where, e.g., the random variable takes a certain value is called an *event*. To tell us the probability of events, we need a *probability measure*, so for example the probability that a random variable takes a certain value can be written

$$P\Big(\big\{\lambda : A(\phi, \lambda) = -1\big\}\Big) = P\Big(A(\phi) = -1\Big). \tag{1}$$

The notation is often simplified by leaving out the set notation and the λ as shown above. The conditional probability of one outcome given another can be calculated as

$$P\Big(A(\phi) = -1 \Big| B(\psi) = -1\Big) = \frac{P\Big(A(\phi) = -1 \cap B(\psi) = -1\Big)}{P\Big(B(\psi) = -1\Big)}. \tag{2}$$

Missing detections (as in lowered efficiency) can be handled through either assigning value 0 to the outcome, or not defining an outcome value at all [18, 19]. Finally, we need the concept of an *expectation value*, so to speak, the average outcome. For brevity I will only give what we need here, the expectation of the product of ± 1 outcomes,

$$E\Big(A(\phi)B(\psi)\Big) = P\Big(A(\phi) = B(\psi)\Big) - P\Big(A(\phi) \neq B(\psi)\Big) \tag{3}$$

The above would tell us if the outcomes are equal with high probability, which would give a value close to $+1$ or unequal with high probability (close to -1). Using this, Bell could formulate and prove a theorem, later improved by Clauser, Horne, Shimony and Holt [10] into

Theorem 1 (The CHSH inequality, *[10].*) *Under the two assumptions*

Realism: *The outcomes can be described by random variables*

$$A(\phi_i, \psi_j, \lambda),\ B(\phi_i, \psi_j, \lambda); \quad |A| = |B| = 1$$

Locality: *The r.v.s are independent of the remote setting*

$$A_i(\lambda) = A(\phi_i, \psi_j, \lambda);\ B_j(\lambda) = B(\phi_i, \psi_j, \lambda)$$

the following inequality holds:

$$\left|E(A_1B_1) + E(A_1B_2)\right| + \left|E(A_2B_2) - E(A_2B_1)\right| \leq 2.$$

An oversimplified but popular attempt to explain this uses anticorrelated outcomes: one ball in one of two boxes, one box sent to each of two sites, and then opened. If the box at the first site is opened and found to contain a ball, there will be no ball found when the box at the second site is opened. Perfect predictability. The oversimplification is that it uses too few boxes, since the CHSH expression uses four boxes, two sent to one site (A_1 and A_2) and two sent to the second site (B_1 and B_2). The terms $E(A_iB_j)$ tells us how often we can expect to see balls in both or none of the boxes. A negative value tells us that the probability is high that there is a ball in only one of the two boxes. On the other hand, a positive value tells us that the probability is high that there are balls in both or none of them. The CHSH inequality puts bounds on these probabilities: a consequence of the inequality is

$$E(A_2B_2) \leq 2 + E(A_2B_1) + E(A_1B_1) + E(A_1B_2). \tag{4}$$

This means: if the probability is large that there is one and only one ball in the two boxes A_2, B_1; one in A_1, B_1; and one in A_1, B_2, the terms on the right will be negative. If they are negative enough, the bound will tell us that the left-hand side must be negative, so that the probability must be large that there is only one ball in A_2, B_2 (in essence because there are three pairs on the right-hand side, an odd number). In the extreme case, if all three terms on the right are -1, the term on the left is forced to be -1. For our system, we are already able to predict with probability 1, so there is no news there.

But Bell realized that using directions ϕ_2, ψ_1, ϕ_1, and ψ_2 that are close to each other (in that order) but not equal, the bound tells us something new. As indicated above, the most common example is the singlet state, for which the quantum-mechanical prediction is

$$E_{\text{QM}}(A_iB_j) = -\cos(\phi_i - \psi_j), \tag{5}$$

with $\phi_2 = 0$, $\psi_1 = \pi/4$, $\phi_1 = \pi/2$, and $\psi_2 = 3\pi/4$. Then,

$$2 + E_{\text{QM}}(A_2B_1) + E_{\text{QM}}(A_1B_1) + E_{\text{QM}}(A_1B_2) = 2 - \frac{3}{\sqrt{2}} < 0, \tag{6}$$

while

$$E_{\text{QM}}(A_2B_2) = \frac{1}{\sqrt{2}} > 0. \tag{7}$$

In other words, for this quantum-mechanical system, there is a large probability of one and only one ball in the box pair A_2, B_1, one in A_1, B_1, and one in A_1, B_2. And the bound tells us that the probability is larger that there is one and only one ball in A_2, B_2, than two or none (the bound is less than 0). But the quantum-mechanical prediction is instead that it is more likely that there are two or no balls in A_2, B_2, than just one (the quantum prediction is larger than 0). The prediction does not obey the bound, there is a *violation*.

If our desired complete description is local realist, then this gives good support for Bohr's argument: there exists no local realist description that gives the quantum-mechanical predictions. The importance of this result cannot be stressed enough: we have learned something about what possible types of mathematical models that can be used to describe quantum-mechanical systems.

Loopholes

At this point, it is important to make one thing clear: even though a local realist description cannot give the quantum-mechanical predictions, it is still possible that it can give the statistics that we see in nature. It could be the case that the underlying assumptions of Theorem 1 does not apply in nature, or more accurately in our experiment. So our task now is to make sure that the theorem really does apply in our experiment, and as it turns out, this may be difficult. There are several loopholes, problems that could occur in experiment that makes the theorem need additional assumptions. These mainly fall into two classes: efficiency and locality (see e.g., [19]).

Lowered efficiency, that not all single systems give outcomes, is a common problem in experiments to test local realism. This may seem like a small problem, can one not just normalize to the subset of systems that do give outcomes? Well, not really: doing this requires an assumption, the *fair sampling assumption.* To be allowed to normalize to that subset, one needs to be sure that the subset is a fair sample of the full ensemble, in a sense that the subset itself does not depend on the local settings. Many modern experiments only violate the inequality under this assumption. And fair sampling is an assumption because in experiment it is impossible to verify that the sample is fair.

If the sample is not fair, the CHSH inequality changes, something first suggested in [27]. The bound increases when the efficiency decreases, making it possible for a local realist description to reach the quantum value. The crossover for CHSH is 82.84 % [13], and below this efficiency, a local realist description can reach the same value as quantum mechanics. Much work has been put into calculating this and other bounds under increasingly general conditions, and also new inequalities for new systems where the effects are smaller.

Many of these modified inequalities need the number of emitted pairs to be known, so that one can get probabilities from counts (by dividing beneficial counts by the total counts). The below version of the inequality gives a bound for the conditional correlations, and the only extra parameter used is the conditional probability of coincidence, conditioned on single detections. It does not need the number of emissions, but simply the ratio of coincidences to single detections. This data is already available in the experimental record, which simplifies the analysis greatly.

Theorem 2 (The CHSH inequality with inefficiency, *[18].*) *Under the three assumptions*

Realism: *The outcomes can be described by random variables*

$$A(\phi_i, \psi_j, \lambda),\ B(\phi_i, \psi_j, \lambda); \quad |A| \leq 1;\ |B| \leq 1$$

Locality: *The r.v.s are independent of the remote setting*

$$A_i(\lambda) = A(\phi_i, \psi_j, \lambda);\ B_j(\lambda) = B(\phi_i, \psi_j, \lambda)$$

Detection: *Detection is controlled by the local realist model*

$$\eta = \min_{settings} P(remote\ detection|local\ detection)$$

the following inequality holds:

$$\left|E(A_1B_1|coinc.) + E(A_1B_2|coinc.)\right| + \left|E(A_2B_2|coinc.) - E(A_2B_1|coinc.)\right| \leq \frac{4}{\eta} - 2$$

So, if the violation is high enough *and* the efficiency is high enough, then a proper test can be performed. It is important to remember that *efficiency* refers to the total efficiency. There is no reference to detectors or other equipment, the focus is on the output data: coincidences compared to local detections.

The requirements for violation are quite high. Also, we need to take experimental noise into account. If we use the singlet state and take noise into account, the amplitude of the cosine correlation is decreased. This is quantified in terms of *visibility*, a multiplicative factor v in front of the cosine. The inequality now translates into

$$2\sqrt{2}v \leq \frac{4}{\eta} - 2 \tag{8}$$

The region of violation is drawn in Fig. 15.2, the small region on the top right.

It is difficult to reach violation when using photons, the system of choice in long-distance experiments. Efficiency and visibility of four example photonic experiments [2, 14, 15, 32] is shown in Fig. 15.3. The most-cited experiment is [2], for which the main intent was to show that the violation remains even with fast-changing settings at the two sites, fast enough that no light-speed signal could carry information on the setting from one site to the other. This is to avoid disturbance in the spirit of EPR, to ensure that the locality assumption of the inequality holds. And the experiment is conclusive, since the violation does remain. The fair sampling assumption is needed because the experiment has an efficiency of approximately 0.1 % which is not visible to the scale of the figure. One reason for the low efficiency of the Aspect experiment

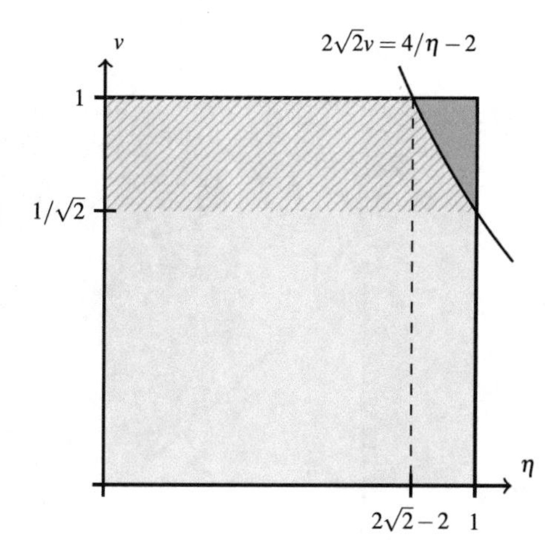

Fig. 15.2 The region of violation of the CHSH inequality for the singlet state, in terms of efficiency η and visibility v. The almost triangular region of violation has corners at visibility 70.71 % and efficiency 82.84 %. The fair-sampling assumption would enable a violation in the entire region above visibility 70.71 %, striped in the figure.

is the wide directional spread of the emissions. More modern sources (for example [17], but there are many improvements after that) give much better collection of photons, and in the experiment of [32] this can be seen, the reported efficiency is 5 %.

That experiment also uses proper random settings rather than the quasiperiodic switching used by Aspect [2]. Another improvement in locality is made in [31], where the random numbers are selected so that no light-speed signal can reach them from the emission event. This requires a large distance resulting in a 35 dB attenuation for the most remote site, or an efficiency of 0.03 %. Impressive for the 144 km distance, but of course lower than our other examples.

At the other end of the scale, we have experiments designed for very high (meaning 100 %) efficiency, these are depicted in the enlarged part of Fig. 15.3. These experiments do not use photons, but instead other more massive systems, or even solid state systems. Our examples are ions in one trap [30], in two separate traps [25], Josephson phase qubits [1], atoms in two separate traps [16], and nuclear spins near a nitrogen-vacancy centre in diamond [28]. In these experiments experimental runs are well defined and do give outcomes, so the efficiency is 100 %, but there is still the locality issue: the systems are not separated enough to guarantee locality. The quotient between the distance you need because of the duration of the relevant measurement, to the distance you have, has improved from 10^9 [30] via 1500 [25], and 300 [1], to 15 [16]. And this is still improving as we speak; the race is really on to achieve and surpass the magic number 1.

Meanwhile, photons are still the system of choice for long-distance quantum experiments, and efficiency numbers have increased steadily over the years. Modern superconducting detectors (e.g., [24] have a very high efficiency, and modern sources (e.g., [29] have very high collection. Coupled with low-loss components for the rest of the setup, an experiment can reach $\eta = 75\,\%$. There were two such exper-

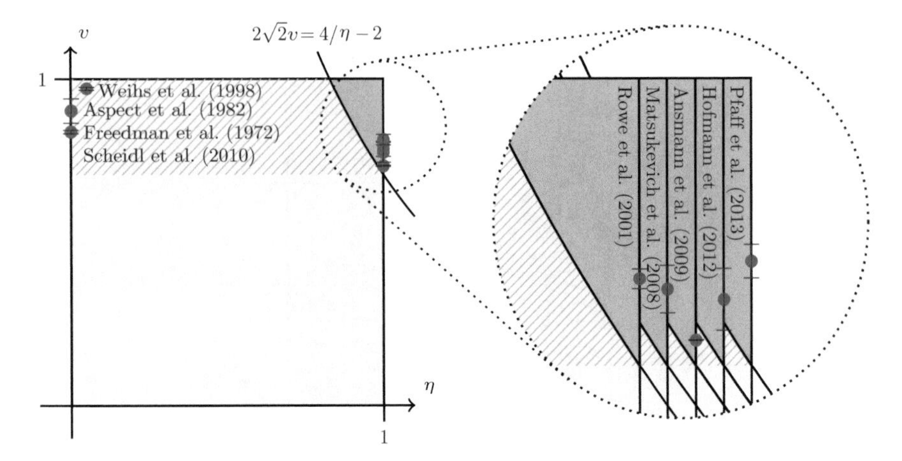

Fig. 15.3 Examples of experimental tests of local realism. The *green dot* shows efficiency and visibility of the experiments. All the papers also report the quality of the visibility estimate in terms of an estimated standard deviation, this is shown as a vertical bar.

iments recently [8, 15], but at this point one may ask: with the bound at 82.84 %, how can an experiment with efficiency 75 % be useful?

As it turns out, Eberhard found that one obtains lower bounds if a different state is used in the test [11]. We need to relax the perfect anticorrelation at equal directions, that we (and Bell, and EPR) thought so important. This is easiest to see if we use a different inequality that relates probabilities directly rather than correlations as in CHSH. The discovery sounds simple but is really profound, so much so that the below inequality sometimes is called *the Eberhard inequality* when used with counts rather than probabilities.

Theorem 3 (The Clauser-Horne (CH) inequality, *[9].*) *Under the two assumptions*

Realism: *The result can be described by random variables*

$$A(\phi_i, \psi_j, \lambda);\ B(\phi_i, \psi_j, \lambda)$$

Locality: *The r.v.s are independent of the remote setting*

$$A_i(\lambda) = A(\phi_i, \psi_j, \lambda);\ B_j(\lambda) = B(\phi_i, \psi_j, \lambda)$$

the following inequality holds:

$$\begin{aligned} P(A_1 = B_2 = 1) + P(A_2 = B_1 = 1) - P(A_2 = B_2 = 1) \\ \leq P(A_1 = 1) + P(B_1 = 1) - P(A_1 = B_1 = 1). \end{aligned}$$

One should be aware that the above is *equivalent* to CHSH, but there is a benefit since the efficiency estimate η does not enter into the calculation. There is no need

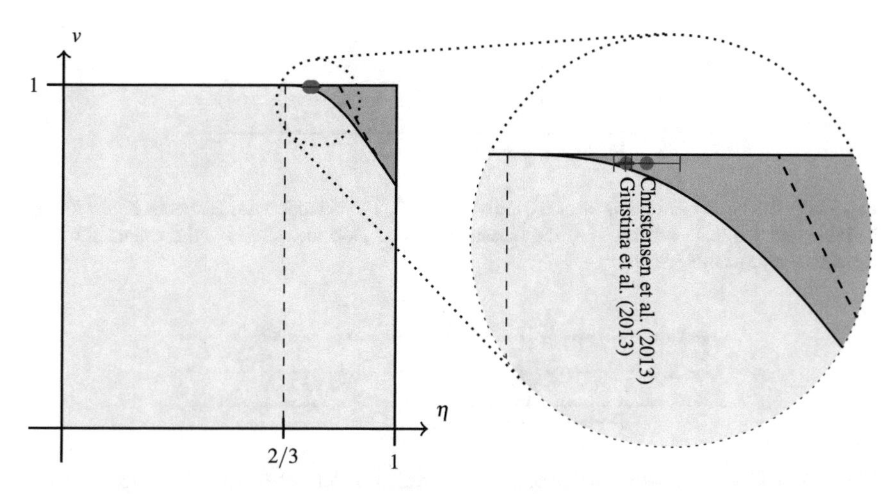

Fig. 15.4 Region of violation for non-maximally entangled quantum states. The region stretches to efficiency 66.67 %, but thins out considerably close to the limit. The two experimental efficiency estimates are qualitatively different and not so easily comparable with the earlier results, or to each other. [15] uses a separate efficiency measurement while [8] uses an estimate from equipment parameters (with a large standard deviation as indicated), and neither of these are necessarily equal to the η of Theorem 2, so the horizontal position of the two dots might not be accurate. On the other hand, the Clauser-Horne inequality does not need an explicit η, and indeed, both datasets violate the bound.

to condition on coincidence here, nor estimate the efficiency of the experiment. We will use this inequality below because of its somewhat simpler structure, since there are fewer outcome combinations that contribute to the expression, but in principle, the same can be done with CHSH.

In any case, for a given efficiency, we can find a non-maximally entangled state (i.e., not our earlier singlet state) and measurement directions that give a violation even if the efficiency is below 82 % but above 66.67 % [11, 21, 23]. The region of violation in terms of visibility and efficiency becomes somewhat larger if this is done, see Fig. 15.4. And the two mentioned experiments are inside the region of violation. This makes photons the only kind of system for which both kinds of loopholes, both efficiency and locality, have been closed, in separate experiments. It remains to close both in the same experiment, but nonetheless, we are gathering evidence that there really exists no local realist description of such a system.

The Coincidence Loophole

There is one problem in photonic experiments that we have not addressed yet: how the experimenter knows which clicks come from the same pair of photons. This is not the same as the efficiency loophole discussed above, which is more the problem of finding photons at all. Typically pairs are identified by *coincidence detection*, by comparing the detection times, either to each other (we'll call this *moving windows*), or to a predetermined reference (which we'll call *fixed time slots*). If the comparison

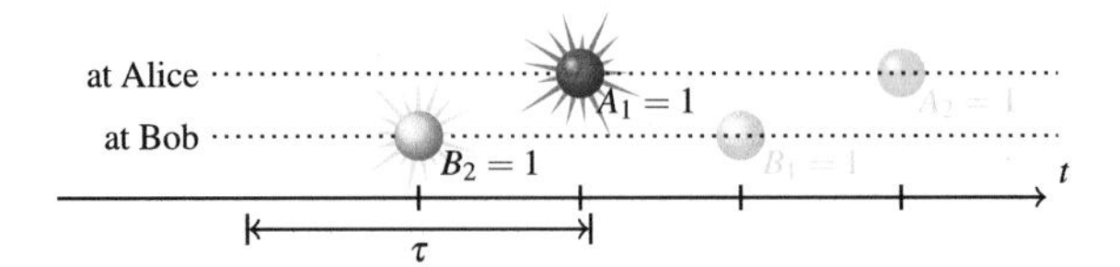

Fig. 15.5 Example local realist model (due to E. Knill). The detection time depends on the hidden variable and the local setting. If Alice measures A_1 and Bob measures B_2, the events are close enough to give a coincidence.

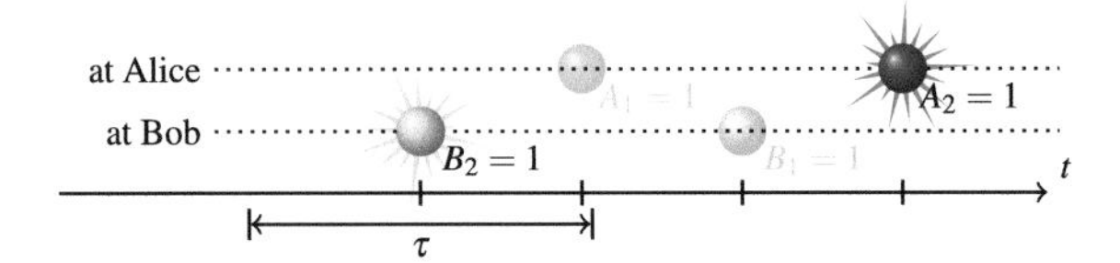

Fig. 15.6 Other choices of setting in the same local realist model will give different detection times. If Alice measures A_2 and Bob measures B_2, the events are *not* close enough to give a coincidence.

is to other events, each event on one side is surrounded with a time interval of length τ, centered on the event, as in Figs. 15.5 and 15.6. If there is a remote event in the time interval, this is considered as belonging to the same pair, or coincident. (In a real experiment there may be several events in the time interval, in which case the closest can be chosen as belonging to the same pair, but we will just consider the simple case here.)

We are considering local realist models, and it is possible that such a model influences the time of detection, similar to influencing (non-)detection. In this situation, the subset of coincidences could change. It could happen that all of the events are shifted in time so that there always are coincidences if A_2, B_1 are measured, or if A_1, B_1 or A_1, B_2 are measured, but never if A_2, B_2 are measured (see Figs. 15.5 and 15.6). This would give

$$P(A_2 = B_1 = 1 \cap \text{coincidence}) = 1 \tag{9a}$$
$$P(A_1 = B_1 = 1 \cap \text{coincidence}) = 1 \tag{9b}$$
$$P(A_1 = B_2 = 1 \cap \text{coincidence}) = 1 \tag{9c}$$
$$P(A_2 = B_2 = 1 \cap \text{coincidence}) = 0 \tag{9d}$$
$$P(A_1 = 1) = P(B_1 = 1) = 1 \tag{9e}$$

so that

$$\begin{aligned}
P(A_1 = B_2 = 1 \cap \text{coinc.}) + P(A_2 = B_1 = 1 \cap \text{coinc.}) - P(A_2 = B_2 = 1 \cap \text{coinc.})& \\
= 1 + 1 - 0 = 2& \\
\not\leq P(A_1 = 1) + P(B_1 = 1) - P(A_1 = B_1 = 1 \cap \text{coinc.})& \\
= 1 + 1 - 1 = 1.&
\end{aligned} \tag{10}$$

The Clauser-Horne inequality does apparently not hold anymore. The apparent violation above is the highest possible, because this reaches the algebraic bound of the CH inequality. But note that the terms containing pairs of outcomes $P(A_i = B_j = 1)$ have been replaced by $P(A_i = B_j = 1 \cap \text{coincidence})$, so it is no longer the same expressions on the right- and left-hand sides. The terms contain an additional restriction in the pair events (that we calculate the probability of), that comes from not knowing beforehand which events belong to the same pair. Clearly, the CH inequality does not apply to the modified expressions, because the left-hand side is larger than the right-hand side.

The effect of this kind of time dependence on the CHSH inequality was studied in [20] (it seems the effect is first discussed in [26]). In [20] a modified CHSH inequality is derived that does apply, and also a local realist model that saturates the inequality (actually containing the above example).

Theorem 4 (The CHSH inequality with coincidence identification, *[20].) Under the three assumptions*

Realism: *The outcomes can be described by random variables*

$$A(\phi_i, \psi_j, \lambda),\ B(\phi_i, \psi_j, \lambda); \quad |A| \le 1;\ |B| \le 1$$

Locality: *The r.v.s are independent of the remote setting*

$$A_i(\lambda) = A(\phi_i, \psi_j, \lambda);\ B_j(\lambda) = B(\phi_i, \psi_j, \lambda)$$

Coincidence: *Coincidence is controlled by local realist time delays and*

$$P(coincidence) = \gamma$$

the following inequality holds:

$$\begin{aligned} &\left|E(A_1B_1|coinc.) + E(A_1B_2|coinc.)\right| \\ &\quad + \left|E(A_2B_2|coinc.) - E(A_2B_1|coinc.)\right| \le \frac{6}{\gamma} - 4 \end{aligned}$$

This is similar to Theorem 2, but there are differences. The expression $4/\eta - 2$ is replaced by $6/\gamma - 4$, giving a higher bound (in γ) than the previous (in η). There is no violation unless $\gamma > 87.87\,\%$, which is visible also in Fig. 15.7. Also, γ is not so simple to estimate from experimental data as η, again the emission rate is needed for the estimate. Since the efficiency on the single-photon level is 100 % in the model example, it has been conjectured that the bound for γ translates directly into an equal bound for η, therefore a comparison with the detection efficiency bound can also be found in Fig. 15.7.

Interestingly, since the model example has single-photon efficiency 100 %, the fair sampling assumption holds. All the photons are registered, so the sample is fair (for the same reason, the *no enhancement* assumption holds, for details see e.g.,

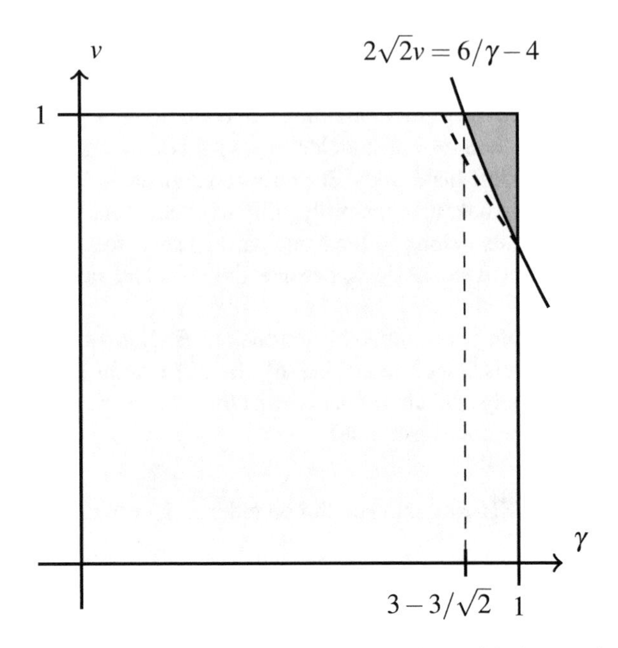

Fig. 15.7 Coincidence probability bound for violation of the modified CHSH inequality, when using a singlet state. The coincidence probability must exceed 87.87 %.

[19]). If anything, it is the *coincidence identification* that is not fair. The subset of coincidences that we register may change with the measurement directions. This happens also when we have lowered efficiency, but there is a difference: detection is a local process, and detection at one site happens on a subset of λs

$$\Lambda_{A_i} = \{\lambda : A_i(\lambda) \text{is detected}\}. \tag{11}$$

A coincidence then occurs on the set

$$\Lambda_{A_i} \cap \Lambda_{B_j}. \tag{12}$$

This is in stark contrast to the coincidence detection we use here, where the detection time T_{A_j} depends on the setting and the hidden variable, and coincidence occurs on the set

$$\Lambda_{A_iB_j} = \left\{\lambda : \left|T_{A_i}(\lambda) - T_{B_j}(\lambda)\right| < \tfrac{1}{2}\tau\right\} \neq \Lambda_{A_i} \cap \Lambda_{B_j} \tag{13}$$

There is even no possibility to rewrite $\Lambda_{A_iB_j}$ so that it can be factorized in this manner. Equivalently, attempting to assign the outcome 0 or ±1 to "missing coincidences" will lead to a non-local hidden-variable model, because even though the time-dependence is local, the newly assigned missing-coincidence-outcome needs to depend on both settings.

At this point, things look bleak. Even 100 % efficient detectors do not help, and the coincidence probability must exceed 87.87 %, considerably higher than the previously known efficiency bound 82.84 %. It is not even enough to assume fair sampling, since it already holds in the model example. It seems that the coincidence loophole is much worse than the efficiency loophole ever was. But fear not, there are quite simple modifications that will re-enable the bounds from before.

How to Avoid the Coincidence Loophole

Of course, one could substitute the fair sampling assumption with a *fair coincidence* assumption, that the subset of *pairs* chosen by our coincidence procedure is a fair sample of the whole ensemble. But we want to avoid that kind of assumptions. Better to check why the inequality does not apply, and attempt find conditions under which it does apply.

The timing of the example is such that all the pairs that increase the violation count as coincidences, while the pairs that decrease the violation does not count. One alternative to re-establish the inequality would therefore be to make sure that the latter pairs do count as coincidences. But ensuring that in general would demand that we know which photons belong together in pairs, and as we already have said, this is not possible. There will always be pairs that we cannot identify, so we need to be less ambitious. What we can aim for is to ensure that a pair that would count as a coincidence for all three measurement combinations that increase the violation, also would count as coincidence for the fourth measurement combination. This can be done by having the time window for A_2, B_2 as wide as *all three* of the other windows taken together, we call this the *window sum method* [22]. The reason that this works is that if λ is such that

$$\begin{cases} |T_{A_2}(\lambda) - T_{B_1}(\lambda)| \leq \tau, \text{and} \\ |T_{A_1}(\lambda) - T_{B_1}(\lambda)| \leq \tau, \text{and} \\ |T_{A_1}(\lambda) - T_{B_2}(\lambda)| \leq \tau, \end{cases} \tag{14}$$

then, for the same λ (see Fig. 15.8), the triangle inequality gives

$$\begin{aligned} &|T_{A_2}(\lambda) - T_{B_2}(\lambda)| \\ &\leq |T_{A_2}(\lambda) - T_{B_1}(\lambda)| + |T_{B_1}(\lambda) - T_{A_1}(\lambda)| + |T_{A_1}(\lambda) - T_{B_2}(\lambda)| \leq 3\tau. \end{aligned} \tag{15}$$

Since this applies to the individual λ values, the subsets (in terms of λ) that give coincidence will obey

$$\Lambda_{A_1B_1} \cap \Lambda_{A_1B_2} \cap \Lambda_{A_2B_1} \subset \Lambda_{A_2B_2}. \tag{16}$$

That this really fixes the issue will of course need a formal proof, so that there is no possibility for another model to evade our construction. The following theorem will do this for us.

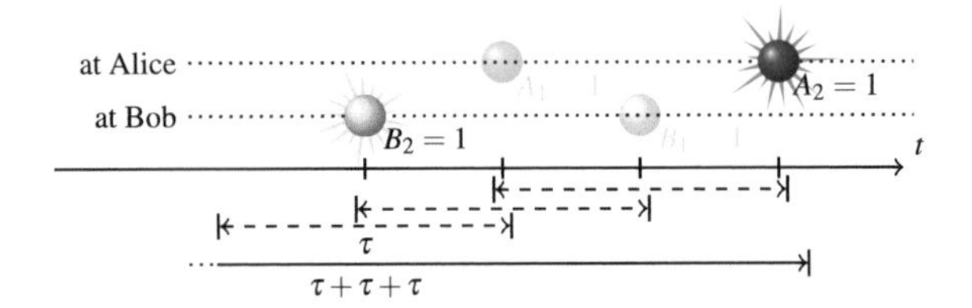

Fig. 15.8 One method to avoid the coincidence loophole is to have longer coincidence windows that decrease the violation (A_2B_2), as long as the sum of the coincidence windows that increase it (A_1B_2, A_1B_1, A_2B_1). This ensures that if there is a chain of possible coincidences that increase the violation ($B_2 \leftrightarrow A_1 \leftrightarrow B_1 \leftrightarrow A_2$), the two endpoints that would decrease the violation ($B_2 \leftrightarrow A_2$) will also be a possible coincidence.

Theorem 5 (The CH inequality, avoiding the coincidence loophole, *[22]*.) *Under the three assumptions*

Realism: *The result can be described by random variables*

$$A(\phi_i, \psi_j, \lambda);\ B(\phi_i, \psi_j, \lambda)$$

Locality: *The r.v.s are independent of the remote setting*

$$A_i(\lambda) = A(\phi_i, \psi_j, \lambda);\ B_j(\lambda) = B(\phi_i, \psi_j, \lambda)$$

Coincidence: *Coincidences are controlled by local realist time delays, are obtained on subsets $\Lambda_{A_1B_1}$; $\Lambda_{A_1B_2}$; $\Lambda_{A_2B_1}$; and $\Lambda_{A_2B_2}$, of Λ, and the last coincidence set contains the intersection of the other three,*

$$\Lambda_{A_1B_1} \cap \Lambda_{A_1B_2} \cap \Lambda_{A_2B_1} \subset \Lambda_{A_2B_2}$$

the following inequality holds:

$$\begin{aligned} P(A_1 = B_2 = 1 \cap \Lambda_{A_1B_2}) &+ P(A_2 = B_1 = 1 \cap \Lambda_{A_2B_1}) \\ &- P(A_2 = B_2 = 1 \cap \Lambda_{A_2B_2}) \\ &\leq P(A_1 = 1) + P(B_1 = 1) - P(A_1 = B_1 = 1 \cap \Lambda_{A_1B_1}). \end{aligned}$$

Another alternative to re-establish the inequality would be to avoid counting some pairs that increase the violation. Here, instead, we ensure that if a pair does not count as a coincidence for the measurement combination that decreases the violation, it is also not counted for at least one of the three measurement combinations that increase the violation. This can be ensured, in general, by introducing fixed points on the time-line that break pairs if the detection times are on different sides of such a point (see Fig. 15.9). And this is what happens when using fixed time slots. Formally, we choose a number of fixed time slots $I_k = \{t : t_k \leq t \leq t_k + \tau\}$ (these must be disjoint), and count a coincidence in slot k when the detection time of both events are in the time

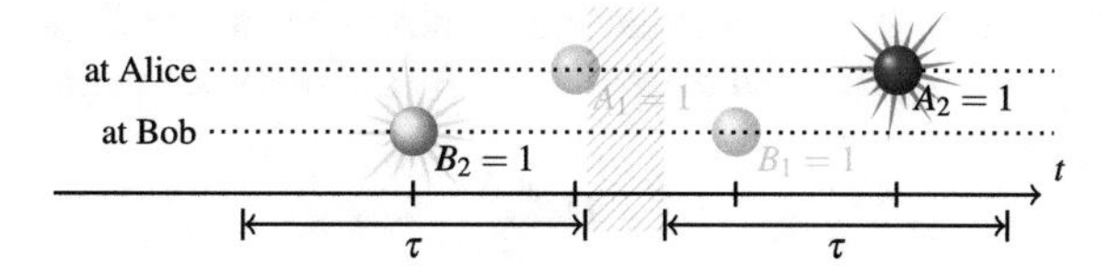

Fig. 15.9 Another method to avoid the coincidence loophole is to have fixed time slots, so that if two events are far enough apart to not give a coincidence ($B_2 \not\leftrightarrow A_2$), there cannot be a chain of possible coincidences from one event to another, that connects the two (here $B_2 \leftrightarrow A_1 \not\leftrightarrow B_1 \leftrightarrow A_2$).

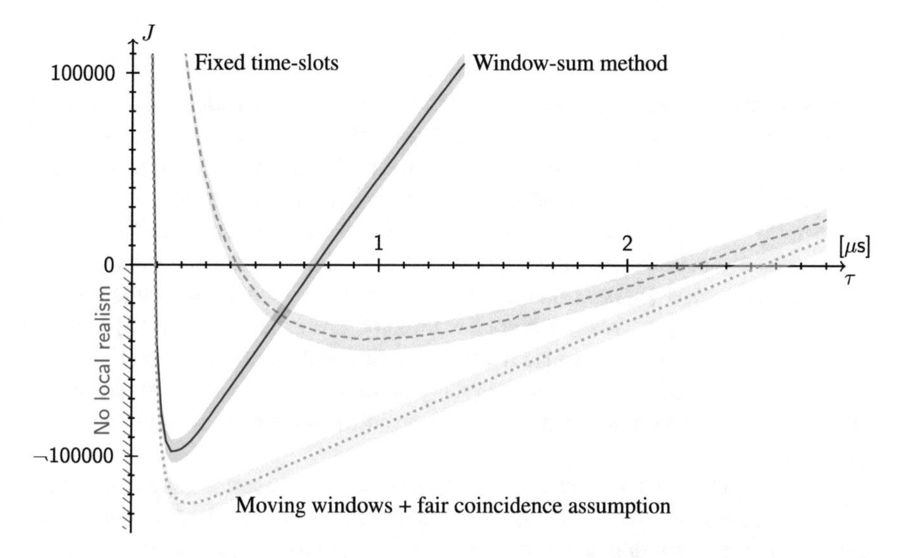

Fig. 15.10 Violation of the Clauser-Horne bound under local realism, using three different methods of handling the coincidence loophole [data from [14]]. The vertical axis scale is in counts rather than probability, a negative value is a violation. The *dotted yellow line* uses standard moving windows and the fair-coincidence assumption. The *green dashed line* uses fixed time slots. The *blue solid line* uses the window-sum method with length τ of the three short windows. The latter two are not vulnerable to the coincidence loophole, and the first one can avoid it by using the fair coincidence assumption. The *shading* corresponds to plus or minus three estimated standard deviations.

slot,

$$\Lambda^{(k)}_{A_iB_j} = \{\lambda : T_{A_i}(\lambda) \in I_k,\ T_{B_j}(\lambda) \in I_k\}. \tag{17}$$

Using these time slots, if λ is such that the events A_2, B_1 are in the same time slot, and also B_1, A_1, and A_1, B_2, then of course A_2, B_2 are in the same time slot, so that

$$\Lambda^{(k)}_{A_1B_1} \cap \Lambda^{(k)}_{A_1B_2} \cap \Lambda^{(k)}_{A_2B_1} \subset \Lambda^{(k)}_{A_2B_2}. \tag{18}$$

Since $\Lambda_{A_iB_j}$ is the disjoint union over k of $\Lambda^{(k)}_{A_iB_j}$, Eq. (16) immediately follows. As does the modified CH inequality.

Conclusion

Thus, there are (at least) two ways to avoid the coincidence loophole, either by using the window-sum method, or by using fixed time slots. This will re-enable the CH inequality, so that the bound for γ (coincidence probability) will be replaced by a bound for η (efficiency, as defined in Theorem 2). Essentially, we have restored the violation region from the small region of Fig. 15.7 to the larger region of Fig. 15.4

This means that both experiments from 2013 do violate the modified CH inequality. The experiment in [8] has a natural time slot structure since the experiment uses a pulsed-pump source, so Theorem 5 applies directly. The experiment in [15] does not have this natural structure because it uses a continuously pumped source, but time slots can still be used. We can also use the window-sum method, and interestingly the best time slot and time window lengths are different for the two methods, see Fig. 15.10.

Even though the coincidence loophole did seem to be very serious, e.g., that 100 % efficiency did not help, we have found two methods that remove it. And both available photonic experiments can use one or both methods, so that there is a violation. In conclusion, we are approaching a proper violation; locality remains to be enforced, but we are getting closer and closer. The importance of performing a conclusive experiment cannot be stressed enough: then we will have learned something fundamental about what possible types of mathematical models that can be used to describe physical reality.

References

1. M. Ansmann, H. Wang, R.C. Bialczak, M. Hofheinz, E. Lucero, M. Neeley, A.D. O'Connell, D. Sank, M. Weides, J. Wenner, A.N. Cleland, J.M. Martinis, Violation of Bell's inequality in Josephson phase qubits. Nature **461**, 504–506 (2009). doi:10.1038/nature08363
2. A. Aspect, J. Dalibard, G. Roger, Experimental test of Bell's inequalities using time-varying analyzers. Phys. Rev. Lett. **49**, 1804–1807 (1982). doi:10.1103/PhysRevLett.49.1804
3. J.E. Baggott, *The Meaning of Quantum Theory* (Oxford University, Press, 1992)
4. J.S. Bell, On the Einstein-Podolsky-Rosen paradox. Physics (Long Island City, N. Y.) **1**, 195–200 (1964). http://philoscience.unibe.ch/documents/TexteHS10/bell1964epr.pdf

5. J.S. Bell, EPR correlations and EPW distributions. Ann. N. Y. Acad. Sci. **480**, 263–266 (1986). doi:10.1111/j.1749-6632.1986.tb12429.x
6. D. Bohm, Quantum Theory. Reprinted by Dover, New York, 1989 (Prentice Hall, New York, 1951). ISBN: 0-486-65969-0
7. N. Bohr, Can quantum-mechanical description of physical reality be considered complete? Phys. Rev. **48**, 696–702 (1935). doi:10.1103/PhysRev.48.696
8. B.G. Christensen, K.T. McCusker, J.B. Altepeter, B. Calkins, T. Gerrits, A.E. Lita, A. Miller, L.K. Shalm, Y. Zhang, S.W. Nam, N. Brunner, C.C.W. Lim, N. Gisin, P.G. Kwiat, Detection-Loophole-free test of quantum nonlocality, and applications. Phys. Rev. Lett. **111**, 130406 (2013). doi:10.1103/PhysRevLett.111.130406
9. J.F. Clauser, M.A. Horne, Experimental consequences of objective local theories. Phys. Rev. D **10**, 526–535 (1974). doi:10.1103/PhysRevD.10.526
10. J.F. Clauser, M.A. Horne, A. Shimony, R.A. Holt, Proposed experiment to test local hidden-variable theories. Phys. Rev. Lett. **23**, 880–884 (1969). doi:10.1103/PhysRevLett.23.880
11. P.H. Eberhard, Background level and counter efficiencies required for a loophole-free Einstein-Podolsky-Rosen experiment. Phys. Rev. A **47**, R747–R750 (1993). doi:10.1103/PhysRevA.47.R747
12. A. Einstein, B. Podolsky, N. Rosen, Can quantum-mechanical description of physical reality be considered complete? Phys. Rev. **47**, 777–780 (1935). doi:10.1103/PhysRev.47.777
13. A. Garg, N.D. Mermin, Detector inefficiencies in the Einstein-Podolsky-Rosen experiment. Phys. Rev. D **35**, 3831–3835 (1987). doi:10.1103/PhysRevD.35.3831
14. S.J. Freedman, J.F. Clauser, Experimental Test of Local Hidden-Variable Theories. Phys. Rev. Lett. **28**, 938–941 (1972). doi:10.1103/PhysRevLett.28.938
15. M. Giustina, A. Mech, S. Ramelow, B. Wittmann, J. Kofler, J. Beyer, A. Lita, B. Calkins, T. Gerrits, S.W. Nam, R. Ursin, A. Zeilinger, Bell violation using entangled photons without the fair-sampling assumption. Nature **497**, 227–230 (2013). doi:10.1038/nature12012
16. J. Hofmann, M. Krug, N. Ortegel, L. Gérard, M. Weber, W. Rosenfeld, H. Weinfurter, Heralded entanglement between widely separated atoms. Science **337**, 72–75 (2012). doi:10.1126/science.1221856
17. P.G. Kwiat, P.H. Eberhard, A.M. Steinberg, R.Y. Chiao, Proposal for a loophole-free Bell inequality experiment. Phys. Rev. A **49**, 3209–3220 (1994). doi:10.1103/PhysRevA.49.3209
18. J.-Å. Larsson, Bell's inequality and detector inefficiency. Phys. Rev. A **57**, 3304–3308 (1998). doi:10.1103/PhysRevA.57.3304
19. J.-Å. Larsson, Loopholes in Bell inequality tests of local realism. J. Phys. A **47**, 424003 (2014). doi:10.1088/1751-8113/47/42/424003
20. J.-Å. Larsson, R.D. Gill, Bell's inequality and the coincidence-time loophole. Europhys. Lett. **67**, 707–713 (2004). doi:10.1209/epl/i2004-10124-7
21. J.-Å. Larsson, J. Semitecolos, Strict detector-efficiency bounds for nsite Clauser-Horne inequalities. Phys. Rev. A **63**, 022117 (2001). doi:10.1103/PhysRevA.63.022117
22. J.-Å. Larsson, M. Giustina, J. Kofler, B. Wittmann, R. Ursin, S. Ramelow, Bell-inequality violation with entangled photons, free of the coincidencetime loophole. Phys. Rev. A **90**, 032107 (2014). doi:10.1103/PhysRevA.90.032107
23. G. Lima, E.B. Inostroza, R.O. Vianna, J.-Å. Larsson, C. Saavedra, Optimal measurement bases for Bell tests based on the Clauser-Horne inequality. Phys. Rev. A **85**, 012105 (2012). doi:10.1103/PhysRevA.85.012105
24. A.E. Lita, A.J. Miller, S.W. Nam, Counting near-infrared singlephotons with 95 (2008)
25. D.N. Matsukevich, P. Maunz, D.L. Moehring, S. Olmschenk, C. Monroe, Bell inequality violation with two remote atomic qubits. Phys. Rev. Lett. **100**, 150404 (2008). doi:10.1103/PhysRevLett.100.150404
26. S. Pascazio, Time and Bell-type inequalities. Phys. Lett. A **118**, 47–53 (1986). doi:10.1016/0375-9601(86)90645-6
27. P. Pearle, Hidden-variable example based upon data rejection. Phys. Rev. D **2**, 1418–1425 (1970). doi:10.1103/PhysRevD.2.1418

28. W. Pfaff, T.H. Taminiau, L. Robledo, H. Bernien, M. Markham, D.J. Twitchen, R. Hanson, Demonstration of entanglement-by-measurement of solidstate qubits. Nat. Phys. **9**, 29–33 (2013). doi:10.1038/nphys2444
29. S. Ramelow, A. Mech, M. Giustina, S. Gröblacher, W. Wieczorek, J. Beyer, A. Lita, B. Calkins, T. Gerrits, S.W. Nam, A. Zeilinger, R. Ursin, Highly efficient heralding of entangled single photons. Opt. Express **21**, 6707–6717 (2013). doi:10.1364/OE.21.006707
30. M.A. Rowe, D. Kielpinski, V. Meyer, C.A. Sackett, W.M. Itano, C. Monroe, D.J. Wineland, Experimental violation of a Bell's inequality with efficient detection. Nature **409**, 791–794 (2001). doi:10.1038/35057215
31. T. Scheidl, R. Ursin, J. Kofler, S. Ramelow, X.-S. Ma, T. Herbst, L. Ratschbacher, A. Fedrizzi, N.K. Langford, T. Jennewein, A. Zeilinger, Violation of local realism with freedom of choice. Proc. Nat. Acad. Sci. **107**, 19708–19713 (2010). doi:10.1073/pnas.1002780107
32. G. Weihs, T. Jennewein, C. Simon, H. Weinfurter, A. Zeilinger, Violation of Bell's inequality under strict Einstein locality conditions. Phys. Rev. Lett. **81**, 5039–5043 (1998). doi:10.1103/PhysRevLett.81.5039

Chapter 16
Bringing Bell's Theorem Back to the Domain of Particle Physics and Cosmology

Beatrix Hiesmayr

Abstract John St. Bell was a physicist working most of his time at CERN and contributing intensively and sustainably to the development of Particle Physics and Collider Physics. As a hobby he worked on so-called "*foundations of quantum theory*", that was that time very unpopular, even considered to be scientifically taboo. His 1964-theorem, showing that predictions of local realistic theories are different to those of quantum theory, initiated a new field in quantum physics: quantum information theory. The violation of Bell's theorem, for instance, is a necessary and sufficient criterion for generating a secure key for cryptography at two distant locations. This contribution shows how Bell's theorem can be brought to the realm of high energy physics and presents the first conclusive experimental feasible test for weakly decaying neutral mesons on the market. Strong experimental and theoretical limitations make a Bell test in weakly decaying systems such as mesons and hyperons very challenging, however, these systems show an unexpected and puzzling relation to another big open question: why is our Universe dominated by matter, why did the antimatter slip off the map? This long outstanding problem becomes a new perspective via quantum information theoretic considerations.

Introduction

Only, since 2012 a promising proposal for testing Bell's 1964-theorem for systems usually produced at accelerator facilities, so called neutral $\mathcal{K}$-mesons, is on the market [1]. Putting it to reality in current or upcoming accelerator facilities will be for several reasons challenging but feasible. However, from the theoretical point of view these systems at high energies are of great interest since—as will be presented in this essay—a puzzling relation between the information theoretic content and the violation of discrete symmetries was found [2–4]. Discrete symmetries and their violation play an important role in the understanding of the four forces ruling our Universe. In

B. Hiesmayr (✉)
Faculty of Physics, University of Vienna, 1090 Vienna, Austria
e-mail: Beatrix.Hiesmayr@univie.ac.at

R. Bertlmann and A. Zeilinger (eds.), *Quantum [Un]Speakables II*,
The Frontiers Collection, DOI 10.1007/978-3-319-38987-5_16

particular, they may be the key observation in finding an explanation to the question why the antimatter did slip off in our Universe.

This essay starts with an introduction to the cosmological question of the imbalance of matter and antimatter and proceeds by discussing discrete symmetries such as parity $\mathcal{P}$, charge-conjugation $\mathcal{C}$ and time reversal $\mathcal{T}$, their breaking and its implementations. After that Bell's theorem and in particular its application to quantum cryptographic protocols is reviewed. Herewith, all key ideas have been gather to relate the security of cryptographic protocols to the difference between a world of matter and antimatter. Last but not least half integer spin particles decaying weakly, so called hyperons, are discussed from an information theoretic perspective.

How Did the Antimatter Slip Off the Map of Our Universe?

The Russian mathematician Alexander Friedmann solved Einstein's equations of general relativity showing that these equations predict a continually expanding Universe. Since an expansion can not go on for ever sometime in the past it must have started out from a tiny spot, an explosion leading to our expanding Universe. In the aftermath of this Big Bang particle and antiparticle pairs were generated. As the Universe cooled, less and less particle-antiparticle pairs formed, while those already existing could have annihilated with each other to produce photons. If this picture is correct, there should be as many particles in the Universe as antiparticles.

Considering on the composition of an average cubic meter of our Universe we find 10^9 photons, 1 proton and no antiprotons. Tracing back to just after the Big Bang the same cubic metre should have had 10^9 photons, 10^9 antiprotons and $10^9 + 1$ protons. This sounds odd, why should there be slightly more baryonic matter particles than baryonic antimatter particles in our visible Universe? Where does this observed imbalance of matter and antimatter come from?

There are two main obvious explanations for this asymmetry: either our Universe began with a small preference for matter, i.e. the total baryonic number of the universe is non-zero, or the universe was originally perfectly symmetric, but somehow a set of phenomena contributed to a small imbalance in favor of matter over time. This hypothetical physical processes that produces an asymmetry between baryons and antibaryons in the very early universe is dubbed baryogenesis. The second point of view is generally preferred, although there is no clear experimental evidence to favor one over the other one. Except the argument that an initial asymmetry should most likely have quickly eliminated due to thermodynamic considerations.

Short after 1964 the Russian nuclear physicist Andrei Dmitrievich Sakharov put forward three necessary conditions for the baryogenesis to occur, regardless of the exact mechanism.[1] Firstly, obviously the baryon number $\mathcal{B}$ must have been violated in at least one process generating baryons and antibaryons. Thus there needs to exist processes of the type

[1] In his paper "*Violation of $\mathcal{CP}$ invariance, $\mathcal{C}$, and baryon asymmetry of the universe*" in JETP Letters 5, 24 (1967), re-published in Soviet Physics Uspekhi 34, 392 (1991), he did not explicitly list the three conditions.

$$a \longrightarrow b + c\,, \tag{1}$$

where a, c are particles with baryon number $\mathcal{B} = 0$ and b a baryon ($\mathcal{B} = 1$). If the discrete symmetry charge-conjugation $\mathcal{C}$, which transforms a particle to its antiparticle or vice versa, is a symmetry of the universe, then the mirrored $\mathcal{C}$ reaction $\bar{a} \longrightarrow \bar{b} + \bar{c}$ has to have the same rate Γ, i.e.

$$\Gamma(a \longrightarrow b + c) = \Gamma(\bar{a} \longrightarrow \bar{b} + \bar{c})\,. \tag{2}$$

Then even if the baryon number $\mathcal{B}$ is violated a conservation of charge-conjugation symmetry $\mathcal{C}$ does not produce an advantage of matter over antimatter for a long period of time. Thus one needs a violation of this symmetry as well. This is not quite enough, since if we consider a (hypothetical) $\mathcal{B}$ symmetry violating process $a \longrightarrow b_l + b_l$, which generates left-handed baryons, then if $\mathcal{CP}$ is a symmetry of Nature ($\mathcal{P}\ldots$parity), the mirrored $\mathcal{CP}$ process $\bar{a} \longrightarrow \bar{b}_r + \bar{b}_r$ ($r\ldots$right handed) exists and hence we have

$$\Gamma(a \longrightarrow b_l + b_l) + \Gamma(a \longrightarrow b_r + b_r) = \Gamma(\bar{a} \longrightarrow \bar{b}_l + \bar{b}_l) + \Gamma(\bar{a} \longrightarrow \bar{b}_r + \bar{b}_r)\,. \tag{3}$$

Thus, though the symmetry $\mathcal{C}$ leads to different rates, the combined symmetry $\mathcal{CP}$ acts to conserve the total baryon number $\mathcal{B}$. Indeed, before 1964 physicists assumed that antiparticles are just there to explain the unexpected violation of the parity symmetry with $\mathcal{P}$, discovered 1956, that was seen in the experiments of madame Wu (see next Sect. "Broken Mirrors and the Absolute Definition of Charges"). Consequently, the symmetry $\mathcal{CP}$ needs to be violated as well such that more baryons are generated.

One striking fact is that the masses of particles and their antiparticles are identical. At the thermal equilibrium Boltzmann distribution law dictates any existing baryon asymmetry to turn back to the same amount. Hence there must have been a period, i.e. outside of the equilibrium, where different regions existed, some conserving, some violating the $\mathcal{B}, \mathcal{C}$ and $\mathcal{CP}$ symmetries. After baryogenesis has taken place, the universe should have turned to equilibrium and no other process should have appeared reversing this asymmetry.

All of these conditions are compatible with today's observations, particulary with the predictions of the Standard Model. In particular we want to draw our attention in the next section to the peculiar violation of the $\mathcal{P}$ symmetry and the $\mathcal{CP}$ symmetry, found in the year 1956 and 1964, respectively.

Broken Mirrors and the Absolute Definition of Charges

Weak interactions, one out of four fundamental forces, are responsible for the decay of massive quarks and leptons into lighter quarks and leptons. When fundamental particles decay, it is very astonishing: we observe the particle vanishing and being replaced by two or more different particles. Although the total of mass and energy is conserved, some of the original particle's mass is converted into kinetic energy, and the resulting particles always have less mass than the original particle that decayed. The only matter around us that is stable is made up of the smallest quarks and leptons, which cannot decay any further. When a quark or lepton changes type (a strange quark changing to down quark, for instance) it is said to change flavor. All flavor changes are due to the weak interaction.

Until the mid-1950s, physicists thought that handedness, i.e. a reflection of right to left or vice versa, does not change subatomic processes. Similar like a right-handed screwdriver would work equally well if they were manufactured to be left-handed, quantum physicists took for guaranteed that if a handed particle or process exists then its mirror-image should also exist. A left-handed particles would reflect via a left-right mirror transformation as a left-handed particle, but otherwise all other properties would happen in the same way and with the same rate. Also the behaviour would not essentially be altered by a mirror that transforms particles into antiparticles and vice versa.

Perhaps stimulated by the discovery of molecules that had due to their handedness very different behaviours Lee and Yang reexamined the evidence of the conservation of these two transformations, parity $\mathcal{P}$ and charge-conjugation $\mathcal{C}$. They found that it has not rigorously been proven. Hearing about that Madame Chein-Shiung Wu set up an experiment with radioactive cobalt in a strong magnetic field. If the symmetry $\mathcal{P}$ holds as many electrons from the β-decay should spray out in direction of the field as in opposite direction. Indeed, it was even not a small effect, a 40 % asymmetry was seen! This was indeed overlooked since no one before Lee and Yang dared to question this discrete symmetry!

Then it was obvious to look also closer to the second discrete symmetry $\mathcal{C}$! Indeed the experiments revealed that also this particle-antiparticle symmetry was flawed.

Wolfgang Pauli, in a letter to Victor Weisskopf [5], has written "*Now the first shock is over and I begin to collect myself again (as one says in Munich).... It is good that I did not make a bet. I would have resulted in a heavy loss of money (which I cannot afford); I did make a fool of myself, however (which I think I can afford to do) incidentally, only in letters or orally and not in anything that was printed. But the others now have the right to laugh at me. What shocks me is not the fact that "God is just left-handed" but the fact that in spite of this He exhibits Himself as left/right symmetric when He expresses Himself strongly. In short, the real problem now is why the strong interaction are left/right symmetric. How can the strength of an interaction produce or create symmetry groups, invariances or conservation laws? This question prompted me to my premature and wrong prognosis. I don't know any good answer to that question...*"

Somehow the idea established that the antiparticles come to rescue the broken parity symmetry, meaning that at least the combined symmetry $\mathcal{CP}$ should be unbroken. Getting familiar with the non-conservation of these discrete symmetries the long outstanding so called "*τ-θ puzzle*" found a solution. The neutral particles named τ and θ surprisingly having the same mass, however, decaying in strongly different ways with opposite parities turned out indeed to be the same particle, nowadays known by the name neutral $\mathcal{K}$-meson or kaon. Kaons are mesons that are composed by a quark and an antiquark, in particular the neutral kaon's quark content is a strange quark s and a down quark d. The new quantum number, strangeness S, was introduced by Kazuhiko Nishijima, Tadao Nakano and Murray Gell-Mann in 1953, to explain the "strange" behaviour of these particles being produced quite often but decaying rather slowly: the strong force responsible in production process conserves the strangeness quantum number S, whereas the weak interaction is the one in charge for the subsequent decay which violates the strangeness conservation. The two different decay times with the two different decay channels, 2 pions having $\mathcal{CP} = +1$ and 3 pions having $\mathcal{CP} = -1$, of a single neutral $\mathcal{K}$-meson are thus naturally explained by the maximal violation of $\mathcal{C}$ and $\mathcal{P}$, respectively. In modern formalism one would say that a neutral kaon strangeness state $|\mathcal{K}^0\rangle$ is a superposition of a short-lived state $|\mathcal{K}_S\rangle$ (decaying into 2 pions) and the long lived state $|\mathcal{K}_L\rangle$ (decaying into 3 pions),

$$|\mathcal{K}^0\rangle = \frac{1}{\sqrt{2}} \{|\mathcal{K}_S\rangle + |\mathcal{K}_L\rangle\} \tag{4}$$

and an antikaon state $|\overline{\mathcal{K}}^0\rangle$ is consequently defined by (where one puts the minus sign is not of physical importance)

$$|\overline{\mathcal{K}}^0\rangle = \frac{1}{\sqrt{2}} \{-|\mathcal{K}_S\rangle + |\mathcal{K}_L\rangle\} \ . \tag{5}$$

Since both the kaon and its antimatter state decay into the same products, one cannot distinguish via the daughter particles the strangeness number, i.e. being a particle or an antiparticle state. This means that if in a process , for instance, an antikaon is generated, it can oscillate in its particle state and vice versa, this is known under the term "*strangeness oscillations*": $s \to \bar{s} \to s \to \ldots$. Solving the Schrödinger equation for the two particle scenario one finds the following time evolution for an initial $|K^0\rangle$

$$|\mathcal{K}^0(t)\rangle = \frac{1}{\sqrt{2}} \left\{ e^{-\frac{\Gamma_S t}{2}} \cdot e^{-i m_S t} \, |\mathcal{K}_S\rangle + e^{-\frac{\Gamma_L t}{2}} \cdot e^{-i m_L t} \, |\mathcal{K}_L\rangle \right\} \ , \tag{6}$$

where $m_{S,L}$ and $\Gamma_{S,L}$ corresponds to the masses and decay constants of the short/long lived states. Note that between the decay rate Γ_S and Γ_L is a factor of about 600. So typically, a short lived state decays after 4 cm and thus the long lived state decay peaks only after about 2.4 m! Was this combined symmetry $\mathcal{CP}$ really conserved or also flawed?

In 1963, James Cronin and Val Fitch set up an experiment at Brookhaven to have a closer look. Concentrating their data analyzes on the long lived kaons end of October 1963, they expected to see only 3 pions, however, for one event out of thousand decays they did not find the third pion. Since obviously such a small effect could have many reasons, they put all their effort into the careful analyzes. Due to a formal mistake in the submission process they could not present their experiment at the Washington meeting in April 1964 giving them several more months to find the error.

After trying everything to explain these 2 pion decays they decided to go public. Indeed no error could be found, also the combined symmetry $\mathcal{CP}$ is broken, not maximally as the other ones, only by a tiny amount!

Consequently, the mass eigenstates become non-orthogonal, i.e.

$$\langle \mathcal{K}_S | \mathcal{K}_L \rangle = \frac{2Re\{\varepsilon\}}{1 + |\varepsilon|^2}, \tag{7}$$

where ε is the $\mathcal{CP}$ violating parameter and of the order 10^{-3}.

The electric charge gets absolutely defined by a neutral particle: Because of $\mathcal{CP}$ violation, Nature allows for an ultimate way of defining the charge. Indeed, the electrically neutral kaon distinguishes positive charges from negative charges! E.g. in semileptonic decays ($\mathcal{K}^0 \longrightarrow \pi^- + l^+ + \nu_l$ and $\overline{\mathcal{K}}^0 \longrightarrow \pi^+ + l^- + \overline{\nu}_l$ with l being an electron e^- or a muon μ) the kaon state decays slightly more often into a positive charged lepton than the antikaon state in a negative charged lepton state. Before meeting an alien from a distant world one certainly should find out whether they consist of antimatter or matter since shaking hands with antimatter would result in annihilation. Thus a kaon experiment is highly recommended before scheduling any meeting!

Considering symmetries of Nature is a powerful tool to explain empirical observations in physics, in particular in Particle Physics. There are three discrete symmetries: The spatial reflection that leads to parity $\mathcal{P}$ conservation if and only if there is no distinction between left and right. Discovering antiparticles another discrete symmetry, charge-conjugation $\mathcal{C}$, connects the partner with its antipartner and its conservation reflects the non-existence of an absolute electric charge. Last but not least there is the time reversal symmetry $\mathcal{T}$, stating that there should be no preferences for a forward or backward direction in time. So far, no violations of the combined symmetry $\mathcal{CPT}$ has been found, but experimental investigations are ongoing (see e.g. Refs. [6, 7]). Since the combined symmetry $\mathcal{CP}$ has been found to be broken in meson-antimeson systems, also $\mathcal{T}$ symmetry violations have to exist to compensate such that the symmetry $\mathcal{CPT}$ is conserved (only recently violation of the $\mathcal{T}$ symmetry have been measured without referring indirectly to $\mathcal{CP}$ symmetry violations [8]). Now let us turn to quantum information theory to discuss how distant parties can generate a key for sending secret messages among them.

Bell's Theorem and the Security of Quantum Cryptography

John St. Bell was a physicist working at CERN and contributing intensively and sustainably to the development of Particle Physics and Collider Physics. His famous 1964-theorem, known nowadays as Bell's theorem, shows that predictions of local realistic theories are different to those of quantum theory. As, e.g. one application it has been found that the violation of Bell's theorem is a necessary and sufficient criterion for generating a secure key for cryptography at two distant locations, which we will show in some detail in the following.

In the famous Einstein-Podolsky-Rosen scenario a source produces two particles, which are separated and independently measured by Alice and Bob. Both parties can choose among two different measurement alternatives $i = n, n'$ for Alice and $j = m, m'$ for Bob. These settings yield either the outcomes $k, l = +1$ or $k, l = -1$. Any classical or quantum correlation function can be defined, e.g., by

$$E_{AB}(i,j) = \sum_{k,l}(k \cdot l)\, P_{AB}^{kl}(i,j)\;, \tag{8}$$

where $P_{AB}^{kl}(i,j)$ is the joint probability for Alice obtaining the outcome k and Bob obtaining the outcome l, when they chose measurements i and j, respectively. For local realistic theories Bell's locality assumption imposes a factorization of the joint probabilities. Bell inequalities are tests for correlations that can be simulated using only local resources and shared randomness (a modern terminology for local hidden variables) and have, therefore, at hitherto nothing to do with quantum theory. Inserting the probabilities derived by the rules of quantum mechanics, however, in some cases lead to a violation of the inequality, i.e. to a contradiction between predictions of local hidden variable theories and quantum theory. For bipartite entangled particles with two degrees of freedom a tight Bell inequality is the famous Clauser-Horne-Shimony-Holt (CHSH) Bell inequality [9], i.e.

$$-2 \leq S(n,m,n',m') := E_{AB}(n,m) - E_{AB}(n,m') + E_{AB}(n',m) + E_{AB}(n',m') \leq 2\;. \tag{9}$$

In quantum mechanics the $S(n,m,n',m')$-function is given by deriving the four quantum mechanical expectation values $E_{AB}^{QM}(n,m')(\rho) = Tr(O_n \otimes O_{m'}\rho)$ (where O_i are appropriate operators and ρ is the density matrix of the bipartite state). It is straightforward to prove that only entangled states can violate CHSH-Bell inequality, but not all entangled states violate the inequality. Moreover, the maximum violation $S^{QM} = 2\sqrt{2}$ is reached for a maximally entangled state, e.g., the antisymmetric Bell state

$$|\psi^-\rangle = \frac{1}{\sqrt{2}}\left\{ |\Uparrow\rangle \otimes |\Downarrow\rangle - |\Downarrow\rangle \otimes |\Uparrow\rangle \right\}. \tag{10}$$

Let us extent the Einstein-Podolsky-Rosen scenario such that Alice and Bob measure randomly and independently one out of three specifically chosen observables each, where two of those observables are equal. Moreover, we assume that the source produces without loss of generality the maximally entangled antisymmetric Bell state (10). Via a fully public open channel Alice and Bob announce their observable choices but not there measurement outcomes. Then we have two cases: Alice and Bob have chosen by chance the same or unequal observables. In case, Alice and

Bob have chosen the same observable, since the source produces the antisymmetric Bell state, their measurement outcomes are perfectly anti-correlated and they can use these outcomes to both obtain a fully identical and random string of "0" and "1" (before, they have decided which measurement outcome is labeled "0" and which one is labeled "1"). In the remaining case, they announce their outcomes (even in public) and use this data to compute the four quantum mechanical correlations functions of the CHSH-Bell inequality. If there is no eavesdropping, the CHSH-Bell inequality should be maximally violated. As proven in detail in Ref. [10] any violation of Bell's theorem guarantees that an attack by an eavesdropper, even including the manipulation of the source, cannot reveal enough bit's of the string of the sifted key of Alice and Bob. Consequently, Alice and Bob can be sure based on the quantum laws that there generated key is secure! These correlations that violate Bell's theorem are stronger than anything that can be generated with classical physics!

Now we are prepared to connect both results, the existence of correlation stronger than those by classical physics and the violation of the discrete symmetries.

Bell's Theorem and the Violation of the $\mathcal{CP}$ Symmetry

The main problems in testing Bell's theorem conclusively are limitations that arise from the experimental side. These are in particular that only the antisymmetric Bell state (compare with Eq. (10)),

$$|\psi^-\rangle = \frac{1}{\sqrt{2}}\left\{ |\mathcal{K}^0\rangle \otimes |\overline{\mathcal{K}}^0\rangle - |\overline{\mathcal{K}}^0\rangle \otimes |\mathcal{K}^0\rangle \right\}, \tag{11}$$

is typically produced with high enough intensity and, secondly, only the strangeness content of neutral $\mathcal{K}$-mesons can be measured by an "*active*" measurement procedure. Active measurements are a crucial requirement for any conclusive test of Bell's theorem since obviously if Alice and Bob have no control over their measurements it is straightforward to construct a local realistic theory resulting in the observed correlation. In particular, a decay event is a "*passive*" measurement procedure, i.e. no experimenter has control over into which particles the meson will decay nor at which time this decay will occur. Though—as shown in Ref. [11]—an decay can be viewed as an open quantum system, in particular modeled by a Markovian Lindblad master equation, the decay states have to be included. This, in particular, means that one is not allowed to normalize to only surviving pairs. These are all subtle points that need to be taken into account for testing Bell's theorem conclusively in the domain of high energy systems. The requirement of an "*active*" measurement procedure rules out all other meson system due to short decay constants, except the neutral $\mathcal{K}$-meson system. The second requirement that all information available has to be considered makes it hard to find a Bell inequality that is violated for the observed constants in the $\mathcal{K}$-meson system.

Actively measuring the strangeness content of neutral $\mathcal{K}$-mesons: The experimenter places at a certain distance from the source a piece of matter that forces the incoming neutral $\mathcal{K}$-meson beam to interact with the material and to reveal the strangeness content, i.e. being at that distance in the state $|\mathcal{K}^0\rangle$ or in the state $|\overline{\mathcal{K}}^0\rangle$. Since Bell's theorem tests against all local realistic theories one is not allowed to ignore the fact that the neutral kaon could have decayed before. Therefore the question that one has to raise has to include that information, i.e. one has to ask: "*Are you at a certain distance from the source in the state* $|\overline{\mathcal{K}}^0\rangle$ *or not?*", which is obviously different to the question "*Are you at a certain distance from the source in the state* $|\overline{\mathcal{K}}^0\rangle$ *or in the state* $|\mathcal{K}^0\rangle$*?*".

To test Bell's theorem given by the S-function, Eq. (9), one has to compute four expectation values for such active measurements of strangeness given for the antisymmetric Bell state for different distances (that one can always convert in proper times since the velocity for a given experimental setup is known). Surprisingly, an optimization over all possible distances (times) does not show any value higher than 2 and -2, i.e. no contradiction to local realistic theories. Why is this the case?

The point is that the oscillation in comparison to the two decay constants is too slow or, equivalently, the decay is too fast in comparison to the oscillation. Since we cannot obviously change the natural constants of elementary particles, we have to search for a different Bell inequality. Unfortunately, the CHSH-Bell version is already the most tight one. In Ref. [1] the authors derived a new type of Bell's inequality for decaying system by including the decay property into the derivations of the bounds from local realistic theories. They assumed that any local realistic theory must also describe the well experimentally tested time evolution of single mesons correctly. This is the more striking since in a typical accelerator experiment $\mathcal{K}$-mesons are only generated in pairs, in huge contrast to typical photon experiments. Observing only a single event on one side, one knows with very high probability that the other one existed but due to purely experimental reasons was not detected.

With this new Bell inequality [1] taking the decay property into account without spoiling the conclusiveness, the authors show which time regions have to be investigated experimentally to reveal correlations that are stronger than those allowed by classical physics. Surprisingly, though investigating strangeness oscillation the $\mathcal{CP}$ symmetry violation plays a crucial role! Asking the question "*Are you at a certain distance from the source in the state* $|\mathcal{K}^0\rangle$ *or not?*" or "*Are you at a certain distance from the source in the state* $|\overline{\mathcal{K}}^0\rangle$ *or not?*" makes the difference, i.e. leading in one case to a violation in the other one not!

Consequently, the security of cryptography protocols depends in a given setup on analyzing the particle or the antiparticle content! How odd is Nature!

Bell's Theorem and the Violation of the $\mathcal{P}$ Symmetry

In the last section we discussed neutral mesons which are spinless particles. Hyperons are half-integer spin particles that are baryons containing in addition to up or down quarks also one or more strange quarks. They decay via the weak interaction violating the $\mathcal{P}$ symmetry. The Standard Model of elementary particles predicts also tiny contribution of $\mathcal{CP}$ violating processes, however, no violation of the $\mathcal{CP}$ symmetry has been up to now experimentally found. In this section we discuss the quantum information theoretic content of weakly decaying hyperons and discuss whether Bell's theorem can be tested for these weakly decaying systems.

Any closed quantum system's dynamic is given by the Schrödinger equation, i.e. a unitary evolution. Some times one is only interested in a part of the closed quantum system or has only access to a part of the system, for instance a spin in a heat bath. The dynamics of the system of interest, the open quantum system, can be derived by the unitary evolution of the total system, system of interest plus environment, and taking the partial trace of the environmental degrees of freedom (for an introduction to open quantum systems consult e.g. Refs. [12, 13]). On the other hand, if the total Hamiltonian is not known, one can study the dynamics of open quantum systems by a proper parametrization of the dynamical map. Any kind of time evolution of a quantum state ρ can always be written in the form [14]

$$\rho(t) = \sum_i \mathcal{K}_i(t, t_0, \rho(t_0))\; \rho(t_0)\; \mathcal{K}_i^\dagger(t, t_0, \rho(t_0)) \tag{12}$$

where the operators $\mathcal{K}_i$ are in general dependent on the initial time t_0 and state $\rho(t_0)$ and are often called the Kraus operators. In particular, the dynamical map defines a universal dynamical map if it is independent of the state it acts upon. This is only the case if and only if the map is induced from an extended system with the initial condition $\sigma_{\text{total}}(t_0) = \rho(t_0) \otimes \rho_{\text{environment}}(t_0)$ where $\rho_{\text{environment}}(t_0)$ is fixed for any $\rho(t_0)$. This is exactly the above described scenario.

In Ref. [15] it has been shown that any hyperon decay process can be modeled efficiently by an open quantum formalism, i.e. via Kraus operators. Typically, the directions of the momentum of the daughter particles of a hyperon are measured. This distribution is connected to the initial spin state of the hyperon. In the weakly decay process there are two interfering amplitudes, one conserves and one violates the parity $\mathcal{P}$ symmetry. The momentum distribution computes to (θ, ϕ are the angular coordinates of the momentum direction of one daughter particle and ρ_{spin} the density operator corresponding to the spin degrees of freedom of the decaying hyperon)

$$I(\theta, \phi) = Tr_{spin}(K_+\; \rho_{spin}\; K_+ + K_-\; \rho_{spin}\; K_-)$$

where the Kraus operators have the conceptually simple form ($\omega_\pm > 0$)

$$K_\pm = \sqrt{\omega_\pm}\; |\vec{\omega}_1 \pm \vec{\omega}_2\rangle\langle\vec{\omega}_1 \pm \vec{\omega}_2| \;:=\; \sqrt{\omega_\pm}\; \Pi_{\vec{\omega}_1 \pm \vec{\omega}_2}$$

with $\omega_+ + \omega_- = 1$. The two Blochvectors $\vec{\omega}_{1,2}$ have to be orthogonal, $\vec{\omega}_1 \cdot \vec{\omega}_2 = 0$, since the transition is completely positive and are chosen such that they have maximal length $|\vec{\omega}_1 \pm \vec{\omega}_2|^2 = s(2s+1)$ ($s\ldots$spin number).

A Blochvector expansion of a density matrix is generally given by $\rho = \frac{1}{d}\{\mathbb{1}_d + \vec{b}\cdot\vec{\Gamma}\}$ where d is the dimension of the system [16]. Since we are dealing with spin-degrees of freedom we have $d = 2s + 1$ and we can choose as a set of orthonormal basis the generalized Hermitian and traceless Gell-Mann matrices $\vec{\Gamma}$ (for $s = \frac{1}{2}$ they correspond to the Pauli matrices). Given this structure we can reinterpret the weak decay process as an incomplete spin measurement of the decaying particle

$$\begin{aligned} I(\theta,\phi) &= \omega_+\, Tr(\Pi_{\vec{\omega}_1+\vec{\omega}_2}\,\rho_{spin}) + \omega_-\, Tr(\Pi_{\vec{\omega}_1-\vec{\omega}_2}\,\rho_{spin}) \\ &= \frac{1}{(2s+1)}\left\{1 + \left(\vec{\omega_1} + (\omega_+ - \omega_-)\vec{\omega_2}\right)\cdot\vec{s}\right\} \end{aligned} \tag{13}$$

where $\vec{s}$ is the Bloch vector representation of ρ_{spin}, i.e. $\vec{s} = Tr(\vec{\Gamma}\,\rho_{spin})$. With probability ω_+ the spin state of the hyperon is projected onto direction $\vec{\omega}_1 + \vec{\omega}_2$ or with the remaining probability ω_- the initial spin state is measured along the direction $\vec{\omega}_1 - \vec{\omega}_2$. Thus the weak process can be associated to a spin measurement with an imperfect Stern-Gerlach apparatus (switching with probability $\omega_\pm$ the magnetic field). The imperfection has two causes: Firstly, the difference $(\omega_+ - \omega_-)$ equals an asymmetry (denoted in the following by α) and is a typical measurable constant for each hyperon. The asymmetry corresponds to interference contrast (visibility) times the cosine of the phase shift of the two interfering amplitudes, one is parity $\mathcal{P}$ conserving and one violates the symmetry $\mathcal{P}$. Secondly, the two directions $\vec{\omega}_1 \pm \vec{\omega}_2$ are typical for the spin number s. Indeed, for $s = \frac{1}{2}$ the Blochvector $\vec{\omega}_1$ is zero, thus only two directions, $\pm\vec{\omega}_2$, are chosen by Nature.

Entangled hyperons can be produced, e.g. by proton-antiproton annihilations. The introduced open quantum formalism allows for a straightforward extension by the tensor product of the Kraus operators [15]. Let us assume that (i) there is no initial correlation between the momentum degrees of freedom and the spin degrees of freedom and (ii) there is no entanglement between the momentum degrees of freedom. Experiments [17–19], e.g. for the spin-$\frac{1}{2}$ Λ hyperon and $\bar{\Lambda}$ anti-hyperon, suggest that the initial spin state is a maximally entangled Bell state (except for backward scattering angles). Therefore without loss of generality we can assume that (iii) the spin degrees of freedom of the particle and antiparticle are produced in the antisymmetric Bell state (compare with Eq. (10))

$$|\psi^-\rangle = \frac{1}{\sqrt{2}}\left\{|\Uparrow_\Lambda\rangle\otimes|\Downarrow_{\bar{\Lambda}}\rangle - |\Downarrow_{\bar{\Lambda}}\rangle\otimes|\Uparrow_\Lambda\rangle\right\}. \tag{14}$$

Then the computation of the angular distribution of the momenta of the two daughter particles of the Λ and $\bar{\Lambda}$ results in

$$I(\theta_\Lambda,\phi_\Lambda;\theta_{\bar{\Lambda}},\phi_{\bar{\Lambda}}) = \frac{1}{4}\left\{1 - \alpha_\Lambda\alpha_{\bar{\Lambda}}\;\vec{n}_\Lambda\cdot\vec{n}_{\bar{\Lambda}}\right\}\;. \tag{15}$$

Since the Bloch vectors

$$n_{\Lambda/\bar{\Lambda}} = \begin{pmatrix} \sin\theta_{\Lambda/\bar{\Lambda}} \cos\phi_{\Lambda/\bar{\Lambda}} \\ \sin\theta_{\Lambda/\bar{\Lambda}} \sin\phi_{\Lambda/\bar{\Lambda}} \\ \cos\theta_{\Lambda/\bar{\Lambda}} \end{pmatrix}$$

are multiplied, $\vec{n}_\Lambda \cdot \vec{n}_{\bar{\Lambda}}$, by the constants $\alpha_\Lambda \cdot \alpha_{\bar{\Lambda}}$, Törnqvist [20] concluded that the hyperon Λ decays "*as if it had a polarization α_Λ tagged in the direction of the π^+ (coming from the $\bar{\Lambda}$) and vice versa*". The knowledge of how one of the $\Lambda's$ decayed—or shall decay (since time ordering is not relevant)—reveals the polarization of the second Λ. He concludes that this is the well-known Einstein-Podolsky-Rosen scenario.

Does the imperfection of the spin measurement allow for detection of entanglement?

In general entanglement is detected by a certain observable that can witness the entanglement content, i.e. a Hermitian operator $\mathcal{W}$ for which holds $Tr(\mathcal{W}\rho) < 0$ for at least one state ρ and $Tr(\mathcal{W}\rho_{sep}) \geq 0$ for all separable states ρ_{sep}. For the antisymmetric Bell state such an optical entanglement witness is given by $\mathcal{W} = \frac{1}{3}(\mathbb{1} \otimes \mathbb{1} + \sum_i \sigma_i \otimes \sigma_i)$ (any other witness can be obtained by local unitary transformations). Since the weak interaction only allows for an imperfect spin measurement we have to multiply the spin part by $\alpha_\Lambda \alpha_{\bar{\Lambda}}$. Thus the entanglement witness for the $\Lambda\bar{\Lambda}$ system results in

$$\frac{1}{3} - \alpha_\Lambda \alpha_{\bar{\Lambda}} \geq 0 \; \forall \; \rho_{sep} \,, \tag{16}$$

which is clearly violated since $\alpha_\Lambda \alpha_{\bar{\Lambda}} = 0.46 \pm 0.06$ [21]. Therefore, the measurement of the correlation functions $\langle \sigma_i \otimes \sigma_i \rangle$ in x, x and y, y and z, z directions of the Λ and $\bar{\Lambda}$ reveals entanglement. Let us here emphasize that a re-normalization (dividing by $\alpha_\Lambda \alpha_{\bar{\Lambda}}$) is not proper since also an mixed separable state may give the value up to $\frac{1}{3}$. Generally, one can say that the asymmetries lead to imperfect spin measurements which shrink the observable space. Equivalently, we can say that the given interferometric device leads to a shrinking of the Hilbert space of the accessible spin states.

However, does the imperfection of the spin measurement allow for detection of correlations stronger than those of classical physics?

For that we have to investigate Bell's inequalities and in principle all its variants. The CHSH-Bell type one, Eq. (9), leads to [15]

$$\alpha_\Lambda \alpha_{\bar{\Lambda}} \overset{\text{for all local realistic theories}}{\leq} \frac{1}{\sqrt{2}} \,. \tag{17}$$

This is clearly not violated since $\alpha_\Lambda \alpha_{\bar{\Lambda}} \approx (0.46 \pm 0.06)$! However, here we anyway missed a requirement for any conclusive test of Bell's theorem: *active* measurements! The weak interaction chooses the quantization axes $\pm\vec{\omega}_2$ spontaneously, we just know the probabilistically which one (with the probabilities $\omega_\pm$). Thus Bell's theorem cannot be tested in this way!

Outlook

We discussed Bell's theorem in the realm of high energy physics. We have seen that a conclusive version that can be experimentally be put to reality is very involved, however, both mesons and hyperons offer a new theoretical perspective since their properties are connected to violations of the discrete symmetry parity $\mathcal{P}$ and the combined discrete symmetry charge-conjugation-parity $\mathcal{CP}$. This surprising relation between Bell's theorem and discrete symmetry breaking, in turn, can be attributed to the unsolved problem of why we live in a universe dominated by matter, a problem which itself links to distinct and fundamental questions in physics. This discovery may pave the way to provide a first understanding for a more fundamental rule in Nature and to whether information theoretic considerations played a key role in the development of our universe (or may play).

Acknowledgments The author gratefully acknowledges the Austrian Science Fund projects FWF-P26783 and wishes to thank the COST Action MP1006 "Fundamental Problems in Quantum Physics".

References

1. B.C. Hiesmayr et al., Revealing Bell's nonlocality for unstable systems in high energy physics. Eur. Phys. J. C **72**, 1856 (2012)
2. R.A. Bertlmann, B.C. Hiesmayr, Bell inequalities for entangled kaons and their unitary time evolution. Phys. Rev. A **63**, 062112 (2001)
3. B.C. Hiesmayr, Nonlocality and entanglement in a strange system. Eur. Phys. J. C **50**, 73 (2007)
4. B.C. Hiesmayr, A generalized Bell inequality and decoherence for the K0 anti-K0 system. Found. Phys. Lett. **14**, 231 (2001)
5. W. Pauli, Wissenschaftlicher Briefwechsel, vol. I, 1919–1929 (Springer, 1979), p. 505. (Translation of the letter by L. O'Raifeartaigh) or vol. II: Part IV, A: 1957 (Springer, 2005), p. 121. (Translation of the letter by V.F. Weisskopf)
6. KLOE-2 Collaboration, Test of CPT and Lorentz symmetry in entangled neutral kaons with the KLOE experiment, Phys. Lett. B **730**, 89 (2014)
7. A. Di Domenico, K.L.O.E. Collaboration, CPT symmetry and quantum mechanics tests in the neutral kaon system at KLOE. Found. Phys. **40**, 852 (2010)
8. J. Bernabeu, A. Di Domenico, P. Villanueva-Perez, Direct test of time reversal symmetry in the entangled neutral kaon system at a ϕ-factory. Nucl. Phys. B **868**, 102 (2013)
9. J.F. Clauser, M.A. Horne, A. Shimony, R.A. Holt, Proposed experiment to test local hidden-variable theories. Phys. Rev. Lett. **23**, 880 (1969)

10. A. Ekert, Quantum cryptography based on Bell's theorem. Phys. Rev. Lett. **67**, 661 (1991)
11. R.A. Bertlmann, W. Grimus, B.C. Hiesmayr, Open-quantum-system formulation of particle decay. Phys. Rev. A **73**, 054101 (2006)
12. H.-P. Breuer, F. Petruccione, *The Theory of Open Quantum Systems* (Oxford, 2002)
13. A. Rivas, S.F. Huelga, *Open Quantum Systems: An Introduction*. Springer Briefs in Physics (2012)
14. D. Salgado, J.L. Sanchez-Gomez, M. Ferrero, Evolution of any finite open quantum system always admits a Kraus-type representation, although it is not always completely positive. Phys. Rev. A **70**, 054102 (2004)
15. B.C. Hiesmayr, Sci. Rep. **5**, 11591 (2015)
16. R.A. Bertlmann, Ph. Krammer, Bloch vectors for qudits. J. Phys. A: Math. Theor. **41**, 235303 (2008)
17. B. Bassalleck et al., (PS185 Collaboration), Measurement of spin-transfer observables in $p\bar{p} \longrightarrow \bar{\Lambda}\Lambda$ at 1.637 GeV/c. Phys. Rev. Lett. **89**, 212302 (2002)
18. K.D. Paschke et al., (PS185 Collaboration), Experimental determination of the complete spin structure for $p\bar{p} \longrightarrow \bar{\Lambda}\Lambda$ at $p_{\bar{p}} = 1.637$ GeV/c. Phys. Rev. C **74**, 015206 (2006)
19. T. Johansson, Antihyperon-hyperon production in antiproton-proton collisions, in *AIP Conference Proceedings of Eigth International Conference on Low Energy Antiproton Physics* (2003)
20. N.A. Törnqvist, Suggestion for Einstein-Podolsky-Rosen experiments using reactions like $e^+e^- \rightarrow \Lambda\bar{\Lambda} \rightarrow \pi^- p \pi^+ \bar{p}$. Found. Phys. **11**, 171 (1981)
21. K.A. Olive et al., (Particle Data Group). Review of particle physics. Chin. Phys. C **38**, 090001 (2014)

Part V
Quantum Topics

Chapter 17
Black Box Quantum Mechanics

Antonio Acín and Miguel Navascués

Abstract There is no doubt that Bell's theorem [1] is a fundamental result for our understanding of quantum physics and its relation with classical physics. Before Bell, the possibility that an intuitive classical model could exist with the same predictive power as quantum physics was valid and, in a sense, justified in view of the arguments by Einstein, Podosky and Rosen (EPR) on the incompleteness of quantum physics [2]. After Bell's work, a classical model for quantum physics is still possible but, as discussed below, requires breaking some very natural assumptions that, in a way, make it as counter-intuitive as quantum physics. In the last decade, our understanding of Bell's theorem, for instance of the assumptions required for its derivation and its implications, has significantly improved using concepts and ideas borrowed from quantum information theory. At the same time, concepts from foundations of quantum physics have opened new approaches to quantum information applications, especially in the so-called device-independent scenario. The purpose of this text is to provide an overview over this new research direction merging quantum foundations and information theory, with an emphasis on the motivations and some of the obtained results. Our text, however, should not be understood as a review paper, but more as a rather personal selection of results in the field, unavoidably biased to some of our works. The structure of the essay is as follows: we start by presenting the assumptions required in the derivation of Bell's theorem and its implications. We move on and show how ideas from Bell's arguments can be used for quantum information purposes: we introduce the device-independent approach to quantum information theory and argue that it can be interpreted as a form of Bell-type quantum information theory. Then, we reverse this direction and show how ideas from information theory help us to understand quantum physics.

A. Acín (✉)
ICFO–Institut de Ciencies Fotoniques, Barcelona Institute of Science and Technology, 08860 Castelldefels, Barcelona, Spain
e-mail: antonio.acin@icfo.es

A. Acín
ICREA–Institucio Catalana de Recerca i Estudis Avançats, 08010 Barcelona, Spain

M. Navascués
Institute for Quantum Optics and Quantum Information (IQOQI), Austrian Academy of Sciences, 1090 Vienna, Austria
e-mail: miguel.navascues@oeaw.ac.at

R. Bertlmann and A. Zeilinger (eds.), *Quantum [Un]Speakables II*,
The Frontiers Collection, DOI 10.1007/978-3-319-38987-5_17

Bell Scenario

One of the main strengths of Bell's theorem is its simplicity and level of abstraction. It just involves two (or more) different systems given to two different observers, named in what follows Alice and Bob, who can perform $m > 1$ possible different actions on it and get $r > 1$ possible results. What the observers precisely do with their systems is irrelevant for the derivation of the theorem. For instance, in a standard quantum optics experiment, a pair of entangled photons is prepared and one photon is sent to each of the two observers. These two photons are the two systems to be measured. The observers perform polarization measurements on their corresponding photon along m different directions, in the simplest case $m = 2$. As a result, the photon may impinge on two possible detectors with different probabilities,[1] defining the two possible results, $r = 2$. However, this is just one of the possible practical realizations of the previous theoretical scenario: any situation in which two observers can perform m different measurements of r possible results is equally valid. In fact, the Bell scenario can just be described, see Fig. 17.1, by a very minimalistic setup in which Alice and Bob have access to a completely uncharacterized device, represented by a box, on which they can perform m different actions, represented by classical variables $x, y = 1, \dots, m$, and get r possible results, denoted by classical variables $a, b = 1, \dots, r$. Each round of the experiment is simply described by a choice of inputs and the corresponding outputs. By repeating the experiment many times,[2] it is possible to describe it by means of a conditional probability distribution $P(ab|xy)$ specifying the probabilities of getting outputs a and b when choosing inputs x and y. The experiment is arranged so that each party has a well-defined statistical description independent of the other. This requirement is also known as the no-signalling condition and states that the marginal probability distribution seen by one party does not depend on the choice of setting by the other, $\sum_a P(ab|xy) = P(b|y)$ and $\sum_b P(ab|xy) = P(a|x)$, for all x, y. Note that the validity of this statement can be tested, up to statistical uncertainties, as it is expressed only in terms of the observed probabilities. This is all what is needed to derive, and to experimentally test, Bell's theorem.

Bell's theorem is a statement about the possible correlations between the measurement results observed in the two systems, described by the conditional probability distribution $P(ab|xy)$. The genius of Bell was to realize that assumptions on the physical behavior of the systems had consequences on the observed correlations. While Bell stuck to the terms *locality* and *realism* introduced by EPR [2], here we will slightly modify the terminology and name the assumptions in his theorem as follows:

---

[1] In fact, in a realistic experiment, there are 4 possible results: no detector clicks, only detector 1 clicks, only detector 2 clicks, and both detectors click. However, here we are considering an idealized scenario where photons are always detected and only one photon is sent to each observer.

[2] There is an implicit assumption when writing this conditional probability distribution, namely that all the rounds of the experiment represent independent and identically distributed (iid) realizations of $P(ab|xy)$. It is however possible to derive a form of Bell's theorem valid without the iid assumption, see for instance [3]. Here, for the sake of simplicity, we work under the iid assumption.

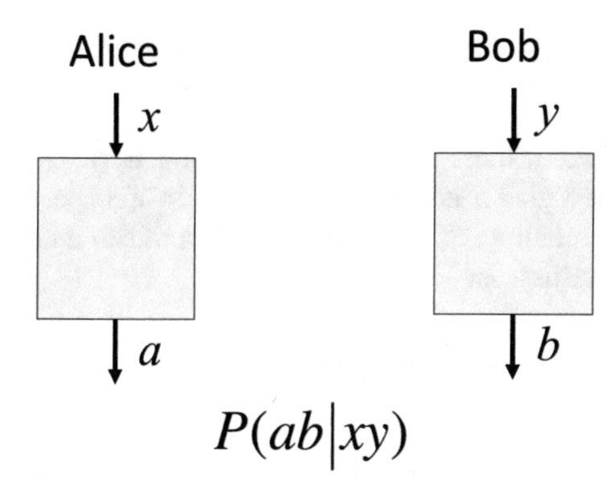

Fig. 17.1 *Bell scenario*: two distant observers, Alice and Bob, have access to two uncharacterised quantum systems. These systems can be seen as *black boxes* that take a classical input, x for Alice and y for Bob, and produce a classical output, a and b. The correlations observed between these variables are described by a joint probability distribution $P(ab|xy)$. This distribution can be statistically estimated by repeating the experiment.

1. *Determinism*: The possible randomness or probabilistic character observed in the experiment is just a consequence of ignorance. In other words, it should be possible to describe the observed correlations as a mixture of deterministic assignments specifying the outputs given the inputs. Formally, this means that it should be possible to write the observed correlations as:

$$P(ab|xy) = \sum_{\lambda} P(\lambda)D_{\lambda}(ab|xy), \tag{1}$$

where $P(\lambda)$ is a probability distribution over a classical random variable λ, usually known as the hidden variable, and $D_{\lambda}(ab|xy)$ denote different deterministic functions that given a value for the inputs x and y return 1 for just one of the r^2 possible configurations of (a, b) and zero otherwise. Note that this assumption alone does not impose any limitation on the set of possible correlations, as any conditional probability distribution $P(ab|xy)$ can be easily decomposed in terms of deterministic assignments. The intuition behind the hidden variable is that, as it happens in classical physics, any source of observed randomness in the setup should be attributed to ignorance or lack of information on the deterministic values of the outputs, encapsulated by the probability distribution $P(\lambda)$.

2. *No-signalling*: The observers should not be able to infer the choice of input by the other observer, even if they had access to a possible hidden variable. Formally, this means that it should be possible to write the observed correlations as:

$$P(ab|xy) = \sum_{\lambda} P(\lambda)P_{\lambda}(ab|xy), \tag{2}$$

where $P(\lambda)$ is a probability distribution over a classical random variable λ, again a hidden variable, and $P_{\lambda}(ab|xy)$ are non-signalling probability distributions, that is $\sum_a P_{\lambda}(ab|xy) = P_{\lambda}(b|y)$ and $\sum_b P_{\lambda}(ab|xy) = P_{\lambda}(a|x)$, for all x, y, λ. Again, this assumption is trivially satisfied in the previous scenario, as $P(ab|xy)$ is already non-signalling.

While, as mentioned, these two assumptions are trivially satisfied separately, when combined, they give a non-trivial structure to the space of correlations. In fact, some given correlations $P(ab|xy)$ satisfy determinism and no-signalling if and only if they can be written as:

$$P(ab|xy) = \sum_{\lambda} P(\lambda) D_{\lambda}^{A}(a|x) D_{\lambda}^{B}(b|y). \tag{3}$$

This easily follows from the fact that a deterministic assignment is compatible with the no-signalling principle if and only if it has a product form. Equation (3) are nothing but the standard EPR-correlations. In fact, all correlations obtained in classical physics are of this form. Bell was the first to notice this non-trivial structure and derived a Bell inequality, linear on the correlations $P(ab|xy)$, that is satisfied by EPR-correlations. Then, Bell showed that these inequalities can be violated by performing local measurements on entangled quantum particles. This phenomenon is often named quantum non-locality, mostly because of historical reasons, and it is the terminology we follow here. Leaving aside whether the term non-locality fully captures the implications of the theorem, here it simply means that the correlations of a theory cannot be written as in Eq. (3).

Before concluding this part, which is rather introductory, we would like to mention several important points. First of all, Bell's Theorem is a mathematical theorem. As such, it cannot be wrong: given the assumptions, or hypotheses, the result just follows after some simple algebra. In this sense, it is desirable to state assumptions that have a physical motivation but, also, can be easily formalized in mathematical terms, as above.

It is of course licit to discuss about the validity of the assumptions, either from a theoretical perspective or in concrete experimental setups. From a theoretical perspective, the discussions about the assumptions is non-scientific and mostly based on personal taste. For instance, in Bohm's model, assumption 2 is abandoned, as the model is signaling at the hidden-variable level. There is no scientific argument against this logical possibility and, in fact, Bohm's model represents a valid and intrinsically deterministic model for quantum physics. The discussion about the assumptions is possibly more justified, yet still subjective, when discussing concrete setups. This is for instance relevant when considering experimental loopholes, especially in the context of the free-will and locality loophole. In this sense, the detection loophole is not really a loophole: whenever it appears, it just means that it is possible to give an EPR model (3) for the observed correlations without affecting assumptions 1 and 2. Of course, this EPR model breaks the fair-sampling assumption, which states that the observed statistics when the particles (usually photons) are detected is a fair representation of the statistics one would observe in the case of perfect detection. In other words, it is impossible to find a model for the correlations that is compatible with assumptions 1 and 2 *plus* the fair-sampling assumption. The locality or the free-will loophole share some similarities and are from a logical point of view completely different from the detection loophole, as they shed doubts on the validity of assumption 2. Arranging the measurements so that they define space-like separated events, or selecting the inputs of the experiments using a proper random number generator or the Geneva phone book [4], will only increase our confidence that assumption 2 is satisfied in the experiment. But, strictly speaking, and contrary to the detection loophole, it will never be possible to close the locality and free-will loophole.

Finally, it is worth recalling that there is nothing quantum in the derivation of Bell's Theorem. What is quantum is the violation of Bell inequalities. In fact, in our opinion, Bell inequalities are nothing but the application of our intuition to explain the correlations that are possible among separate devices. It is not surprising that some Bell inequalities already appeared in texts by Boole in 1854, much before the birth of quantum physics.

Device-Independent Quantum Information Processing

Bell inequalities prove the existence of correlations in quantum physics that do not have a classical analogue: they are satisfied by classical physics, which is deterministic and non-signalling, but violated by quantum physics. It is somehow natural to expect that quantum non-locality plays an important role in quantum information theory. In fact, the main goal of this field is to design protocols for information processing that are impossible using classical resources. For quantum information purposes, the more quantum the better. However, before the advent of the device-independent approach, described in what follows, no quantum information protocol was based on non-locality.

In the case of quantum computation, this is perhaps less surprising. Bell non-locality tells us about something which is possible in quantum physics and impossible in classical physics. In quantum computation, the main objective is to design quantum algorithms that solve a relevant computational problem much faster than any classical algorithm. It is not expected that classical resources cannot solve the problem, but "just" that they do it in a significantly less efficient way, possibly with an exponential overhead. In the case of quantum cryptography, Artur Ekert, in 1991, proposed a protocol for cryptography based on the violation of a Bell inequality [5]. But soon after Eker's proposal, Bennett, Brassard and Mermin wrote an article [6] showing that the role played by Bell's inequalities in the security of Ekert's protocol was marginal and security can equally be guaranteed by the non-orthogonality of quantum superpositions, as in previous quantum cryptographic protocols à la BB84 (for Bennett and Brassard in 1984 [7]). Yet, Ekert's work was the first connecting Bell's Theorem with quantum cryptography.

A few years later, Mayers and Yao [8] realized that quantum non-local correlations were useful as an information resource in a scenario where devices are not characterized. The idea is that if some devices provide an almost maximal quantum violation of a Bell inequality, they should come from measurements on a highly entangled state and security follows. This gave rise to the concept of self-testing, as the protocol certifies that the state and measurements implemented by the uncharacterized devices are the expected ones. Somehow unrelated, another work connecting quantum cryptography and Bell violation was by Barrett, Hardy and Kent [9], who studied whether it is possible to have secure key distribution in a scenario where quantum physics is not assumed, but the no-signalling principle, that is, the impossibility of faster-than-light communication, holds. The authors of [9] showed how to establish a secret bit (not a secret key) when the maximal violation of some Bell inequalities is observed.[3] Finally, in [10], the two approaches were merged and it was shown how the techniques developed in [9] for security could be adapted to present a quantum key distribution protocol using uncharacterized devices, the new approach being named device-independent quantum key distribution (DIQKD).[4]

The general goal of DIQKD is to construct protocols that allow two parties, Alice and Bob, to establish a secure secret key without making any assumption on the internal working of the devices used in the protocol. In other words, the devices are just seen as black boxes, as in a standard Bell test (see Fig. 17.1) and the key should be established from the observed correlations between the measurement results. Alice and Bob don't need to have any characterization of the local processes producing the classical outputs a and b given the classical inputs x and y, but they have to assume that these processes are quantum. On the contrary Eve may a have a complete knowledge of this process, that is, contrary to the honest parties, she perfectly knows which quantum state is prepared and which measurements are performed on it.

In any cryptographic problem, one has to assume a paranoid perspective and work in the worst-case scenario. In the device-independent approach, this means that, given some correlations observed between their devices, Alice and Bob must assume that the actual preparation, unknown to them, is the worst compatible with these correlations. This immediately explains why non-locality is necessary to have a secure DIQKD protocol. In fact, if the observed data are compatible with an EPR model, they are of deterministic nature and can be perfectly predicted by the eavesdropper Eve with a perfect knowledge of the hidden variable λ. In other words, the observed correlations can be reproduced by classical information, which can be copied and therefore be in Eve's hands.

The parallelism between the device-independent approach to QKD and the Bell scenario becomes even clearer when analyzing the assumptions needed in the definition of DIQKD. We can identify three main assumptions:

---

[3]It is at the moment an open problem whether a secret key, and not just a single bit, can be distributed in a realistic noisy scenario only under the assumption of the no-signalling principle.

[4]Strictly speaking, the security proof in [10] was valid only under a specific class of attacks, known as collective attacks. A general security proof has later been established in [11].

1. Quantum physics is correct.
2. There is no information leakage from the honest parties' locations.
3. Alice and Bob have trusted randomness to choose their classical inputs.

It should be now clear that assumptions 2 and 3 are the cryptographic analogs of the locality and free-will loopholes. If these assumptions cannot be met (the loopholes cannot be closed), a secure key distribution (proper Bell violation) cannot be guaranteed.

DIQKD is a formalism that supersedes traditional quantum key distribution. In recent years, several works, see for instance [12], reported the successful hacking of some standard QKD protocols. In all these attacks, the hackers did not break the principle but the implementation. The security proofs of standard QKD protocols build on several assumptions about the states and measurements that are crucial for the security proof but very hard to meet in practice. The unavoidable mismatch between theoretical security proofs and practical implementations was the loophole exploited by the quantum hackers. Of course, it is always possible to put in place patches solving the problems used in the attacks. This is in fact a possible approach to the problem: trying to experimentally guarantee that the states and measurements required in the protocol are correctly implemented. But it is unclear whether a complete solution to this problem would ever be possible, due to the unavoidable presence of noise and imperfections in any implementation. DIQKD offers a completely different solution to this problem. In DIQKD security is guaranteed only from the observed statistics and does not rely on any detailed modelling of the internal working of the devices. Any implementation is good provided that it produces the needed correlations between the honest parties, something that can be easily tested in an implementation. Thus, there is no mismatch between the theoretical requirements in the security proof and the actual implementation and the previous hacking attacks become impossible.

We conclude this part by mentioning that the device-independent scenario, originally introduced in the context of QKD, has been generalized in recent years to other quantum information problems. We now have, for instance, schemes for device-independent randomness generation [13] or entanglement detection [14]. All this effort is sometimes known as device-independent quantum information processing and can be understood as a form of quantum information theory with black boxes and without any explicit mention to any Hilbert space. As for DIQKD, any quantum information protocol in this scenario must be built from quantum correlations violating a Bell inequality.

From Information Principles to Quantum Foundations

The device-independent approach to quantum information theory exploits one of the most fundamental quantum properties, namely quantum non-locality, to design information protocols which do not have any classical analogue. However, it is possible to reverse this direction and borrow concepts from quantum information theory to gain a better understanding of quantum foundations.

In 1992, Popescu and Rohrlich proposed to classify physical theories according to their ability to produce correlations between distant observers [15]. Each experimental set of correlations $P(ab|xy)$ is viewed as a point in a probability space, and the goal is to study the shape of the set of all accessible points of the physical theory under consideration. In this framework, classical physics is represented by flat-sided convex figures (polytopes), with facets corresponding to Bell inequalities and positivity conditions of the sort $P(ab|xy) \geq 0$. Quantum mechanics is described by a much more convoluted object, combining both flat and curved surfaces [16]. In fact, to this day we ignore how to characterize the quantum set explicitly even in the simplest Bell scenarios, and there are arguments which hint that the problem may be actually undecidable [17]. A natural question in this context is: why quantum correlations? Which physical principles are behind this set?

Popescu and Rohrlich ventured that perhaps the shape of the quantum set could be entirely explained by the no-signalling conditions [15]. Inspired by Einstein's notion of causality, the no-signalling conditions represent an intuitive restriction on the transmission of information between separate parties. Most importantly, they can be formulated in terms of correlations between measurement devices, without making any reference to states, observables or any other element of the actual physical theory behind the experiment. As such, the no-signaling conditions can be regarded as the first *information-theoretic physical principle.*

Unfortunately, as the authors discovered, the no-signalling conditions are not strong enough to recover the set of quantum correlations [15]. This state of affairs led to a race to come up with an information-theoretic physical principle which did the job. Soon many candidates appeared in the literature: Non-Trivial Communication Complexity [18], No-Advantage for Nonlocal Computation [19], Information Causality [20], Macroscopic Locality [21] and Local Orthogonality [22]. Let us briefly review them:

- Non-Trivial Communication Complexity (NTCC) considers a communication scenario where two separated parties receive the strings of bits $\bar{x}$, $\bar{y} \in \{0, 1\}^n$, respectively, and try to compute the function $f(\bar{x}, \bar{y})$ with just one bit of communication. If they succeed in doing so with probability greater than $p > \frac{1}{2}$ for all possible functions and input sizes n, then we say that the world where they live has trivial communication complexity. In analogy with the quantum mechanical case, NTCC postulates the impossibility of such a state of affairs. In [18] it is proven that the NTCC principle restricts nonlocality further than the no-signalling condition alone. The authors of [18] leave open, though, whether NTCC is enough to single out the quantum set.
- Similarly to NTCC, No-Advantage for Nonlocal Computation (NANLC) [19] posits constraints in a communication scenario where two parties are asked to compute a function in a distributed way. This time, the n-bit string $\bar{z}$ is sampled from the probability distribution $p(\bar{z})$, while $\bar{x}$ is completely random. Both parties are handed the bit strings $\bar{x}$, $\bar{c}$, respectively, with $c_i = x_i + z_i$ (mod 2) and asked

to generate bits a, b such that $a + b \pmod 2) = f(\bar{z})$, for some Boolean function f. Note that, taken separately, the strings $\bar{x}$, $\bar{c}$ contain no information about the variable $\bar{z}$. However, if both parties were allowed to share information, they could recover the value of $\bar{z}$ and output the bits $a = 0, b = f(\bar{z})$, hence winning the game with certainty. If they are not allowed to communicate, however, they can only expect to guess the value of $f(\bar{z})$ with some probability, that will depend on both f and the distribution of the variable $\bar{z}$. Interestingly, quantum nonlocality does not offer any advantage towards winning this family of games: namely, if Alice and Bob share quantum correlations prior to being distributed $\bar{x}, \bar{c}$, they cannot expect to win the game more often than by applying the best classical strategy. The authors of [19] postulate that future physical theories should also satisfy this nonlocality constraint. NANLC allows recovering important limits on the strength of quantum nonlocality [19].

- In Information Causality (IC) [20], Alice is distributed an n-bit string $\bar{x}$, while Bob receives a number $k \in \{1, \ldots, n\}$. Bob's task is to guess Alice's kth bit. In order to increase his chances, we allow Alice to transmit him m bits. Intuitively, one would not expect Alice and Bob to succeed with high probability if $m \ll n$. IC tries to formalize this intuition by demanding that the sum of the *mutual informations*[5] between the correct answer x_k and Bob's guess b for each value k does not exceed the number of transmitted bits. That is,

$$\sum_{k=1}^{n} H(x_k : b|k) \leq m. \tag{4}$$

IC has been shown to induce severe constraints over the set of possible correlations [20, 23–27].

The last two principles have a different flavor, in that their original formulation does not make any reference to communication games.

- Roughly speaking, Macroscopic Locality (ML) [21] states that, even though certain microscopic experiments cannot be understood classically, their natural macroscopic counterparts should admit a classical description. Think about the Physics behind a typical experiment of nonlocality: some event somewhere in-between Alice's and Bob's labs produces a pair of correlated particles. After a time, Alice (Bob) receives a particle, which she (he) subjects to a given interaction x (y). Frequency analysis then allows one to estimate the probability $P(ab|xy)$ that the particles respectively impinge on detectors a, b when interactions x, y were switched on. Now, consider an experiment where $N \gg 1$ independent particle pairs with distribution $P(ab|xy)$ are produced. This time Alice (Bob) will receive not just one particle, but a beam of them. As before, Alice and Bob can interact with each particle beam; each interaction, however, will affect all the particles of

[5]The mutual information between two random variables A and B is defined as $H(A : B) = \sum_{A,B} P(A, B) \log_2 \left(\frac{P(A,B)}{P(A)P(B)} \right)$.

the beam at the same time. As a consequence, in Alice's (Bob's) lab, a number of particles I_a (J_b) of the incident beam will impinge against detector a (b). ML postulates that, under low resolution detectors, the 'macroscopic' (Gaussian) probability distribution $P(I_1, \ldots, I_r, J_1, \ldots, J_r)|x, y)$ observed by Alice and Bob should be local and not violate any Bell inequality, that is, it should be of the form (3). The authors of [21] show that many quantum limits in nonlocality follow straightforwardly from ML.

- Local Orthogonality (LO) is a genuinely multipartite principle; as such, in order to formulate it, we need to go beyond the bipartite scenario. Suppose, then, that n distant parties are conducting experiments simultaneously. Denoting by x_k (a_k) the experimental input (output) of the kth party, the full statistics of the multipartite experiment are given by $P(a_1 \ldots a_n|x_1 \ldots x_n)$. Any pair of vectors $e \equiv (\bar{x}, \bar{a})$, representing n inputs and outputs, will be called *event*; and by $P(e)$ we will denote the corresponding conditional probability $P(\bar{a}|\bar{x})$. Two events $(\bar{a}, \bar{x})$, $(\bar{a}', \bar{x}')$ are *locally orthogonal* if one of the parties performs the same measurement in both events ($x_k = x'_k$), but obtains different outcomes ($a_k \neq a'_k$). The principle of LO states that, for any set E of pair-wise locally orthogonal events, $\sum_{e \in E} P(E) \leq 1$. LO can be shown to limit nonlocality non-trivially even in the bipartite case [22]. Most notably, LO allows precluding the existence of tripartite supraquantum correlations which nonetheless satisfy all bipartite physical principles [28].

The above principles highlight previously unnoticed features of quantum mechanics. Taken together, these works hint that the nonlocality of any future physical theory cannot be much different from that of quantum mechanics. However, the question of whether all those principles are enough to single out the quantum set *exactly* remains open.

In [29], it was argued that no set of reasonable physical principles would suffice. There the authors propose a set of correlations, dubbed *almost quantum*, that contains strictly the quantum set. The authors of [29] prove that, with the possible exception of Information Causality [20] for which no proof could be obtained, the almost quantum set satisfies all the aforementioned physical principles. Furthermore, they provide compelling numerical evidence that the IC axiom is satisfied as well. The findings of this work suggest that singling out quantum correlations by means of device-independent principles, that is principles stated only in terms of the observed correlations, may be much subtler than initially expected.

In addition, the almost quantum set has a number of interesting features:

1. The almost quantum set can be shown to be physically consistent, in the sense that natural operations over a number of almost quantum correlations cannot create effective correlations outside the set. Mathematically, this is an extremely non-trivial property, and fairly natural sets of correlations fail to comply with it [30–32].
2. In any Bell scenario, the almost quantum set admits an efficient characterization. In contrast, characterizing the quantum set in two-party Bell scenarios is, at best, computationally hard [33] and, at worst, an undecidable problem [17].

3. As shown in [34], the set of almost quantum correlation naturally emerges when one tries to extend the consistent histories approach to physical theories beyond quantum mechanics *à la* Sorkin [35].

The above properties led the authors of [29] to pose the question of whether the almost quantum set corresponds to the set of correlations of a yet-to-be-discovered consistent physical theory. Finding such a theory could be an important step. In fact, quantum theory has been proven to a rather robust theoretical framework, in the sense that the proposed alterations of the theory, even if a priori simple, have been problematic for fundamental reasons. For instance, they have been shown to imply a significant increase of the computational power of the theory [36] or violations of the no-signalling principle [37, 38]. These results, among others, have led researchers to question whether "quantum mechanics is special" [39] and "an island in theory space"[40]. Finding a modification of quantum theory that does not lead to violations of the no-signalling principle, as a theory compatible with the set of almost quantum correlations would imply, would shed light on how a theory alternative to quantum physics could look like. Moreover, a theory for almost quantum correlations would be operationally equivalent to quantum physics for all the physical principles proposed so far.

Conclusion

In this essay, we have tried to give the reader a brief account of the ongoing interaction between quantum information science and the foundations of quantum mechanics. On one hand, Bell's ideas have inspired the very recent device-independent approach to quantum information theory, with DIQKD [10] and randomness generation [13] as its maximum exponents. In this framework, the security of quantum communication protocols relies on minimal assumptions, such as the validity of quantum theory, and does not require any modelling of the devices used in the protocol. Hence device-independent solutions provide an unprecedented degree of security.

On the other hand, ideas from quantum information science have allowed us to deepen our understanding of quantum nonlocality: now we know that quantum nonlocality does not represent a significant advantage for a number of communication games [18–20]; that it vanishes in natural macroscopic experiments [21] and that it constrains the statistics of locally orthogonal events [22]. Furthermore, we have strong reasons to believe that future theories beyond quantum mechanics cannot exhibit correlations much stronger than what quantum theory allows. Finally, ideas from quantum information science have pointed out new promising directions for future generalizations of quantum theory [29].

We would like to conclude this essay with an interesting remark. Note that, seminal papers aside, most of the works we have cited in this review were published in the past decade. It is certainly impressive how this synergy between quantum information and the foundations of physics has evolved in such a short period of time! We nonetheless feel that we are not at the apex of the hill, but barely have begun to explore the bottom: the best is yet to come.

Acknowledgments This work is supported by the ERC CoG QITBOX, the AXA Chair in Quantum Information Science, Spanish MINECO (FOQUS FIS2013-46768-P and SEV-2015-0522), Fundación Cellex, the Generalitat de Catalunya (SGR 875), the John Templeton Foundation and the FQXi grant "Towards an almost quantum physical theory".

References

1. J. Bell, On the Einstein Podolsky Rosen paradox. Physics **1**, 195 (1964)
2. A. Einstein, B. Podolsky, N. Rosen, Can quantum-mechanical description of physical reality be considered complete? Phys. Rev. **47**, 777 (1935)
3. R.D. Gill, Time, Finite Statistics, and Bell's Fifth Position, Preprint at http://arxiv.org/abs/quant-ph/0301059 (2003); J. Barrett et al. Quantum nonlocality, Bell inequalities and the memory loophole. Phys. Rev. A **66**, 042111 (2002); Y. Zhang, S. Glancy, E. Knill, Efficient quantification of experimental evidence against local realism. Phys. Rev. A **88**, 052119 (2013)
4. S. Pironio, arXiv:1510.00248
5. A. Ekert, Quantum cryptography based on Bell's theorem. Phys. Rev. Lett. **67**, 661 (1991)
6. C.H. Bennett, G. Brassard, N.D. Mermin, Quantum cryptography without Bell's theorem. Phys. Rev. Lett. **68**, 557 (1992)
7. C.H. Bennett, G. Brassard, Quantum cryptography: Public key distribution and coin tossing, in *Proceedings of IEEE International Conference on Computers, Systems and Signal Processing*, vol.175, 1984, p. 8
8. D. Mayers, A. Yao, Self testing quantum apparatus. Found. Phys. **24**, 379 (1994)
9. J. Barrett, L. Hardy, A. Kent, No signalling and quantum key distribution. Phys. Rev. Lett. **95**, 010503 (2005)
10. A. Acín et al., Device-independent security of quantum cryptography against collective attacks. Phys. Rev. Lett. **98**, 230501 (2007)
11. U. Vazirani, T. Vidick, Fully device independent quantum key distribution. Phys. Rev. Lett. **113**, 140501 (2014)
12. L. Lydersen et al., Hacking commercial quantum cryptography systems by tailored bright illumination, Nature Phot. **4**, 686 (2010); I. Gerhardt et al., Full-field implementation of a perfect eavesdropper on a quantum cryptography system. Nat. Comm. **2**, 349 (2011)
13. R. Colbeck, Quantum and Relativistic Protocols for Secure Multi-Party Computation, Ph.D. thesis, University of Cambridge, also available at http://arxiv.org/abs/0911.3814; S. Pironio et al., Random numbers certified by Bell's theorem, Nature **464**, 1021 (2010); R. Colbeck, A. Kent, Private Randomness Expansion With Untrusted Devices, J. Phys. A: Math. Th. **44**(9), 095305 (2011)
14. J.D. Bancal, N. Gisin, Y.C. Liang, S. Pironio, Device-independent witnesses of genuine multipartite entanglement. Phys. Rev. Lett. **106**, 250404 (2011)
15. S. Popescu, D. Rohrlich, Nonlocality as an axiom. Found. Phys. **24**, 379 (1994)
16. B. Tsirelson, Quantum analogues of the Bell inequalities. The case of two spatially separated domains. J. Soviet Math. **36**(4), 557 (1987); B. Tsirelson, Some results and problems on quantum Bell-type inequalities. Hadronic J. Suppl. **8**, 329 (1993)
17. T. Fritz, T. Netzer, A. Thom, Can you compute the operator norm? Proc. Am. Math. Soc. **142**, 4265 (2014)
18. W. van Dam, *Nonlocality and communication complexity*, Ph.D. thesis, University of Oxford (2000); G. Brassard et al., Limit on nonlocality in any world in which communication complexity is not trivial. Phys. Rev. Lett. **96**, 250401 (2006)

19. N. Linden et al., Quantum nonlocality and beyond: limits from nonlocal computation. Phys. Rev. Lett. **99**, 180502 (2007)
20. M. Pawłowski et al., Information causality as a physical principle. Nature **461**, 1101 (2009)
21. M. Navascués, H. Wunderlich, A glance beyond the quantum model. Proc. R. Soc. A **466**, 881 (2010)
22. T. Fritz et al., Local orthogonality as a multipartite principle for quantum correlations. Nat. Comm. **4**, 2263 (2013)
23. J. Allcock, N. Brunner, M. Pawłowski, V. Scarani, Recovering part of the boundary between quantum and nonquantum correlations from information causality. Phys. Rev. A **80**, 040103 (2009)
24. A. Ahanj, Bound on Hardy's nonlocality from the principle of information causality. Phys. Rev. A **81**(3), 032103 (2010)
25. D. Cavalcanti, A. Salles, V. Scarani, Macroscopically local correlations can violate information causality. Nat. Comm. **1**, 136 (2010)
26. Y. Xiang, W. Ren, Bound on genuine multipartite correlations from the principle of information causality. Quant. Inf. Comp. **11**, 0948 (2011)
27. T.H. Yang et al., Information-causality and extremal tripartite correlations. New J. Phys. **14**, 013061 (2012)
28. R. Gallego, L.E. Würflinger, A. Acín, M. Navascués, Quantum correlations require multipartite information principles. Phys. Rev. Lett. **107**, 210403 (2011)
29. M. Navascués, Y. Guryanova, M. Hoban, A. Acín, Almost quantum correlations. Nat. Comm. **6**, 6288 (2014)
30. J. Allcock, N. Brunner, M. Pawlowski, V. Scarani, Closed sets of non-local correlations. Phys. Rev. A **80**, 062107 (2009)
31. B. Lang, M. Navascués, T. Vertesi, Closed sets of correlations: answers from the zoo. J. Phys. A: Math. Theor. **47**, 424029 (2014)
32. B. Salman, A.A. Gohari, Monotone measure for non-local correlations, Preprint at http://arxiv.org/abs/1409.3665 (2014)
33. T. Ito, H. Kobayashi, K. Matsumoto, Oracularization and Two-Prover One-Round Interactive Proofs against Nonlocal Strategies, Preprint at http://arxiv.org/abs/0810.0693 (2008)
34. F. Dowker, J. Henson, P. Wallden, A histories perspective on characterising quantum non-locality. New J. Phys. **16**, 033033 (2014)
35. R. Sorkin, Quantum mechanics as a quantum measure theory. Mod. Phys. Lett. **9**, 3119 (1994)
36. D. Abrams, S. Lloyd, Nonlinear quantum mechanics implies polynomial-time solution for NP-complete and $\#P$ problems. Phys. Rev. Lett. **81**, 3992 (1998)
37. N. Gisin, Stochastic quantum dynamics and relativity. Helvetica Physica Acta **62**, 363 (1989)
38. J.D. Bancal et al., Quantum non-locality based on finite-speed causal influences leads to superluminal signalling. Nat. Phys. **8**, 867 (2012)
39. S. Popescu, Nonlocality beyond quantum mechanics. Nat. Phys. **10**, 264 (2014)
40. S. Aaronson, Quantum computing, postselection, and probabilistic polynomial-time. Proc. R. Soc. A **461**, 2063 (2005)

Chapter 18
Quantum Measurement of Spins and Magnets, and the Classical Limit of PR-Boxes

Nicolas Gisin

Abstract Using weak quantum measurements one can determine the direction in which a large ensemble of spins, as in a classical magnet, points. Assume Alice and Bob share a large ensemble of N pairs of spin-$\frac{1}{2}$. If Alice measures all her spins, all along the same direction, she prepares at a distance an ensemble of spins for Bob which, because of statistical fluctuations, have a magnetic moment of the order $\sqrt{N}$. By making N large enough, this magnetic moment can be made arbitrarily large, indicating an apparent possibility to signal. However, we show that an arbitrarily large magnetic moment is not necessarily classical in the sense that it might be fundamentally impossible to determine in which direction it points. We also consider stronger than quantum correlations and show that Tsirelson's bound follows from the physical assumption that in the macroscopic limit all measurements are compatible and that this should not lead to signaling.

Introduction

The question of how one should apply quantum theory to our macroscopic world, and even the big question whether quantum theory applies at all scales, have been with us since the inception of quantum theory. To illustrate the question, let as study the following little conundrum.

First, consider a single spin $\frac{1}{2}$. When measured the result is probabilistic and the quantum state perturbed, except if the spin was in a state without quantum indeterminacy. Next, consider a large ensemble of N spins, as in a magnet. Then there is no doubt that one can measure the global magnetization essentially without any disturbance. Now, if Alice and Bob share 2 spins in the singlet state and if Alice measures her spin in a direction we label z, then she will get as result ± 1 (assuming her measurement is described by the Pauli operator σ_z). This prepares Bob's spin at a distance in the state $|\mp z\rangle$. Since the mixture of these two states is independent

N. Gisin (✉)
Group of Applied Physics, University of Geneva, 1211 Geneva 4, Switzerland
e-mail: Nicolas.Gisin@unige.ch

R. Bertlmann and A. Zeilinger (eds.), *Quantum [Un]Speakables II*,
The Frontiers Collection, DOI 10.1007/978-3-319-38987-5_18

of Alice's measurement direction (and equal to Bob's state obtained by tracing out Alice), there is no signaling from Alice to Bob, as is well known.

But consider now the case where Alice and Bob share a large number N of pairs of spins, each in the singlet state. If Alice measures all of them, individually, in the z-direction and adds all her results, then she will get a positive or negative fluctuation around zero of the order $\mu \approx \pm\sqrt{N}$. Because of the quantum correlation, Bob will also get a fluctuation of the order μ, i.e. a magnetization of about $\mp\mu$ in the z-direction. By making N large enough, Bob's magnetization μ can be made arbitrarily large. But then, if the magnetization is arbitrarily large, it may seem that Bob can measure it without significantly perturbing it. Obviously, the same should hold if Alice chooses to measure her spins in the x-direction. But then, it seems that Bob could determine the direction in which his magnetization points, either $\pm z$ or $\pm x$. Bob could thus deduce from his magnetization the measurement direction chosen by Alice, without anything carrying this information from Alice to Bob; this would be signaling. Moreover, by enlarging the distance between Alice and Bob and assuming Bob's measurement takes a finite time, this signaling would lead to faster than light communication. But that is impossible. Hence, there must be something wrong in the above story.

In this paper we apply standard quantum measurement theory to resolve this conundrum, i.e. we couple the spin system under investigation to the pointer of the measuring device, treated quantum mechanically. We shall see that the size of the system, here the size of the magnetization μ doesn't suffice to characterize systems that can be measured "classically". In our example, the background noise of the randomly oriented spins, although averaging to zero magnetization, can't be ignored.

Next, we investigate what happens if one replaces the quantum singlet state by stronger than quantum correlations, such as the so called PR-boxes. Following Navascues and Wunderlich [1] and Rochlich [2], we argue that any physical box, when there are large ensembles of them, should admit "classical" measurements. We show that isotropic noisy PR-boxes [3] satisfy this highly plausible physical constrain if and only if the noise is large enough for the correlation to be quantum. We thus recover Tsirelson's bound from a physical assumption, in contrast to previous derivations based on more information theoretical arguments [4–6].

Weak Measurements as Classical Measurements

The standard description of quantum measurements goes as follows. First, one couples the system to be measured to an auxiliary system called the pointer. The later is initially in the state $|q = 0\rangle$, i.e. it points to zero. The coupling between the system and the pointer is assumed so strong and brief that, during that short time, one may safely ignore all Hamiltonian evolutions, except the one that describes the coupling [7, 8]:

$$H_c = g(t)A \otimes p \tag{1}$$

where A is the operator describing the physical quantity to be measured and p is the translation operator acting on the pointer's position; g(t) is a function with non-zero values only during the short system-pointer interaction time, normalized such that $\int g(t)dt = 1$.

Let us illustrate this in the case of N spin $\frac{1}{2}$, all in the state $|\vec{m}\rangle$, with measurement $A = \sum_{j=1}^{N} \sigma_z^j$, where σ^j acts on the j-th spin. Assume that the pointer's initial state is $|q = 0\rangle = \Phi(x)$, with, for example, the function $\Phi(x)$ a Gaussian [9]:

$$\Phi(x) = (2\pi\Delta^2)^{-1/4} exp\{-x^2/4\Delta^2\}, \tag{2}$$

where Δ is the mean square deviation of the pointer's position. After the interaction (1) the initial state $|\vec{m}\rangle^{\otimes N} \otimes |q = 0\rangle$ evolves to:

$$\Psi_{SP} = \sum_{k=0}^{N} \langle\, k, N \mid \vec{m}^{\otimes N}\,\rangle \cdot |k, N\rangle \otimes |q = 2k - N\rangle \tag{3}$$

where $|k, N\rangle$ is the normalized and symmetrized state of N spins with k pointing up z and $N - k$ pointing down in the z-direction, so that the magnetization is $\mu = 2k - N$. The pointer's state is thus merely its initial state displaced by μ: $|q = 2k - N\rangle = \Phi\big(x - (2k - N)\big)$.

The usual quantum measurement story goes then on as follows. The pointer being macroscopic, one can directly look at it. If one finds it at position x_p, then the state of the measured system collapses to the unnormalised state (its norm square being the probability of finding x_p):

$$\Psi_{S|x_p} = \sum_{k=0}^{N} \langle\, k, N \mid \vec{m}^{\otimes N}\,\rangle \Phi\big(x_p - (2k - N)\big) \cdot |k, N\rangle \tag{4}$$

A measurement is strong if the pointer has a well defined position, i.e. if its mean square deviation is small with respect to the distance between the eigenvalues of the measured operator: $\Delta \ll 1$. In this case the sum in (4) reduces to a single value $k \approx x_p$, because for all other values of k, $\Phi\big(x - (2k - N)\big)$ is (practically) zero. This corresponds to the standard textbook measurement process.

A measurement is weak if, on the contrary, the pointer's position has a quantum uncertainty much larger than the distance between the eigenvalues of the measured operator: $\Delta \gg 1$. In this case, many terms in (4) remain. Actually, for the most likely results x_p all significant terms remain quasi unchanged. Hence, weak measurements practically don't perturb the N-spin system. This is how one can measure the magnetization of magnets.

As a first example, assume $\vec{m} = \vec{e}_z$, i.e. all N spins are up in the z-direction. Then $\langle\, k, N \mid \vec{m}^{\otimes N}\,\rangle$ vanishes for all k except $k = N$, hence (3) simplifies to:

$$\Psi_{SP} = |\vec{e}_z^{\;\otimes N}\rangle \otimes |q = N\rangle \tag{5}$$

In this case the spin system is not perturbed at all and the pointer moves N steps to the right.

As second example, consider a magnet in the x-direction, i.e. $\vec{m} = \vec{e}_x$, and a weak measurement with $\Delta \geq \sqrt{N}$. In this case the scalar product $\langle\, k, N \mid \vec{m}^{\otimes N}\, \rangle$almost vanishes except for $k \approx N/2 \pm \frac{1}{2}\sqrt{N}$ in which cases $\Phi\big(x_p - (2k - N)\big) \approx \Phi\big(x_p \mp \sqrt{N}\big)$ is essentially independent of k. Hence, the first terms in Eq. (3) are non-negligible only when the second term is independent of k. Consequently, the pointer's central position doesn't move, but merely broadens a bit. Its mean square deviation after the interaction is the convolution of the initial spread Δ and the square root of the number of spins: $\sqrt{\Delta^2 + N}$. Again, the state of the N spins is almost not perturbed.

In summary, weak measurements, as we recalled their formalization, allow one to discriminate magnets pointing to any of the 4 directions $\pm z$ or $\pm x$ without significantly perturbing their quantum state.

Weak Measurements on *N* Half Singlets

Let Alice and Bob share N pairs of spins, each in the singlet state. Alice can chose between measuring all her spins individually in the z- or in the x-directions, i.e. measure σ_z or σ_x on each spin. Adding all her results, on average she should find zero. But in any run (a run consists of N measurement, one on each of her spins), she will find a fluctuation, typically $\pm\sqrt{N}$. Hence, Bob's N spins will result in a magnetization of about $\mu = \pm\sqrt{N}$ in either the z- or the x-direction, depending on Alice's choice. If Bob could use weak measurements to determine this direction, there would be signalling. How is it that signalling is impossible, despite the fact that Alice's measurement does indeed trigger an arbitrarily large magnetization on Bob's side?

If μ is the magnetization, i.e. the difference between the number of spins up and down along any direction, then one has $\frac{N+\mu}{2}$ spin up and $\frac{N-\mu}{2}$ spin down along that direction. Assume first that this direction is the z-direction. Then, according to the formalism recalled in the previous section "Weak Measurements as Classical Measurements", the pointer will move a distance μ without broadening and without perturbing the state of the N spins. Hence, the probability distribution of the pointer's position, condition on a magnetization μ, reads:

$$\rho^z(x_p|\mu) = |\Phi(x_p - \mu)|^2 \tag{6}$$

where the suffix z recalls that Alice measured her spins along the z-direction.

Since the probability of a magnetization μ is binomial: $2^{-N}\begin{pmatrix} N \\ j \end{pmatrix}$, with $j = \frac{N+\mu}{2}$, Bob's pointer distribution reads:

$$\rho^z(x_p) = 2^{-N} \sum_{j=0}^{N} \binom{N}{j} |\Phi\big(x_p - (2j - N)\big)|^2 \tag{7}$$

Next, assume that Alice chooses the x-direction, hence that the magnetization is along the x-direction:

$$\Psi_{in}^x = |+x\rangle^{\otimes \frac{N+\mu}{2}} \otimes |-x\rangle^{\otimes \frac{N-\mu}{2}} \tag{8}$$

where[1]

$$|\pm x\rangle^{\otimes n} = 2^{-n/2} \sum_{j=0}^{n} \sqrt{\binom{n}{j}} (\pm 1)^{n-j} |j, n\rangle \tag{9}$$

Accordingly, in the z-basis Ψ_{in}^x reads:

$$\Psi_{in}^x = \sum_{j=0}^{\frac{N+\mu}{2}} \sum_{k=0}^{\frac{N-\mu}{2}} c_{jk} |j, \frac{N+\mu}{2}\rangle \otimes |k, \frac{N-\mu}{2}\rangle \tag{10}$$

where

$$c_{jk} = 2^{N/2} \sqrt{\binom{\frac{N+\mu}{2}}{j} \binom{\frac{N-\mu}{2}}{k}} (-1)^{\frac{N-\mu}{2}-k} \tag{11}$$

The unitary system-pointer interaction results in:

$$U(\Psi_{in}^x \otimes |q = 0\rangle) = \tag{12}$$
$$\sum_{j,k} c_{jk} \cdot |j, \frac{N+\mu}{2}\rangle \otimes |k, \frac{N-\mu}{2}\rangle \otimes |q = 2k + 2j - N\rangle$$

Accordingly, the pointer's position probability distribution obtains by tracing out the N-spin system reads:

$$\rho^x(x_p|\mu) = \sum_{j=0}^{j_m} \sum_{k=0}^{k_m} c_{jk}^2 |\Phi\big(x_p - (2k + 2j - N)\big)|^2 \tag{13}$$

where $j_m = \frac{N+\mu}{2}$, $k_m = \frac{N-\mu}{2}$ and the suffix x recalls that Alice measured her spins along the x-direction.

Note that the $(-1)^{\frac{N-\mu}{2}-k}$ sign in the expression of c_{jk} in (13) cancels because only the square of c_{jk} appears in $\rho(x_p)$. Furthermore, the double sum in (13) can be reduced

[1]Note that it is not necessary to symmetrize Ψ_{in}^x; indeed, the system-pointer interaction being symmetric, a symmetrized Ψ_{in}^x would lead to the same effect.

to a single sum by using the identity $\sum_{j=0}^{s} \binom{\frac{N+\mu}{2}}{j} \binom{\frac{N-\mu}{2}}{s-j} = \binom{N}{j}$ and the convention $\binom{k}{j} = 0$ for all $j > k$. For this purpose introduce the variable $s = k + j$ and rewrite the double $\sum_{j=0}^{j_m} \sum_{k=0}^{k_m} = \sum_{s=0}^{j_m+k_m} \sum_{j=0}^{s}$:

$$\begin{aligned} \rho^x(x_p|\mu) &= \\ &= 2^{-N} \sum_{s=j}^{j_m+k_m} \sum_{j=0}^{s} \binom{j_m}{j} \binom{k_m}{s-j} |\Phi(x_p - (2s-N))|^2 \\ &= 2^{-N} \sum_{s=0}^{N} \binom{N}{s} |\Phi(x_p - (2s-N))|^2 \end{aligned} \tag{14}$$

Consequently, Bob's pointer position distribution doesn't depend on the magnetization μ and is rigorously equal to the case Alice measured along the z-direction; this holds for all pointer's state $\Phi(x)$, see (7).

This proves that Bob can't get any information about Alice's choice of measurement direction. The reason is that when Alice choses the z-direction, Bob's pointer moves without any deformation by a random distance depending on Alice's result, i.e. depending on the magnetization μ. If, on the other hand, Alice chooses the x-direction, then the pointer's central position doesn't move, but the noise due to the background spins broadens its distribution by precisely the amount required to make it indistinguishable from the case of a z-direction measurement. In other words, Bob's magnetization $\mu \approx \sqrt{N}$ doesn't consist of $\sqrt{N}$ spins in the direction corresponding to Alice's measurement, but is smeared in a bath of N random spins with a $\sqrt{N}$ fluctuation in a the direction chosen by Alice. The large bath of random spins in which Bob's magnetization exists hides the information about Alice's direction.

Note that this result is exact for any number N of spins and for any strength of the measurement, i.e. any function $\Phi(x)$, in particular any Δ.

In summary, an arbitrarily large magnetic moment is not necessarily classical in the sense that it might be fundamentally impossible to determine in which direction it points.

Macroscopic Limit of Isotropic PR-Boxes

We just saw that quantum entanglement doesn't allow for signaling, even in the case when it allows one to prepare arbitrarily large magnetic moments at a distance. The inavoidable noise is precisely sufficient to prevent any information transfer, just as in quantum cloning [10] and general quantum dynamics [11, 12]. This raises the question whether stronger than quantum correlations would lead to signaling when large ensemble are considered.

Let Alice and Bob share N noisy PR-boxes [13], with isotropic noise [3]. Denote the inputs $x, y \in \{0, 1\}$ and outcomes $a, b = \pm 1$. Hence, each PR-box has random marginals and correlation

$$P(a \cdot b = (-1)^{x \cdot y} | x, y) = V \tag{15}$$

where the "visibility" V is the "pure PR-box weight", $V \in [0, 1]$.

As in the case of N singlets, consider the case where Alice measures all her boxes with either the setting $x = 0$ or with all $x = 1$ (i.e. she inputs into all her boxes either $x = 0$ or $x = 1$). In this way she prepares Bob's ensemble of boxes at a distance. If Bob measures all his boxes with the same input y and sums up all his outcomes, he finds a fluctuation of the order $\pm\sqrt{N}$ around zero. If $x \cdot y = 0$, then Alice and Bob's fluctuations are likely to be of the same sign; however, if $x \cdot y = 1$, then they are likely to be of opposite signs. So far, this is very similar to the N singlet case. But if the noise is small enough for the correlations to be stronger than quantum, then one may wonder whether signaling is still excluded.

At this point one would like to define weak-measurements for large ensemble of PR-boxes. Indeed, as emphasized by Rohrlich [2], any physical box must be such that when large ensembles are considered, then collective measurements of their global "magnetization" should be feasible. Unfortunately, at present one doesn't know how to define the analog of weak measurements for ensembles of PR-boxes, a clear weakness of today's concept of PR-boxes. Nevertheless, it makes good sense to assume that in the macroscopic limit of large enough N, the following two quantities on Bob's side can both be measured[2]:

$$B_y = \sum_{j=1}^{N} b_{j|y} \tag{16}$$

where $b_{j|y}$ is the outcome of Bob's j-th PR-box when it gets the input y.

If the PR-boxes are noise-free, i.e. $V = 1$, when Bob could read Alice's input x from B_0 and B_1. Indeed, if $x = 0$, then $B_0 = B_1$, while if $x = 1$, then $B_0 = -B_1$.

But clearly, if the PR-boxes are noisy enough to be realizable with quantum entanglement, i.e. if $V \leq \frac{1}{2}(1 + \sqrt{\frac{1}{2}}) \approx 0.85$, then, as we have seen in the previous section "Weak Measurements on N Half Singlets", the assumption that B_0 and B_1 are jointly measurable doesn't lead to signaling. Hence the natural question: "How much noise should PR-boxes have to avoid signaling in the macroscopic limit?".

We shall consider the limit of infinitely many PR-boxes and assume that, in this limit, all 4 quantities A_0, A_1, B_0, B_1 can be measured simultaneously, where

---

[2]Note that according to quantum theory B_0 and B_1 (i.e. $\sum_j \sigma_z^j$ and $\sum_j \sigma_x^j$) can be measured simultaneously. Indeed, a weak measurement of B_0 with a pointer' spread of the order $\sqrt{N}$ essentially doesn't perturb the quantum state, hence can be followed by a similar weak measurement of B_1. Both measurements provide pretty good information about B_0 and B_1, respectively. Note furthermore that this process can be extended to series of measurements, similarly to [9].

$$A_x = \sum_{j=1}^{N} a_{j|x} \tag{17}$$

with similar notations for Alice's $a_{j|x}$.

Hence there exist a well defined (i.e. non-negative) probability distribution $P(A_0, A_1, B_0, B_1)$. This implies that the possibility for Bob to measure simultaneously B_0 and B_1 doesn't lead to signaling. Indeed, the existence of a global probability distribution excludes violation of a Bell inequality, hence guarantees the existence of a local model [14]. This also establishes the connection with the concept of macroscopic-locality [1].

In the limit of many PR-boxes, thanks to the central limit theorem, the probability distribution $P(A_0, A_1, B_0, B_1)$ is Gaussian, with zero mean:

$$\begin{aligned} &P(A_0, A_1, B_0, B_1) = \\ &exp\{-(A_0, A_1, B_0, B_1)K^{-1}(A_0, A_1, B_0, B_1)^t\} \end{aligned} \tag{18}$$

where the suffix t indicates the transpose and the correlation matrix is defined as follows:

$$K \equiv \begin{pmatrix} \langle A_0A_0\rangle & \langle A_0A_1\rangle & \langle A_0B_0\rangle & \langle A_0B_1\rangle \\ \langle A_1A_0\rangle & \langle A_1A_1\rangle & \langle A_1B_0\rangle & \langle A_1B_1\rangle \\ \langle B_0A_0\rangle & \langle B_0A_1\rangle & \langle B_0B_0\rangle & \langle B_0B_1\rangle \\ \langle B_1A_0\rangle & \langle B_1A_1\rangle & \langle B_1B_0\rangle & \langle B_1B_1\rangle \end{pmatrix} \tag{19}$$

with $\langle A_0A_1\rangle$ the correlation between A_0 and A_1 and similarly for all entries. K is clearly symmetric.

The first entry is easy to evaluate: $\langle A_0A_0\rangle = \langle\sum_{i,j=1}^{N} a_{i|0} \cdot a_{j|0}\rangle$. If $i = j$ one has $a_{i|0} \cdot a_{j|0} = 1$. If $i \neq j$, in the limit of large N the average vanishes. Hence $\langle A_0A_0\rangle = N$, and similarly for all 4 diagonal terms of K.

The second entry $\langle A_0A_1\rangle$ can't be evaluated without further assumptions. Hence, let's move to the next entry: $\langle A_0B_0\rangle = \langle\sum_{i,j=1}^{N} a_{i|0} \cdot b_{j|0}\rangle$. If $i = j$, $P(a_{i|0} \cdot b_{j|0} = (-1)^{0\cdot 0} = +1) = V$, hence $\langle\sum_{j=1}^{N} a_{j|0} \cdot b_{j|0}\rangle = N(2V - 1) \equiv Nv$. If $i \neq j$, in the limit of large N the average vanishes. Hence $\langle A_0B_0\rangle = Nv$. Similarly $\langle A_0B_1\rangle = \langle A_1B_0\rangle = Nv$ and $\langle A_1B_1\rangle = -Nv$.

Consequently, the correlation matrix reads:

$$K = N\begin{pmatrix} 1 & s & v & v \\ s & 1 & v & -v \\ v & v & 1 & s \\ v & -v & s & 1 \end{pmatrix} \tag{20}$$

where we assume $\langle A_0A_1\rangle = \langle B_0B_1\rangle$ and define $s \equiv \langle A_0A_1\rangle/N = \langle B_0B_1\rangle/N$. Note that one can prove that this symmetry assumption is not necessary to derive our conclusion, though it is a very natural assumption.

Now, the Gaussian probability $P(A_0, A_1, B_0, B_1)$ is non-negative if and only if the correlation matrix K is non-negative. The eigenvalues of K/N are:

$$1 + \sqrt{2v^2 + s^2 + 2vs} \tag{21}$$

$$1 - \sqrt{2v^2 + s^2 + 2vs} \tag{22}$$

$$1 + \sqrt{2v^2 + s^2 - 2vs} \tag{23}$$

$$1 - \sqrt{2v^2 + s^2 - 2vs} \tag{24}$$

These must be non negative. Adding (22) and (24) one gets $2 \geq 4v^2 + 2s^2$. Hence, $1 - 2v^2 \geq s^2 \geq 0$, thus $v^2 \leq \frac{1}{2}$, i.e.

$$v \leq \sqrt{1/2} \tag{25}$$

which is Tsirelson's bound (recall $V = \frac{1+v}{2}$) [15].

Hence, Tsirelson's bound follows from the physical assumption that in the macroscopic limit all measurements are compatible and that this should not lead to signalling.

Extension to Asymmetric Noisy PR-Boxes

The result of the previous section can easily be extended to asymmetric non-signalling boxes with arbitrary noise. It suffices to replace (18) by:

$$P(A_0, A_1, B_0, B_1) = \\ exp\{-(\bar{A}_0, \bar{A}_1, \bar{B}_0, \bar{B}_1)\bar{K}^{-1}(\bar{A}_0, \bar{A}_1, \bar{B}_0, \bar{B}_1)^t\} \tag{26}$$

where $\bar{A}_i \equiv A_i - \langle A_i \rangle$ and $\bar{B}_i \equiv B_i - \langle B_i \rangle$ and the correlation matrix $\bar{K}$ is constructed as (19), but using the $\bar{A}_i$ and $\bar{B}_j$ instead of the A_i and B_j.

The non-negativity of $\bar{K}$ is then equivalent to the first step in the hierarchy [16] characterizing quantum correlations. It is known that, in general, this first step is not sufficient to single out quantum correlations, hence—surprisingly—there are stronger than quantum correlations that have a macroscopic non-signalling limit, as emphasized in [17].

Conclusion and Open Problems

Large ensembles of small systems should be jointly measurable in some sort of a macroscopic or classical limit. If not, they are not physical [2]. This is true as well for quantum systems as for systems described by any post-quantum theory. In section "Weak Measurements on N Half Singlets" we illustrated this for large ensembles of spin-$\frac{1}{2}$ and showed that indeed, the quantum formalism of weak measurement provides the tool to describe collective measurements and how they carefully are just at the border of not violating the no-signaling principle. In the following section we considered noisy PR-boxes, that is hypothetical boxes with stronger than quantum correlations. In the case of isotropic noise and in the limit of infinitely many boxes, we found that the assumption that all collective measurements are compatible leads to non-physical signaling whenever the noise is weak enough for the boxes to share correlations stronger than possible according to quantum theory; that is we recovered Tsirelson's bound. This is physically very nice, however one should be able to get to this result without the N to infinity limit. Furthermore, in the case of non-isotropic noise one doesn't recover the quantum boundary (even in the limit $N \to \infty$), as already emphasized in [17]. This is absolutely remarkable and deserves deeper investigation. In particular, there is an urgent need for a model of collective measurements of large-but-finite ensembles of noisy PR-boxes.

Acknowledgments This work profited from numerous comments from colleagues from Bristol and Barcelona and discussions after my presentation at the DIQIP meeting in May 2013. I am especially in debt to Sandu Popescu for numerous stimulating discussions and debates and to Daniel Rohrlich for sending me his draft [2] before publication. Financial support by the European projects ERA-NET DIQIP and ERC-AG MEC are gratefully acknowledged.

References

1. M. Navascues, H. Wunderlich, Proc. R. Soc. A **466**, 881–890 (2010)
2. D. Rohrlich, PR-box correlations have no classical limit, in *Quantum Theory: A two Time Success Story* (Springer, NY, 2013). arXiv:1407.8530 and arXiv:1408.3125
3. L. Masanes, A. Acn, N. Gisin, Phys. Rev. A **73**, 012112 (2006)
4. G. Brassard, H. Buhrman, N. Linden, A.A. Methot, A. Tapp, F. Unger, Phys. Rev. Lett. **96**, 250401 (2006)
5. N. Linden, S. Popescu, A.J. Short, A. Winter, Phys. Rev. Lett. **99**, 180502 (2007)
6. M. Pawlowski, T. Paterek, D. Kaszlikowski, V. Scarani, A. Winter, M. Zukowski, Nature **461**, 1101 (2009)
7. J. von Neunam, *Mathematical Foundations of Quantum Mechanics* (Princeton University Press, Princeton, 1955)
8. Y. Aharonov, D. Rohrlich, *Quantum Paradoxes* (Wiley-VCH, 2005)
9. R. Silva, N. Gisin, Y. Guryanova, S. Popescu, Phys. Rev. Lett. **114**, 250401 (2015)
10. N. Gisin, Phys. Lett. A **242**, 1 (1998)
11. N. Gisin, Helv. Phys. Acta **62**, 363 (1989)
12. V. Buzek, N. Gisin, Ch. Simon, Phys. Rev. Lett. **87**, 170405 (2001)
13. S. Popescu, D. Rohrlich, Found. Phys. **24**(3), 379385 (1994)
14. A. Fine, Phys. Rev. Lett. **48**, 291 (1982)
15. B.S. Cirel'son, Lett. Math. Phys. **4**, 93 (1980)
16. M. Navascues, S. Pironio, A. Acín, Phys. Rev. Lett. **98**, 010401 (2007). New J. Phys. **10**, 073013 (2008)
17. M. Navascues, Y. Guryanova, M. Hoban, A. Acin, Nat. Commun. **6**, 6288 (2015)

Chapter 19
The Dynamical Roles Played by Mass and Proper Time in Physics

Daniel M. Greenberger

Introduction

If EPR (Einstein, Podolsky, and Rosen) had been correct in 1935, and "elements of reality" actually existed, the subsequent history of quantum theory would have been completely different, and much more conservative. But Bohr was right on this issue, and Bell's Theorem made the subject experimentally accessible. Bohr's answer was quite unintuitive, and so it took many experiments to convince most people. But even then, it took a long time before it led to the modern revolution of information technology. But after the initial challenge of EPR, who would have bet on how the controversy would turn out? Furthermore, almost nobody cared, even though the result is crucial to the interpretation and meaning of quantum mechanics.

I am going to present to you today another challenge to the usual theory, that most people also won't care about. If it is confirmed, it will have enormous consequences, but, as Carl Sagan was fond of saying, "extraordinary claims require extraordinary proof." Only experiment can give you that. I can only offer plausibility arguments at the present time. (I hope you won't put it in the category "silly claims require silly proofs.")

One problem that I have always had with the standard physics formalism is with the roles that are played by mass and proper time within it. Mass plays a dynamical role in many problems, and yet it enters the formalism as a passive parameter. You can't do anything with it. In the formula $E = mc^2$, E is an operator, and energy states can be superposed, while m has no dynamical role at all.

If an atom decays from an excited state to its ground state, the energy of the atom changes, and the interaction provides the mechanism for this to happen. But nothing happens to the mass. If you need to keep track of the changing mass in this case,

D.M. Greenberger (✉)
City College of New York (CCNY), City University of New York (CUNY), New York, NY 10031, USA
e-mail: dan.greenbergernyc@gmail.com

R. Bertlmann and A. Zeilinger (eds.), *Quantum [Un]Speakables II*,
The Frontiers Collection, DOI 10.1007/978-3-319-38987-5_19

you can enter it by hand. But there ought to be some kind of interaction mechanism that automatically keeps track of the changing mass. It should enter the formalism as a dynamical variable, and there should be an equation of motion for it, to keep track of these things [1–9].

For this to be so, the mass should be defined as the energy in the center of mass (barycentric) frame. That way, it will keep track of changes in binding energies. Also, if the energy is uncertain, and has a spread ΔE, then the mass will have a spread Δm. So even if you know, say, that the particle is a free electron, the mass will not necessarily be m_e, the mass that appears in the free particle Dirac Equation. We call that the "nominal mass" of the particle. The actual mass will include binding energies and uncertainties.

If m plays the role of a dynamical variable, then what will be its conjugate variable? The obvious candidate is the proper time of the particle. In the standard formalism the proper time is defined geometrically, by

$$d\tau^2 = dt^2 - dx^2,$$

within special relativity (throughout the paper we will take c = 1, unless otherwise stated), and there is no room for it to be a dynamical variable. So if it is to be considered as a dynamical variable, this equation, or equivalently,

$$\dot{\tau} = \sqrt{1 - v^2},$$

must follow as an equation of motion, which can be changed with an appropriate interaction. We will consider the proper time of a particle to be the time read by a clock in the rest frame of the particle, and it will be subject to uncertainties.

The Hamiltonian Formalism

The Hamiltonian formalism is perfectly suited to bring out the meaning of this new addition to dynamics. A particle normally is described by a set of dynamical variables, x, the position, and p, the momentum, the conjugate variable. Generally the Hamiltonian will be the energy of the particle. Then the Hamiltonian will be

$$H = H(x, p).$$

The equations of motion are

$$\dot{x} \equiv v = \frac{\partial H}{\partial p}, \quad \dot{p} = -\frac{\partial H}{\partial x}.$$

The meaning of these equations are that if space is homogeneous, i.e., the Hamiltonian does not depend on x, then p will be conserved. But if there is a potential that depends on x, it will induce forces that will change the momentum.

Similarly, we now include the two new dynamical variables τ and m, and the Hamiltonian and equations of motion become

$$
\begin{aligned}
H &= H(x, p; \tau, m), \\
\dot{x} &= \frac{\partial H}{\partial p}, \quad \dot{p} = -\frac{\partial H}{\partial x}, \\
\dot{\tau} &= \frac{\partial H}{\partial m}, \quad \dot{m} = -\frac{\partial H}{\partial \tau}.
\end{aligned}
$$

In a likewise manner, if H does not depend on τ, space is homogeneous in τ and the mass will not change. This is the usual situation in mechanics and in special relativity.

However if H does depend on τ, then the particle can decay, and change its mass, even classically. So here is a natural extension of the classical theory, and the theory automatically accommodates to a changeable mass. This is a totally new role for dynamics, and with the correct Hamiltonian the particle can react to the physical situation and, for example, pick up an addition to its mass that responds to, say, a change in the binding energy. Also, the particle can decay into a lighter one. And it becomes possible to have coherent superpositions of masses, which can happen in relativity, but again it does not automatically get produced by the formalism. But also, there will be an equation of motion for $\dot{\tau}$, which then becomes more general than the geometrical prescription it usually has, and it can display more physical properties.

As a very simple example, consider the case of a free particle. Here, the Hamiltonian will be the usual one

$$
\begin{aligned}
H &= H(p; m) = \sqrt{p^2 + m^2}, \\
\dot{x} &= v = \frac{\partial H}{\partial p} = \frac{p}{\sqrt{p^2 + m^2}}, \quad \text{or} \quad \mathrm{p} = \frac{mv}{\sqrt{1 - v^2}}; \\
\dot{\tau} &= \frac{\partial H}{\partial m} = \frac{m}{\sqrt{p^2 + m^2}} = \sqrt{1 - v^2}.
\end{aligned}
$$

So we see that $\dot{\tau}$ obeys the usual equation, not as a constraint, but as an equation of motion, that comes from the form of the Hamiltonian. In this simple case, m is constant, and there is no modification to the behavior of τ. But notice that m no longer plays the role of an external parameter, but is as intrinsically dynamical as is p.

To see how a particle can classically decay, take the simple case where a particle decays from some initial m_O to some final m_f, by an exponential decay, $m = m_f + \Delta e^{-\gamma\tau}$, where $\mathrm{m_f}$ is a stable state, and $\Delta \ll m_f$. The decay is a $f(\tau)$, not a $f(t)$, because it is known experimentally that if it decays in a moving system, there will be a time dilatation effect. This decay is a low energy one, like an atom decaying into its ground state. Imagine that the decay is caused by some potential, $\varphi(\tau)$. Then

$$H = m + m\varphi(\tau) = m_f.$$

H will be a constant, since there is no direct t dependence. So we have

$$m = \frac{m_f}{1+\varphi} = m_f + \Delta e^{-\gamma\tau},$$
$$\varphi = \frac{1}{1+\frac{\Delta}{m_f}e^{-\gamma\tau}} - 1 \approx -\frac{\Delta}{m_f}e^{-\gamma\tau}.$$

We can solve for τ as a function of t, and we get

$$\dot{\tau} = \frac{\partial H}{\partial m} = 1 + \varphi = \frac{1}{1+\frac{\Delta}{m_f}e^{-\gamma\tau}},$$
$$\tau + \frac{\Delta}{m_f\gamma}(1 - e^{-\gamma\tau}) = t.$$

Here we have set τ = 0 at t = 0, and we see that as

$$t \to \infty, \quad t - \tau \to \frac{\Delta}{m_f\gamma}, \quad t > \tau.$$

So even though the particle is at rest, $\tau \neq t$. So there is a decay time delay on τ. One can make this plausible by making a model of the particle as a damped oscillator which, even though it is at rest on average, is moving back and forth while decaying, and so its proper time is slowly falling behind (one gets the same order of magnitude effect). But here it is a general phenomenon associated with all decays, and one need not concoct a specific model.

However we see here a general and surprising phenomenon. The potential φ acts here just like a gravitational potential, and so there is a second and extremely important role for gravity. When the gravitational potential depends on τ as well as, or instead of, on x, it acts as a mechanism causing particles to decay. Thus besides its usual role of bending space to cause the usual gravitational attraction, it also distorts proper time, and causes particles to decay. This is the unexpected connection of gravity to the other forces of physics, and it implies that one will have to rethink and extend general relativity, especially if one wants to quantize it.

The Uncertainty Principle Between Mass and Proper Time

Once one quantizes a system of particles, one is confronted by the fact that there must be an uncertainty principle between conjugate variables. This would imply that one cannot measure both the proper time (the reading of a clock in the rest system of the particle) and its mass at the same time. This shows that the mass

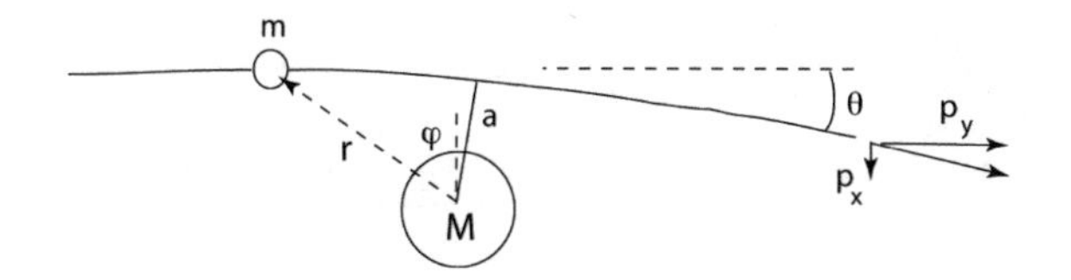

Fig. 19.1 *Mass-Proper Time Uncertainty*. The mass m is measured by gravitationally scattering it off a much larger mass M. The distance of closest approach is a, mass m is located at r, and θ is the scattering angle.

cannot mean the nominal mass of the particle, but the dynamical mass associated with the given experimental arrangement. In fact, this is true, and we can illustrate it with countless examples, both gravitational and non-gravitational. A simple gravitational example would be the following.

One wants to measure the mass of a light particle by gravitationally scattering it off a much heavier particle. The situation is shown in Fig. 19.1.

The mass m is located at r, and the distance of closest approach is a. The gravitational force on m is $F = GmM/r^2$, and the momentum p_x it picks up will be $\int F_x dt = p_x$. Most of the force will be exerted in the region where $|\varphi| < 45°$, and so

$$p_x \sim \frac{GMm}{a^2}\frac{2a}{v}, \qquad \Delta p_x \sim \frac{2GM}{av}\Delta m.$$

Even if the velocity is perfectly known asymptotically, a clock sitting on m will be uncertain, because if the distance a is not perfectly known, then the gravitational potential on m, φ, will also be unknown to that extent, and will affect the accuracy of the clock on m. So

$$\Delta\tau \sim \frac{\varphi}{c^2}\Delta t \sim \frac{\varphi}{c^2}\frac{2\Delta a}{v} \sim \frac{GM}{ac^2}\frac{2\Delta a}{v}.$$

So that finally, one has

$$\Delta p_x \Delta x \sim \frac{2GM}{av}\Delta m \frac{ac^2 v}{2GM}\Delta\tau \sim c^2 \Delta m \Delta\tau \sim \hbar.$$

If one believes in the uncertainty relation for the other conjugate variables, one cannot avoid it for m and τ.

A second simple example, which does not depend on gravity, is the attempt to measure the mass of a charged particle by passing it through a thin slit, and into an opposing electric field, and measuring how far it travels before coming to rest. The situation is shown in Fig. 19.2.

A particle with known velocity v_0 and mass m, charge e, enters a slit, and is brought to rest by an opposing field E, in distance L. The distance it takes to come to rest is

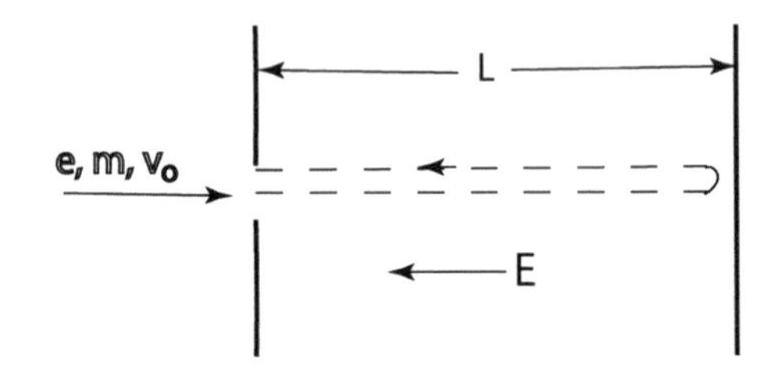

Fig. 19.2 *Mass-Proper Time Uncertainty Without Gravity.* A particle enters the slit, encounters an electric field E, and is brought to rest at distance L.

$$L=\frac{eE}{2m}T^2, \quad \Delta L=\frac{eE}{m}T\Delta T=\Delta x.$$

Since v_0 is accurately given, the momentum will be

$$p=v_0 m, \quad \Delta p=v_0\Delta m,$$

and so

$$\Delta x\Delta p\sim\hbar\sim\frac{eE}{m}T\Delta T v_0\Delta m.$$

Now $v_0 = aT = (eE/m)T$, and

$$\tau=\sqrt{1-v^2/c^2}T, \quad \Delta\tau\sim\frac{v^2}{c^2}\Delta T, \quad \text{and}$$
$$\Delta p\Delta \mathrm{x}\sim \mathrm{v}_0^2\Delta m\Delta T\sim c^2\Delta m\Delta\tau\sim\hbar.$$

In this case, τ is determined by special relativity, while in the previously case it was determined by the gravitational potential of general relativity.

These examples depend on the usual uncertainty principles, but it would be nice to find an example that could show the effect of the mass and proper time as dynamic operators in a much more spectacular and convincing way. Our candidate for such an experiment is to find a system that can be broken into N parts, with equally spaced energies, like a spin in a magnetic field, and to break the system up, and after some time T to coherently recombine it. Then the masses of each part would be

$$c^2 m_n=E_0+n\varepsilon, \quad n=0,\ 1,\ \ldots, N-1.$$

The proper time, if the particle states are separated such that each part has the same momentum (i.e., each has the same probability of absorbing the momentum of an incident exciting photon),

$$\tau_n = t - \frac{v_n^2}{2c^2}t, \quad v_n = \frac{p}{m_n} = \frac{c^2 p}{E_0}\left(1 - \frac{n\varepsilon}{E_0}\right),$$
$$\tau_n = t\left(1 - \frac{c^2p^2}{E_0^2}\left(1 - \frac{2n\varepsilon}{E_0}\right)\right) = \tau_0 + n\kappa t, \quad \kappa = \frac{2c^2p^2\varepsilon}{E_0^3}.$$

So the proper times are equally spaced, and when the beams are brought together, the energy differences will be removed, and the wave function will become

$$\psi = \sum \psi_0 e^{-iE_0\tau_n/\hbar} = \psi_0 e^{-iE_0\tau_0/\hbar} \sum e^{-niE_0\kappa\tau_0/\hbar}$$
$$= \psi_0 e^{-iE_0\tau_0/\hbar} \sum e^{-in\omega t} = \psi_0 e^{-iE_0\tau_0/\hbar} \frac{\sin^2 N\omega t/2}{\sin^2 \omega t/2}.$$

where we have $\omega = E_0\kappa\tau_0/\hbar$. So we have in effect a diffraction pattern in proper time.

Now if the original set of states were slowly decaying, due to some other cause, the graph of the envelope of $|\Psi(\mathrm{t})|^2$ will look like that below (Fig. 19.3).

So experimentally, one would only see decays during the brief periods when all the different segments of ψ were in phase, a pretty convincing demonstration of the dynamical properties of proper time.

There are many other aspects to the dynamical behavior of proper time and mass, but one hopes that this shows that the inclusion of these variables into the dynamical aspects of the theory would prove to introduce a whole range of unexpected behavior to the mechanics of particles. We also believe that the implications of gravity playing a dynamical role, by being responsible for particle decays, will provide a big clue for how to include gravity into the general scheme of

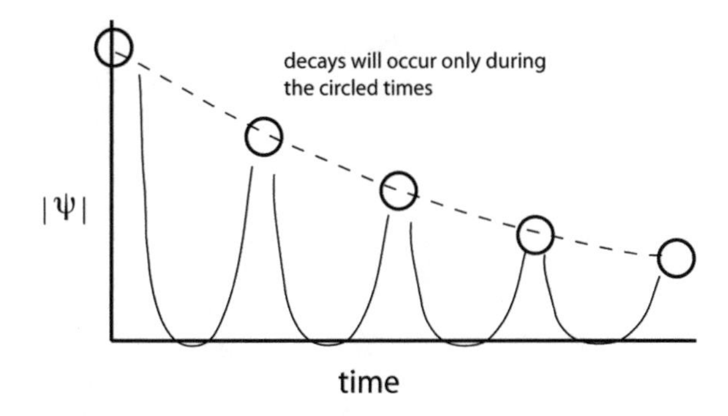

Fig. 19.3 *Truncated decay times due to coherent proper time interference*. Due to the coherent adding of wave-functions with different proper times having elapsed, the overall effect is that of a diffraction pattern in proper time, and the particle can decay only in those brief periods when the various parts are in phase, during the *circled* periods in the graph. Without this coherence, the decay graph would be the *dotted line*.

the other forces. Our contention is that the failure to find a straightforward quantized theory of gravity is not due to the mathematical complexities of the problem, such as non-linearities, but rather to the failure to recognize this as a physics problem, even at the simple level of the equivalence principle.

References

Here are some articles I have written that are relevant to the subject.

1. D. M. Greenberger, The role of equivalence in quantum mechanics. Ann. Phys. (N. Y.) **47**, 116–126 (1968)
2. D. M. Greenberger, Theory of particles with variable mass, I – Formalism. J. Math. Phys. **11**, 2329–2340 (1970)
3. D. M. Greenberger, Theory of particles with variable mass, II – Some Physical Consequences. J. Math. Phys. **11**, 2341–2347 (1970)
4. D. M. Greenberger, Some useful properties of a theory of variable-mass particles. J. Math. Phys. **15**, 395–405 (1974)
5. D. M. Greenberger, Wave packets for particles of indefinite mass. J. Math. Phys. **15**, 406 (1974)
6. D. M. Greenberger, The proper role of proper time, in *Experimental Metaphysics: Quantum Mechanical Studies for Abner Shimony*, vol. I, ed. by R.S. Cohen, M. Horne, J. Stachel. Kluwer, Dordrecht (1997)
7. D. M. Greenberger, The inadequacy of the Galilean transformation in quantum mechanics. Phys. Rev. Lett. **87**(10), 100405–1 (2001)
8. D. M. Greenberger, Conceptual problems related to time and mass in quantum theory, Quant-ph. arXiv:1109.3709m (2010)
9. D. M. Greenberger, The Disconnect Between Quantum Mechanics and Gravity, Quant-ph. arXiv:1011.3719 (2010)

Chapter 20
On Spatial Entanglement Wave Functions

Michael Horne

There exists tension between two of Bell's papers collected in *Speakable and Unspeakable in Quantum Mechanics* [1]. Paper 2, the 1964 inequality paper [2], employs Bohm's 1951 entanglement of spin states [3] instead of EPR's original 1935 entanglement of spatial wave functions [4]. Moreover, earliest [5], improved [6], and most other experimental violations of Bell's inequality, have employed spin-like entanglements, specifically polarization-entangled photon pairs. But in the "Generalization" section of paper 2 Bell argues that the shift from EPR to Bohm is of no significance because "we can always consider two-dimensional subspaces" in each particle's Hilbert space. In 1986, in paper 21, Bell returns to the EPR state and argues [7] that "there is no non-locality problem" with this state because it has a non-negative Wigner distribution. Does the EPR state pay inadequate attention to suitable two-dimensional subspaces? Is the non-negativity of a Wigner distribution an inadequate test of the absence of a locality problem? I haven't resolved this tension, but I do wish to propose it as a research project.

Bell's motivation for shifting from Bohm back to EPR comes partly from a 1985 explicit, but still gedanken proposal [8] for a Bell experiment with linear momentum entanglement instead of polarization entanglement. Such gedanken arrangements, known as two-particle interferometers [9], became realizable as down-conversion replaced atomic cascades as the primary source of entangled photon pairs; absence of the residual atom provides strong correlation in the magnitude and/or direction of the pair's momenta. Prototypical two-particle interferometers are either two-particle Mach-Zehnder type or two-particle double-slit type, and hence emphatically require two spatial dimensions to set up. (The one violation of Bell's inequality produced via spatial entanglement employed a two-particle Mach-Zehnder-type setup [10].) The original EPR spatial entanglement is emphatically in one spatial dimension. Does the number of spatial dimensions

M. Horne (✉)
Department of Physics and Astronomy, Stonehill College, Easton, MA 02357, USA
e-mail: mhorne@stonehill.edu

R. Bertlmann and A. Zeilinger (eds.), *Quantum [Un]Speakables II*,
The Frontiers Collection, DOI 10.1007/978-3-319-38987-5_20

enter into creating (and resolving) the tension? On the one hand, interferences made possible by having two spatial dimensions play a pivotal role in violating Bell's inequality, with the relative phase of two paths playing the role of the orientation of the Stern-Gerlach magnet [linear polarizer] in a spin [polarization] experiment. On the other hand, couldn't one contemplate a two-particle Fabry-Perot type setup and hence work in one spatial dimension?

It is noteworthy in paper 21 that Bell only shows positiveness of the Wigner distribution for the raw or initial EPR state. He doesn't address what effects complicated manipulations of the EPR state might have on the initially positive Wigner distribution, specifically, manipulations with beamsplitters and mirrors which bring two of the EPR momenta of each particle to interference. However, in the single spatial dimension of the EPR state, the only two different momenta for a particle are two that differ only in *magnitude* (or sign). Therein may lay the importance of having at least two spatial dimensions: the existence of two momenta of equal magnitude but different *direction* (other than simply opposite).

To prepare for an investigation of the behavior of Wigner's distribution in two-particle interferometry, first review (or create) the basic elements (i.e. definition) of two-particle Wigner distributions [11]. To expose the behavior and evolution of the distribution in the context of, say, a double-slit-type interferometer, first compute the time-independent Wigner distribution in the region between the source and the two double slits, then in the region close behind the double slits where the wave functions haven't yet crossed, and finally in the region where they do cross, e.g. on the surfaces of the two receiving planes where the two-particle fringes reside. Which, if any, of these Wigner distributions are everywhere positive? Which, if any, are not? Did the double double-slits change the former into the latter?

To summarize, spatial entanglements that violate Bell's inequality confirm the "Generalization" section of Bell's paper 2. But these spatial entanglement experiments also suggest that paper 21 is misleading, or incomplete, or both. An investigation is called for.

References

1. J.S. Bell, *Speakable and Unspeakable in Quantum Mechanics* (Cambridge University Press, Cambridge, 1987)
2. J.S. Bell, On the Einstein–Rosen–Podolsky paradox. Physics **1**, 195–200 (1964)
3. D. Bohm, *Quantum Theory*, (Prentice-Hall, Englewood Cliffs, 1951), pp. 611–23
4. A. Einstein, B. Podolsky, N. Rosen, Can quantum-mechanical description of physical reality be considered complete? Phys. Rev. **47**, 777–80 (1935)
5. S.J. Freedman, J. Clauser, Experimental test of local hidden-variable theories. Phys. Rev. Lett. **28**, 938 (1972)
6. A. Aspect, J. Dalibard, G. Roger, Experimental test of Bell's inequalities using time-varying analyzers. Phys. Rev. Lett. **49**, 1804–07 (1982)

7. J.S. Bell, EPR correlations and EPW distributions in *New Techniques and Ideas in Quantum Measurement Theory,* vol. 480, ed. by D. Greenberger (New York Academy of Sciences, New York, 1986)
8. M.A. Horne, A. Zeilinger, A Bell-type EPR experiment using linear momenta, in *Symposium on the Foundations of Modern Physics*, *Joensuu, 1985*, ed. by P. Lahti, P. Mittelstaedt (World Scientific, Singapore, 1985). See also M. Horne, A. Zeilinger, A possible spin-less experimental test of Bell's inequality in *Microphysical Reality and Quantum Formalism,* vol. 2, ed. by A. van der Merwe, et al. (Kluwer, Dordrecht, 1988), p. 401
9. M.A. Horne, A. Shimony,A. Zeilinger, Two-particle interferometry. Phys. Rev. Lett. **62**, 2209 (1989). See also D. Greenberger, M.A. Horne, A. Zeilinger, Multiparticle interferometry and the superposition principle. Phys. Today 22 (1993)
10. J.G. Rarity, P.R. Tapster, Experimental violation of Bell's inequality based on phase and momentum. Phys. Rev. Lett. **64**, 2495–98 (1990)
11. The one-particle distribution is introduced in E. Wigner. Phys. Rev. **40**, 749 (1932)

Part VI
Entanglement Features

Chapter 21
Analysing Multiparticle Quantum States

Otfried Gühne, Matthias Kleinmann and Tobias Moroder

Abstract The analysis of multiparticle quantum states is a central problem in quantum information processing. This task poses several challenges for experimenters and theoreticians. We give an overview over current problems and possible solutions concerning systematic errors of quantum devices, the reconstruction of quantum states, and the analysis of correlations and complexity in multiparticle density matrices.

Introduction

The analysis of quantum states is important for the advances in quantum optics and quantum information processing. Many experiments nowadays aim at the generation and observation of certain quantum states and quantum effects. For instance, in quantum simulation experiments thermal or ground states of certain spin models should be observed. Another typical problem is the demonstration of advanced quantum control by preparing certain highly entangled states using systems such as trapped ions, superconducting qubits, nitrogen-vacancy centers in diamond, or polarized photons.

All these experiments require a careful analysis in order to verify that the desired quantum phenomenon has indeed been observed. This analysis does not only concern the final data reported in the experiment but in fact, many more questions have to be considered in parallel. Did the experimenter align the measurement devices correctly? Have the count rates been evaluated properly in order to obtain the mean values of the measured observables? Such questions are relevant and, as we demonstrate below, ideas from theoretical physics can help the experimenters answer them.

O. Gühne (✉) · T. Moroder
Faculty IV for Science and Technology, University of Siegen,
57068 Siegen, Germany
e-mail: otfried.guehne@uni-siegen.de

M. Kleinmann
Department of Theoretical Physics, University of the Basque Country UPV/EHU,
48080 Bilbao, Spain

R. Bertlmann and A. Zeilinger (eds.), *Quantum [Un]Speakables II*,
The Frontiers Collection, DOI 10.1007/978-3-319-38987-5_21

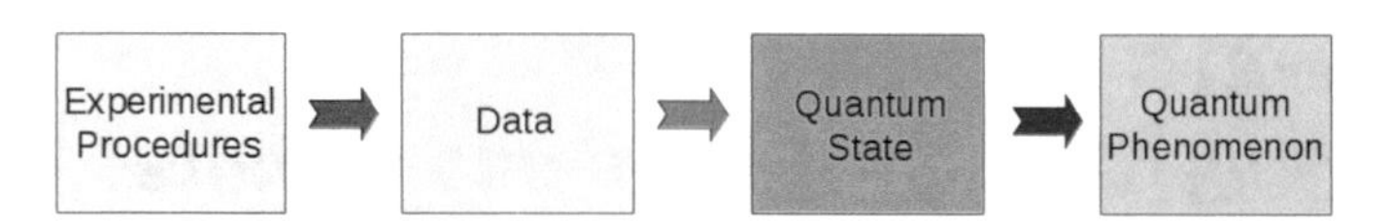

Fig. 21.1 The analysis of many experiments in quantum physics can be divided into several steps, from the experimental procedures to the verification of quantum mechanical properties of the generated states.

Many experiments in quantum optics can be divided in several steps (see also Fig. 21.1). In the beginning, some experimental procedures are carried out and measurements are taken. The results of the measurements are collected as data. These data are then processed to obtain a quantum state or density matrix ϱ, which is often viewed as the best description of the "actual state" generated in the experiment. This quantum state can then be further analysed, for instance, its entanglement properties may be determined.

In this article, we show how ideas from statistics and entanglement theory can be used for analysing the transitions between the four building blocks in Fig. 21.1. First, we consider the transition from the experimental procedures to the data. We show that applying statistical tests to the data can be used to recognize systematic errors in the experimental procedures, such as a misalignment of the measurement devices. Then, we consider the reconstruction of a quantum state from the experimental data. We explain why many frequently used state reconstruction schemes, such as the maximum-likelihood reconstruction, lead to a bias in the resulting state. This can, for instance, result in a fake detection of entanglement, meaning that the reconstructed state is entangled, while the original state, on which the measurements were carried out, was not entangled. We also show how such a bias can be avoided. Finally, we discuss the characterization of quantum states on a purely theoretical level. Assuming a multiparticle density matrix ϱ we show how its entanglement can be characterized and how the complexity of the state can be quantified using tools from information geometry and exponential families.

Systematic Errors in Quantum Experiments

In this first part of the article, we discuss what assumptions are typically used in quantum experiments. The violation of these assumptions leads to systematic errors and we show how these systematic errors can be identified using statistical methods and hypothesis tests.

Assumptions Underlying Quantum Experiments

Before explaining the assumptions, it is useful to discuss a simple example. Consider a two-photon experiment, where a quantum state should be analysed by performing state tomography. For that, Alice and Bob have to measure all the nine possible combinations of the Pauli matrices $\sigma_i \otimes \sigma_j$ for $i, j \in \{x, y, z\}$. In practice, this can be done as follows: Alice and Bob measure the three Pauli matrices $\sigma_x, \sigma_y, \sigma_z$ by measuring the polarization in different directions, getting the possible results $+$ and $-$. These results correspond to the projectors on the eigenvectors of the observable. By combining the results, they obtain one of four possible outcomes from the set $\{++, +-, -+, --\}$. The measurement is repeated N times on copies of the state, where the outcome $++$ occurs $N_{++|ij}$ times etc. From that, one can obtain the relative frequencies $F_{++|ij} = N_{++|ij}/N$ and estimate the expectation values as $\langle\sigma_i \otimes \sigma_j\rangle_{\exp} = (N_{++|ij} - N_{+-|ij} - N_{-+|ij} + N_{--|ij})/N$. In addition, the expectation values of the marginals $\sigma_i \otimes \sigma_0$ (here and in the following, we set $\sigma_0 = \mathbb{1}$) can be determined from the same data. Given all the experimental results, Alice and Bob may then reconstruct the quantum state via the formula

$$\hat{\varrho} = \frac{1}{4} \sum_{i,j \in \{0,x,y,z\}} \lambda_{ij} \sigma_i \otimes \sigma_j, \text{ where } \lambda_{ij} = \langle\sigma_i \otimes \sigma_j\rangle_{\exp}. \tag{1}$$

This simple quantum state reconstruction scheme is often called linear inversion. It assumes that the observed frequencies equal the probabilities, we will discuss its advantages and disadvantages below. For the moment, we just use it as an example to illustrate the definitions and discussion concerning systematic errors in experiments.

Now we can formulate the assumptions that lead to the statistical model typically used in quantum experiments. We consider a scenario where one actively chooses between different measurements (e.g., the $\sigma_i \otimes \sigma_j$), each having a finite number of results. We use the label s to denote the measurement setting and r to denote the result. It is important to note that, if in an experiment using the setting s one registers the result r, then this outcome $r|s$ is not just treated as a classical result. In addition, each outcome is tied to an operator $M_{r|s}$ (e.g., the projectors onto the eigenstates corresponding to the results $\{++, +-, -+, --\}$ of $\sigma_i \otimes \sigma_j$) that serves as the object to compute probabilities within quantum mechanics: If the underlying quantum state is characterized by the density operator ϱ, then the probability to observe $r|s$ is given by $P(r|s; \varrho) = \mathrm{tr}(\varrho M_{r|s})$. Therefore, this quantum mechanical description is one of the essential ingredients to connect the observed samples with the parameters of the system that one likes to infer. Knowledge about this description can come from previous calibration measurements or from other expertise that one has acquired with the equipment. But one thing should be obvious: If one assumes a description $M_{r|s}$, which deviates from the true description in the experiment $\tilde{M}_{r|s}$, then things can go terribly wrong and these type of errors are the ones that we like to address in the following.

Clearly, such deviations are presumably present in any model, but they are usually assumed to be small. However, considering the increased complexity of present experiments, one can ask the question, whether or not these deviations show up significantly in the data. Well known examples, like different detection efficiencies or dark-count rates in photo-detectors or non-perfect gate fidelities for single-qubit rotations preceding the readout of a trapped ion, could support this scepticism. However, these effects are hardly ever considered in the description of $M_{r|s}$.

Let us complete the list of assumptions. Most often each experiment of setting s is repeated N times, which are assumed to be independent and identically distributed trials. So one further assumes that one always prepares the same quantum state ϱ, measures the same observables $M_{r|s}$, and that both are completely independent.[1] Clearly, also in all these steps there can be errors, for instance, due to drifts in the measuring devices or dead-times in detectors coming from previous triggering events. However, if everything works as planned, then it is not necessary anymore to keep track of the individual measurement results, since every information that can be inferred about the state parameters is already included in the count rates $N_{r|s}$ of the individual measurement results $r|s$. Their probability is then given by a multinomial distribution $\mathrm{Mult}[N, P(r|s;\varrho)]$ for each setting, which is the distribution characterizing N repetitions of independent trials. Here, the single event probabilities $P(r|s;\varrho) = \mathrm{tr}(\varrho M_{r|s})$ are calculated according to quantum mechanics and these are the only parameters that depend on the quantum state.

Finally, the whole collection of distributions for all measurement settings is the exact parametric model used for most quantum experiments. These distributions are given by the set

$$\mathcal{P}_{\mathrm{QM}} = \left\{ P(\{N_{r|s}\}_{r,s};\varrho) = \prod_s \mathrm{Mult}\left[N, \mathrm{tr}(\varrho M_{r|s})\right], \text{ for all } \varrho \text{ with } \varrho \geq 0, \mathrm{tr}(\varrho) = 1 \right\}, \tag{2}$$

and the observed probabilities are assumed to be an element of this set. In the following, we discuss how the validity of this model can be tested.

Testing the Assumptions

How can one test in this framework whether the assumed measurement description is correct for the experiment? As a first try, we could intersperse the experiment with test measurements, in which one prepares previously characterized states. But such an option seems very cumbersome, independent of problems like how to characterize the test states in the first place and to ensure that they are well prepared in between

[1]This means that both, measurements and states are described by the corresponding N-fold tensor products. While such a property can be inferred for the states with the help of the de Finetti theorem [1], one should be aware that its exchangeability requirements do not apply to experiments where one actively measures first all the $s = 1$ measurements, followed by all $s = 2$ measurements and so on.

the true experiment. In contrast, we want to do it more directly and this becomes possible, at least partially, by exploiting that quantum states only allow a restricted set of event probabilities.

Let us first discuss the idea for the case where one has access to the true event probabilities $P_0(r|s)$ which can be attained from the relative frequencies $F_{r|s} = N_{r|s}/N \to P_0(r|s)$ in the limit $N \to \infty$. We want to know whether these observed probabilities are at all compatible with the assumed measurement description. This boils down to the question whether there exists a quantum state ϱ_0 with $P_0(r|s) = \text{tr}(\varrho_0 M_{r|s})$ for all r, s. Since quantum states must respect the positivity constraint $\varrho \geq 0$, not all possible probabilities are accessible: For instance, if one measures a qubit along the x, y, z directions, its corresponding probabilities will be constrained by the requirement that the Bloch vector must lie within the Bloch ball. To make this more general, assume that we have a certain set of numbers $w_{r|s}$ such that the observable $\sum_{r|s} w_{r|s} M_{r|s} \geq 0$ has no negative eigenvalues and is, therefore, positive semidefinite. If the probabilities $P_0(r|s)$ can indeed by realized by a quantum state, one has

$$w \cdot P_0 \equiv \sum_{r,s} w_{r|s} P_0(r|s) = \sum_{r,s} w_{r|s} \text{tr}(\varrho_0 M_{r|s}) = \text{tr}[\varrho_0 \big(\sum_{r,s} w_{r|s} M_{r|s} \big)] \geq 0, \tag{3}$$

where the inequality holds because both operators are positive semidefinite. Thus, if everything is correct one must get a non-negative value for $w \cdot P_0 \geq 0$. Consequently, whenever one observes $w \cdot P_0 < 0$, one knows that something must be wrong and that the description of the measurements $M_{r|s}$ has some flaws. This type of inequalities is similar in spirit to Bell inequalities for local hidden variable models or entanglement witnesses for separable states [2, 3]. Let us point out that the above inequalities are necessary and sufficient. So, indeed any $P_0(r|s)$ which cannot originate from a quantum state, can be detected by appropriately chosen coefficients $w_{r|s}$ by $w \cdot P_0 < 0$ [4]. Finally, we add that besides the positivity, some other constraints for the measurement description are conceivable. For instance, in the example of state tomography from above, the marginal $\langle \sigma_x \otimes \sigma_0 \rangle$ should not depend on whether it has been derived from the measurement $\sigma_x \otimes \sigma_x$ or $\sigma_x \otimes \sigma_y$. This can be formulated as a linear dependency of the form $\sum_{r,s} w_{r|s} M_{r|s} = 0$ and the corresponding constraint even becomes an equality $w \cdot P_0 = 0$.

Note that, with this test we ask the question whether the data $P_0(r|s)$ *fit at all* to the assumed measurement model $M_{r|s}$. But it should be clear that this approach can never serve as a proof that everything is correct in the experiment. For example, one can consider again the Bloch ball, where the measurement model assumes perfectly aligned measurements in the x, y, z directions, but in the true experiment one measures in slightly tilted directions which distorts the resulting Bloch ball. All states from this tilted Bloch ball, which lie outside the standard sphere, will be detected by the above method as being incompatible with the assumed model. For all other states, however, we do not see the difference because they are still consistent with the model.

Finally, let us address the point that we only collect count rates in the experiment. Since the relative frequencies $F_{r|s}$ are only approximations to the true probabilities $P(r|s)$, it is clear that a similar inequality as Eq. (3) does not need to hold anymore for $w \cdot F$, even if everything is correct. One would expect, however, that larger negative values are much less likely. This is indeed the case and is made more quantitative via Hoeffding's inequality [5].

This inequality states the following: Consider N independent, not necessarily identically distributed, bounded random variables $X_i \in [a_i, b_i]$. Then the sample mean $\bar{X} = \sum_i X_i/N$ satisfies

$$\text{Prob}[\bar{X} - \mathbb{E}(\bar{X}) \leq -t] \leq \exp\left(\frac{-2t^2N^2}{\sum_i (b_i - a_i)^2}\right) \tag{4}$$

for all $t > 0$, where $\mathbb{E}(\bar{X})$ denotes the expectation value of $\bar{X}$. In practice, the main statement of this inequality is that for N independent repetitions of an experiment, the probability of deviations from the mean value by a difference t scales like $\exp(-t^2N)$. It is important to stress that this result uses no extra assumptions, like N being large, at all.

For our case, we can use Hoeffding's inequality to bound the probability of observing data that violate positivity constraints as in Eq. (3). More precisely, we can derive the following statement [6]: For all distributions compatible with quantum mechanics, the probability to observe frequencies $\{F_{r|s}\}_{r,s}$ such that $w \cdot F < -\varepsilon$ is bounded by

$$\text{Prob}_P\big[w \cdot F < -\varepsilon\big] \equiv \sum_{\substack{\{n_{r|s}\}_{r,s}:\\ w\cdot F<-\varepsilon}} P(\{n_{r|s}\}_{r,s}) \leq \exp\left(-\frac{2\varepsilon^2 N}{C_w^2}\right), \text{ for all } P \in \mathcal{P}_{QM} \tag{5}$$

with $C_w^2 = \sum_s (w_{\max|s} - w_{\min|s})^2$ and $w_{\max|s}$, $w_{\min|s}$ being the extreme values of $\{w_{r|s}\}_r$. Again, this can be interpreted as showing that if everything is correct, then the probability of finding a violation of the positivity constraint is exponentially suppressed.

We can use this statement as follows: Suppose that we should reach a conclusion whether the observed data are "compatible" or "incompatible" with our assumed model. Of course, if we say "incompatible", we do not want to reach this conclusion too often, if indeed everything is perfect. For definiteness, we may assume that the probability of claiming incompatibility if everything is correct should be at maximum $\alpha = 1\,\%$. We then use Eq. (5) to deduce the threshold value that we need to beat, $\varepsilon_\alpha = \sqrt{C_w^2|\log(\alpha)|/2N}$. If we now carry out the experiment and register click rates with $w \cdot F < -\varepsilon_\alpha$, we know that there was at most a $\alpha = 1\,\%$ chance that we would have registered such badly looking data, if everything is correct. Since this would be really bad luck we would rather say "incompatible", and assume that some systematic error was present.[2]

In practice, this test can be used to detect systematic errors in various scenarios: In ion trap experiments, a typical systematic error comes from the cross talk between the ions, i.e. the fact that a laser focused on one ion also influences the neighbouring ions. This phenomenon can be detected with the presented method [6]. The second application are Bell experiments: In these experiments, the choice of the measurements on one party should ideally not influence the results of the other party and a violation of this condition completely invalidates the result of a Bell test. Again, this non-signalling condition can be formulated as linear constraints on the probabilities and this can be tested with the presented method. In all these applications, the determination of the vector w characterizing the positivity constraint or the linear constraint can be done as follows: One splits the observed data into two parts. From the first part one determines the w leading to the maximal violation of the respective constraint for the first half. Then, one applies this w as a test to the second part of the data. If the violations of the constraint are only due to statistical fluctuations, the respective w for the two parts of the data are uncorrelated and the test will not find a significant violation of the constraint.

Let us point out that the mathematical framework just described is called a hypothesis test [7], in which one tests the null-hypothesis N_0: "compatible", against the alternative A: "incompatible". The special property of such a test is that there is an asymmetry about the two types of errors that can occur. As already explained, our concern is that, when saying "incompatible", then this statement is more or less correct. The other error can occur when we respond "compatible" to incompatible data. Naturally, this error characterizing the detection strength of our test, ideally, should also be made small. However, it is not possible to reduce both errors equally simultaneously. Nevertheless, since we cannot detect all possible systematic deviations from the assumed model, anyway, one should not be too euphoric about the statement "compatible" in this sense.

Note that, while the presented test has been build up by first deriving specific inequalities for event probabilities and then equipping it with the necessary statistical rigour to arrive at an hypothesis test, one can also take the other direction, by using techniques which are known to be good for hypothesis tests and apply them to the special statistical model of the quantum experiments. We have done this for the so-called generalized likelihood-ratio test [7] and details can be found in Ref. [6]. Finally, other tests for systematic errors can be found in Refs. [8–10].

---

[2]Since one typically likes to leave the choice of appropriate levels of α to the reader one can also report the p-value [7] of the observed data: It is the smallest α with which we would have still said "incompatible" with the test.

Performing State Tomography

In the previous section, we have seen that care has to be taken when making the measurements on the quantum system. In this section, we show that the interpretation of tomographic data, such as the reconstruction of the quantum state, has to be done with care, too. Otherwise, one introduces yet another class of systematic errors.

Problems with State Estimates

We are used to summarize experimental data by an estimate together with an error margin. In quantum state tomography this corresponds to an estimate for the density matrix together with an error region. So, the first question is how one can obtain an estimate $\hat{\varrho}$ for the experimentally prepared density matrix ϱ_{exp} from the observed frequencies $F_{r|s}$. The simplest approach is to use linear inversion, that is, the method given in Eq. (1). This has, however, at first sight some disadvantages: Due to statistical fluctuations the observed frequencies are not equal to the true probabilities and this leads to the consequence that the reconstructed "density matrix" will typically have some negative eigenvalues. This makes the further analysis of the experiment, e.g. the evaluation of entanglement measures, not straightforward. In order to circumvent this, one often makes a density matrix reconstruction by setting

$$\hat{\varrho} = \arg\max_{\sigma \geq 0} \mathcal{F}(F_{r|s}, \sigma). \tag{6}$$

Here, one optimizes a target function $\mathcal{F}$ over all density matrices σ and the optimal σ will obviously be a valid density matrix. Examples for this type of state reconstruction are the maximum-likelihood reconstruction or the least-squares reconstruction, both are frequently used for experiments in quantum optics.

An important property of such an estimator is the question whether it is biased or unbiased. This means the following: The underlying state ϱ_{exp} leads via the multinomial distribution to a probability distribution over the frequencies $F_{r|s}$. The estimator $\hat{\varrho}$ is a function from the observed data (the frequencies $F_{r|s}$) to the state space. In this way, the original state ϱ_{exp} induces a probability distribution over the estimators $\hat{\varrho}$, and one can ask whether the expectation value of this equals the original state, $\mathbb{E}[\hat{\varrho}] \stackrel{?}{=} \varrho_{\text{exp}}$. If this is the case, the estimator is unbiased, otherwise it is biased. It must be stressed, however, that biased estimators are not necessarily useless or bad, as it all depends on the purpose the estimator is used for.

For quantum state reconstruction one can prove the following: Any state reconstruction scheme that yields a density matrix from experimental data will be biased, i.e., on average, the reconstructed state $\hat{\varrho}$ will not be the state used in the experiment, $\mathbb{E}[\hat{\varrho}] \neq \varrho_{\text{exp}}$. A proof of this statement was given in Ref. [11], but the following example demonstrates that the problem in finding an unbiased estimator comes from the fact that the quantum mechanical state space is bounded by the positivity constraint.

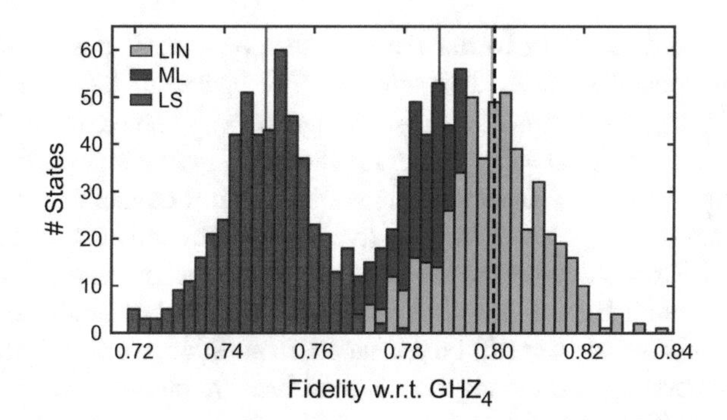

Fig. 21.2 Demonstration of the bias for different state estimators. A given state ϱ_{exp} having 80 % fidelity with the GHZ state was used to sample the distribution of the estimator $\hat{\varrho}$ in state space, and from these samples the fidelities with the GHZ state were computed. The maximum-likelihood (ML) and least-square (LS) estimators clearly underestimate the fidelity, while the linear inversion (LIN) is unbiased. The figure is taken from Ref. [11].

Consider a coin toss where we are interested in the modulus of the difference between the probability of obtaining heads or tails, $\Delta = |p_h - p_t| = 1 - 2\min\{p_h, 1 - p_h\}$. This quantity cannot be negative, so also an estimator should not be negative. Let us assume that an estimator $\hat{\Delta}$ is unbiased, $\mathbb{E}[\hat{\Delta}] = \Delta_{\text{exp}}$. Then, for any experimental data that could come from a fair coin ($\Delta_{\text{exp}} = 0$) we cannot have $\hat{\Delta} > 0$ since this would imply $\mathbb{E}[\hat{\Delta}] = \sum_k (1/2)^n \binom{n}{k} \hat{\Delta}(k) > 0$, where k denotes the number of occurrences of heads. On the other hand, any possible number of heads and tails is compatible with a fair coin. So, the estimate $\hat{\Delta}$ for *any* data must be 0. Then in particular $\mathbb{E}[\hat{\Delta}] = 0$, which means that $\hat{\Delta}$ is a biased estimator whenever the coin is *not* fair.

Apart from this theoretical argument, the question arises whether this effect plays a significant role in practical quantum state reconstruction. Unfortunately, this is the case and this effect can causes substantial fidelity underestimation or spurious entanglement detection in realistic scenarios [11]. This problem applies to the established schemes for reconstructing a density matrix, in particular to the maximum-likelihood method [12] and the constrained least-squares method [13]. So, how large is the bias? For example, in a tomography of a four-qubit GHZ state with fidelity 0.8, when reconstructing from a total number of 8100 samples, the state from a maximum-likelihood estimate has a fidelity of 0.788 ± 0.010 [11], i.e., the fidelity is systematically underestimated (see also Fig. 21.2). Such an underestimation may be considered to be unfortunate, but acceptable. However, it was also demonstrated that maximum-likelihood and least-square methods tend to *overestimate* the entanglement. In fact, for a clearly separable state the reconstructed states can be always entangled, thus leading to spurious entanglement detection [11]. This is not acceptable for many experiments.

A way to avoid the bias is to accept that the reconstructed density matrix is not always a valid quantum state and can have negative eigenvalues. The simplest unbiased method is linear inversion explained above. More generally, if $M_{r|s}$ are the operators corresponding to the measurement outcomes in a complete tomographic measurement scheme, then one can find operators $X_{r|s}$ such that for all states $\varrho = \sum_{r|s} X_{r|s} P(r|s; \varrho)$ holds, generalizing Eq. (1).[3] The estimate given by linear reconstruction is then $\hat{\varrho} = \sum_{r|s} X_{r|s} F_{r|s}$, where $F_{r|s}$ are the relative frequencies of the result r for setting s. This estimate is unbiased but it comes with the price that in all realistic scenarios $\hat{\varrho}$ has some negative eigenvalues and hence it is not a valid density matrix. Depending on the intended use of the reconstructed density matrix this may be problematic, but it was shown in Ref. [11] that entanglement measures or the Fisher information can still be estimated. In addition, we stress that the eigenvectors corresponding to these negative eigenvalues are randomly distributed in the following sense: If we choose a rank one projection $|\alpha\rangle\langle\alpha|$ independently of the data, then the probability that $\mathrm{tr}(\hat{\varrho}|\alpha\rangle\langle\alpha|) < -\epsilon$ is exponentially suppressed, as can be seen from the inequality in Eq. (3).

Problems with Error Regions

Any report of an experiment has to equip the reported results with error bars. In the case of a density matrix, this will be a high-dimensional error region. When specifying an error region, one first has to decide between the Bayesian framework and the frequentistic framework. A Bayesian analysis gives a credible region, which has the property that with high probability the actual state is in this region. A frequentist's analysis gives a confidence region, which is a map from the data to a region in state space such that with high probability the region contains the actual state. There is a long debate in mathematical statistics which method is appropriate, but most of the subsequent discussion is independent of this dispute.

Before discussing the advantages and disadvantages of an error region, it is important to remember, that the variance does in general not give an appropriate error region. This occurs in particular if the underlying distributions are far from being Gaussian. But for state tomography, the data is sampled from a multinomial distribution, typically with a very low number of events. Indeed, in many experiments the number of clicks per measurement outcome is about ten, but sometimes even below one. Also the method of bootstrapping may yield an inappropriate error region. In bootstrapping, one uses an estimate $\hat{\varrho}$ for the state (parametric bootstrapping) or the empirical distributions of the outcomes of the measurements $F_{r|s}$ (non-parametric bootstrapping) in order to estimate the variance of the estimate. This estimate is usually obtained by Monte Carlo sampling from the corresponding distributions. There is no particular reason that this should be a good error region, and it was also demonstrated that the most commonly used schemes yield invalid error regions.

Methods to obtain valid error regions both in the Bayesian [14] and in the frequentistic framework [15] have been suggested, however, they turn out to be notoriously difficult to compute. But even when it is possible to achieve a proper error region,

---

[3]The new operators $X_{r|s}$ may be necessary, since the $M_{r|s}$ can be overcomplete or not orthogonal.

one has to keep in mind that the size of the error region scales with the dimension of the underlying Hilbert space, i.e., exponentially with the number of qubits. This makes it very difficult to perform state tomography of a large system with a reasonable sized error region. Fortunately, in many situations the error region for the state is not of uttermost importance. Often one is only interested in certain scalar quantities like a measure of entanglement or the fidelity with the (pure) target state. In this cases it is possible to infer an appropriate confidence region directly from the data, without taking the detour over an error region for the density operator. This is particularly simple, if the quantity of interest is linear in the density matrix, e.g., the fidelity with a pure state $F = \langle\psi|\varrho|\psi\rangle$. One can again use Hoeffding's tail inequality in order to obtain a lower bound $\hat{F}_l$ on the fidelity. The promise is then that $P(\hat{F}_l > \langle\psi|\varrho_{\text{exp}}|\psi\rangle) < 1\,\%$ for any state ϱ_{exp}. A general method to provide such confidence regions for convex functions, like the bipartite negativity or the quantum Fisher information, has been introduced in Ref. [11].

Analysing Density Matrices

In the last section of this article, we assume that a valid multiparticle density matrix ϱ is given and the task is to analyse its properties. Naturally, many questions can be asked about a density matrix, but we concentrate on two of them. First, we consider the question whether the state is genuinely multiparticle entangled or not. We explain a powerful approach for characterizing multiparticle entanglement with the help of so-called PPT mixtures and semidefinite programming. Second, we consider the problem of characterizing the complexity of a given quantum state and explain an approach using exponential families. For example, in this approach a state that is a thermal state of a Hamiltonian with two-body interactions only, is considered to be of low complexity and the distance to these thermal states can be considered as a measure of complexity. The underlying techniques also allow to characterize pure states which are not ground states of a two-body Hamiltonian.

Characterizing Entanglement with PPT Mixtures

Notions of Entanglement

Before explaining the characterization of multiparticle entanglement, we have to explain some basic facts about entanglement on a two-particle system. The definition of entanglement is based on the notion of local operations and classical communication (LOCC). If a quantum state can be prepared by LOCC, it is called separable, otherwise it is entangled. For pure states, this just means that product states of the form $|\phi\rangle = |\alpha\rangle \otimes |\beta\rangle$ are separable and all other states (e.g. the singlet state

$|\psi^-\rangle = (|01\rangle - |10\rangle)/\sqrt{2})$ are entangled. If mixed states are considered, a density matrix ϱ is separable, if it can be written as a convex combination of product states,

$$\varrho = \sum_k p_k \varrho_A^k \otimes \varrho_B^k, \tag{7}$$

where the p_k form a probability distribution, so they are non-negative and sum up to one. Physically, the convex combination means that Alice and Bob can prepare the global state by fixing the joint probabilities with classical communication and then preparing the states ϱ_k^A and ϱ_k^B separately. The question whether or not a given quantum state is entangled is, however, in general difficult to answer. This is the so-called separability problem [2, 3].

Many separability criteria have been proposed, but none of them delivers a complete solution of the problem. The most famous separability test is the criterion of the positivity of the partial transpose (PPT criterion) [16]. For that, one considers the partial transposition of a density matrix $\varrho = \sum_{ij,kl} \varrho_{ij,kl}|i\rangle\langle j| \otimes |k\rangle\langle l|$, given by

$$\varrho^{T_A} = \sum_{ij,kl} \varrho_{ij,kl}|j\rangle\langle i| \otimes |k\rangle\langle l|. \tag{8}$$

In an analogous manner, one can also define the partial transposition ϱ^{T_B} with respect to the second system. The PPT criterion states that for any separable state ϱ the partial transpose ϱ^{T_A}, (and consequently also $\varrho^{T_B} = (\varrho^{T_A})^T$) has no negative eigenvalues and is therefore positive semidefinite. So, if one finds a negative eigenvalue of ϱ^{T_A}, then the state ϱ must necessarily be entangled. The PPT criterion solves the separability problem for low dimensional systems (that is, two qubits or one qubit and one qutrit) [2], but in all other cases the set of separable states is a strict subset of the PPT states. The entangled states which are PPT are of great theoretical interest: It has been shown that their entanglement can never be distilled to pure state entanglement, even if many copies of the state are available. This weak form of entanglement is then also called *bound entanglement* and bound entangled states are central for many challenging questions in quantum information theory.

The characterization of entanglement becomes significantly more complicated, if more than two particles are involved. Let us consider three particles (A, B, C). First, a state can be fully separable, meaning that it does not contain any entanglement and is of the form $|\phi^{\text{fs}}\rangle = |\alpha\rangle \otimes |\beta\rangle \otimes |\gamma\rangle$. If a state is entangled, one can further ask whether only two parties are entangled or all three parties. For instance, in the state $|\phi^{\text{bs}}\rangle = |\psi^-\rangle_{AB} \otimes |\gamma\rangle_C$ the parties A and B are entangled, but C is not entangled with A or B, therefore the state is called biseparable. Alternatively, if all parties are entangled with each other, the state is called genuine multipartite entangled [3]. For the simplest case of three two-level systems (qubits) it has been shown that even the genuine multipartite entangled states can be divided into two subclasses, represented by the GHZ state $|GHZ\rangle = (|000\rangle + |111\rangle)/\sqrt{2}$ and the W state $|W\rangle = (|001\rangle + |010\rangle + |100\rangle)/\sqrt{3}$. These subclasses are distinguished by the fact that a single copy of a state in one class cannot be converted via LOCC into a state in the other class, even if this transformation is not required to be performed with probability one [3].

The classification of entanglement for pure states can be extended to mixed states by considering convex combinations as in Eq. (7). First, a mixed state is fully separable, if it can be written as a convex combination of fully separable states

$$\varrho^{\mathrm{fs}} = \sum_k p_k \varrho_A^k \otimes \varrho_B^k \otimes \varrho_C^k, \tag{9}$$

and a state is biseparable, if it can be written as a mixture of biseparable states, which might be separable with respect to different partitions,

$$\varrho^{\mathrm{bisep}} = p_1 \varrho_{A|BC}^{\mathrm{sep}} + p_2 \varrho_{B|AC}^{\mathrm{sep}} + p_3 \varrho_{C|AB}^{\mathrm{sep}}. \tag{10}$$

The different notions of entanglement in the multipartite case and the different bipartitions that have to be taken into account imply that the question whether a given mixed multipartite state is entangled or not is extraordinarily complicated.

The Approach of PPT Mixtures

A systematic approach for characterizing genuine multiparticle entanglement makes use of so-called PPT mixtures [17]. Instead of asking whether a state is a mixture of separable states with respect to different partitions as in Eq. (10), one asks whether it is a mixture of states which are PPT for the bipartitions

$$\varrho^{\mathrm{pptmix}} = p_1 \varrho_{A|BC}^{\mathrm{ppt}} + p_2 \varrho_{B|AC}^{\mathrm{ppt}} + p_3 \varrho_{C|AB}^{\mathrm{ppt}}. \tag{11}$$

Since the separable states are a subset of the PPT states, any biseparable state is also a PPT mixture. This means that if a state is no PPT mixture, then it must be genuine multipartite entangled (see also Fig. 21.3).

At first, it is not clear what can be gained by this redefinition of the problem. First, the condition for PPT mixtures is a relaxation of the definition of biseparability and it might be that the conditions are relaxed too much, implying that not many states can be detected by this method. Second, it is not clear how the criterion for PPT mixtures can be evaluated in practice and whether this is easier than evaluating the conditions for separability directly. In the following, however, we will see that the question whether a state is a PPT mixture or not can directly be checked with a technique called semidefinite programming. Furthermore, the approximation to the biseparable states is rather tight, and for many families of states the property of being a PPT mixture coincides with the property of being biseparable.

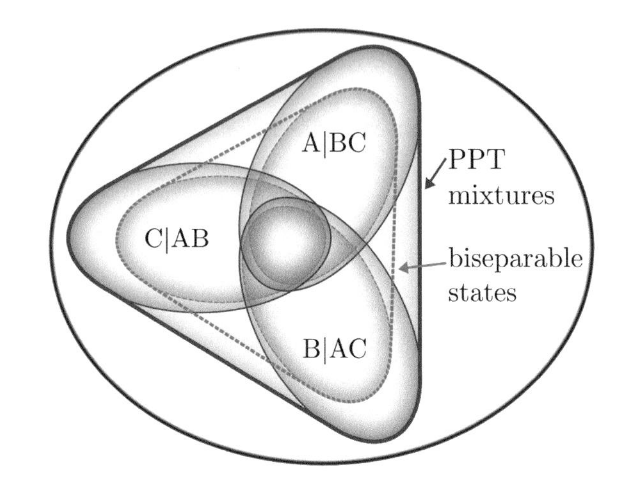

Fig. 21.3 Schematic view of the states which are PPT mixtures and the biseparable states for three particles. There are three possible bipartitions, and the corresponding sets of states which are separable or PPT for the bipartition. The figure is taken from Ref. [17].

Evaluation of the Criterion

Let us discuss the evaluation of the condition for PPT mixtures. For that, we need to introduce the notion of entanglement witnesses. In the two-particle case, an entanglement witness $\mathcal{W}$ is an observable with the property that the expectation value is positive for all separable states, $\text{tr}(\varrho^{\text{sep}}\mathcal{W}) \geq 0$. This implies that a measured negative expectation value signals the presence of entanglement. In this way, the concept of an entanglement witness bears some similarity to a Bell inequality, where correlations are bounded for classical states admitting a local hidden variable model, while entangled states may violate the bound.

How can entanglement witnesses be constructed? For the two-particle case a simple method goes as follows: Consider an observable of the form

$$\mathcal{W} = P + Q^{T_A}, \tag{12}$$

where $P \geq 0$ and $Q \geq 0$ are positive semidefinite operators. Using the fact that $Tr(XY^{T_A}) = Tr(X^{T_A}Y)$ for arbitrary operators X, Y, we find that for a separable state $Tr(\mathcal{W}\varrho^{\text{sep}}) = Tr(P\varrho^{\text{sep}}) + Tr(Q(\varrho^{\text{sep}})^{T_A}) \geq 0$, since ϱ^{sep} has to be PPT. Therefore, the observable $\mathcal{W}$ is an entanglement witness, which may be used to detect the entanglement in states that violate the PPT criterion.

This construction can be used to decide whether a given three-particle state is a PPT mixture or not. For that, consider the optimization problem

$$\begin{aligned} \underset{\mathcal{W},P_i,Q_i}{\text{minimize}} \quad & Tr(\varrho\mathcal{W}) \\ \text{subject to:} \quad & \mathcal{W} = P_1 + Q_1^{T_A} = P_2 + Q_2^{T_B} = P_3 + Q_3^{T_C} \text{ and} \\ & P_i \geq 0 \text{ for } i = 1, 2, 3 \text{ and} \\ & \mathbb{1} \geq Q_i \geq 0 \text{ for } i = 1, 2, 3. \end{aligned} \tag{13}$$

The constraints guarantee that the observable $\mathcal{W}$ is of the form as in Eq. (12) for any of the three bipartitions. This means, that if a state is a PPT mixture as in Eq. (11), the expectation value $Tr(\varrho\mathcal{W})$ has to be non-negative. On the other hand, one can show that if a state is not a PPT mixture, then the minimization problem will always result in a strictly negative value [17]. In this way, the question whether a state is a PPT mixture or not, can be transformed into a optimization problem under certain constraints.

The point is that the optimization problem belongs to the class of semidefinite programs (SDP). An SDP is an optimization problem of the type

$$\begin{aligned} \underset{x_i}{\text{minimize}} \quad & \sum_i c_i x_i \\ \text{subject to:} \quad & F_0 + \sum_i x_i F_i \geq 0, \end{aligned} \tag{14}$$

where the c_i are real coefficients defining the target function, the F_i are hermitean matrices defining the constraints and the x_i are real coefficients which are varied. This type of optimization problem has two important features [18]. First, using the so-called dual problem one can derive a lower bound on the solution of the minimization, which equals the exact value under weak conditions. This means that the optimality of a solution found numerically can be demonstrated. In this way, one can prove rigorously by computer whether a given state is a PPT mixture or not. Second, for implementing an SDP in practice there are ready-to-use computer algebra packages available and therefore the practical solution of the SDP is straightforward.

Results

Concerning the characterization of PPT mixtures, the following results have been obtained:

- First, the practical evaluation of the SDP in Eq. (13) can be carried out easily with standard numerical routines. A free ready-to-use package called `PPTMixer` is available online [19], and it solves the problem for up to six qubits on standard computers. For a larger number of particles, the numerical evaluation becomes difficult, but analytical approaches are also feasible [17, 20].
- For many families of states, the approach of the PPT mixtures delivers the strongest criterion of entanglement known so far. For many cases it even solves the problem of characterizing multiparticle entanglement. For instance, three-qubit permutation-invariant states are biseparable, if and only if they are PPT mixtures [21]. The same holds for states with certain symmetries, like GHZ diagonal states or four-qubit states diagonal in the graph-state basis [20, 22].
- Nevertheless, the approach of PPT mixtures can not detect all multiparticle entangled states. There are examples of genuinely entangled three-qubit and three-qutrit states, which are PPT mixtures [23, 24]. For an increasing dimension and number

of particles one can even show that the probability that a given multiparticle entangled state can be detected by the PPT mixture approach decreases [25]. This finding, however, is in line with the observation that also in bipartite high-dimensional systems no single entanglement criterion detects a large fraction of states [26].
- The value $\mathcal{N}(\varrho) = -Tr(\varrho\mathcal{W})$, that is, the amount of violation of the witness condition is a computable entanglement monotone for genuine multiparticle entanglement [20]. It can be called the genuine multiparticle negativity, as it generalizes the entanglement measure of bipartite negativity.
- An interesting feature of the PPT mixer approach is that it can also be evaluated, if only partial information on the state ϱ is available. Namely, if only the expectation values $\langle A_i \rangle$ of some observables A_i are known, one can add in the SDP in Eq. (13) that the witness $\mathcal{W}$ should be a linear combination of the measured observables $\mathcal{W} = \sum_i \lambda_i A_i$. It can be shown that this is then still a complete solution of the problem, meaning that the SDP returns a negative value, if and only if all states that are compatible with the data $\langle A_i \rangle$ are not PPT mixtures.

Characterizing the Complexity of Quantum States

Besides the question whether a given multiparticle quantum state is entangled or not, one may also be interested in other questions about a reconstructed quantum state ϱ. For instance, one may ask: Is the given state is a ground state or thermal state of a simple Hamiltonian? In the following, we will explain how this question can be used to characterize the complexity of a many-body quantum state.

Exponential Families

First, one can consider the set of all possible two-body Hamiltonians. For multi-qubit systems they are of the form

$$H_2 = \sum_{i,\alpha} \lambda_\alpha^{(i)} \sigma_\alpha^{(i)} + \sum_{i,j,\alpha,\beta} \mu_{\alpha\beta}^{(ij)} \sigma_\alpha^{(i)} \sigma_\beta^{(j)} + \nu \mathbb{1}, \tag{15}$$

where $\sigma_\alpha^{(i)}$ is the Pauli matrix σ_α acting on the ith qubit. This Hamiltonian H_2 contains, apart from the identity, single-particle terms and two-particle interactions. However, no geometrical arrangement of the particles is assumed and the two-particle interactions are between arbitrary particles and not restricted to nearest-neighbour interactions. We also denote the set of all two-particle Hamiltonians by $\mathcal{H}_2$, and in a similar manner one can define the sets of k-particle Hamiltonians $\mathcal{H}_k$.

Given the set of k-particle Hamiltonians, we can define the so-called exponential family of all thermal states

$$\mathcal{Q}_k = \{\exp\{H_k\} \text{ with } H_k \in \mathcal{H}_k\}, \tag{16}$$

where the normalization of the state has been included into the Hamiltonian via the term $\nu\mathbb{1}$.

If a given quantum state is in the family $\mathcal{Q}_k$ for small k, then one can consider it to be less complex, since only a simple Hamiltonian with few parameters are required to describe the interaction structure. One the other hand, if a state is not in the exponential family, one can consider the distance

$$D_k(\varrho||\mathcal{Q}_k) := \inf_{\eta \in \mathcal{Q}_k} D(\varrho||\eta) \tag{17}$$

with $D(\varrho||\eta) = \mathrm{tr}[\varrho \log(\varrho)] - \mathrm{tr}[\varrho \log(\eta)]$ being the relative entropy, as a measure of the complexity of the quantum state. The optimal η is also called the information projection $\tilde{\varrho}_k$ and one can show that this $\tilde{\varrho}_k$ is the maximum likelihood approximation of ϱ within the family $\mathcal{Q}_k$ [27]. Below, we will explain several further equivalent characterizations which can help to solve the underlying minimization problem.

This type of complexity measure has been first discussed for the case of classical probability distributions in the context of information geometry [28]. The measure D_1 is also known as the multi-information in complexity theory [29]. For classical complex systems, these quantities have been used to study the onset of synchronization and chaos in coupled maps or cellular automata [27]. For the quantum case, this measure and its properties have been discussed in several recent works [30–33].

At this point, it is important to note that in the quantum case as well as in the classical case the quantity D_k does not necessarily decrease under local operations [31, 32]. Simple examples for this fact follow from observation that taking a thermal state of a two-body Hamiltonian and tracing out one particle typically leads to a state that is not a thermal state of a two-body Hamiltonian anymore. Therefore, the quantity D_k should not be considered as a measure of correlations in the quantum state, it is more appropriate to consider it as a measure of the complexity of the state.

Characterizing the Approximation

For the characterization of the information projection $\tilde{\varrho}_k$, the following result is quite helpful [31]. First, let ϱ be an arbitrary quantum state, and $\tilde{\varrho}_k$ be the information projection onto the exponential family $\mathcal{Q}_k$. Furthermore, let M_k be the set of all quantum states that have the same k-body marginals as ϱ. M_k is, contrary to $\mathcal{Q}_k$, a linear subspace of the space of all density matrices (see Fig. 21.4). Then, the following statements are equivalent:

(a) The state $\tilde{\varrho}_k$ is the closest state to ϱ in $\mathcal{Q}_k$ with respect to the relative entropy.
(b) The state $\tilde{\varrho}_k$ has the maximal entropy among all states in M_k.
(c) The state $\tilde{\varrho}_k$ is the intersection $\mathcal{Q}_k \cap M_k$.

This equivalence can be used for many purposes. For example, it is useful for developing an algorithm for computing the information projection [33, 34]. Instead of minimizing the relative entropy as a highly nonlinear function over $\mathcal{Q}_k$, one can do

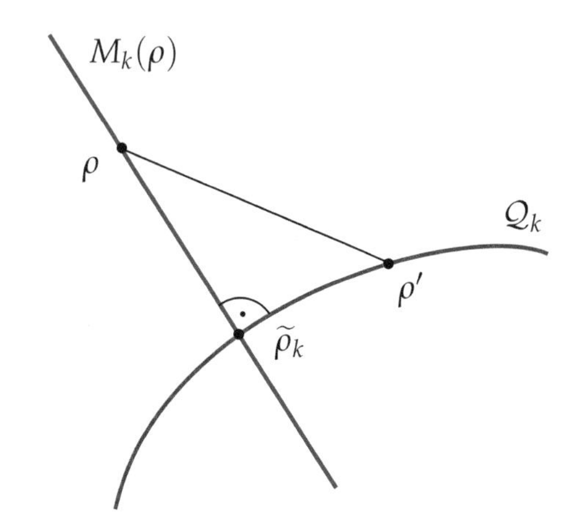

Fig. 21.4 The information projection $\tilde{\varrho}_k$ of the state ϱ is the closest state to ϱ within the exponential family $\mathcal{Q}_k$. M_k denotes the set of all quantum states that have the same k-body marginals as ϱ, and can also be used to characterize $\tilde{\varrho}_k$. For arbitrary states ϱ' within $\mathcal{Q}_k$ the relation $D(\varrho||\varrho') = D(\varrho||\tilde{\varrho}_k) + D(\tilde{\varrho}_k||\varrho')$ holds, which resembles the Pythagorean Theorem. The figure is taken from Ref. [33].

the following: One optimizes over all states in $\mathcal{Q}_k$ with the aim to make the k-body marginals the same as for the state ϱ. The resulting algorithm converges well and allows the computation of the complexity measure D_k for up to six qubits [33].

Second, from the equivalences it follows that the multi-information D_1 can directly be calculated, since the closest state to ϱ in the family $\mathcal{Q}_1$ is the product state $\tilde{\varrho}_1 = \varrho_1 \otimes \varrho_2 \otimes ... \otimes \varrho_N$ built out of the reduced single-particle density matrices of ϱ. Clearly, $\tilde{\varrho}_1$ has the same marginals as ϱ and maximizes the entropy.

A Five-Qubit Example

As a final example, let us discuss how the notion of exponential families can help to characterize ground states of two-body Hamiltonians. For that, consider the five-qubit ring-cluster state $|R_5\rangle$. This state is defined to be the unique eigenstate fulfilling

$$|R_5\rangle = g_i|R_5\rangle, \tag{18}$$

where $g_1 = \sigma_x\sigma_z\mathbb{1}\mathbb{1}\sigma_z, g_2 = \sigma_z\sigma_x\sigma_z\mathbb{1}\mathbb{1}, g_3 = \mathbb{1}\sigma_z\sigma_x\sigma_z\mathbb{1}, g_4 = \mathbb{1}\mathbb{1}\sigma_z\sigma_x\sigma_z$, and $g_5 = \sigma_z\mathbb{1}\mathbb{1}\sigma_z\sigma_x$. Here, the tensor product symbols have been omitted. After appropriate local transformations, the ring-cluster state can also be written as

$$|R_5\rangle = \frac{1}{\sqrt{8}}\big[|00000\rangle + |00110\rangle - |01011\rangle + |01101\rangle + |10001\rangle - |10111\rangle + |11010\rangle + |11100\rangle\big]. \tag{19}$$

The ring-cluster state is an example of a so-called graph state, and plays an important role in quantum error correction as a codeword of the five-qubit Shor code. It was known before that the state $|R_5\rangle$ cannot be the unique ground state of a two-body Hamiltonian [35]. This, however, leaves the question open whether it can be approximated by ground states of two-body Hamiltonians. For instance, for three qubits it was shown that not all pure states are ground states of two-body Hamiltonians, but all pure states can be approximated arbitrarily well by such ground states [36].

The characterization of the exponential families from the previous section can indeed help to prove that the state has $|R_5\rangle$ has finite distance to all thermal states of two-body Hamiltonians. For that, first note that the two-body marginals of the state $|R_5\rangle\langle R_5|$ are all maximally mixed two-qubit states. Then, one can directly find states which have the same two-body marginals, but their entropy is larger than the entropy the state $|R_5\rangle\langle R_5|$. This last property is, of course, not surprising, since $|R_5\rangle\langle R_5|$ has as a pure state the minimal possible entropy. According to the previous section, this already implies that $|R_5\rangle\langle R_5|$ cannot be the thermal or ground state of any two-body Hamiltonian.

Furthermore, if an arbitrary state ϱ has a high fidelity with $|R_5\rangle$ then the two-body marginals will be close to the maximally mixed states, and in addition the entropy of ϱ will be small. This implies that one can find again states with the same marginals and higher entropy. Using these ideas and some detailed calculations one can prove that if a state fulfils

$$F = \langle R_5|\varrho|R_5\rangle \geq \frac{31}{32} \approx 0.96875 \tag{20}$$

then it cannot be a thermal state of a two-body Hamiltonian [37]. This shows that the state $|R_5\rangle$ cannot be approximated by thermal states of two-body Hamiltonians. In principle, this bound can also be used to prove experimentally that a given state is not a thermal state of a two-body Hamiltonian.

Conclusion

In conclusion we have explained several problems occurring in the analysis of multiparticle quantum states, ranging from systematic errors of the measurement devices to the characterization of ground states of two-body Hamiltonians. We believe that several of the explained topics are important to be addressed in the future. First, since the current experiments in quantum optics are getting more and more complex, advanced statistical methods need to be applied in order to reach solid conclusions. Second, the analysis of ground states and thermal states of simple Hamiltonians is relevant for quantum simulation and quantum control, so direct characterizations would be very helpful.

Acknowledgments We thank Rainer Blatt, Tobias Galla, Bastian Jungnitsch, Martin Hofmann, Lukas Knips, Thomas Monz, Sönke Niekamp, Daniel Richart, Philipp Schindler, Christian Schwemmer, and Harald Weinfurter for discussions and collaborations on the presented topics. Furthermore, we thank Mariami Gachechiladze, Felix Huber, and Nikolai Miklin for comments on the manuscript. This work has been supported by the EU (Marie Curie CIG 293993/ENFOQI, ERC Starting Grant GEDENTQOPT, ERC Consolidator Grant 683107/TempoQ), the FQXi Fund (Silicon Valley Community Foundation), and the DFG (Forschungsstipendium KL 2726/2-1).

References

1. R. Renner, Nat. Phys. **3**, 645 (2007)
2. R. Horodecki, P. Horodecki, M. Horodecki, K. Horodecki, Rev. Mod. Phys. **81**, 865 (2009)
3. O. Gühne, G. Tóth, Phys. Rep. **474**, 1 (2009)
4. T. Moroder, M. Keyl, N. Lütkenhaus, J. Phys. A: Math. Theor. **41**, 275302 (2008)
5. W. Hoeffding, J. Am. Stat. Assoc. **58**, 301 (1963)
6. T. Moroder, M. Kleinmann, P. Schindler, T. Monz, O. Gühne, R. Blatt, Phys. Rev. Lett. **110**, 180401 (2013)
7. A.F. Mood, *Introduction to the Theory of Statistics* (McGraw-Hill Inc., 1974)
8. L. Schwarz, S.J. van Enk, Phys. Rev. Lett. **106**, 180501 (2011)
9. N.K. Langford, New J. Phys. **15**, 035003 (2013)
10. S.J. van Enk, R. Blume-Kohout, New J. Phys. **15**, 025024 (2013)
11. C. Schwemmer, L. Knips, D. Richart, H. Weinfurter, T. Moroder, M. Kleinmann, O. Gühne, Phys. Rev. Lett. **114**, 080403 (2015)
12. Z. Hradil, Phys. Rev. A **55**, 1561(R) (1997)
13. D.F.V. James, P.G. Kwiat, W.J. Munro, A.G. White, Phys. Rev. A **64**, 052312 (2001)
14. J. Shang, H.K. Ng, A. Sehrawat, X. Li, B.-G. Englert, New J. Phys. **15**, 123026 (2013)
15. M. Christandl, R. Renner, Phys. Rev. Lett. **109**, 120403 (2012)
16. A. Peres, Phys. Rev. Lett. **77**, 1413 (1996)
17. B. Jungnitsch, T. Moroder, O. Gühne, Phys. Rev. Lett. **106**, 190502 (2011)
18. L. Vandenberghe, S. Boyd, SIAM Rev. **38**, 49 (1996)
19. See the program PPTmixer. http://mathworks.com/matlabcentral/fileexchange/30968
20. M. Hofmann, T. Moroder, O. Gühne, J. Phys. A: Math. Theor. **47**, 155301 (2014)
21. L. Novo, T. Moroder, O. Gühne, Phys. Rev. A **88**, 012305 (2013)
22. X. Chen, P. Yu, L. Jiang, M. Tian, Phys. Rev. A **87**, 012322 (2013)
23. G. Tóth, T. Moroder, O. Gühne, Phys. Rev. Lett. **114**, 160501 (2015)
24. M. Huber, R. Sengupta, Phys. Rev. Lett. **113**, 100501 (2014)
25. C. Lancien, O. Gühne, R. Sengupta, M. Huber, J. Phys. A: Math. Theor. **48** 505302 (2015)
26. S. Beigi, P.W. Shor, J. Math. Phys. **51**, 042202 (2010)
27. T. Kahle, E. Olbrich, J. Jost, N. Ay, Phys. Rev. E **79**, 026201 (2009)
28. S. Amari, I.E.E.E. Trans, Inf. Theory **47**, 1701 (2001)
29. N. Ay, A. Knauf, Kybernetika **42**, 517 (2007)
30. D.L. Zhou, Phys. Rev. Lett. **101**, 180505 (2008)
31. D.L. Zhou, Phys. Rev. A **80**, 022113 (2009)
32. T. Galla, O. Gühne, Phys. Rev. E **85**, 046209 (2012)
33. S. Niekamp, T. Galla, M. Kleinmann, O. Gühne, J. Phys. A: Math. Theor. **46**, 125301 (2013)
34. D.L. Zhou, Commun. Theor. Phys. **61**, 187 (2014)
35. M. Van den Nest, K. Luttmer, W. Dür, H.J. Briegel, Phys. Rev. A **77**, 012301 (2008)
36. N. Linden, S. Popescu, W.K. Wootters, Phys. Rev. Lett. **89**, 207901 (2002)
37. F. Huber, O. Gühne, Phys. Rev. Lett. **117**, 010403 (2016)

Chapter 22
Few-Body Entanglement Manipulation

C. Spee, J.I. de Vicente and B. Kraus

Abstract In order to cope with the fact that there exists no single maximally entangled state (up to local unitaries) in the multipartite setting, we introduced in [1] the maximally entangled set of n-partite quantum states. This set consists of the states that are most useful under conversion of pure states via Local Operations assisted by Classical Communication (LOCC). We will review our results here on the maximally entangled set of three- and generic four-qubit states. Moreover, we will discuss the preparation of arbitrary (pure or mixed) states via deterministic LOCC transformations. In particular, we will consider the deterministic preparation of arbitrary three-qubit (four-qubit) states via LOCC using as a resource a six-qubit (23-qubit) state respectively.

Introduction

Bell's theorem states that the predictions of quantum mechanics are not compatible with any local realistic theory (i.e. any local hidden variable model) [2]. This result has deep conceptual implications and constitutes a milestone in our understanding of quantum mechanics. However, 50 years after its discovery, the relevance of this result goes beyond foundational issues. The development of quantum information theory [3] in the last decades has taught us that the counter-intuitiveness of quantum mechanics can be exploited to devise technologies beyond what can be reached classically. Interestingly, it is the correlations present in entangled quantum states that allow to violate Bell inequalities and it is entanglement what is considered to be the fundamental ingredient behind many of the applications of quantum information processing. Nowadays, entanglement is regarded as a resource and a big effort has been put up in the last twenty years to develop a theory of entanglement, which aims at its characterization, manipulation and quantification. Central to this theory is

C. Spee · B. Kraus (✉)
Institute for Theoretical Physics, University of Innsbruck, 6020 Innsbruck, Austria
e-mail: Barbara.Kraus@uibk.ac.at

J.I. de Vicente
Departamento de Matemáticas, Universidad Carlos III de Madrid, 28911 Leganés, Madrid, Spain

R. Bertlmann and A. Zeilinger (eds.), *Quantum [Un]Speakables II*,
The Frontiers Collection, DOI 10.1007/978-3-319-38987-5_22

the paradigm of Local Operations assisted with Classical Communication (LOCC). These are the most general operations allowed by the rules of quantum mechanics to spatially separated parties: each subsystem can undergo any possible form of quantum dynamics at will and the parties can correlate these local protocols via classical communication. The restriction to LOCC is, thus, a very natural one and at the heart of many applications. Moreover, as entanglement cannot be created by LOCC alone, it can be considered to be a resource to overcome the limitations of state manipulation restricted to this class of operations.

Unfortunately, the set of LOCC transformations is mathematically subtle and to characterize the possible conversions among entangled states is in general a very hard problem. Entanglement theory is much better established in the bipartite case than in the multipartite one. This is partly because entanglement manipulation via LOCC is better understood in the former setting. This has allowed to identify the bipartite maximally entangled state as the most useful state under LOCC transformations, to introduce the scenario of entanglement distillation and to define several operational entanglement measures [4]. On the other hand, not surprisingly, most bipartite quantum information protocols such as teleportation [5] and cryptography schemes [6] are optimally implemented by means of the bipartite maximally entangled state. Although several applications are also known for multipartite entanglement (e.g. quantum secret sharing [7] and one-way quantum computation [8]), a more systematic and profound study of LOCC manipulation in this realm appears to be of paramount importance. In addition to playing a pivotal role in the theory of entanglement, this would identify the most useful multipartite states and could lead to new applications of quantum information in many-body scenarios.

In this paper we review our recent results on the characterization of LOCC transformations over few-body pure states [1]. As a unique maximally entangled state does not exist in this situation, we introduced the notion of the maximally entangled set of states, as those that are most useful under LOCC conversions. We characterized the maximally entangled set for systems of three and four qubits. In addition to this, we present LOCC protocols that allow to obtain arbitrary (pure or mixed) bipartite, three or four-qubit states from a single multipartite state with more subsystems. In particular, we provide a specific six-qubit state that allows to prepare any three-qubit state by LOCC and the analogous for any four-qubit state with a given 23-qubit state.

Preliminaries

Entanglement is well understood in the bipartite pure state case. In this case it is well known that any state can be written in Schmidt decomposition [3], i.e. it can be written as $U_A \otimes U_B \sum_{i=1}^{d} \sqrt{\lambda_i}\,|ii\rangle$, where U_A and U_B are Local Unitaries (LUs), d is the dimension of the smaller subsystem, $\lambda_i \geq 0$, $\sum_i \lambda_i = 1$ and $|i\rangle$ are computational basis states. The coefficients $\sqrt{\lambda_i}$ are called Schmidt coefficients.

The entanglement properties of a bipartite pure state are completely characterized by its Schmidt coefficients, as states that are equivalent up to LUs have the same entanglement. In order to see this consider two states, $|\Psi\rangle$ and $|\Phi\rangle$, that are LU-equivalent, i.e. $|\Psi\rangle = V_1 \otimes \ldots \otimes V_n |\Phi\rangle$. By applying LUs (which corresponds to a deterministic LOCC protocol) one can convert $|\Psi\rangle$ to $|\Phi\rangle$ and vice versa. As entanglement is non-increasing under LOCC, one obtains that $E(\Phi) \geq E(\Psi)$ and $E(\Phi) \leq E(\Psi)$ for any entanglement measure E, i.e. these states have the same entanglement ($E(\Phi) = E(\Psi)$). Interestingly, the converse is also true. It has been shown that if one can transform a pure state $|\psi\rangle$ via LOCC deterministically to a pure state $|\phi\rangle$ and vice versa, then the two states are LU-equivalent [9]. As applying LUs does not change the entanglement of a state, the entanglement properties of a bipartite pure state are completely determined by the Schmidt coefficients. Moreover, in this case it can be shown that any LOCC protocol can be simulated by a (simple) LOCC transformation where one party performs a POVM measurement and the other party applies depending on the outcome a LU [10]. As only one-way communication is necessary this simplifies the study of LOCC transformations and entanglement for the bipartite case. In fact, LOCC transformations among bipartite pure states have been characterized completely by Nielsen [11]. In particular, it has been shown in [11] that a state $|\psi\rangle$ can be transformed via LOCC deterministically to state $|\phi\rangle$ iff $\vec{\lambda}_\phi \succ \vec{\lambda}_\psi$, i.e. $\vec{\lambda}_\phi$ majorizes $\vec{\lambda}_\psi$. The vectors $\vec{\lambda}_{\phi(\psi)}$ correspond to d-dimensional vectors containing the squares of the Schmidt coefficients of $|\phi\rangle$ ($|\psi\rangle$) respectively. A vector $\vec{y} = (y_1, \ldots, y_d)$ is said to majorize $\vec{x} = (x_1, \ldots, x_d)$ (i.e. $\vec{y} \succ \vec{x}$) if $\forall k \in \{1, \ldots, d-1\}$

$$\sum_{i=1}^{k} x_i^{\downarrow} \leq \sum_{i=1}^{k} y_i^{\downarrow} \tag{1}$$

and

$$\sum_{i=1}^{d} x_i = \sum_{i=1}^{d} y_i. \tag{2}$$

Note that $\downarrow$ denotes that the components are ordered in non-increasing order (e.g. $x_d^{\downarrow}$ is the smallest and $x_1^{\downarrow}$ is the largest component of $\vec{x}$). Note further that in the two qubit case this result implies that LOCC imposes a total order on the states with respect to their entanglement. In the case of two d-dimensional systems with $d > 2$ the situation changes, as there are states that are incomparable. More precisely, there exist states, $|\Psi\rangle$ and $|\Phi\rangle$, for which it holds that one can neither transform $|\Psi\rangle$ into $|\Phi\rangle$ via deterministic LOCC transformations, i.e. $\vec{\lambda}_\Phi \not\succ \vec{\lambda}_\Psi$ nor $|\Phi\rangle$ into $|\Psi\rangle$, i.e. $\vec{\lambda}_\Psi \not\succ \vec{\lambda}_\Phi$. This shows a clear difference between the two-qubit case and the case of two d-dimensional systems with $d > 2$. Nevertheless, in both cases there exists a single state (up to LUs) that allows to obtain any other state via deterministic LOCC transformations, namely the state $|\Phi^+\rangle_d = 1/\sqrt{d}\sum_{i=1}^{d} |ii\rangle$.

As knowing which LOCC transformations are possible allows to impose a (partial) order on the states with respect to their entanglement it allows to identify entanglement measures. To be more precise, the condition that an entanglement measure E has to fulfill is that it is non-increasing under LOCC. Therefore, any function E (defined on bipartite pure states) that respects the ordering due to the majorization condition, i.e. for $\vec{\lambda}_\phi \succ \vec{\lambda}_\psi$ it holds that $E(\psi) \geq E(\phi)$, is an entanglement measure for pure states. Moreover, using the majorization condition it is easy to identify the maximally entangled states. As entanglement is non-increasing under LOCC, a maximally entangled state can not be obtained via LOCC (excluding LUs) deterministically from any other state. Thus, the condition is that its Schmidt vector does not majorize any other Schmidt vector. It is easy to see that up to LUs this property is fulfilled by a single state, namely $|\Phi^+\rangle_d = 1/\sqrt{d}\sum_{i=1}^{d}|ii\rangle$. Note that any other state can be reached via LOCC from this state, as its Schmidt vector is majorized by any other Schmidt vector. Hence, this state is the most useful one concerning applications. If a protocol is based on another bipartite state with smallest dimension of the subsystems smaller than or equal to d, it can be also implemented using $|\Phi^+\rangle_d$ as a resource by applying the corresponding LOCC protocol beforehand. On the other hand, if a protocol requires that the parties share a maximally entangled state, then there exists no other (LU-inequivalent) state that allows to perform the same task. For example, faithful teleportation [5] requires a maximally entangled state as a resource.

Characterizing LOCC transformations in the multipartite setting is hard, as one has to deal with several rounds of communication. It has even been proven that some tasks require infinitely many rounds [12]. Moreover, LOCC is not closed, i.e. there exist sequences of LOCC protocols $\{\Lambda_1, \Lambda_2, \ldots\}$ such that $\lim_{n\to\infty}\Lambda_n$ is not a LOCC transformation [13].

Due to these difficulties multipartite LOCC transformations have only been characterized for a few classes of states [1, 14]. Other classifications of multipartite states with respect to entanglement have been studied such as LU-equivalence [15] and SLOCC-equivalence [16, 17]. As already argued before, states that are in the same LU-equivalence class have the same entanglement properties. The problem of deciding whether two pure n-qubit states are LU-equivalent has been solved in [15]. By applying LUs a generic state can be brought into its unique standard form. Two generic states are then LU-equivalent iff their standard forms are the same. For non-generic states there exists an algorithm that allows to decide whether two states are LU-equivalent or not [15]. Another classification can be established by considering Stochastic LOCC (SLOCC) transformations. Two states, $|\Psi\rangle$ and $|\Phi\rangle$, are said to be in the same SLOCC class, if one can transform $|\Psi\rangle$ via LOCC with non-zero probability into $|\Phi\rangle$ and vice versa. Mathematically, this implies that $|\Psi\rangle$ can be written as $g|\Phi\rangle$, where $g \in \mathcal{G}$ with $\mathcal{G}$ being the set of local invertible operators, i.e. $g = g^1 \otimes \ldots \otimes g^n$ and $g^i \in GL(2)$ [16]. Note that in contrast to deterministic LOCC transformations the classification according to SLOCC (or LU) corresponds to an equivalence relation. States in different SLOCC classes have different kinds of entanglement but SLOCC does not impose any order with respect to their entanglement. For three qubits there are two different truly tripartite entangled SLOCC classes, the W-class and the GHZ-class [16]. As representatives for the GHZ- and the W-class one can choose the GHZ-state [18], i.e. $1/\sqrt{2}(|000\rangle + |111\rangle)$ and the W-state [16], i.e. $1/\sqrt{3}(|100\rangle + |010\rangle + |001\rangle)$ respectively. In the four-qubit case it has been shown that there exist infinitely many SLOCC classes which can be grouped into 9 different families [17].

As the mathematical study of multipartite LOCC transformations is formidably hard, other approaches towards the characterization of entanglement have been pursued. They consist of enlarging the set of allowed operations such that one obtains classes of operations that are easier to deal with mathematically as for example PPT preserving maps [19] or separable transformations (SEP) (see [20] and references therein). As it will be important for our discussion we will focus here on SEP. LOCC is strictly included in SEP [21], i.e. any LOCC transformation corresponds to a completely positive trace-preserving map Λ whose action on any density matrix ρ can be written as $\Lambda(\rho) = \sum_i X_i \rho X_i^\dagger$ where $X_i = x_i^{(1)} \otimes x_i^{(2)} \ldots \otimes x_i^{(n)}$, $x_i^{(j)}$ is a complex matrix and $\sum_i X_i^\dagger X_i = \mathbb{1}$. However, there exist SEP transformations that can not be implemented via LOCC deterministically [22]. Note that in order for Λ to correspond to a LOCC protocol the operators X_i have to fulfill some condition apart from being local, i.e. each X_i denotes the operation corresponding to one branch of the LOCC protocol. To be more precise, $x_i^{(j)}$ is a product of operators where each of these operators originates from the implementation of a POVM and in general there is a dependence between the implemented POVM and previous measurement outcomes. Obviously, SEP transformations are mathematically much easier to characterize. Unfortunately, however, they lack an operational meaning. Nevertheless, the result of [20] on convertibility via SEP transformations has been crucial for determining the maximally entangled set for three and four qubits as we will explain in the next section.

The Maximally Entangled Set

As argued in the previous section, the entanglement properties of multipartite states are hard to characterize. This is also reflected in the fact that there exist several different notions of maximally entangled multipartite states in the literature (see e.g. [23, 24]). Why is it of importance to know which states are maximally entangled? Apart from fundamental interest, these states are the most useful ones concerning application. Any protocol that might rely on the parties sharing a state that is not maximally entangled can also be performed using as resource a maximally entangled state (by applying the LOCC protocol beforehand that transforms the corresponding maximally entangled state deterministically into the state that the protocol is based on), but the converse is not true. Moreover, these states are promising candidates to discover new applications for multipartite entanglement, as also most bipartite applications like teleportation [5] and cryptography [6] rely on the parties sharing a maximally entangled state. In order to characterize maximal entanglement one has to develop a deeper understanding of multipartite LOCC transformations. Knowing which LOCC transformations are possible allows to define new (operational) entanglement measures. This is due to the fact that, as mentioned before, the condition an entanglement measure has to obey is that it is non-increasing under deterministic LOCC transformations.

As already discussed, it is well known which states are maximally entangled in the bipartite case. They are those states which correspond up to LUs to $|\Phi^+\rangle_d \propto \sum_{i=1}^{d} |ii\rangle$, where d denotes the dimension of the smaller subsystem. As the application of LUs does not change the entanglement properties of a state and therefore two states in the same LU-equivalence class are equally useful, we will in the following only consider one representative per LU-equivalence class. Hence, $|\Phi^+\rangle_d$ is *the* maximally entangled state. Recall that the important property of $|\Phi^+\rangle_d$ (and the reason why it is maximally entangled) is that it can not be reached via deterministic LOCC transformations from any other state (excluding LU) and that any state can be obtained deterministically via LOCC from this state [11].

In the multipartite setting there exists no single state (up to LUs) with this property. This can already easily be seen in the three-qubit case. Recall that there are two different truly three-partite entangled SLOCC classes, the W-class and the GHZ-class. If there would exist a state $|\Psi\rangle$ that can be transformed via a deterministic LOCC transformation to a state in the GHZ-class $|\Psi_{GHZ}\rangle$ as well as to a state in the W-class $|\Psi_W\rangle$, then it would hold that there exist complex 2×2 matrices M_k^i, N_k^i such that $M_k^1 \otimes M_k^2 \otimes M_k^3 |\Psi\rangle = |\Psi_{GHZ}\rangle$ and $N_k^1 \otimes N_k^2 \otimes N_k^3 |\Psi\rangle = |\Psi_W\rangle$. As we only consider transformations between truly tripartite entangled states this implies that the matrices M_k^i and N_k^i have to be invertible for all i and therefore $|\Psi_{GHZ}\rangle$ and $|\Psi_W\rangle$ would be in the same SLOCC class.[1] Thus, deterministic LOCC transformations among fully-entangled three-qubit state are only possible between states in the same SLOCC class and so there exists no single three-qubit state that allows to reach any other state via deterministic LOCC transformations.

In order to cope with the fact that there exists no single maximally entangled state in the multipartite setting, we introduced in [1] the Maximally Entangled Set (MES) of n-partite states. The MES of n-partite states, MES_n, is defined as the minimal set of n-partite pure states such that any other truly n-partite entangled pure state can be reached via LOCC deterministically from one of the states in MES_n.[2] To state it differently, MES_n is the set of n-partite states characterized by the following two properties:

1. No state in MES_n can be reached via LOCC (excluding LU) deterministically from any other n-partite state.
2. For any truly n-partite entangled state $|\Psi\rangle \notin MES_n$ there exists a state $|\Phi\rangle \in MES_n$ such that $|\Psi\rangle$ can be obtained via a deterministic LOCC transformation from $|\Phi\rangle$.

---

[1]Note that this line of argumentation can be easily generalized to the n-qubit case. Thus, deterministic LOCC transformations among fully entangled n-qubit states are only possible between states in the same SLOCC class.

[2]Note that MES_n is unique (up to LUs).

Hence, MES_n is the minimal set of states that are most useful with respect to any application. As already explained before, any application that uses a state that is not in the MES can also be performed by using a state that is in the MES (but not vice versa). Note that the MES for bipartite d-level systems is given by $\{|\Phi^+\rangle_d\}$.

Another concept that is important for our study is the one of isolation [1]. Isolated states are defined as truly n-partite entangled states that can neither be reached nor can they be converted to any other truly n-partite entangled pure state via LOCC (excluding LU) in a deterministic way. It is obvious from this definition that isolated states have to be contained in the MES. The subset of the MES of LOCC convertible states is of particular interest, as it is the only relevant set of states concerning deterministic entanglement manipulation and therefore will most probably play an important role in discovering new applications of multipartite entanglement.

In [1] we determined the MES for three-qubit and generic four-qubit states. In contrast to the two-qubit case where the MES is given by $\{|\Phi^+\rangle_2\}$ the MES for three qubits, MES_3, contains infinitely many states. It is characterized by 3 parameters, whereas an arbitrary three-qubit state is characterized by 5 parameters (up to LUs) [25, 26]. Therefore, MES_3 is of measure zero. Interestingly, no state in MES_3 is isolated (see Fig. 22.1). The picture changes again drastically when going from the three-qubit to the four-qubit case. For four qubits MES_4 is of full measure. This is due to the fact that almost all states are isolated, i.e. deterministic LOCC transformations are hardly ever possible among fully entangled four-qubit states. Interestingly, the subset of non-isolated states in the MES is of measure zero (see Fig. 22.1). As already mentioned, these states are the only relevant ones for entanglement manipulation.

Let us now present our results on MES_3, the MES of three qubits, in more detail. Due to the existence of two truly tripartite-entangled SLOCC classes for three qubits [16], the W-class and the GHZ-class, MES_3 has to contain at least two states. Recall that if $|\Psi\rangle$ and $|\Phi\rangle$ are in the same SLOCC class, then $|\Psi\rangle$ can be written as $g\,|\Phi\rangle$, where $g = g^1 \otimes \ldots \otimes g^n$ and $g^i \in GL(2)$. Thus, we can write any state in the GHZ-class as $g\,|GHZ\rangle$. Note that the local operator g is not unique, as the two states $g\,|GHZ\rangle$ and $(gS)\,|GHZ\rangle$ coincide if S is a local symmetry of the GHZ-state. Here and in the following, we denote by $S(\Psi) = \{S \in \mathcal{G} : S\,|\Psi\rangle = |\Psi\rangle\}$ the local symmetry group of the state $|\Psi\rangle$. In order to get rid of this ambiguity, we defined in [1] a standard form for states in the GHZ-class. In particular, we showed that any state in the GHZ-class can be written up to LUs as

$$g_x^1 \otimes g_x^2 \otimes g_x^3 P_z\,|GHZ\rangle\,, \tag{3}$$

where $P_z = diag(z, 1/z)$ with $z \in \mathbb{C}\backslash 0$ and $g_x^i = \sqrt{G_x^i}$ with $G_x^i = (g_x^i)^\dagger g_x^i = 1/2\mathbb{1} + \gamma_x^i \sigma_x$ and $0 \leq \gamma_x^i < 1/2$. Here and in the following we denote by $\mathbb{1}$ and σ_w where $w \in \{x, y, z\}$ the identity operator and the Pauli matrices. Note that the restriction $\gamma_x^i \in [0, 1/2)$ ensures that the operators g_x^i are invertible, as otherwise entanglement would be destroyed.

It is well known that any state in the W-class can be written (up to LUs) as $g_1 \otimes g_2 \otimes \mathbb{1} |W\rangle$ [16], where $|W\rangle \propto |001\rangle + |010\rangle + |100\rangle$,

$$g_1 = \begin{pmatrix} 1 & 0 \\ 0 & x_1/x_3 \end{pmatrix} \quad \text{and} \quad g_2 = \begin{pmatrix} x_3 & x_0 \\ 0 & x_2 \end{pmatrix}, \tag{4}$$

with $x_1, x_2, x_3 > 0$ and $x_0 \geq 0$, i.e. $x_0 |000\rangle + x_1 |100\rangle + x_2 |010\rangle + x_3 |001\rangle$. Using this notation we can now present the MES for three qubits [1].

Theorem 1 *The MES of three qubits, MES_3, is given by*

$$MES_3 = \{g_x^1 \otimes g_x^2 \otimes g_x^3 P_z |GHZ\rangle, g_1 \otimes g_2 \otimes \mathbb{1} |W\rangle\}, \tag{5}$$

where $z \in \{1, i\}$, no $g_x^i \propto \mathbb{1}$ (except for the GHZ-state) and g_1 and g_2 are diagonal.

The general idea of how to obtain this result will be presented below. Due to Theorem 1 a state in the GHZ-class is in MES_3 iff it is either the GHZ-state or in its corresponding standard form $z \in \{1, i\}$ and $\gamma_x^i \neq 0 \; \forall i$. States in the W-class which are in MES_3 have the property that $x_0 = 0$ (see Eq. (4)), i.e. their standard form corresponds to $x_1 |100\rangle + x_2 |010\rangle + x_3 |001\rangle$. All three-qubit states that are not in MES_3 can be reached via a LOCC protocol deterministically from some state in MES_3. Interestingly, the states in the set MES_3 have a simple description in terms of the decomposition presented in [26]. Any state in MES_3 can be written up to LUs as

$$|\Psi\rangle = |0\rangle |\Psi_s\rangle + |1\rangle Y(\beta') \otimes Y(\beta) |\Psi_s\rangle\}, \tag{6}$$

where $|\Psi_s\rangle = a|00\rangle + \sqrt{1-a^2}|11\rangle$ and $a, \beta, \beta' \in \mathbb{R}$. In this decomposition it is easy to see that MES_3 is characterized by 3 parameters. As three-qubit states up to LUs are characterized by 5 parameters (see Eq. (3) and e.g. [25, 26]), MES_3 is of measure zero (see Fig. 22.1). Moreover, one can easily show (by constructing the corresponding LOCC protocol) that all states in MES_3 are non-isolated.

Let us proceed by presenting the results on the MES for generic four-qubits states[3] [1]. A generic four-qubit state belongs to one of the SLOCC classes denoted by G_{abcd} in [17]. Its representatives can be chosen to be

$$\begin{aligned}|\Psi\rangle = &\frac{a+d}{2}(|0000\rangle + |1111\rangle) + \frac{a-d}{2}(|0011\rangle + |1100\rangle) \\ &+ \frac{b+c}{2}(|0101\rangle + |1010\rangle) + \frac{b-c}{2}(|0110\rangle + |1001\rangle),\end{aligned} \tag{7}$$

where $a, b, c, d \in \mathbb{C}$ with $b^2 \neq c^2 \neq d^2 \neq b^2, a^2 \neq b^2, c^2, d^2$ and the parameters fulfill the condition that there exists no $q \in \mathbb{C}\backslash 1$ such that $\{a^2, b^2, c^2, d^2\} = \{qa^2, qb^2, qc^2, qd^2\}$. The non-trivial symmetries of these states are given by $\sigma_w^{\otimes 4}$ where $w \in \{x, y, z\}$. As for the three-qubit case one can define a unique standard form (by sorting the

[3]Note that for the non-generic cases similar results can be obtained [28].

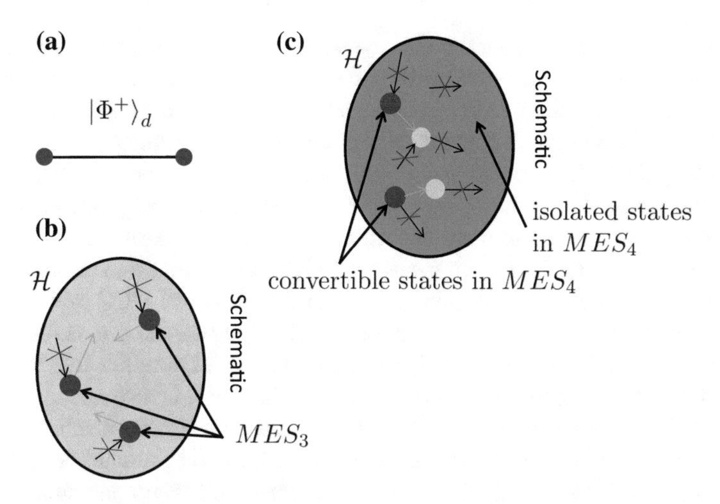

Fig. 22.1 This graphic shows schematically the MES for the bipartite-, three- and four-qubit case. In (**b**) and (**c**) the reachable states are indicated in *green*, the isolated states in the MES in *light red* and the convertible states in the MES in *dark red*. **a** Bipartitie case. **b** Three-qubit case. **c** Four-qubit case.

parameters in Eq. (7), using the symmetries to uniquely define $g^\dagger g$ and fixing the LUs). In the following we use the notation $g_w^i = \sqrt{G_w^i}$ where $w \in \{x, y, z\}$ and $G_w^i = (g_w^i)^\dagger g_w^i = 1/2\mathbb{1} + \gamma_w^i \sigma_w$ with $|\gamma_w^i| < 1/2$. Using this notation we can state our result on which fully-entangled four-qubit states can be reached via a deterministic LOCC transformation from some other state.

Theorem 2 *A generic state, $h\,|\Psi\rangle$, is reachable via LOCC from some other LU-inequivalent state iff (up to permutations) either $h = h^1 \otimes h_w^2 \otimes h_w^3 \otimes h_w^4$, for $w \in \{x, y, z\}$ where $h^1 \neq h_w^1$ or $h = h^1 \otimes \mathbb{1}^{\otimes 3}$ with $h^1 \not\propto \mathbb{1}$ arbitrary.*

Note that whereas the reachable states are characterized by 12 parameters a generic state is (up to LUs) described by 18 parameters. Hence, the set of reachable states is of measure zero. This implies that the MES for four qubits is of full measure. That is almost all states are in MES_4. By investigating which states are convertible via LOCC it becomes apparent that almost all states are isolated. In the following theorem we state which states are convertible via deterministic LOCC transformations to some other state.

Theorem 3 *A generic state $g\,|\Psi\rangle$ is convertible via LOCC to some other LU-inequivalent state iff (up to permutations) $g = g^1 \otimes g_w^2 \otimes g_w^3 \otimes g_w^4$ with $w \in \{x, y, z\}$ and g^1 arbitrary.*

Combining Theorems 2 and 3, one obtains that the non-isolated states in MES_4 are given by

$$g_w^1 \otimes g_w^2 \otimes g_w^3 \otimes g_w^4\,|\Psi\rangle\,, \tag{8}$$

where $w \in \{x, y, z\}$ and excluding the case where $\gamma_w^i \neq 0$ for exactly one i. Note that as these states are characterized by 10 parameters, the subset of LOCC-convertible states in MES_4 is of measure zero. These states are particularly interesting as they are in the MES and can be converted deterministically to some other states. Thus, investigating these states could lead to the discovery of new applications.

We will proceed by presenting the general idea of how to prove which states are reachable (or convertible) via deterministic LOCC transformations. As already pointed out, it is very hard to characterize LOCC transformations in the multipartite setting. In fact, in the four-qubit case a classification of LOCC protocols is not known. Nevertheless, the characterization of MES can be achieved as follows. The fact that LOCC protocols are strictly included in SEP [21] implies that if a state is not reachable (or convertible) deterministically via SEP it is also not reachable (or convertible) deterministically via LOCC. Hence, a state that is not reachable via SEP has to be in the MES. In order to characterize the states that can not be reached (or converted) via SEP we used the result of [20] where necessary and sufficient conditions for the deterministic convertibility of pure states via SEP have been derived. With this, we could characterize all three- and generic four-qubit states that are not reachable (or convertible) via LOCC. Moreover, we showed that all other states can be reached (or converted) via LOCC by constructing the corresponding LOCC protocols. Interestingly, these protocols turned out to be very simple. Any reachable three- or generic four qubit state can be obtained from some other state via a LOCC protocol were just one party performs a non-trivial measurement and the other parties apply, depending on the measurement outcome, a LU. The detailed proof of which states are in the MES in the three-qubit and generic four-qubit case can be found in [1].

We will in the following present the result of [20] and discuss some technical details. In [20] it has been shown that a state $|\Psi_1\rangle = g\,|\Psi\rangle$ can be transformed via SEP to $|\Psi_2\rangle = h\,|\Psi\rangle$ iff there exists a $m \in N$ and a set of probabilities, $\{p_k\}_1^m$ ($p_k \geq 0, \sum_{k=1}^m p_k = 1$) and local symmetries, $S_k \in S(\Psi)$ such that

$$\sum_k p_k S_k^\dagger H S_k = rG. \tag{9}$$

Here, we use the notation $H = h^\dagger h$, $G = g^\dagger g$ and $r = n_{\Psi_2}/n_{\Psi_1}$ with $n_{\Psi_i} = ||\,|\Psi_i\rangle\,||^2$. The POVM elements that allow to do this transformation are of the form $M_k = \frac{\sqrt{p_k}}{\sqrt{r}} h S_k g^{-1}$. Note that M_k, H, G and S_k are local operators. From Eq. (9) it becomes apparent that the symmetries of the chosen representative of the SLOCC class play an important role for the convertibility via deterministic SEP transformations. To be more precise, deterministic SEP transformations only become possible if there exists at least one non-trivial symmetry and the symmetries specify which transformations are possible. Thus, in order to characterize SEP convertibility it is crucial to know the symmetries of a representative of the corresponding SLOCC class. Moreover, it is important to note that H and G are not uniquely defined by the states, as the state $g\,|\Psi\rangle$ is the same as $\tilde{g}\,|\Psi\rangle = gS\,|\Psi\rangle$ for $S \in S(\Psi)$. Thus, G and $\tilde{G} = S^\dagger G S$ correspond to the same state. This makes it more complicated to characterize which SEP transformations are possible as the only information about the initial and the final state that enters Eq. (9) is (apart from the normalization factor) H and G. In order to get rid of this ambiguity, we defined for each SLOCC class a standard form, which also takes into account that we only consider one representative per LU-equivalence class. In order to get a unique correspondence (up to LUs) between the operator G and the state $g|\Psi\rangle$, we choose symmetries $S_i \in S(\Psi)$ such that $\tilde{G} = S_i^\dagger G S_i$ is of a specific form which ensures that there exists no symmetry $S_j \in S(\Psi)$ such that $S_j^\dagger \tilde{G} S_j$ is also of this form. Obviously, it is sufficient to consider only transformations between states in standard form. As outlined before, using this standard form and Eq. (9) allowed us to characterize the states that can not be reached (or converted) via LOCC.

Up to now, we discussed deterministic LOCC transformations among pure states. In the following section we will extend our study to the preparation of mixed states.

Deterministic State Preparation of an Arbitary Two-, Three- and Four Qubit State (Pure or Mixed)

In this section, we will show that in the bipartite- and three-qubit case having access to all states in the MES allows to obtain not only all pure fully entangled states but an arbitrary state (pure or mixed) of the same dimension via deterministic LOCC transformations. Moreover, we will present a six-qubit state that allows the local preparation of an arbitrary three-qubit state. Finally, we will discuss the differences that arise if one tries to extend our discussion to the four-qubit case.

In the bipartite case MES_2 contains a single state, namely $|\Phi^+\rangle_d$. Interestingly, this state is not only a resource to prepare an arbitrary pure states but also to obtain any mixed bipartite state, $\rho = \sum_i p_i |\Psi_i\rangle\langle\Psi_i|$. The protocol that allows to achieve this task is the following. With probability p_i apply the LOCC protocol that deterministically transforms $|\Phi^+\rangle_d$ into $|\Psi_i\rangle$ (see Fig. 22.2b). Thus, $|\Phi^+\rangle_d$ is not only the most useful state among pure states but also among mixed bipartite states.

In the three qubit case the MES, MES_3, contains infinitely many states. Having simultaneous access to all of the states in MES_3 allows to obtain any pure or mixed three-qubit state deterministically via LOCC as we will show in the following. It is clear from the definition of the MES that any pure fully entangled three-qubit state can be obtained via LOCC from one of the states in MES_3. Moreover, any bipartite entangled three-qubit state can be obtained from the GHZ-state by performing a projective measurement in the σ_x-basis on the appropriate qubit (to obtain up to LUs $|0\rangle|\Phi^+\rangle_2$) and then applying the corresponding LOCC protocol to reach the desired state. It is clear that any product state can be reached from an arbitrary state in MES_3 via projective measurements. Thus, having access to all of the states in MES_3 allows to prepare any pure three-qubit state. Interestingly, it also allows to obtain any mixed three-qubit state as we will show in the following. In order to obtain a mixed three-qubit state, $\rho = \sum_i p_i |\Psi_i\rangle\langle\Psi_i|$, choose with probability p_i the state in MES_3 that allows to reach $|\Psi_i\rangle$ deterministically and apply the corresponding LOCC protocol to obtain $|\Psi_i\rangle$. Thus, being able to prepare any state in MES_3 and to perform LOCC transformations does not only allow to obtain any pure but also any mixed three-qubit state deterministically. This implies that MES_3 contains also the most useful states in the mixed state case.

Interestingly, there exists also a single six-qubit state that allows to obtain an arbitrary three-qubit state (pure or mixed) via deterministic LOCC transformations as we will explain in the following. Thus, this six-qubit state is a resource for three-qubit state preparation. It has been introduced as a resource state in a different context, namely Remote Entanglement Preparation (REP) [27]. In REP the aim is that party A (which can be split into several spatially separated parties) is able to deterministically provide the spatially separated parties B_i with arbitary multipartite entanglement. In the three qubit case it has been shown that a specific eight-qubit state has to be shared between the parties to provide the parties B_i with an arbitary state up to LUs. The scenario that we are interested in here is the deterministic preparation of an arbitrary state in MES_3. The corresponding resource has been shown to be the following six-qubit stabilizer state [27] (see Fig. 22.2a)

$$|\Phi_3\rangle = Z_3(\tfrac{\pi}{4})H_3Z_2(\tfrac{\pi}{2})Z_6(-\tfrac{\pi}{4})Z_5(-\tfrac{\pi}{4})Z_4(-\tfrac{\pi}{4})H_1S_{46}S_{56}S_{45}S_{15}S_{35}S_{34}S_{12}S_{23}\,|+\rangle^{\otimes 6}, \tag{10}$$

where $S_{ij} = |0\rangle\langle 0| \otimes \mathbb{1} + |1\rangle\langle 1| \otimes \sigma_z$ and here and in the following we denote by $Z_{i_1\ldots i_m}(\alpha)$ a phase gate $e^{i\alpha\sigma_z^{i_1}\otimes\ldots\otimes\sigma_z^{i_m}}$ acting non-trivially on parties $i_1 \ldots i_m$.

By performing projective local measurement on the qubits 4, 5 and 6 one prepares the state for qubits 1, 2 and 3. The choice of the measurement basis decides on the values of the parameters in the following decomposition

$$|\Psi_3\rangle = Z_{13}(\alpha_4)Z_{12}(\alpha_5)(T_2 \otimes T_3)Z_{23}(\alpha_6)\,|+\rangle^{\otimes 3}, \tag{11}$$

where $T_3 = e^{-i\frac{\pi}{4}\sigma_x}Z(-\frac{\pi}{4})H$ and $T_2 = e^{i\frac{\pi}{4}\sigma_y}Z(\frac{\pi}{4})H$. Note that any state that can be given (up to LUs) in the decomposition of Eq. (6) can be written up to LUs also in this form.[4] Thus, via choosing the measurement basis accordingly one can prepare any state in MES_3 (up to LUs) on the qubits 1, 2 and 3. Let us now present some details of the protocol. In order to specify the parameters in Eq. (11) qubit 4, 5 and 6 have to be measured in the basis $\{\sigma_z^{k_i}Z(-\theta_i)\,|+\rangle\}_{k_i=0,1}$ with $\theta_i = \alpha_i$ for $i = 4, 6$ and $\theta_5 = \pm\alpha_5$ for $i = 5$ where the sign depends on the outcome of the measurement on qubit 6. Thus, the measurement on qubit 6 that determines the parameter α_6 has to be performed before the measurement on qubit 5 and depending on the measurement outcome the basis of the measurement that determines α_5 might have to be adjusted (for outcome $k_6 = 1$ one has to choose $-\alpha_5$ instead of α_5). Moreover, the state on the qubits 1, 2 and 3 is prepared up to Pauli operators which depend on the measurement outcomes.

[4]Note that also any product or biseparable state can be written up to LUs in the decomposition given in Eq. (11).

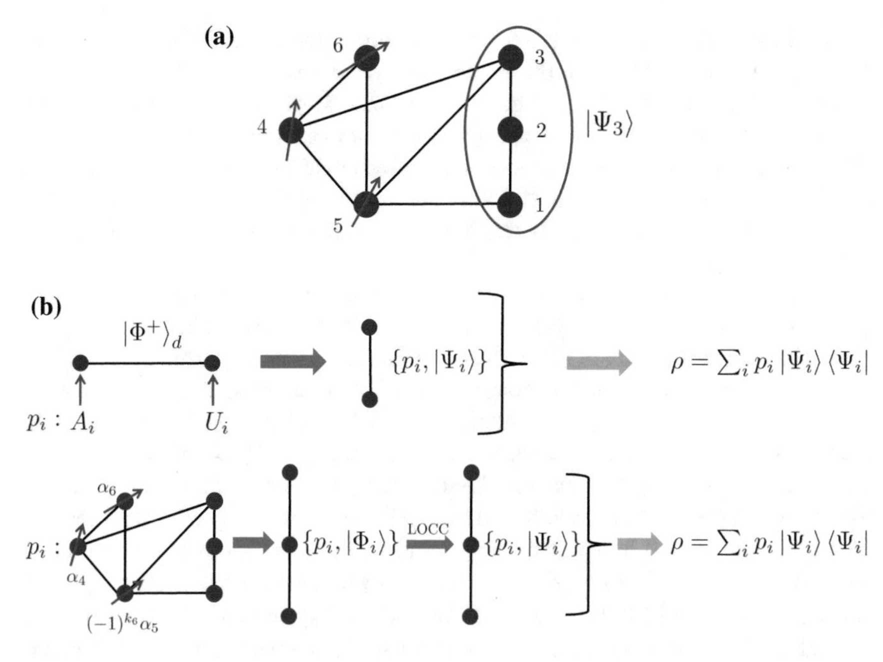

Fig. 22.2 **a** This graphic shows the graph state that is LU-equivalent to the six-qubit state in Eq. (10). **b** In this graphic we indicate how any pure or mixed bipartite (three-qubit) state can be obtained via deterministic LOCC transformations from $|\Phi^+\rangle_d$ (the six-qubit state in Eq. (10)) respectively.

More precisely, one obtains the following state $(\sigma_z^{k_4+k_5} \otimes \sigma_z^{k_5}\sigma_y^{k_6} \otimes \sigma_z^{k_4+k_6})\,|\Psi_3\rangle$. The origin of the importance of the order of the measurements, as well as of the fact that the state is prepared up to Pauli operators is that we used deterministic gate implementation in order to construct the resource state (for details see [27]). As the six-qubit state in Eq. (10) allows to obtain any state in MES_3, it is a resource for the preparation of any pure or mixed three-qubit state via deterministic LOCC transformations (see Fig. 22.2b).

Note that there exists a different six-qubit state, namely $|\Phi^+\rangle_2^{\otimes 3}$ which is shared among A (who possesses 4 qubits), B (1 qubit) and C (1 qubit), that allows to obtain any state in MES_3 via a projective measurement and application of LUs, i.e. by teleporting the corresponding state in MES_3 [5]. In fact, any pure (or mixed) three-qubit state can be obtained from $|\Phi^+\rangle_2^{\otimes 3}$ by performing (with some probability) a corresponding measurement and applying depending on the measurement outcome some Pauli operators. The important difference to our scheme is that except for fully-separable states the projective measurement is non-local and therefore it does not correspond to a LOCC protocol on the 6 qubits.

In the four-qubit case the 23-qubit state that is a resource state for REP can be used to obtain any pure or mixed four-qubit state, as it allows to prepare, by performing local projective measurements, an arbitrary pure four-qubit state up to LUs [27].

Unfortunately, the number of qubits of the resource state for REP in the four qubit case can not be reduced by requiring only the preparation of states in the MES. The reason for this is the following. The fact that MES_4 is of full measure implies that (independent of the chosen decomposition) the number of parameters that describe MES_4 and the number of parameters that describe an arbitrary four-qubit state up to LUs have to be the same. By construction the number of qubits of the resource state depends on the number of parameters in some decomposition of specific form (see [27] for details).

Moreover, there is another clear difference to the three-qubit case. In the four-qubit case having access to all states in the MES does not allow to obtain all biseperable states via deterministic LOCC transformations. States of the form $|\phi\rangle\,|\psi\rangle$, where $|\phi\rangle$ and $|\psi\rangle$ are truly bipartite entangled, can not be obtained from some state in MES_4. This can be seen as follows. Applying a LOCC protocol to a state $|\Psi\rangle \in MES_4$ leads in any branch of the protocol to a state $M_k^1 \otimes M_k^2 \otimes M_k^3 \otimes M_k^4\,|\Psi\rangle$ where M_k^i are 2×2 matrices. Clearly, states of the form $|\phi\rangle\,|\psi\rangle$, where $|\phi\rangle$ and $|\psi\rangle$ are truly bipartite entangled, can not be obtained from $|\Psi\rangle$ via a local projective measurement, i.e. all the matrices M_k^i have to have rank 2 and are therefore invertible. This implies that $|\Psi\rangle$ and $M_k^1 \otimes M_k^2 \otimes M_k^3 \otimes M_k^4\,|\Psi\rangle$ have to be in the same SLOCC class, which is not the case for $|\Psi\rangle \in MES_4$ and $|\phi\rangle\,|\psi\rangle$, where $|\phi\rangle$ and $|\psi\rangle$ are truly bipartite entangled. Hence, having access to all state in MES_4 and being able to perform any LOCC protocol does not allow to obtain all biseparable states and therefore not all four-qubit states. As the 23-qubit state that is a resource state for REP allows to obtain any four-qubit state (including also all biseparable four-qubit states) by performing the corresponding local measurements this state allows to prepare any mixed or pure four-qubit state via LOCC.

In summary, we have shown that in the bipartite- and three qubit case the states that are contained in the MES are the most useful ones among all states (pure or mixed) of the same dimension. We presented a single six-qubit state from which any three-qubit state (pure or mixed) can be obtained deterministically by applying the corresponding LOCC protocol. Moreover, there exists a 23-qubit state that allows to obtain any pure or mixed four-qubit state via deterministic LOCC transformations.

Summary and Outlook

The characterization of the possible LOCC transformations among quantum states is a very relevant problem, both in the foundations of entanglement theory and to identify the most useful states in order to look for new applications of quantum information. Although this is in general a very hard task, we have shown that it is in principle possible to address this question using the tools that characterize the conversions under the mathematically more tractable set of SEP operations. Following this idea, we have characterized the MES of truly entangled three-qubit states and generic four-qubit states. Interestingly, it turned out that there is a subset of states of measure zero which is clearly more relevant for state manipulation and these families should therefore be good candidates to look for practical applications. MES_n allows to obtain by LOCC any possible n-qubit truly entangled pure state. However, we have shown that the power of MES_3 extends to the preparation of all possible (pure or mixed) tripartite qubit states but that this is not the case for MES_4. As a complement, we have considered LOCC protocols that allow to obtain all few-body states from a single state of more parties. We have provided a specific six-qubit (23-qubit) state that allows to prepare any form of three-qubit (four-qubit) entanglement.

Our results open the door to a systematic and general study of LOCC manipulation in multipartite systems. In the future we will investigate extensions to systems of higher dimension and/or more parties. Moreover, we will study the case of more copies and also the more general case of approximate LOCC transformations in which, contrary to exact LOCC, one allows for a certain error in the output state of the protocol. Furthermore, the investigation of possible LOCC transformations will allow, similarly to the bipartite case, to introduce operational entanglement measures.

This research was funded by the Austrian Science Fund (FWF): Y535-N16. JdV further acknowledges the Spanish MINECO through grants MTM2014-54692 and MTM2014-54240-P and the CAM regional research consortium QUITEMAD+CM S2013/ICE-2801.

References

1. J.I. de Vicente, C. Spee, B. Kraus, Phys. Rev. Lett. **111**, 110502 (2013)
2. J.S. Bell, Physics **1**, 195 (1964)
3. See e. g. M.A. Nielsen, I.L. Chuang, *Quantum Computation and Quantum Information* (Cambridge University Press, 2000)
4. R. Horodecki et al., Rev. Mod. Phys. **81**, 865 (2009)
5. C.H. Bennett, G. Brassard, C. Crépeau, R. Jozsa, A. Peres, W.K. Wootters, Phys. Rev. Lett. **70**, 1895 (1993)
6. A.K. Ekert, Phys. Rev. Lett. **67**, 661 (1991)
7. M. Hillery, V. Bužek, A. Berthiaume, Phys. Rev. A **59**, 1829 (1999). R. Cleve, D. Gottesman, H.-K. Lo, Phys. Rev. Lett. **83**, 648 (1999)
8. R. Raussendorf, H.J. Briegel, Phys. Rev. Lett. **86**, 5188 (2001)
9. R.M. Gingrich, Phys. Rev. A **65**, 052302 (2002)
10. H.-K. Lo, S. Popescu, Phys. Rev. A **63**, 022301 (2001)
11. M.A. Nielsen, Phys. Rev. Lett. **83**, 436 (1999)
12. E. Chitambar, Phys. Rev. Lett. **107**, 190502 (2011)
13. E. Chitambar, D. Leung, L. Mancinska, M. Ozols, A. Winter, Commun. Math. Phys. **328**(1), 303–326 (2014). and references therein
14. S. Turgut, Y. Gül, N.K. Pak, Phys. Rev. A **81**, 012317 (2010). S. Kintas, S. Turgut, J. Math. Phys. **51**, 092202 (2010)
15. B. Kraus, Phys. Rev. Lett. **104**, 020504 (2010). Phys. Rev. A **82**, 032121 (2010)
16. W. Dür, G. Vidal, J.I. Cirac, Phys. Rev. A **62**, 062314 (2000)
17. F. Verstraete, J. Dehaene, B. De Moor, H. Verschelde, Phys. Rev. A **65**, 052112 (2002)
18. D. M. Greenberger, M. Horne, A. Zeilinger, *Bell's Theorem, Quantum Theory, and Conceptions of the Universe*, ed. by M. Kafatos (Kluwer, Dordrecht, 1989), p.69

19. S. Ishizaka, M.B. Plenio, Phys. Rev. A **71**, 052303 (2005)
20. G. Gour, N.R. Wallach, New J. Phys. **13**, 073013 (2011)
21. C.H. Bennett, D.P. DiVincenzo, C.A. Fuchs, T. Mor, E. Rains, P.W. Shor, J.A. Smolin, W.K. Wootters, Phys. Rev. A **59**, 1070 (1999)
22. E. Chitambar, R. Duan, Phys. Rev. Lett. **103**, 110502 (2009)
23. G. Gour, N.R. Wallach, J. Math. Phys. **51**, 112201 (2010)
24. P. Facchi, G. Florio, G. Parisi, S. Pascazio, Phys. Rev. A **77**, 060304(R) (2008)
25. A. Acin, A. Andrianov, L. Costa, E. Jané, J.I. Latorre, R. Tarrach, Phys. Rev. Lett. **85**, 1560 (2000)
26. J.I. de Vicente, T. Carle, C. Streitberger, B. Kraus, Phys. Rev. Lett. **108**, 060501 (2012)
27. C. Spee, J.I. de Vicente, B. Kraus, Phys. Rev. A **88**, 010305(R) (2013)
28. C. Spee, J.I. de Vicente, B. Kraus, J. Math. Phys. **57**, 052201 (2016)

Part VII
Neutron Interferometry

Chapter 23
Search for Hidden Observables in Neutron Experiments

Helmut Rauch

Here, hidden observables and not hidden variables will be discussed. Neutrons are proper tools for testing basic laws of quantum physics since they are massive and can be handled and measured with high efficiency. Suitable post-selection experiments demonstrate coherence features of sub-ensembles even when the whole ensemble seems to have lost its coherence. All experiments have the capacity to explain more details by additional pre- and post-selection methods. It will be shown that specific losses are unavoidable in any interaction. Coherence and decoherence are intrinsic quantum effects and can shed light on the measurement problem. Quantum contextuality is a consequence of the entanglement of different degrees of freedom. This makes quantum phenomena more strongly correlated than classical ones. Most experiments have been performed with perfect neutron interferometers and some others by using ultra-cold neutrons and spin-echo systems. An event by event based interpretation can also be brought into agreement with the experimental results. In many cases parasitic beams carry the same information as the main beam and this relates such measurements to "weak" measurements. The coupling and entanglement of various parameter spaces guide us to a more elaborate discussion of quantum effects.

Introduction

The ongoing discussion about the interpretation of quantum physics which oscillates between very mystic and more rational interpretations. John Bell contributed to this discussion and made very useful proposals for related test measurements [2]. In this paper it will be shown that in all experiments there are some kinds of hidden

H. Rauch (✉)
Atominstitut, Vienna University of Technology, 1020 Vienna, Austria
e-mail: rauch@ati.ac.at

R. Bertlmann and A. Zeilinger (eds.), *Quantum [Un]Speakables II*,
The Frontiers Collection, DOI 10.1007/978-3-319-38987-5_23

observables mainly because not all pre- and post-selection possibilities are utilised and average values are measured where more detailed investigations would be possible. In this respect we discuss neutron experiments where the particle and wave features are essential and where one can show that much more information can be extracted by using more sophisticated measurement methods. This will be demonstrated by means of spatial-, momentum-, time- and polarization post-selection experiments. We will also demonstrate that unavoidable quantum losses may play an important role in the interpretation of the quantum to classic transition. These losses can be used as weak measurements to obtain basic information about the main object without interacting with it. In some sense the analysis follows the pragmatic access of [7] who showed how quantum phenomena can be explained when one follows strictly the state of knowledge one gains from an experiment starting from a few up to a high number of events and where the averaging procedure becomes more reliable. In this respect one includes the measurement time, or the number of particles used, into the analysis.

Our approach uses neutron interferometer experiments where two coherent neutron beams are produced by dynamical Bragg diffraction from a perfect silicon crystal and they are superposed at the exit crystal plate and exhibit all well-known interference phenomena (Fig. 23.1; e.g. [14]). Since one deals with a stationary situation one can use the time-independent Schrödinger equation and gets for the beams behind the interferometer a superposition of beam I and II which are transmitted-reflected-reflected ψ_{trr} and reflected-reflected-transmitted ψ_{rrt} respectively. From symmetry follows that they are equal in intensity and phase. A phase shift χ can be applied, which is given by the index of refraction of any material and related to a spatial shift $\Delta(\chi = \Delta.\vec{K})$.

$$I_0 = |\psi_I + \psi_{II}|^2 = |\psi_{trr} + \psi_{rrt} e^{i\chi}|^2 = |\psi_0|^2 \left(1 + |\Gamma(\vec{\Delta})| \cos\chi\right) \tag{1}$$

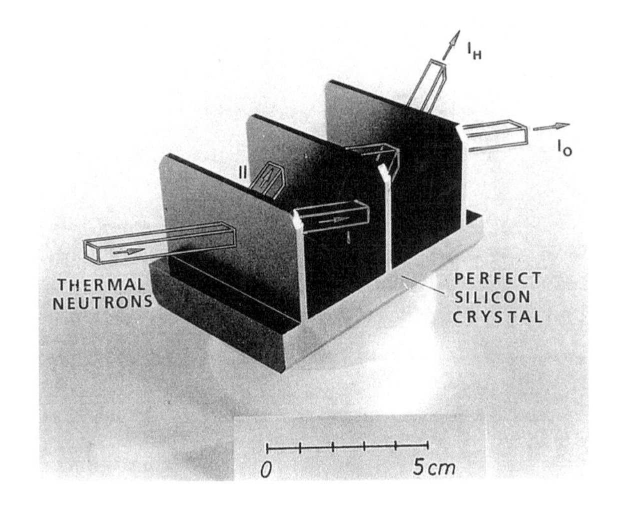

Fig. 23.1 Photo of a perfect crystal neutron interferometer.

$\Gamma(\vec{\Delta})$ denotes the coherence function, which defines the coherence lengths $\vec{\Delta}_c$ as its characteristic dimension and is related to the size of the wave packets involved

$$\left|\Gamma(\vec{\Delta})\right| = \left|\langle\psi(0)\psi(\vec{\Delta})\rangle\right| \propto \left|\int g(\vec{k})e^{i\Delta\vec{k}}d\vec{k}\right| \tag{2}$$

For Gaussian beams with momentum widths δk one obtains

$$|\Gamma(\Delta)| = \exp\left[-(\Delta.\delta k)^2/2\right] \tag{3}$$

where $g(\vec{k})$ is the momentum distribution. In an experiment Eq. (1) can be approximated by

$$I_{\text{exp}} = A(1 + V\cos(\chi + \phi)) \tag{4}$$

V denotes the visibility of the interference pattern and ϕ an internal phase caused by some small deviations from the perfectness of the crystal or due to external effects like gravity or magnetic fields. In practice high visibilities (up to 95 %) and high order interferences (up to 200th) have been observed.

Analysis of the Parameters in Eqs. (1)–(4)

All parameters of Eqs. (1)–(4) are average values over the beam cross section, the momentum distribution and the measurement time. In the following sections we intend to analyse the various parameters of the equations shown above. We try to show in the following sections how they can be specified by various post-selection experiments.

Wave Function—Momentum Post-Selection

The wave function follows from the solution of the time-independent Schrödinger equation and can be written for free space motion as a wave packet centred around k_0 with a width δk. For Gaussian shaped beams this can be written as

$$-\frac{\hbar^2}{2m}\Delta\psi = E\psi \tag{5}$$

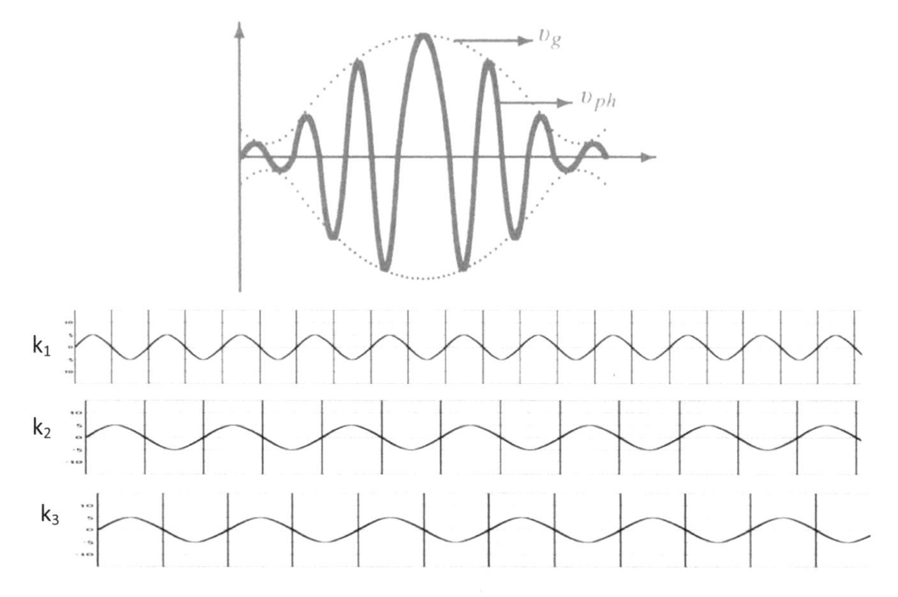

Fig. 23.2 Wave packet (*above*) and partial waves (*below*).

$$\psi(r) \propto \int a(k) e^{ikr} dk \propto \int e^{-(k-k_0)^2/2\delta k^2} e^{ikr} dk \tag{6}$$

which is an eigenvalue solution for the freely moving particle. $a(k)$ denotes the amplitudes of the coherently superposed partial waves and $|a(k)|^2$ the momentum distribution function as used in Eq. (2). Whereas the wave packet is localized within a region compatible with its coherence length ($\Delta_c \delta k \geq 1/2$) the partial waves are arbitrarily widely spread as shown in Fig. 23.2. That means that information about the particle exists even far away from the packet. This has to be taken into account when non-locality effects are discussed e.g. [13]. Proper momentum post-selection measurements have made these partial waves visible by applying additional monochromatization [12, 16]. This shows how more information can be extracted by means of more sophisticated experimentation. Figure 23.3 shows an arrangement where some of these far reaching components of the wave functions can be analysed and fringe visibility can be preserved although the beam I_0 without additional monochromator crystals and the artificially summed up intensity ($\sum_1^n I_n$) of all measuring channels do not show interference features at all. This means that one can decide after the measurement whether one is more interested in wave features (individual channel intensities) or particle features (summed up intensity). This has to be discussed on the basis of the Greenberger-Englert duality relation which treats wave and particle properties as a duality system where the particle

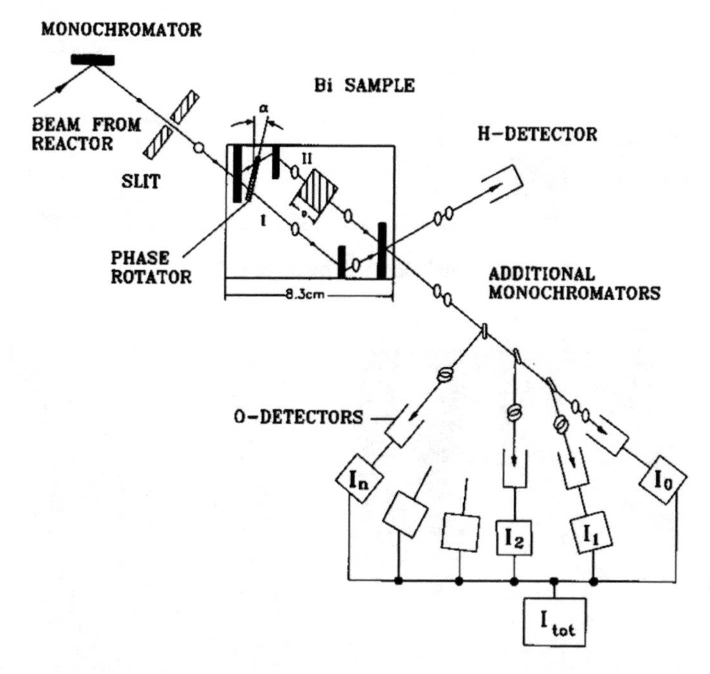

Fig. 23.3 Sketch of a feasible momentum post-selection experiment using various monochromaters (1, … n) to observe wave features and where the intensities can be summed up to reach I_0, i.e. a situation which permits beam path (particle) information.

feature is determined by the path distinguishability (P_D) and the wave features by the visibility (V) of the interference pattern [9, 6]. This indicates that a single particle system can exhibit far reaching features outside its wave packet range and that a two or many body system never separates when it has a common origin.

$$P_D^2 + V^2 \geq 1 \tag{7}$$

Beam Cross Section and Wave Packets—Position Post-Selection

As mentioned above the size of the wave packet is given by the Fourier transform of the momentum distribution function (Eq. 6) and can be determined from the measurement of the coherence function (Eq. 2), i.e. from the decrease of the visibility at high interference order [16]. For a neutron interferometer situation the coherence lengths are different for the various directions due to the different momentum distributions $\left(\Delta_{longitudinal} \approx 100\ \text{A},\ \Delta_{vertical} \approx 50\ \text{A},\ \Delta_{transvers} \approx 20\ \mu\text{m}\right)$, which are much smaller than a typical beam cross section ($\approx 1 \times 2\,\text{cm}$). When one

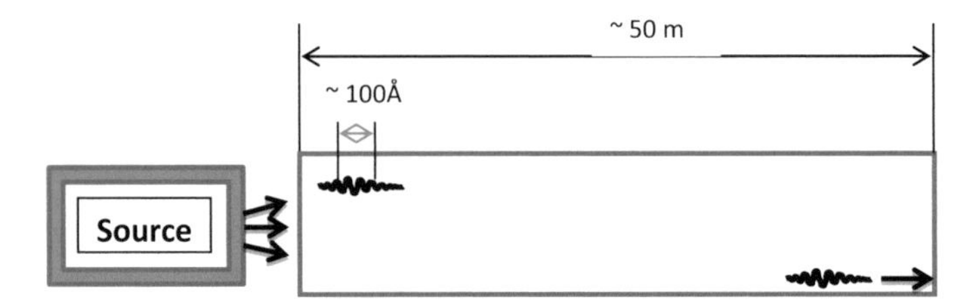

Fig. 23.4 Sketch of wave packets within the neutron flight path.

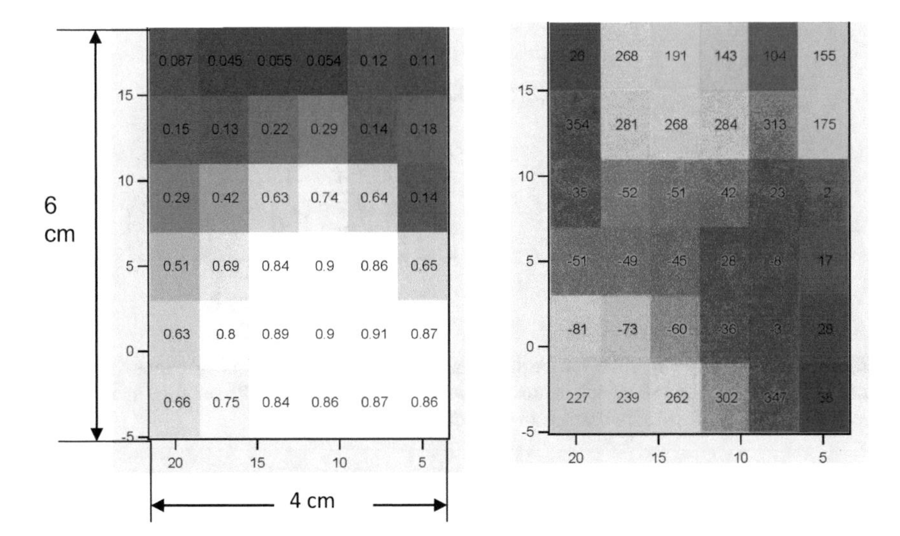

Fig. 23.5 Result of a position post-selection experiment showing the contrast (*left*) and the internal phase (*right*) across the beam cross section.

considers the available intensity of about 10^4 n/s one notices that one deals with single particle interference and there is no interaction between successive neutrons besides the fact that they are shaped by the same monochromators, collimators and crystal reflections. Thus a situation as shown in Fig. 23.4 exists where the wave packets of the individual neutron do not overlap at all.

The intensity and the coherence features vary across the beam cross section and therefore the parameters of Eq. (4) vary as well, as shown in Fig. 23.5. This is a position post-selection result when a position sensitive detector is used to scan the intensity, the contrast $(V\, or\, |\Gamma(\Delta)|)$ and the internal phase (ϕ) of the interference pattern. One notices that the measured interference pattern depends on the size and position of the aperture used. One can also measure the momentum distribution at any position and will notice some differences. This means that the wave-functions are different at any position of the aperture, but they are similar to each other which permits the definition of a mean wave-function for any beam. One should keep in mind that in any experiment an average is taken over aperture area and various momentum distributions at any position within the beam cross section. This underlines that the wave packet features are determined by the apparatus only but each neutron of the beam experience a similar history which causes common features and determine the coherence properties of the beam.

Time—Post-Selection

Neutron choppers can be used to measure the velocity distribution of the beam by means of time-of-flight methods. These methods are similar to the momentum post-selection methods discussed in Sect. "Wave Function—Momentum Post-Selection" [15, 10]. The energy-resolved interference pattern show a higher visibility than the full beam. A classical analysis of the data is possible as long as the resolution time of the chopper is larger than the coherence time of the beam (5 μs compared to 10 ns).

Another method measures the arrival time of each neutron. For a thermal (Poissonian) beam the probability to measure a neutron within a time interval t after another neutron has arrived is given by [8]

$$W(t) = Ie^{-It} \tag{8}$$

This is shown in Fig. 23.6 [18]. When measuring two neutrons arriving within short or long time intervals one can achieve a considerably higher contrast and phase sensitivity than analysing the full beam only. Since arrival time

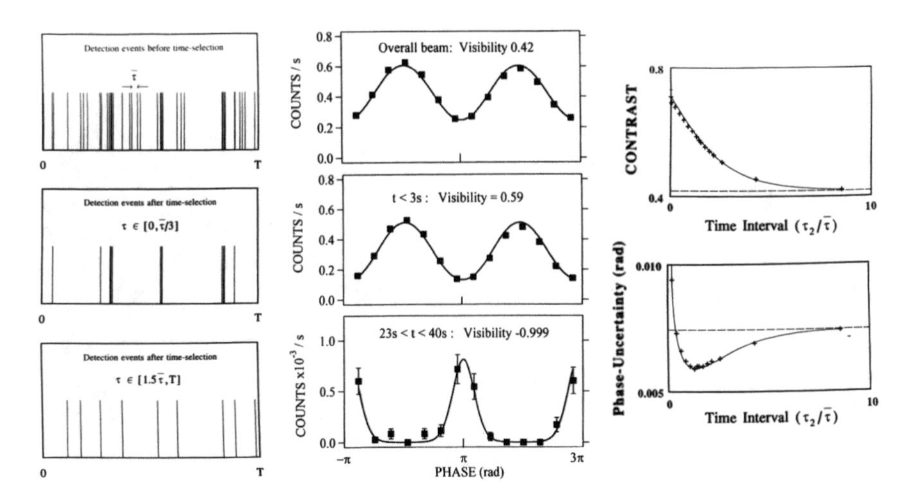

Fig. 23.6 Arrival times of neutrons and short pair and long pair arrivals (*left*), measured interference pattern for the overall beam and the short and long pair arrivals (*middle*) and the measured and calculated contrast and sensitivity (*right*).

measurements can be implemented quite easily this opens a new possibility to improve experimental results. A more complete analysis has been given on the basis of a Bayes estimation by [17]. The mean time interval between two arriving neutrons is given by $\bar{\tau} = 1/I$. One notices that for "long" pairs the contrast always reaches nearly 100 % and that the interference pattern becomes shifted by π.

Unavoidable Losses

In many cases losses during a quantum measurement are neglected and treated as caused by experimental imperfections only. Here we deal with unavoidable losses caused by the theory itself. Such losses may become important for the understanding of quantum decoherence and the quantum measurement process [19, 20, 11]. It will be shown that not only dissipative interactions cause an irreversible change of the wave-function, but deterministic ones can cause such a change too. We start with the phase echo experiment where a large positive phase shift should be compensated by a large negative one. This has been verified to a high degree [3], but a closer look shows that each phase shifter causes additional back and forth reflections as indicated in Fig. 23.7. When the energy of the particle E is much larger than the height of the barrier $(\bar{V})$ and its thickness produces phase shifts larger than the coherence length of the beam the reflectivity can be written as e.g. [4]

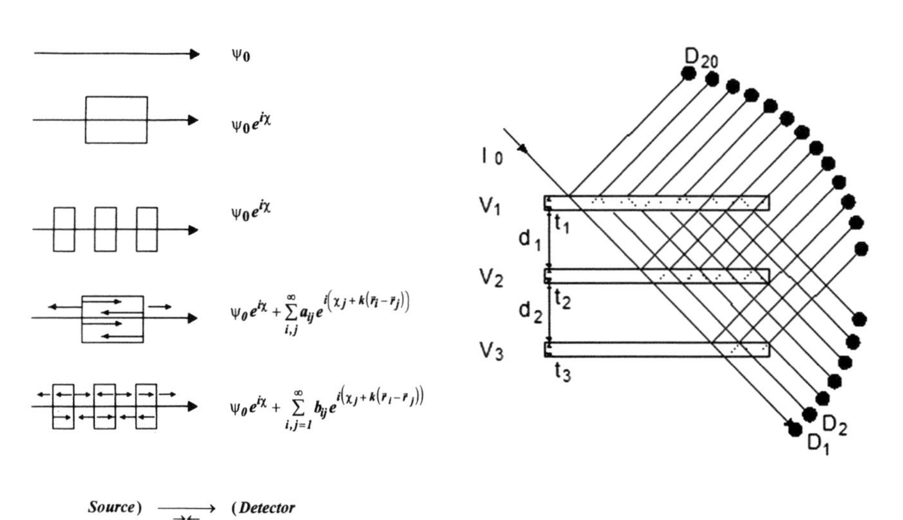

Fig. 23.7 Approximate and complete wave-functions when differently shaped phase shifters are used (*left*) and a method to measure these parasitic beams (*right*).

$$\bar{R} \cong \frac{1}{2}\left(\frac{\bar{V}}{2E}\right)^2. \tag{9}$$

For thermal neutrons this is very small (10^{-10}) but unavoidable even when specially shaped barriers are taken into account. It should be mentioned that the same information which exists in the main beam (D_1) is available in the parasitic beams (D_2–D_{20}). This relates such measurements to "weak" measurements [1, 5]. The measurement period may be much longer but that does not enter the analysis and there is no method to retrieve all components into the original beam.

Discussion

These measurements and their analysis have shown that many coherence properties of a beam can be retrieved by various post-selection methods. Nevertheless a complete retrieval cannot be achieved due to unavoidable losses. These losses need not be caused by dissipative forces but can also be caused by unavoidable quantum losses. Thus irreversibility seems to start with the first interaction the system experiences and can take place at any time scale. In this respect irreversibility and the related measurement process seems to be part of the quantum formalism as stated by [7]. Wave and particle features (interference and beam path) are related according to Eq. (7) and they can be taken as a two level system which can be entangled in a contextual sense. This shows that quantum physics involves more entanglements and makes the world and the human being more correlated than classical physics.

References

1. Y. Aharonov, D.Z. Albert, L. Vaidman, Phys. Rev. Lett. **60**, 1351 (1988)
2. J.S. Bell, *Speakable and unspeakable in quantum mechanics* (Cambridge University Press, 1987)
3. R. Clothier, H. Kaiser, S.A. Werner, H. Rauch, H. Wölwitsch, Phys. Rev. A **44**, 5357 (1991)
4. C. Cohen-Tannoudji, B. Diu, F. Laloe, *Quantum Mechanics*, vol. 1 (Wiley, N.Y., 1977)
5. J. Dressel, M. Malik, F.M. Miatto, A.N. Jordan, Rev. Mod. Phys. **86**, 307 (2014)
6. B.-G. Englert, Phys. Rev. Lett. **77**, 2154 (1996)
7. B.-G. Englert, Eur. Phys. J. **67**, 238 (2013)
8. R.J. Glauber, *Fundamental Problems in Statistical Mechanics*, ed. by E.G.D. Cohen (North Holland, 1968)
9. D.M. Greenberger, A. Yasin, Phys. Lett. **128**, 391 (1988)
10. D.L. Jacobson, B.E. Allman, M. Zawisky, S.A. Werner, H. Rauch, J. Phys. Soc. Jpn. **A65**, 94 (1996)
11. E. Joos, H.D. Zeh, C. Kiefer, D. Giulini, J. Kupsch, I.-O. Stamatescu, *Decoherence and the Appearance of a Classical World*, 2nd edn. (Springer, Berlin, 2003)

12. H. Kaiser, R. Clothier, S.A. Werner, H. Rauch, H. Wölwitsch, Phys. Rev. A **45**, 31 (1992)
13. H. Rauch, Phys. Lett. A **173**, 240 (1993)
14. H. Rauch, S.A. Werner, *Neutron Interferometry* (Oxford University Press, Oxford, 2015)
15. H. Rauch, H. Wölwitsch, R. Clothier, H. Kaiser, S.A. Werner, Phys. Rev. A **46**, 49 (1992)
16. H. Rauch, H. Wölwitsch, H. Kaiser, R. Clothier, S.A. Werner, Phys. Rev. A **53**, 902 (1996)
17. J. Rehacek, Z. Hradil, M. Zawiska, S. Pascazio, H. Rauch, J. Perina, Phys. Rev. A **60**, 473 (1999)
18. M. Zawisky, H. Rauch, Y. Hasegawa, Phys. Rev. A **50**, 5000 (1994)
19. H.D. Zeh, Found. Phys. **1**, 69 (1970)
20. W.H. Zurek, Phys. Rev. D **24**, 1516 (1981)

Chapter 24
What Does Quantum Theory Tell Us? A Matter-Wave Approach

Yuji Hasegawa

It is my great pleasure and honor to present here a review of our recent experimental achievements investigating foundations of quantum mechanics with neutrons. We are doing quantum-optical experiments with neutrons. Neutrons are massive particles with ½-spin. Matter-wave experiments such as interferometer experiments with neutrons have been established as an almost ideal tool for tests of quantum mechanics for the last several decades. In addition, the coupling of neutrons with the electromagnetic field allows us to manipulate neutron's spin degrees of freedom with extremely-high precision. Stationary as well as time-dependent manipulation scheme is accomplished: Larmor precession is actually involved in the former case and (total) energy degree of freedom can be affected in the latter case. Here, we concentrate on three major topics, i.e., entanglement achieved in a single-particle (neutron) system, quantum Cheshire-Cat and a new error-disturbance uncertainty relation, which are very great concerns from the fundamental view point of quantum theory.

Introduction

From the earlier stage of the development of quantum theory, the double–slit experiment, in particular with single-particles has been serving as the best example to view central mystery in quantum mechanics [1]. In classical physics, after lots of trials with particles, the obtained distribution at the screen behind the two opening slits will be given by the sum of two distributions separately obtained by individual single-slit openings. This undoubtedly understood by presuming the fact that each particle can go through merely either one or the other slit so that the final distri-

Y. Hasegawa (✉)
Atominstitut, Vienna University of Technology, 1020 Vienna, Austria
e-mail: Hasegawa@ati.ac.at

R. Bertlmann and A. Zeilinger (eds.), *Quantum [Un]Speakables II*,
The Frontiers Collection, DOI 10.1007/978-3-319-38987-5_24

bution is the sum of those obtained from two sub-ensembles, one with particle going through one slit and the other through the other slit. In contrast, in quantum version of double-slit experiments, when both slits are open and particles like neutrons, electrons, molecules and so forth are sent, there appears "a fine structure" in the final distribution at the screen, This fine structure exposed more frequent and even less frequent occurrences in some places than in cases of individual single-slit openings: this is explained due to a constructive/destructive interference effect from a wave model in classical physics. It sounds reasonable if one accepts the situation where particles, which cannot be divided into pieces, hit both slits simultaneously and "an effect" from both opening with wave-like property reaches the screen, exhibiting fringe pattern due to constructive and destructive interference. It is worth noting here that non-local effects, not in a sense of quantum kinematics observed in two-particle correlation but in a sense of quantum dynamics described by quantum equation of motion [2], are clearly observed in quantum version of the double-slit experiments.

Optical experiments with massive particles such as neutrons, electrons, atoms, molecules are playing a significant role for tests of peculiar phenomena predicted by quantum theory: the first and, in some cases, even the second order quantum interference effects are observed [3]. It is worth noting that such an important physical issue is sometime easily forgotten but lies beneath that the (non-relativistic) Schrödinger equation can be directly applied in describing the time-evolution of quantum state of massive quantum system: de Broglie wave with the wavelength $\lambda = h/mv$, spin quantum numbers with its norm given by $\|\mathbf{s}\| = \sqrt{s(s+1)}\hbar$, canonical commutation relation such as $[x_i, p_j] = i\hbar\delta_{ij}$, *et cetra* are purely quantum mechanical features and cannot be found in classical mechanics.

In interferometer experiments, separated coherent beams are produced typically by a wave-front division, e.g., in Young type, or by an amplitude division, e.g., in Mach-Zehnder type. These beams are recombined and superposed coherently after propagating through some regions of space, where phase as well as amplitude can be manipulated by various interactions. The advent of the perfect crystal neutron interferometer in 1974 opened up a new era of fundamental studies of quantum mechanics with matter-waves [4]. Coherent beam split is carried out in the manner of the amplitude division and optical elements such as beam-splitter/mirror/analyzer, are attained by the use of (dynamical) diffraction at the perfect crystal slab. A schematic view of the skew-symmetric neutron interferometer made of a Si perfect-crystal and an intensity modulation as a function of the relative phase tuned by a phase shifter are depicted in Fig. 24.1. A monolithic construction allows us alignment-free structure. Typical beam separation is several centimeters, which enables quantum mechanical studies in a macroscopic scale. It is to be mentioned here that all perfect-crystal neutron interferometer experiments up to now concern with self-interference experiment, where only one particle (neutron) is inside the interferometer at a time, which goes through two split beam paths, and interference fringes are observed for a certain ensemble: $I_0 \propto \left| |\Psi_I\rangle + e^{i\chi}|\Psi_{II}\rangle \right|^2 \propto (1 + \cos\chi)$.

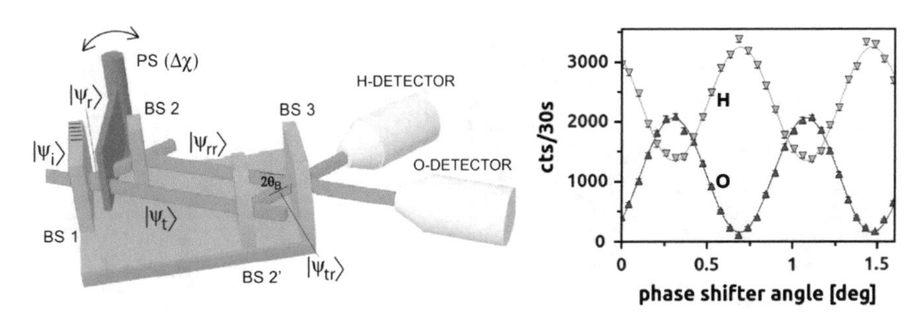

Fig. 24.1 Schematic view of the perfect crystal neutron interferometer (*left*). A typical interferogram obtained by rotating phase shifter (*right*).

In typical perfect crystal neutron interferometers, interference patterns between two coherent sub-beams, Ψ_I and Ψ_I, can be observed by tuning the relative phase. Exploiting ½-spin of neutrons, another interference effects between the spin eigenstates, e.g., $|\Uparrow\rangle$ and $|\Downarrow\rangle$ (spin eigenstates parallel and anti-parallel to the quantization axis), can be observed in a so-called neutron polarimeter: the polarimeter is a device, typically where a polarized incident neutron beam is sent through the sample and the spin vector of the exit beam is analyzed subsequently. For quantum-mechanical investigations, neutron polarimeter is used in the following manner. (i) A coherent superposition of the up- and the down-spin eigenstate is created by turning the spin vector of the up polarized incident beam by $\pi/2$, $|\Uparrow\rangle \mapsto 1/\sqrt{2}(|\Uparrow\rangle + |\Downarrow\rangle)$. (ii) A relative phase shift between the spin eigenstates is given: $1/\sqrt{2}(|\Uparrow\rangle + |\Downarrow\rangle) \mapsto 1/\sqrt{2}(|\Uparrow\rangle + e^{i\alpha}|\Downarrow\rangle)$. (iii) Finally, the direction of the spin vector is measured: the direction depends on the relative phase α, e.g., $P_{\Uparrow} \propto (1 + \cos\alpha)$. This sinusoidal modulation dependent on the relative phase α has exactly the same origin, i.e., quantum interference, as the intensity modulation observed in the interferometer experiments. A schematic view of a typical neutron polarimeter experiment and an intensity modulation as a function of the relative phase α are depicted in Fig. 24.2. Neutron polarimetry has several advantages compared to perfect crystal neutron interferometry: we mention here, in particular, higher phase-stability due to insensitivity to ambient disturbances, larger available space for inserting diverse optical instruments and apparatus into the superposing beam, wider angular acceptance range of the beam bringing higher intensity, and higher (>99 %) efficiency of spin manipulation attaining higher contrast (>98 %) of the final interference fringes.

The main goal of this article is to present tutorial review of the recent fundamental studies of quantum mechanics explored with neutron's matter-waves: some recent experiments with the neutron interferometer and the polarimeter are presented. We exploited entanglements between degrees of freedom in a single-particle system: bi-partite and even tri-partite entangled quantum systems are investigated [5–7]. Furthermore, following a recent theory suggesting a separation of the photon and its polarization in the curious way of quantum mechanics [8], we carried out an experiment where so-called a quantum Cheshire-cat is observed in neutron interferometer setup: the

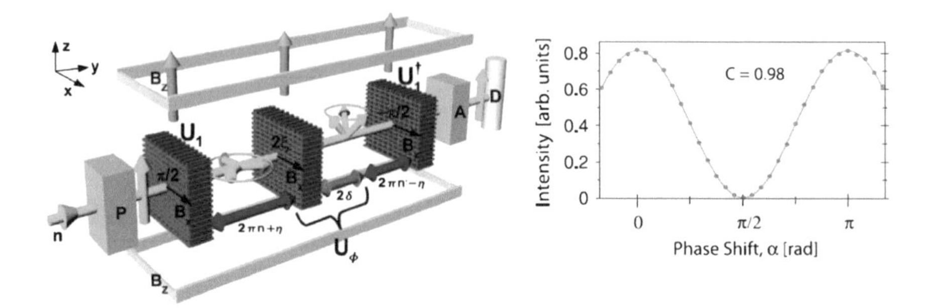

Fig. 24.2 Schematic view of a typical neutron polarimeter experiment (*left*). Sinusoidal intensity modulation is obtained by tuning the relative phase between the two spin eigenstates (*right*).

experimental results suggest that the system as a whole behaves as if neutron and its magnetic moment are disembodied and they travel along different beam paths [9]. In the last part, a new theory concerning the error-disturbance uncertainty relation is tested by modifying a conventional neutron polarimeter setup [10, 11]: successive measurements of neutron's spin exhibits the violation of the naïve error-disturbance relation suggested by Heisenberg and confirms the validity of a new universally valid error-disturbance uncertainty relation by Ozawa [12, 13]. A more detailed summary of recent neutron interferometer and polarimeter experiments is published elsewhere [14].

Entanglement Achieved in a Single-Particle (Neutron) System

It was Einstein, Podolsky, and Rosen (EPR) [15] and afterwards Bell [16] who shed light on the non-local properties between subsystems in quantum mechanics (QM). Bell inequalities are constraints imposed by local hidden-variable theories (LHVTs) on the results of spacelike separated experiments on distant systems. The conflict between LHVTs and QM is even more apparent in tri- or multi-partite quantum systems as analyzed by Greenberger, Horne and Zeilinger (GHZ) [17]: the contradiction arises in contrast to statistical violation in common with Bell-inequalities.

The neutron optical studies exploit properties of bi-partite and even tri-partite entangled quantum system [5–7]. In our experiments entanglements are achieved not between particles but between degrees of freedom: this allows investigations properties not due to quantum non-locality but due to quantum contextuality. It is worth noting here the fact that local hidden variable theories are a subset of a larger class of hidden variable theories known as non-contextual hidden variable theories (NCHVTs). By definition, NCHVTs assume that the result of a measurement of an observable is predetermined and independent of a previous or simultaneous measurement of any other compatible (commeasurable) observables.

The concept of contextuality in quantum mechanics was first analyzed by Kochen and Specker, which suggests certain constraints on the allowed hidden variable theories [18]: there appears a contradiction between predictions by non-contextual hidden variable theories and quantum mechanics. Experimental tests of quantum contextuality with neutrons started with the study of Bell-like inequality [5]. In this experiment, the path and the spin degrees of freedom of neutrons in the interferometer are employed to accomplish the Bell-like state such as $|\Psi_{Bell}\rangle = 1/\sqrt{2}(|\Uparrow\rangle|\Psi_I\rangle + |\Downarrow\rangle|\Psi_{II}\rangle)$, where $|\Uparrow\rangle$ and $|\Downarrow\rangle$ denote the up-spin and down-spin states, and $|\Psi_I\rangle$ and $|\Psi_{II}\rangle$ denote the two beam paths in the interferometer, respectively. The Bell-like inequality is obtained with parameters of the relative phase χ between the two beams and the direction α of the final spin analysis as $S \equiv E(\alpha_1, \chi_1) + E(\alpha_1, \chi_2) - E(\alpha_2, \chi_1) + E(\alpha_2, \chi_2)$ and $-2 \leq S \leq 2$. In this experiment the expectation values of the joint measurement of the path and the spin $E(\alpha_j, \chi_k)$ (j, k = 1, 2) are determined by a combination of count rates with appropriate adjustments of α and χ and given by

$$E(\alpha_j,\chi_k) = \frac{N(\alpha_j,\chi_k) + N(\alpha_j^\perp,\chi_k^\perp) - N(\alpha_j^\perp,\chi_k) - N(\alpha_j,\chi_k^\perp)}{N(\alpha_j,\chi_k) + N(\alpha_j^\perp,\chi_k^\perp) + N(\alpha_j^\perp,\chi_k) + N(\alpha_j,\chi_k^\perp)}, \tag{1}$$

with $\alpha_j^\perp = \alpha_j + \pi$ and $\chi_j^\perp = \chi_j + \pi$. A schematic view of the experiment and the sinusoidal intensity modulations used to determine $E(0, 0.79\pi)$, $E(\pi, 0.79\pi)$, $E(0, 1.79\pi)$, and $E(\pi, 1.79\pi)$ are shown in Fig. 24.3. A final value of $S_{\text{exp}} \equiv E(0, 0.79\pi) + E(0, 1.79\pi) - E(\pi, 0.79\pi) + E(\pi, 1.79\pi) = 2.051 \pm 0.019 \nleq 2$ was obtained, which violates the Bell-like inequality by almost three standard deviations. Although this first experiment exhibits only a small violation of the Bell-like inequality, the violations are more explicit as $S_{\text{exp}} = 2.365 \pm 0.013 \nleq 2$ in the

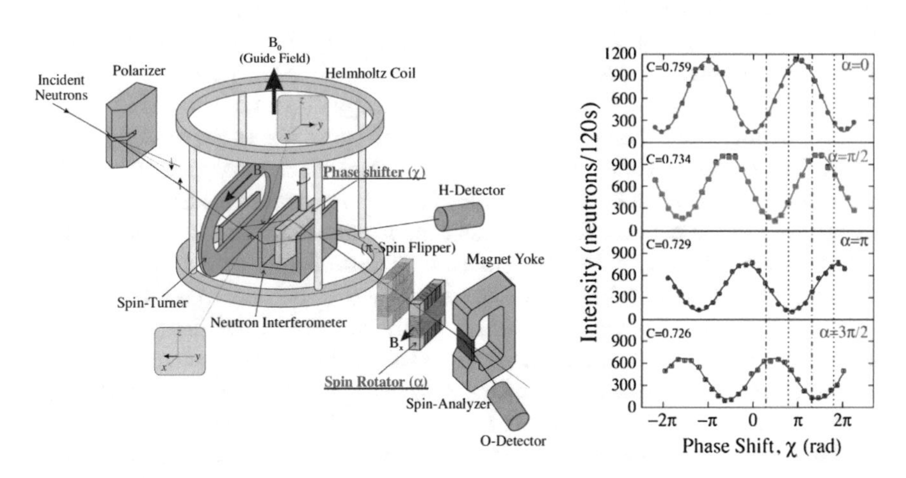

Fig. 24.3 Schematic view of the experimental setup to demonstrate a violation of the Bell-like inequality (*left*). Sinusoidal intensity modulation obtained by tuning the relative phase for various directions (α) of the final spin analysis [5].

recent experiments with neutron interferometer with the use of spin-path entanglement [19] and as as $S_{\exp} = 2.781 \pm 0.015 \nleq 2$ the polarimeter with spin-energy entanglement [7]. By the use of the bi-partite spin-path entanglement in a single-neutron system, Kochen-Specker-like contradictions between NCHVTs and the quantum-mechanical predictions are studied with neutron interferometric setup [20–22]. All these experiments confirm quantum contextuality: the result of a certain measurement in quantum mechanics cannot be ascribed to a definite value before the measurement but actually depends on, for instance, what is measured together.

In the first experiment studying the Bell-like inequality, the spin and the path degrees of freedom in a single-neutron system are entangled. For further investigations, other degrees of freedom of neutrons were pursued to be manipulated with high fidelity and entangled. It was already known and used in the neutron optics community that the total energy of neutrons can be shifted through the spin-flip by the interaction of neutron's magnetic moment with a sinusoidally oscillating magnetic field: an interaction represented by a time-dependent Hamiltonian can influence the total energy of the system. A neutron with the up-spin is suffered by the sinusoidally oscillating magnetic field $B_1(\omega t)$ situated in the region with a guide magnetic field B_0. By tuning the strengths of the two magnetic fields and the frequency ω of the oscillation, the neutron's spin can be flipped: the neutron with the down-spin leaves the instrument. At the entrance (and exit) of the guide magnetic field, the kinetic energy and the potential energy of the neutron decrease and increases, respectively so that the total energy, which is given by the sum of the kinetic and potential energies, is preserved. By the interaction with the oscillating magnetic field, the neutron's spin is flipped so that the potential energy changes with the unaffected kinetic energy: the total energy of the neutron is shifted through this interaction. (Since typical frequency of the oscillating magnetic field is some tens of kHz, i.e. a radio frequency, the spin-flipper of this type is called a radio-frequency/RF spin-flipper.) This interaction can be viewed by the James-Cummings model [23]: the shift of the total energy of the neutron is attributed to the photon exchange, i.e. photon emission/absorption [24, 25]. A schematic view of the spin-flip by the oscillating field together with an energy diagram is shown in Fig. 24.4.

Interferometer experiment accompanied by RF spin-flippers are reported already in the 1980s: one intended to observe a time-dependent superposition of spinor wavefunction [26] and a kind of a double-slit experiment with marking the paths by energy shift of the neutron [27]. In these experiments, however, the contrast of the interferogram was pretty low due to thermal disturbance to the perfect-crystal interferometer: a temperature difference even by 0.1° between the interferometer crystal and the spin-flip device, which is easily and almost inevitably attained by heat-loads due to electric currents, can considerably reduce the contrast of the inteferogram. For advanced experiments, a special apparatus was to be developed: a small box, e.g., several cm^3, made of acrylic glass, which is a thermal and electrical insulator, is fed with temperature-controlled water. A (mini) guide-field and a RF coil are buried in the cooling-water. An illustration and a picture of this apparatus are shown in Fig. 24.5.

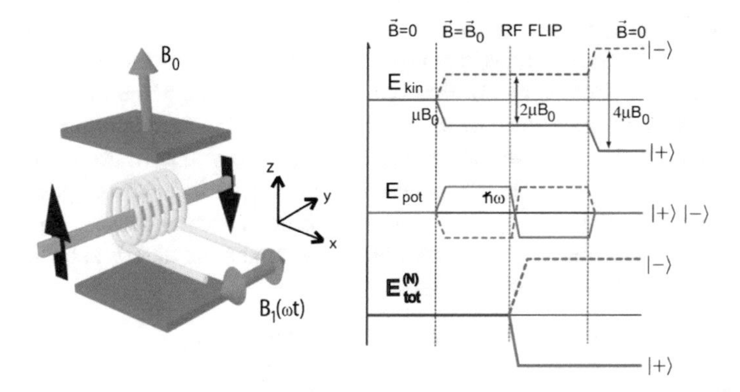

Fig. 24.4 Spin-flip accomplished by the interaction between a sinusoidally oscillating magnetic field and neutron's spin (*left*). An energy diagram of the neutron: the total energy is shifted by going through the RF spin flipper.

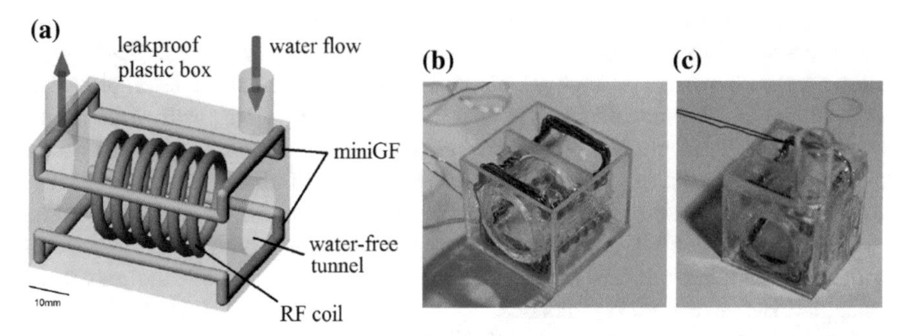

Fig. 24.5 An illustration of the RF spin-flipper box equipped with water cooling system and a mini guide field (**a**). A picture of the spin-flipper box (**b**, **c**).

By using this new apparatus, we have attained to manipulate (total) energy degree of freedom coherently with high fidelity [28]. This achievement enabled us to study properties of tripartite-entangled system, e.g., between spin-path-energy degrees of freedom. The Greenberger-Horne-Zeilinger-like (GHZ-like) state, given by $|\Psi_{GHZ}\rangle = 1/\sqrt{2}(|\Uparrow\rangle|\Psi_I\rangle|E_0\rangle + |\Downarrow\rangle|\Psi_{II}\rangle|E_0 - \hbar\omega\rangle)$ is generated to test quantum contextuality [6] and the W-like state given by the superposition of $|\Downarrow\rangle|\Psi_I\rangle|E_0 - 2\hbar\omega\rangle$, $|\Uparrow\rangle|\Psi_I\rangle|E_0 - \hbar\omega\rangle$, and $|\Downarrow\rangle|\Psi_{II}\rangle|E_0 - \hbar\omega\rangle$ is used to study evidence of the distinct types of genuine multi-partite entanglement between three degrees of freedom [29, 30].

The derivation of the Bell's inequality is based on the combination of two assumptions of locality and realism, which is violated with certain circumstances predicted by quantum theory. Taking one step further Leggett proposed in 2003 a class of hidden variable theories, i.e., crypto-nonlocal, that abandons locality [31]. He showed that predictions of a certain class of this kind of non-local theories are in conflict with those of quantum mechanics. The first experimental test was carried out by using a pair of entangled photons [32]: the key issue of the test is correlation measurements of photon polarization lying not in the same plane but out of plane, i.e., correlations in 3D instead of in 2D. It is worth noting here that this experiment demands extremely high contrast of the final interferograms (more than 97.4 %); such a high contrast is only accessible with polarimeter in neutron optics.

In our polarimetric test [7], we followed the criteria used in the first experimental study by Gröblacher et al. [32]. In particular, the contextual theory to be tested here is based on the following assumptions:

(i) All the values of measurements are predetermined.
(ii) States are a statistical mixture of subensembles.
(iii) The expectation values taken for each subensemble obey cosine dependence.

Assumptions (i) and (ii) are common to experimental tests of ordinary non-contextual theories and assumption (iii) is a peculiarity of this model. Here, the result of the final measurement of B [A] depends on the setting of the previous measurement of A [B]: a realistic contextual model is tested in our experiment. The experimental setup is depicted in Fig. 24.6. Passing through a bent Co-Ti super mirror array, the beam is highly polarized. The same technique is employed to analyze the polarization. Two identical radio-frequency (RF) spin rotators are employed. Both are put in a homogeneous and static magnetic guide field. In the present experiment, a maximally spin-energy entangled Bell-like state is generated and affected by successive energy and spin measurements. Initial Larmor-precession scan exhibit sinusoidal intensity modulations with more than 99 % contrast. By tuning the rotation angle of RF spin-flipper 1 to π/2, the maximally entangled Bell-like state is generated.

For the test of alternative model of quantum mechanics à la Leggett, correlation measurements are required between spin-directions outside the single plane. By tuning the angle adequately, the maximum discrepancy was observed, as the obtained value $S_{\text{Exp}} = 3.8387 \pm 0.0061$ at the upper limit $S_{\text{Th}} = 3.7921$: the violation is more than 7.6 standard deviations. In order to see the tendency of the violations, the parameter φ is tuned to 8 different values between 0 and 0.226π.

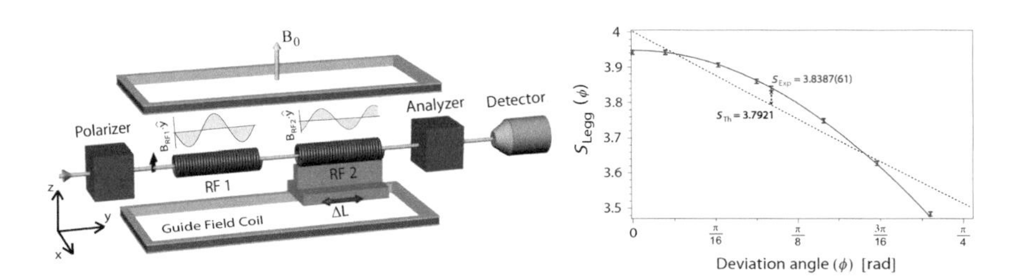

Fig. 24.6 Schematic view of the experimental setup to test alternative model of quantum mechanics à la Leggett (*left*). The final results plotted as a function of an experimental parameter φ: some experimentally determined points, on the *red curve* according to the quantum theory, are clearly above the boundary, i.e. *black curve* predicted by the alternative model (*right*) [7].

Figure 24.6 shows a plot of the experimentally determined S_{Exp} together with the limit of the Leggett's model as well as the quantum mechanical prediction (calculated for the final contrast of 99 %). The experimental values follow the quantum mechanical prediction, and this clearly confirms the violation of Leggett's model.

Neutron optical approach for studies of quantum contextuality has exploited entanglement between several degrees of freedom, i.e. spin, path, and (total) energy. Even though the ½-spin of neutrons is, in fact, a two-level quantum system, the path and energy degrees of freedom can be extended to higher-order multi-level quantum system. For instance, three-level quantum systems, i.e., so-called qutrits, are created in neutron optical experiments and utilized: a three-path neutron interferometer is developed to be actually used in some measurements [33] and three energy-levels are in usage for generations of tripartite entangled systems [30]. It is to be emphasized here that the energy levels (as well as the path) used in the neutron experiments are not of natural origin but artificially created. For instance, gaps between each levels are controlled by the frequency of the time-dependent interaction: applying various frequencies for interactions, one can artificially generate lots of energy levels to be used for, say, quantum information/communication processing. In this case, the properties of neutrons such as storability in a "bottle", less fine-structure and lower decoherence-rate by avoiding electric interactions will be of help.

Cheshire-Cat in Quantum Mechanics

Quantum theory is one of the most successful theories in physics and its predictions are verified with high precision in a wide range of the field by experiments. From the beginning, however, quantum mechanics is providing extraordinary and strange view of nature, which is different from that in classical physics. For instance, a particle such as electron, neutron and positron can behave non-locally located waves: wave-particle duality postulates that all particles exhibit both wave and particle properties. Furthermore, another paradox known as a Schödinger's cat is a good example that quantum mechanics is capable to exhibit counterfactual phenomena. Recently, another counterfactual paradox, called quantum Cheshire-cat, attracted attention: in a pre- and post-selected circumstances, a cat, i.e. a particle, spots in one place and her grin, e.g., a spin, does in the other [9, 34]. This situation is illustrated in Fig. 24.7 on the left side.

The Cheshire-cat featured in a novel "Alice in Wonderland" by Lewis Caroll is a remarkable creature: Alice says that she sees a cat disappears, leaving its grin behind. It is contradictory to our ordinary life, since no grin can be found without a cat. In the quantum world, however, it is possible that an object is separated from its properties. This is a "quantum Cheshire-cat" which is allowed by quantum theory. In more detail, according to the law of quantum physics, particles can be in different states at the same time: this phenomenon is known to be observed in so-called a double-slit experiment. If, for example, a beam of neutrons is divided into two

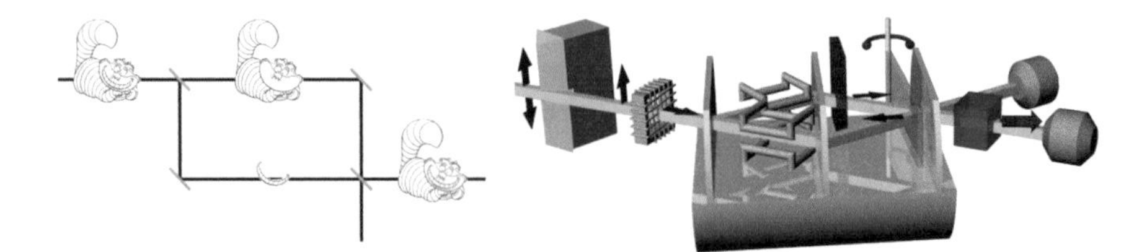

Fig. 24.7 Illustration of quantum Cheshire-cat: a cat is found in the upper beam path, while her grin is in the lower path in the Mach-Zehnder interferometer (*left*). A schematic view of the experiment setup to demonstrate quantum Cheshire-cat phenomenon: with appropriate pre- and post-selected states, the neutron itself is detected in one of the beam paths, while its spin is in the other path in a neutron interferometer (*right*).

beam paths using a silicon crystal, the individual neutrons can travel along both beam paths at the same time in a quantum superposition. Neutrons have no electric charge, but they carry a magnetic moment. They have the spin, which can be influenced by external magnetic fields. A neutron beam is split into two parts in a neutron interferometer. Let us see if neutrons and their spin will be found in difference paths of the neutron interferometer setup depicted in Fig. 24.7 on the right side. Neutrons enter the interferometer: then the spins of the two beams are rotated into different directions. The neutron in the upper beam has a spin parallel to the propagation, the spin of the lower beam points into the opposite direction. After the two beams are recombined, only those neutrons are selected, which have a spin parallel to their stream. These procedures are called the pre- and the post-selection, which are key issues in creation of quantum Cheshire-cat.

For the observation of the quantum Cheshire-cat, two measurements are required: the measurement of neutron's population and the measurement of the location of the neutron's spin. This is not a trivial task, since the quantum system is so fragile that conventional measurement procedure in quantum mechanics unavoidably leads to disturbances on the measured system to prevent further evolution of the system. Therefore, we decided to use an alternative strategy: not a strong but a weak coupling of the system with the measurement apparatus allows a minimum disturbance on the system to extract the information of the system in between. Since the "signal" of this kind of measurement is so low, the price we should have paid was to repeat the measurement until enough reliable results are obtained. In the first measurement, absorbers with pretty high transmissivity (about 80 or 60 %) are inserted in one of the beam paths in the interferometer. Typical results are shown in Fig. 24.8. The absorbers in the beam path I (lower path) do not affect the final intensity of the O-beam with a spin-analysis, while final intensity decreases by inserting the absorbers in the path II (upper path). This suggests that neutrons are traveling through the interferometer, following the beam path I. Things get tricky, when the system is put under the second measurement: the spin in one of the beam paths is rotated slightly ($\sim 15^{\circ}$) by applying a weak magnetic field. Typical results are shown in Fig. 24.9. The magnetic field in the beam path II (upper path) does not affect the final intensity of the O-beam with a spin-analysis, while

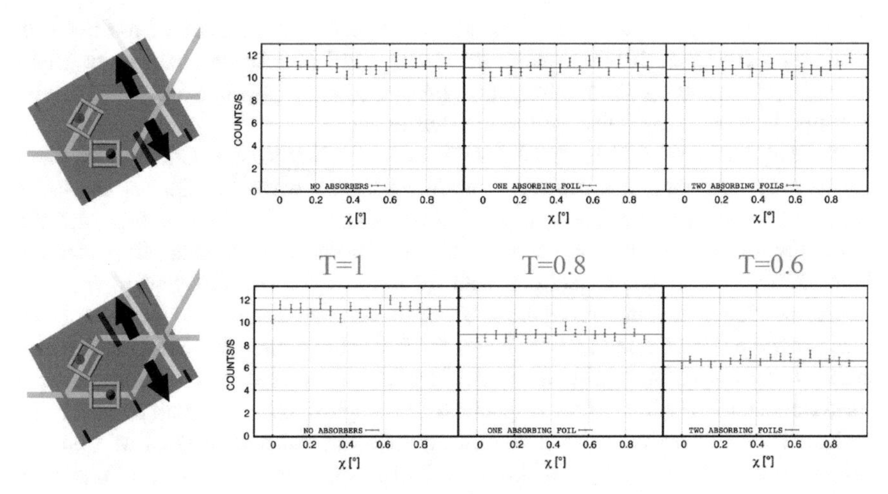

Fig. 24.8 Results of the measurements of neutron's population: absorbers with transmissivity, T = 0.8 and 0.6 are inserted in one of the beam path in the interferometer. The absorbers in the lower path do not affect the final intensity of the O-beam with a spin-analysis (*upper panel*). Intensity decreases by inserting the absorbers in the upper path (*lower panel*) These results suggest that neutrons are travelling through the upper beam path.

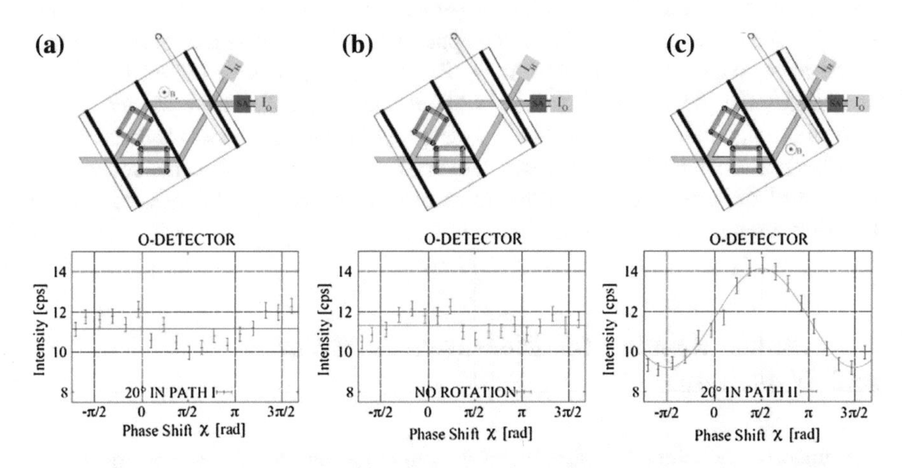

Fig. 24.9 Results of the measurements of the location of the neutron's spin: the spin in the *upper/lower* beam path is rotated slightly (~15°) by a weak magnetic field. The magnetic field in the *upper path* does not affect the final intensity of the O-beam with a spin-analysis. In contrast, sinusoidal intensity modulation appears by applying the magnetic field in the *lower path*—that is the path, which the neutrons are actually never supposed to take.

sinusoidal intensity modulation appears by applying the magnetic field in the path I (lower path): effective is the magnetic field in the beam path I (lower path)—that is the path, which the neutrons are actually never supposed to take. These two measurements together imply that one component of the neutron's spin is travelling through the interferometer following the beam path II in absence of the neutron, while the neutron itself is following the beam path I in absence of the spin. These results are consistent with the theoretical prediction [34]: physical properties can be disembodied from the object in the interferometer.

A keen reader would ask whether one can view the quantum Cheshire-cat alternatively by following the evolution of the neutron's wave function or not. This argument is clarified here. Looking at the experimental setup depicted in Fig. 24.7, the population of the neutrons can be intuitively understood. The spin-analysis behind the interferometer allows only neutrons with the spin in forward direction to transmit. These are exactly the neutrons in the upper beam path: the neutrons in the lower beam path will be filtered out by the spin-analyzer. The first measurement shows exactly this situation. (See Fig. 24.8) How about the location of the neutron's spin? In the spin measurement of the spin, a weak magnetic field is applied: the spin-vector in each beam paths is rotated. That is, the spin in forward/backward direction is deviated from the original direction by the angel α, followed by the spin analysis in forward direction. Since the spin-analysis is described by the projection, the intensities after the spin-analysis on spins deviated from the parallel and the anti-parallel are proportional to $\cos(\alpha/2)$ and $\sin(\alpha/2)$, respectively. (This is analogous to the transmission of the polarized light through a polarizing filter.) Therefore, the differences in intensity with and without the weak magnetic field result in $\cos(\alpha/2) - 1 \approx -\alpha^2/8$ (parallel spin in the upper path) and $\sin(\alpha/2) - 0 \approx \alpha/2$ (anti-parallel spin in the lower path). This means, the influence of the magnetic field in the upper path is in the first order of α, while that in the lower path is in the second order. For the parameter, $\alpha \sim 15°$ realized in the experiment, the former becomes 0.13, while the latter is 0.0085: the former is more than one magnitude larger the latter. This phenomenon is clearly observed in the interferograms obtained in the second measurement. (See Fig. 24.9) A major change in the case of the magnetic field in the lower path for small α is attributed to the spin traveling through the lower path.

Uncertainty Relation for Measurement Error and Disturbance

The uncertainty principle refers to intrinsic indeterminacy of quantum mechanics and ranks among the most famous statements of modern physics. It describes the limitations of simultaneous measurements of certain pairs of observables: there are physical properties, which can be accessed only with limitation. It was Heisenberg who first formulated the uncertainty relation as a limitation of accuracies of position and momentum measurements [35]. He proposed a reciprocal relation for measurement *mean-error* (nowadays referred to as *error*) and *discontinuous-change* (nowadays referred to *disturbance)* using the famous γ-ray microscope thought experiment: "*At the instant when the position is determined—therefore, at the moment when the photon is scattered by the electron—the electron undergoes a discontinuous change in momentum. This change is the greater the smaller the wavelength of the light employed—that is, the more exact the determination of the position... Thus, the more precisely the position is determined, the less precisely the momentum is known, and conversely... Let* q_1 *be the precision with which the value* q *is known (*q_1 *is, say, the mean error of* q*), therefore here the wavelength of the light. Let* p_1 *be the precision with which the value* p *is determinable; that is, here, the discontinuous change of* p *in the Compton effect* [35]." Finally, he gave a rather heuristic estimate for the product of the error of a position measurement q_1 and the disturbance p_1 induced on the particle's momentum, as $p_1 \cdot q_1 \sim h$. (This relation is read in modern treatment as $\varepsilon(\mathrm{Q}) \cdot \eta(\mathrm{P}) \geq \hbar/2$ for the error $\varepsilon(\mathrm{Q})$ of the position measurement Q and the disturbance $\eta(\mathrm{P})$ of the momentum measurement P.)

Later on, the uncertainty relation was reformulated in terms of standard deviations by Kennard [36] and Robertson [37] as $\sigma(\mathrm{Q})) \cdot \sigma(\mathrm{P}) \geq \hbar/2$ for position (Q)/momentum(P) measurements and $\sigma(\mathrm{A})) \cdot \sigma(\mathrm{B}) \geq ½| < \Psi|[\mathrm{A}, \mathrm{B}]|\Psi > |$ for arbitrary A/B measurements. It is worth noting here that the quantities such as $\sigma(\mathrm{A})$ $\sigma(\mathrm{B})$ are logically unrelated with errors/disturbances of measurements; the physical situations concerned are not simultaneous measurements of A and B but rather the measurements of one or the other of these dynamical variables on each independently prepared representative of the particular state being studied [38]. The relation in the form with standard deviations describe the limitation in preparing microscopic objects but no direct relevance to the limitation of precision of quantum measurements. In his famous book, von Neumann wrote "*We shall first derive the most important relation of this type mathematically, and then return to its fundamental meaning, and its connection with experiments...With the foregoing considerations, we have comprehended only one phase of the uncertainty relations, that is, the formal one* (sigma-sigma form obtained by Kennard/Robertson, remark by the author)*; for a complete understanding of these relations, it is still necessary to consider them from another point of view: from that of direct physical experience. For the uncertainty relations bear a more easily understandable and simpler relation* (error-disturbance form suggested by Heisenberg, remark by the author) *to direct experience than many of the facts on which quantum mechanics was originally based, and therefore the above, entirely formal, derivation does not do them full justice* [39]." In lots of textbooks of quantum mechanics, the Heisenberg's uncertainty principle is still referred to be related with the inequality defined for standard deviations [40].

It was known that the validity of Heisenberg's original error-disturbance relation is justified only under limited circumstances; Ozawa introduced the correct form of a generalized error–disturbance uncertainty based on rigorous theoretical treatments of quantum measurements as

$$\varepsilon(A)\cdot\eta(B)+\varepsilon(A)\cdot\sigma(B)+\sigma(A)\cdot\eta(B)\geq\frac{1}{2}|\langle\Psi|[A,B]|\Psi\rangle|, \tag{2}$$

and proved its universal validity [12, 13]. Here, $\varepsilon(A)$ denotes the root-mean-square (r.m.s.) error of the measurement for an observable A, $\eta(B)$ is the r.m.s. disturbance on the measurement of another observable B induced by the measurement A, and $\sigma(A)$ and $\sigma(B)$ are the standard deviations of the measurement A and B for the state $|\psi\rangle$. Note that the additional second and third terms imply a new accuracy limitation, which does not necessarily follow the trade-off relation of error and disturbance.

It took about a decade after the publication of the theoretical woks [12, 13], that an experimental test of the new error-disturbance uncertain relation was performed: a successive spin-measurement of neutrons, i.e. σ_x and σ_y, that allows determining the error of the first x spin-component measurement and the disturbance caused on another y spin-component measurement [10]. The experimental setup is depicted in Fig. 24.10. The initial state of the neutron's spin is tuned to be $|+z\rangle$. The observable for the preceding and the following measurements, carried out by M1 and M2, are set as $A = \sigma_x$ and $B = \sigma_y$. Consequently, the right term of Eq. (2) becomes 1: a maximum uncertainty is expected under these circumstances. Then, in order to investigate behaviors of the error and the disturbance systematically, we took an approach where the observable of the first measurement $O_A(\phi)$ is detuned from the observable-to-be ($A = \sigma_x$) on purpose: in practice, $O_A(\phi)$ is set $O_A(\phi) = \sigma_x \cdot \cos\phi + \sigma_y \cdot \sin\phi$. This detuning turned out to be the origin of the measurement error of the measurement A and was able to reduce the disturbance of the measurement B. In the Bloch sphere description, the vectors representing these

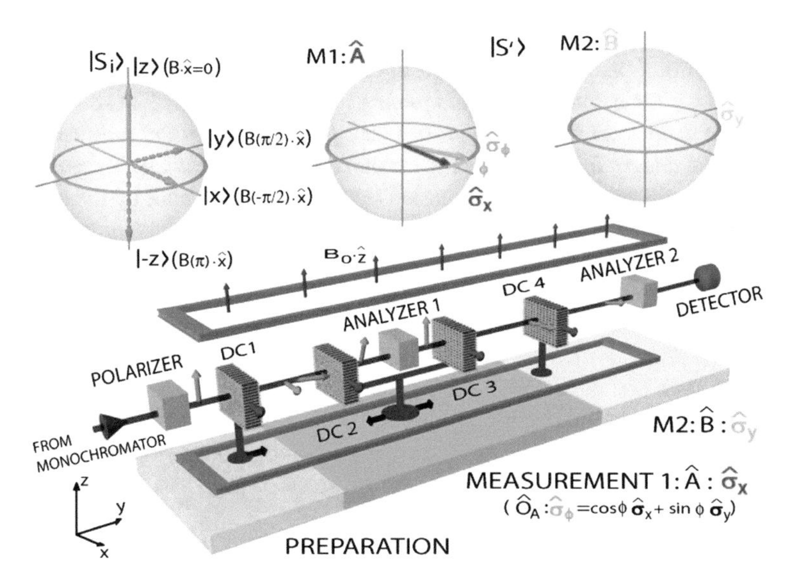

Fig. 24.10 Illustration of the experimental setup to test the new error-disturbance uncertainty relation in the successive spin measurements σ_x and σ_y [10].

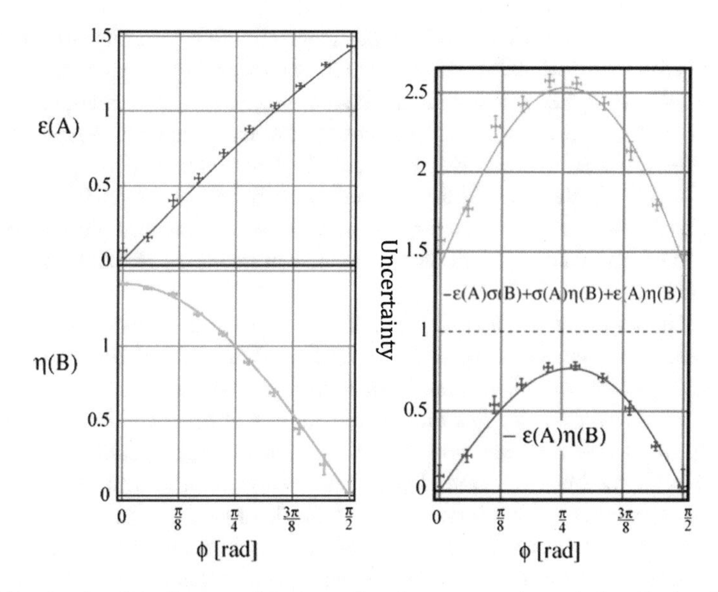

Fig. 24.11 Results of the first test of the error-disturbance uncertainty-relation. Trade-off relation is seen for the error ε(A) and the disturbance error η(B) (*left*). Plot of the error-disturbance uncertainty relations (*right*). The new (*orange*) term is always above the calculated limit, while the old term (*red*) is below the limit.

observables and the state are lying on the equator, which is also depicted in Fig. 24.10.

The experiment was carried out at the research reactor facility TRIGA Mark II of the Vienna University of Technology. The monochromatic neutron beam is polarized crossing a super-mirror polarizer and two other super-mirrors are used as analyzers. The guide field together with four DC spin rotator coils, induces Larmor precession to allow state preparation and projective measurements of $O_A(\phi)$ in M1 and B in M2. To test the error-disturbance uncertainty relation given by Eq. (2), the standard deviations σ(A), σ(B), the error ε(A) and the disturbance η(B) are determined from the experimentally obtained data. The trade-off relation of the error and the disturbance is clearly seen. (See Fig. 24.11 on the left) The error-disturbance uncertainty relations, following the Eq. (2) and the product suggested by Heisenberg, are depicted in Fig. 24.11 on the right: the determined values confirm the fact that the Heisenberg product is always below the calculated limit, and that the universally valid expression in Eq. (2) is always larger than the limit.

The neutron's spin-measurement of ours is the first experimental test of the error-disturbance uncertainty relation. The validity of the new relation (Eq. (2)) proposed as a universally valid error-disturbance relation is demonstrated; moreover the failure of the old relation as a reciprocal relation between the error and the disturbance is also illustrated. This experiment stimulated further studies on the error-disturbance uncertainty relation: in fact, experiments using a photonic system appeared [41–43]. All measurements concern photon's polarization, which is described as a two-level quantum system in the same manner as the ½-spin of neutrons. Afterwards, a remarkable extension in theory was made; a tighter relation was obtained [44] and its validity is confirmed by experiments [45, 46]. Now, uncertainty relations have become again a hot topic in quantum physics, after the publication of the first account of the uncertainty principle by Heisenberg more than four decades ago.

In writing the paper, we actually thought, "*our result demonstrates that the new relation solves a long-standing problem of describing the relation between measurement accuracy and disturbance, and sheds light on fundamental limitations of quantum measurements, for instance on the debate of the standard quantum limit for monitoring free-mass position* [10]". Nevertheless, *ex post facto* critical analysis appeared [47]: for instance, state-dependence of the error and the disturbance by Ozawa is claimed and state-independent definitions of error and disturbance are proposed to reconstruct the error-disturbance uncertainty relation in the same form as the original one proposed by Heisenberg [48]. It turned out that the newly defined error and disturbance there characterize, for instance, the disturbing power of the measurement devices for hypothetical state, somehow in a worst case scenario: this is regarded unsuitable in a sense of the disturbance typically associated with Heisenberg's microscope [49]. Furthermore, a paper is published which deal with "operational constraints" on the measures of the error and disturbance [50]: it is stated that, since "*only the change in the measurement statistics can be detected by the measurement,*" "*a measurement cannot be treated as disturbed if its outcome statistics is identical to the one for the perfect measurement.*" (underlines given by the author) The fact that this view does not accomplish its intended purpose is clearly seen in the first experimental test of ours [10]: as we already emphasized as "*It is worth noting that the mean value of the observable A is correctly reproduced for any detuning angle* ϕ*, that is,* $\langle +z|O_A| + z\rangle = \langle +z|A| + z\rangle$*, so that the projective measurement of* O_A *reproduces the correct probability distribution of A, whereas we can detect the non-zero r.m.s. error* $\varepsilon(A)$ *for* $\phi \neq 0$*.* [10] ", it is *physically* reasonable that the difference of the observable $O_A \neq$ A for $\phi \neq 0$, which is realized in the experiment, leads to the *error* of the measurement, even though the measurement-results are identical. This is unsurprisingly understood when one considers, for instance, an apparatus which (is broken and) always gives the results of the measurement as (+1) and (–1) with a fifty-fifty chance: can one regard this not a *causal* but an *accidental* coincidence as (physically) error-free? As far as *physical* consequences are concerned, *causal* differences, which can appear even in an operational form, are resources of the measurement error/disturbance; modern quantum measurement schemes, i.e. process tomography or that in combination with weak-values, can actually reveal the operational difference. The *functional* differences, emerging only in the final results, can be considered as *informational* aspects of the measurements. Indeed, another form of noise-disturbance uncertainty relation in context of *information-theoretic approach* is derived [51], where the correlations of the measurement results are considered as the resource.

Concluding Remarks

In the above sections, recent neutron interferometer and polarimeter experiments are explained, which exploit fully the properties described in terms of wave-functions and Schrödinger equations. Correlations between measurements of observables belonging to different Hilbert space, thus commuting, are observed, which is stronger than the predictions by models extended simply from classical physics. In addition to the fact that the results of measurements in quantum mechanics can be foreseen only as probabilistic numbers following the Born's rule, quantum contextuality demands that they cannot have definite values before the measurements. So far, (kinematically) non-local and non-classical correlations of the quantum measurements are considered to be a resource for a killer application, e.g. in quantum information and communication technology. Instead, the way how to build a view to obtain results of measurements in quantum mechanics is unknown and the explanation of what is really going on in quantum world is still missing. A serious (re)consideration of quantum *dynamics* (rather than the results of measurements) is one attempt to explore veiled mysteries: for instance, dynamical quantum non-locality appearing in the observation of Aharonov-Bohm (AB) effect is argued [2]. In this procedure, all quantum effects *en route* such as superposition, interaction, phase shift and interference, are to be reexamined. Last two examples explained in this article can be seen from this view-point: quantum Cheshire-cat emerges in superposing quantum states in the interferometer and new (operational) definitions of measurement error and disturbance in quantum mechanics provide new and deeper insights into error-disturbance uncertainty relation.

Acknowledgments This work has been supported by the Austrian Science Foundation (FWF), most recently by the project P25795-N20. The author would like to thank all colleagues who are involved in realizing experiments described here.

References

1. R.P. Feynman, P.B. Leighton, M.L. Sands, *The Feynman Lectures on Physics*, vol. III: Quantum Mechanics (Addision-Wesley, Reading, MA, 1965)
2. S. Popescu, Nature Phys. **6**, 151–153 (2010)
3. M. Arndt, A. Ekers, W. von Klitzing, H. Ulbricht, New J. Phys. **14** 125006 (2012) and references therein

4. H. Rauch, S.A. Werner, *Neutron Interferometry* (2nd Ed., Clarendon Press, Oxford, 2015)
5. Y. Hasegawa, R. Loidl, G. Badurek, M. Baron, H. Rauch, Nature **425**, 45 (2003)
6. Y. Hasegawa, R. Loidl, G. Badurek, K. Durstberger-Rennhofer, S. Sponar, H. Rauch, Phys. Rev. A **81**, 032121 (2010)
7. Y. Hasegawa, C. Schmitzer, H. Bartosik, J. Klepp, S. Sponar, K. Durstberger-Rennhofer, G. Badurek, New J. Phys. **14**, 023039 (2012)
8. Y. Aharonov, S. Popescu, D. Rohrlich, P. Skrzypczyk, New J. Phys. **15**, 113018 (2013)
9. T. Denkmayr, H. Geppert, S. Sponar, H. Lemmel, A. Matzkin, J. Tollaksen, Y. Hasegawa, Nat. Commun. **5**, 4492 (2014). doi:10.1038/ncomms5492
10. J. Erhart, S. Sponar, G. Sulyok, G. Badurek, M. Ozawa, Y. Hasegawa, Nat. Phys. **8**, 185 (2012)
11. G. Sulyok, S. Sponar, J. Erhart, G. Badurek, M. Ozawa, Y. Hasegawa, Phys. Rev. A **88**, 022110 (2013)
12. M. Ozawa, Phys. Rev. A **67**, 042105 (2003); ibid., Phys. Lett. A **318**, 21 (2003)
13. M. Ozawa, Ann. Phys. (N.Y.) **311**, 350 (2004)
14. J. Klepp, S. Sponar, Y. Hasegawa, Prog. Theor. Exp. Phys. 082A01 (2014)
15. A. Einstein, B. Podolsky, N. Rosen, Phys. Rev. **47**, 777 (1935)
16. J.S. Bell, Physics **1**, 195 (1964)
17. D.M. Greenberger, M.A. Horne, A. Shimony, A. Zeilinger, Am. J. Phys. **58**, 1131 (1990)
18. S. Kochen, E.P. Specker, J. Math. Mech. **17**, 59 (1967)
19. H. Geppert, T. Denkmayr, S. Sponar, H. Lemmel, Y. Hasegawa, Nucl. Instr. Methods A **763**, 417 (2014)
20. Y. Hasegawa, R. Loidl, G. Badurek, M. Baron, H. Rauch, Phys. Rev. Lett. **97**, 230401 (2006)
21. A. Cabello, S. Filipp, H. Rauch, Y. Hasegawa, Phys. Rev. Lett. **100**, 130404 (2008)
22. H. Bartosik, J. Klepp, C. Schmitzer, S. Sponar, A. Cabello, H. Rauch, Y. Hasegawa, Phys. Rev. Lett. **103**, 040403 (2009)
23. L. Allen, J.H. Eberly, *Optical Resonance and Two-Level Atoms* (Dover Pub. Inc., NY, 1987)
24. E. Muskat, D. Dubbers, O. Schärpf, Phys. Rev. Lett. **58**, 2047 (1987)
25. J. Summhammer, Phys. Rev. A **47**, 556 (1993)
26. G. Badurek, H. Rauch, J. Summhammer, Phys. Rev. Lett. **51**, 1015 (1983)
27. G. Badurek, H. Rauch, D. Tuppinger, Phys. Rev. A **34**, 2600 (1986)
28. S. Sponar, J. Klepp, R. Loidl, S. Fillipp, G. Badurek, Y. Hasegawa, H. Rauch, Phys. Rev. A **78**, 061604 (2008)
29. M. Huber, F. Mintert, A. Gabriel, B.C. Hiesmayr, Phys. Rev. Lett. **104**, 210501 (2010)
30. D. Erdösi, M. Huber, B.C. Hiesmayr, Y. Hasegawa, New J. Phys. **15**, 0233033 (2013)
31. A.J. Leggett, Found. Phys. **33**, 1469 (2003)
32. S. Gröblacher, T. Paterek, R. Kaltenbaek, Č. Brukner, M. Zukowski, M. Aspelmeyer, A. Zeilinger, Nature (London) **446**, 871 (2007)
33. S. Filipp, Y. Hasegawa, R. Loidl, H. Rauch, Phys. Rev. **A72**, 021602 (2005)
34. Y. Aharonov, S. Popescu, D. Rohrlich, P. Skrzypczyk, New J. Phys. **15**, 113018 (2013)
35. W. Heisenberg, Z. Phys. **43**, 172 (1927)
36. E.H. Kennard, Z. Phys. **44**, 326 (1927)
37. H.P. Robertson, Sitzungsberichte der Preussischen Akademie der Wissenschaften **14**, 296 (1930); ibid. Rev. Mod. Phys. **42**, 358 (1970)
38. L.E. Ballentine, *Quantum Mechanics: A Modern Development* (World Scientific Pub. Co., NJ, 1998)
39. J. von Neumann, *Mathematische Grundlagen der Quantenmechanik* (Springer, Berlin, 1932); *ibid Mathematical Foundations of Quantum Mechanics* (Princeton Univ. Press, NJ, 1955) (Engl. Transl.)
40. L.I. Schiff, *Quantum Mechanics* (McGraw-Hill Inc, NY, 1968)
41. L.A. Rozema, A. Darabi, D.H. Mahler, A. Hayat, Y. Soudagar, A.M. Steinberg, Phys. Rev. Lett. **109**, 100404 (2012)
42. S.-Y. Baek, F. Kaneda, M. Ozawa, K. Edamatsu, Sci. Rep. **3**, 2221 (2013)

43. M.M. Weston, M.J.W. Hall, M.S. Palsson, H.M. Wiseman, G.J. Pryde, Phys. Rev. Lett. **110**, 220402 (2013)
44. C. Branciard, Proc. Nat. Acad. Sci. U.S.A. **110**, 6742 (2013)
45. M. Ringbauer, D.N. Biggerstaff, M.A. Broome, A. Fedrizzi, C. Branciard, A.G. White, Phys. Rev. Lett. **112**, 020401 (2014)
46. F. Kaneda, S.-Y. Baek, M. Ozawa, K. Edamatsu, Phys. Rev. Lett. **112**, 020402 (2014)
47. X.-M. Lu, S. Yu, K. Fujikawa, C.H. Oh, Phys. Rev. A **90**, 042113 (2014). and references therein
48. P. Busch, P. Lahti, R.F. Werner, Phys. Rev. Lett. **111**, 160405 (2013)
49. L.A. Rozema, D.H. Mahler, A. Hayat, A. M. Steinberg, arXiv:1307.3604 [quant-ph]
50. K. Korzekwa, D. Jennings, T. Rudolph, Phys. Rev. A **89**, 052108 (2014)
51. F. Buscemi, M.J.W. Hall, M. Ozawa, M.M. Wilde, Phys. Rev. Lett. **112**, 050401 (2014)

Part VIII
Bell Inequalities—Experiment

Chapter 25
Nonlocality and Quantum Cakes, Revisited

Bradley G. Christensen and Paul G. Kwiat

Abstract Entanglement is a nonintuitive feature of quantum mechanics, leading to various nonlocal phenomena. For example, entangled states can display nonlocal correlations stronger than allowed by any local realistic theory, violating a Bell inequality, assuming various experimental loopholes are addressed. The Hardy paradox allows us to find a more familiar example of the difference between nonlocality and classical expectations. Here, we review the Hardy paradox and the "quantum cakes" example, and present a source of high-quality entangled photons with the best-to-date violation of this paradox.

Introduction

One of the most shocking aspects of quantum mechanics is that of entanglement. In their paper predicting the existence of entanglement, Einstein, Podolsky, and Rosen thought it was so counter-intuitive to how the world should behave that they instead concluded it was quantum mechanics that is wrong rather than our view of reality [1]. Today, entanglement remains difficult to grasp, even for physicists, and even more challenging to describe to non-physicists, despite that it is the quintessential quantum mechanical feature, "the one that enforces its entire departure from classical lines of thought" [2].

The most natural view of reality is arguably that of local realism. That is, we naturally believe any two non-causal events should have no influence on each other (locality), and that the probability of any outcome of a measurement on a system should only depend on the state of that system (realism). However, quantum mechanics does not follow local realism, as noted by Einstein, Podolsky, and Rosen. For some time, this remained a mere philosophical point, until Bell showed that there are statistical

B.G. Christensen · P.G. Kwiat (✉)
Department of Physics, University of Illinois at Urbana-Champaign,
Urbana, IL 61801, USA
e-mail: kwiat@illinois.edu

B.G. Christensen
e-mail: bgchris2@illinois.edu

R. Bertlmann and A. Zeilinger (eds.), *Quantum [Un]Speakables II*,
The Frontiers Collection, DOI 10.1007/978-3-319-38987-5_25

differences between predictions of quantum mechanics and that of a local realistic theory, which is quantified in terms of an inequality that local realism must obey [3]. The most common of these Bell inequalities is the Clauser-Horne-Shimony-Holt Bell inequality [4], which has now been used to verify the existence of entanglement in a plethora of systems [5–9], to the extent that now it is even an undergraduate lab experiment [10]. However, despite the experiments being "simple", until very recently there were "loopholes" in every implementation.

Loopholes

Loopholes in Bell inequalities arise from the need to still make assumptions about the system or source. That is, local realism could still be a viable description if the assumptions are incorrect. The two most common assumptions made are the fair-sampling assumption and the no-signaling assumption. The first assumption is that the detected particles are an accurate representation of all particles emitted from the source. To close this loophole, one needs system efficiencies over 2/3 (with no background noise) [11], which is incredibly challenging experimentally. As an example, when photons are used to test nonlocality, typical detectors will have efficiencies ranging from 20 to 65 %, which by itself is already too low of a system efficiency to close this loophole. Instead, transition-edge-sensor (TES) detectors have been used, as they can have efficiencies exceeding 95 % (at the cost of requiring temperatures down to 100 mK) [12]. In addition to low detection efficiency, there is additional loss in the system from optical elements, as well as loss through filtering the correct wavelengths and spatial modes. The two systems that first closed this loophole with photons had efficiencies barely over the required threshold [13, 14].

The locality loophole is the statement that the two detection events would have time to send signals to each other. That is, to test local realism, one needs to make measurements that are strictly nonlocal. This loophole is challenging to close due to the distance separation that must occur to keep all relevant events (measurement setting choice and detection of the particle) far enough apart to prevent this loophole. While this loophole has also been closed [15, 16], the additional loss incurred by increasing the separation of the detections has made closing both loopholes simultaneous a challenge, and only very recently has this been done.

To prevent either of these loopholes requires an impressive amount of effort (both to collect nearly all of the entangled photons, and to ensure the relevant events are all space-like separated). But recent advances across many different systems have led to four loophole-free Bell test experiments using three different sources of entanglement: spins in nitrogen-vacancy centers [17], polarization of photons [18, 19], and spins of rubidium atoms [20].

Entanglement Source

Our entanglement source consists of a 355 nm pulsed pump laser incident onto two orthogonal nonlinear BiBO crystals to produce polarization-entangled photon pairs at 710 nm, via spontaneous parametric down-conversion [21]: the first crystal has an amplitude to convert a vertically polarized pump photon into a pair of horizontally polarized daughter photons, while the second has an amplitude to convert a horizontally polarized pump photon into a vertically polarized pair. Using wave plates to control the polarization of the pump beam, we can thus control the relative amplitude (and phase) of the $|HH\rangle$ and $|VV\rangle$ terms [22].

Our source achieves extremely high state quality, once we pre-compensate the temporal decoherence from group-velocity dispersion in the down-conversion crystals with a birefringent crystal [23], resulting in an interference visibility of 0.997 ± 0.0005 in all bases. The high state quality (along with the capability of creating the required state) allows us to make measurements very close to the optimal values for standard Bell tests (e.g., CHSH) and for implementing the Hardy paradox, discussed below.

For the Bell tests, the local polarization measurements are implemented using a fixed Brewster angle polarizing beam splitter, preceded by an adjustable half-wave plate, and followed by single-photon detectors to detect the transmitted photons (see Fig. 25.1). We have used this source to measure a CHSH Bell value of $2.8261 \pm$

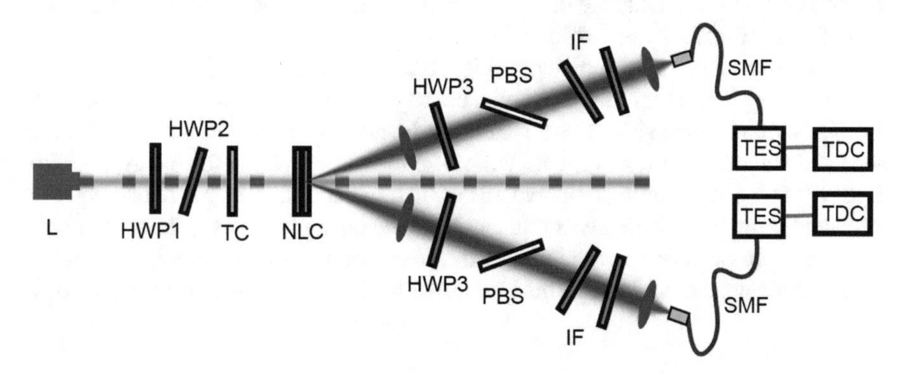

Fig. 25.1 A diagram of the entanglement source. The laser (L) is prepared in a specific polarization state by two half-wave plates (HWP1 and HWP2). We compensate for the temporal decoherence (arising from the group velocity dispersion in the downconversion crystals) by passing through the laser through a crystal (TC) designed to have the opposite group velocity dispersion. Passing the pump through a pair of orthogonal nonlinear crystals (NLC) produces the entangled photons. The measurements are performed using a motorized half-wave plate (HWP3) and a polarizing beam splitter (PBS). We then spectrally filter (IF) the photons to ensure the collected bandwidth is not too large, as well as spatially filter the photons using a single-mode fiber (SMF) to remove any spatial decoherence. Finally, the photons are then detected using transition-edge sensors (TES), the events of which are recorded on a time-to-digital converter (TDC) and saved on a computer for analysis.

0.0004, with the local bound being 2, and the quantum mechanical bound being $2\sqrt{2} \approx 2.8284$.

Quantum Cakes

If the underlying meaning of experiments on nonlocality is difficult to understand, it is even more difficult to explain. An alternative CHSH-like Bell test, the Hardy paradox, is a simpler experiment to explain since it can be seen in terms of an "all-or-nothing" argument [24]. We can write the Hardy paradox as [22]

$$p_{12}(1,1|L_1,R_1) - p_{12}(1,0|L_1,R_0) - p_{12}(0,1|L_0,R_1) - p_{12}(1,1|L_0,R_0) \leq 0, \quad (1)$$

where $p_{1,2}(m,n|L_i,R_j)$ is the probability of a detection event where one side (here called "Lucy") detected outcome m while using setting L_i, and the other side ("Ricardo") detected outcome n while using setting R_j. Any local realistic theory must satisfy the inequality, that is, be less than or equal to 0. However, quantum mechanics (a nonlocal theory) is able to achieve values greater than 0. Here, we are not interested in the maximal quantum violation attainable with this inequality; instead we focus on values which make this inequality easy to understand. We choose to measure the values that are described by the "quantum cakes" example presented in Ref. [25] (Fig. 25.2 and Table 25.1).

Consider the example depicted in Fig. 25.2. Here, there is a single kitchen with two opposing doors, out of which come conveyor belts, and on the belts come pairs of ovens, one to each side. There is an experimenter on each side, Lucy on the left side and Ricardo on the right, who will make measurements on the ovens; later the two will come together to compare their results. In particular, there are two types of (noncommuting) measurements that can be made on a given oven. The tester could wait until the oven reaches the end of the conveyor belt before opening it. Inside, he/she finds a cake, which can then be tested to see whether it tastes good or bad.

Fig. 25.2 Lucy and Ricardo use their knowledge of quantum nonlocality to start up a quantum cakes factory. If Ricardo's cake rose early (*far right*), then Lucy's cake (*far left*) will always taste Good, and vice versa. Despite the fact 9 % of the time both cakes rise early, they never both taste Good! Fig. from [25].

Table 25.1 Raw counts for the Hardy paradox

Probability term	Lucy's singles	Coincidences[a]	Ricardo's singles
p(*Risen*, *Risen*)	62067	16135	58878
p(*Bad*, *Risen*)	150439	205	59481
p(*Risen*, *Bad*)	61154	207	147748
p(*Good*, *Good*)	92945	252	92878

[a]Normalization for the coincidences is 174245; see text for details

This is one observable, the taste of the cake. Alternatively, the tester can open the oven midway on its journey, to see whether or not the batter has risen early or not risen early, the second observable. Assuming we have some sort of soufflé, it is easy to justify why these measurements might be noncommuting, as re-closing the oven in the middle will cause the cake to collapse, and the result will always be a destroyed cake. Hence, only one of these qualities can be measured on a given cake.

Each experimenter will randomly decide which measurement they will make, and record the results obtained. Comparing the records later on will reveal the strangeness which arises if the cakes are made using entangled dough. There are three main classes to consider, depending on whether Lucy and Ricardo both opened their respective ovens in the middle, one waited until the end to do so, or both did. The results of the first two types of measurements are:

(a) In cases where Lucy and Ricardo both checked their ovens midway, they find that 9 % of the time, both cakes rose early.
(b) When Lucy's cake rose early, Ricardo's tasted Good; similarly, when Ricardo's cake rose early, Lucy's tasted Good.

Classically, we can make some inference as to the cake's history, based on the correlations in (b). That is, when one set of dough rose early, the other cake tasted Good, so we can guess that the cakes were made perhaps from the same dough (or the two doughs at least had some interaction in their history) or by the same chef, and so rising early implies that the (not destroyed) cake will taste Good (though a cake still can taste Good if the batter did not rise early). Now, we can make further logically valid predictions of what should happen when both Lucy and Ricardo taste their respective cakes without checking if they rose early. From (a), we know 9 % of the time both cakes rise early, and from (b) we know if one person's cake rises early, that the other person's cake will taste Good; consequently, it follows logically to conclude that at least 9 % of the time both cakes will taste Good. However, if these two cakes are entangled in a particular way, then Lucy and Ricardo *never* find both cakes tasting Good simultaneously, despite the logical conclusion one would reach assuming local realism. The "flaw" in the classical argument was an implicit (but quantum mechanically invalid) assumption that getting a particular result on one side (e.g., cake tastes Good) cannot affect the result on the other side. Such nonlocal correlations are intrinsic to quantum entanglement. Note, however, that there is still no way to control the outcome of the other party—this would enable superluminal communication and violate causality.

While our technology is currently far from making quantum superpositions of cakes, we can still do an equivalent demonstration using entangled photons. For example, we can produce two photons that are nonmaximally entangled with respect to their polarization (it is important that these photons are nonmaximally entangled for the paradox to work, which is an interesting topic on its own). In the cake example we considered measurements of two properties of the cake: the taste and whether it had risen. In the case of a polarization entangled state, these correspond to measurements of polarization in two different bases. Such a measurement can be performed by having a rotatable polarizing beamsplitter with two detectors (in our experiment, and many others, the beamsplitter is "rotated" by rotating the basis it will measure in with a waveplate instead of physically rotating the beamsplitter), one placed in each output port. When this polarizing beamsplitter is oriented at one angle (e.g., 0°), a click at one detector corresponds to the cake tasting Good and a click at the other detector corresponds to the cake tasting Bad. When the polarizing beamsplitter is rotated to another angle, a click at one detector corresponds to the cake having Risen, and a click at the other detector corresponds to the cake Not having Risen. With these correspondences understood, if we identify, for example, "Good" and "Bad" with Horizontal (H) 0° and Vertical (V) 90° polarization, respectively, and "Risen" and "Not Risen" with linear polarizations at 250.8° and 39.2°, respectively, the predictions above will all hold exactly if we use the state

$$|\psi\rangle = 1/2|VV\rangle - \sqrt{3/8}(|VH\rangle + |HV\rangle). \tag{2}$$

Note that there is no $|HH\rangle$ contribution, i.e., no "Good-Good" outcome in the cakes analogy.

Results

Using the source described in Sec. "Entanglement Source", we make the necessary measurements to show the Hardy paradox. To reference back to the quantum cakes example, the measurement of both cakes rising early is the probability term $p_{12}(1, 1|L_1, R_1)$ (in the quantum cakes example, this is 9 %), and the measurement of one cake rising early and the other tasting bad corresponds to the $p_{12}(1, 0|L_1, R_0)$ and $p_{12}(0, 1|L_0, R_1)$ terms (in the example this was 0 % since a Risen cake always implied the other one would taste Good). Finally, both cakes tasting Good is the $p_{12}(1, 1|L_0, R_0)$ term; here we expect this to be 0 % quantum mechanically, but 9 % classically. The classical bound for these measurements is then given by the following inequality:

$$p(Risen, Risen) - p(Risen, Bad) - p(Bad, Risen) - p(Good, Good) \leq 0. \quad (3)$$

As we only detect the transmitted photons, to measure both the 0 and 1 events, we use two different half-wave plate settings (that correspond to orthogonal polarization measurements). To extract probabilities, we measure each possible setting and outcome (i.e., $p(Bad, Risen) + p(Good, Risen) + p(Bad, NotRisen) + p(Good, NotRisen)$), and use this to normalize all events. Our data is given in Table 25.1.

Applying the normalization, we then measure the Hardy paradox of $9.3\,\% - 0.12\,\% - 0.12\,\% - 0.14\,\% \leq 0$, a clear violation of the paradox, and remarkably close to $0\,\%$ on the three terms. However, we can do even better, as our detectors had poor temporal resolution, so it was rather common to detect an "accidental" coincidence count; i.e., there were events where our detector saw photons from two different entangled states at the "same" time (according to the detector). This results in an increase in coincidence counts because uncorrelated photons were detected within the temporal resolution of the detector. To first order, these events are given by $S_L S_R \Delta w/t$, where $S_{L(R)}$ is the singles counts on Lucy's (Ricardo's) side, Δw is the coincidence window size (1 μs), and t is the duration for the measurement (60 s). After accidental correction, we see the Hardy paradox of $9.2\,\% - 0.032\,\% - 0.032\,\% - 0.062\,\% < 0$.

Loopholes and Cakes

The Hardy paradox is nice in terms of the potential to explain the results. As an "all-or-nothing" Bell inequality (that is, one where three of the terms should be 0, which leads to a logical conclusion that the fourth must also be zero), it lends itself to an intuitive description of the oddity of entanglement. So far we have neglected any form of loophole, but it also can be seen from the quantum cakes example how these loopholes can occur. Just as with standard Bell tests, nonlocality is not required as an explanation if there are loopholes. For example, if the chef that made the cakes is spying upon Ricardo, and upon seeing his result has time to alter Lucy's cake before she sees the results (either if the cake is Risen or if it tastes Good), then the chef can make them reach any conclusion he desires. This is the locality loophole. One can also see how, in a real experiment, it is important to look at as many of the "cakes" as possible. In the simplest possible argument, if we were to measure only 91 % of the pairs, and still see no Good-Good events, it is logically possible that the predicted 9 % were just those that we did not measure. Obviously, this would require a rather peculiar sampling of the cakes, so one often makes a "fair-sampling" assumption, which leads to the detection loophole discussed earlier.

Conclusion

We have made measurements of a CHSH Bell inequality, as well as the inequality viewed as the Hardy paradox. By looking at the CHSH Bell inequality in this paradoxical way, we can make "all-or-nothing" type measurements (either there is an probability of a detection event, or there is not). With the source described here, we have realized measurements extremely close to the theoretically predicted values of both the Hardy paradox, as well as the quantum mechanical maximum for the CHSH Bell inequality. In some sense such results have little to do with quantum mechanics per se, since Bell inequalities refer only to the exclusion of local realistic models. As such, Bell tests are not limited to quantum mechanics and can be useful to characterize nonlocality. For a review of some of the interesting searches these Bell test have inspired, see, e.g., Ref. [26], and for an experimental exploration of nonlocality, see, e.g., Ref. [27].

Acknowledgments This research was supported by the NSF grant No. PHY 12-05870.

References

1. A. Einstein, B. Podolsky, N. Rosen, Can quantum-mechanical description of physical reality be considered complete? Phys. Rev. **47**, 777–780 (1935)
2. E. Schrodinger, Proc. Camb. Philos. Soc. **31**, 555 (1935). Translated in *Quantum Theory and Measurement*, ed. by J.A. Wheeler, W.H. Zurek (Princeton, NJ, 1983)
3. J.S. Bell, On the EPR paradox. Physics **1**, 195 (1964)
4. J.F. Clauser, M.A. Horne, A. Shimony, R.A. Holt, Proposed experiment to test local hidden-variable theories. Phys. Rev. Lett. **23**, 880 (1969)
5. A. Aspect, J. Dalibard, G. Roger, Experimental test of Bell's inequalities using time- varying analyzers. Phys. Rev. Lett. **49**, 1804 (1982)
6. P.G. Kwiat et al., New high-intensity source of polarization-entangled photon pairs. Phys. Rev. Lett. **75**, 4337 (1995)
7. M.A. Rowe et al., Experimental violation of a Bell's inequality with efficient detection. Nature **409**, 791 (2001)
8. M. Ansmann et al., Violation of Bell's inequality in Josephson phase qubits. Nature **461**, 504 (2009)
9. J. Hofmann et al., Heralded entanglement between widely separated atoms. Science **337**, 72 (2012)
10. D. Dehlinger, M.W. Mitchell, Entangled photons, nonlocality, and Bell inequalities in the undergraduate laboratory. Am. J. Phys. **70**, 903 (2002)
11. P. Eberhard, Background level and counter efficiencies required for a loophole-free Einstein-Podolsky-Rosen experiment. Phys. Rev. A **47**, R747–R750 (1993)
12. A. Lita et al., Counting near-infrared single-photons with 95% efficiency. Opt. Express **16**, 3032 (2008)
13. Christensen et al., Detection-loophole-free test of quantum nonlocality, and applications. Phys. Rev. Lett. **111**, 130406 (2013)
14. M. Giustina et al., Bell violation using entangled photons without the fair-sampling assumption. Nature **497**, 227–230 (2013)
15. G. Weihs, T. Jennewein, C. Simon, H. Weinfurter, A. Zeilinger, Violation of Bell's inequality under strict Einstein locality conditions. Phys. Rev. Lett. **81**, 5039–5043 (1998)

16. T. Scheidl et al., Violation of local realism with freedom of choice. Proc. Nat. Acad. Sci. **107**, 19708 (2010)
17. B. Hensen et al., Loophole-free Bell inequality violation using electron spins separated by 1.3 kilometers. Nat. **526**, 682 (2015)
18. M. Giustina et al., Significant-loophole-free test of Bell's theorem with entangled photons. Phys. Rev. Lett. **115**, 250401 (2015)
19. L.K. Shalm et al., Strong loophole-free test of local realism. Phys. Rev. Lett. **115**, 250402 (2015)
20. H. Weinfurter, Event-ready loophole free Bell test using heralded atom-atom entanglement, International Conference on Quantum Communication, Measurement and Computing (QCMC), Singapore (2016)
21. P.G. Kwiat, E. Waks, A.G. White, I. Appelbaum, P.H. Eberhard, Ultrabright source of polarization-entangled photons. Phys. Rev. A **60**, 773 (1999)
22. A.G. White, D.F.V. James, P.H. Eberhard, P.G. Kwiat, Nonmaximally entangled states: production, characterization, and utilization. Phys. Rev. Lett. **16**, 3103 (1999)
23. R. Rangarajan, M. Goggin, P. Kwiat, Optimizing type-I polarization-entangled photons. Opt. Express **17**, 18920 (2009)
24. L. Hardy, Nonlocality for two particles without inequalities for almost all entangled states. Phys. Rev. Lett. **71**, 1665 (1993)
25. P.G. Kwiat, L. Hardy, The mystery of the quantum cakes. Am. J. Phys. **68**, 1 (2000)
26. S. Popescu, Nonlocality beyond quantum mechanics. Nat. Phys. **10**, 264270 (2014)
27. B.G. Christensen, Y.-C. Liang, N. Brunner, N. Gisin, P.G. Kwiat, Exploring the limits of quantum nonlocality with entangled photons. Phys. Rev. X. **5**, 041052 (2015)

Chapter 26
An Early Long-Distance Quantum Experiment

Gregor Weihs

In this essay I would like to describe the history of our long-distance test of Bell's inequality experiment [1] from a personal point of view.

For many quantum physicists the series of beautiful experiments by Alain Aspect and coworkers [2–4] put an end to the debates that raged about entanglement and realism in quantum physics since the famous declaration by Einstein, Podolsky and Rosen [5]. For others, Aspect's experiments had the opposite effect and even reinvigorated those debates. Up until the summer of 2015, after the first version of the present article was written, many a (crypto-) realist still hung on to the last straw of an unexpected result in a final, decisive test of Bell's inequality. With the three loophole-free Bell experiments [6–8] we now know the answer that nature violates Bell's inequality and local realism is untenable. Critics are still trying to find flaws in these latest experiments, but given that they were done independently this seems very unlikely.

Among those for which the debate was not settled was my then supervisor Anton Zeilinger, who in one of his lesser known publications [9] pointed out a little flaw in Aspect's beautifully constructed and executed last Bell-type experiment, which involved a distance from the source to each analyzer of 6 m and periodic switching of their orientations with a period of 20 ns. It so had happened that the switching period in Aspect's experiment coincided with the time interval of the photons travelling from the source to the analyzer. Even before that Alain Aspect had pointed out that the ultimate goal would be to randomly switch the analyzers, but that this was not reasonably possible with the technology available at a time.

It was in the early 1990s that I started as a Master's student in Anton Zeilinger's group initially working on Hong-Ou-Mandel interferometry [10]. When I consid-

G. Weihs (✉)
Institute for Experimental Physics, University of Innsbruck,
6020 Innsbruck, Austria
e-mail: Gregor.Weihs@uibk.ac.at

R. Bertlmann and A. Zeilinger (eds.), *Quantum [Un]Speakables II*,
The Frontiers Collection, DOI 10.1007/978-3-319-38987-5_26

ered pursuing a PhD thesis under his supervision, Anton proposed doing a long-distance test of Bell's inequality with randomly switched analyzers. Coincidentally, just in the same year, 1995, Paul Kwiat, a post-doc in Innsbruck at the time, had built a new much brighter and better source of polarization-entangled photon pairs [11]. Some quick estimates of the count rates, entanglement quality, and fiber losses proved encouraging. Switching the analyzers, generating random bits, and recording the data, on the other hand appeared daunting. So in hindsight, my naïvete and excitement about this opportunity made us embark on a project that turned out well, but could have failed in many ways. Apart from my supervisors Anton Zeilinger and Harald Weinfurter this experiment had a lot of help from Thomas Jennewein, Ulrich Achleitner and Christoph Simon. In the rest of this essay I would like to give a somewhat personal account of the realization of the long-distance Bell-experiment, its further uses, subsequent experiments and some consequences

Distance and Channel

The fact that we had a working source of polarization entangled photon pairs at 702 nm wavelength determined many other choices. We never really considered free-space transmission links for lack of experience with the required alignment and for fear that the beam quality at the receiver would not be good enough for polarization modulation. But even optical fibers are not ideal at this wavelength due to their inherent loss of about 4 dB/km. Even more concern came when Nicolas Gisin told me that this would never work because of the polarization mode dispersion (PMD) in optical fiber [12]. This effect is caused by some small, randomly oriented birefringence in single-mode optical fiber. To lowest order this causes a change of the state of polarization of light that we launch into the fiber. At the exit we will find that the state has been rotated to an arbitrary other state on the Poincaré sphere. For a perfectly monochromatic wave this rotation can always be undone, even if it is the result of a concatenation of many statistically distributed small rotations along the fiber. However, if the radiation is not monochromatic, different frequency components will generally undergo different polarization rotations causing incoherence at the fiber output. A typical value of the PMD coefficient for standard SMF-28 fiber is $0.1\ \mathrm{ps}/\sqrt{\mathrm{km}}$, meaning that at a length of 1 km we expect a dispersion of a pulse of about 0.1 ps. Given our relatively broad bandwidth this could have adversely affected the polarization of our photons. Much less was known at the time about the behavior of single-mode fiber in the visible wavelength range. Again, being somewhat naïve and bold helped, because it turned out that statistical PMD was almost no issue for our experiment, except for a slight reduction in the polarization purity after the long distance. The bigger issue was that the polarization rotation introduced by the fiber changed over time and had to be recalibrated frequently.

Many tests involving optical fibers are carried out with the long fiber on a spool. It was clear to us that, in contrast, we would actually have to lay out the optical fiber to establish a "field" test with the highest possible geometric distance. The distances that count are measured along the bee-line, straight distances between the source and

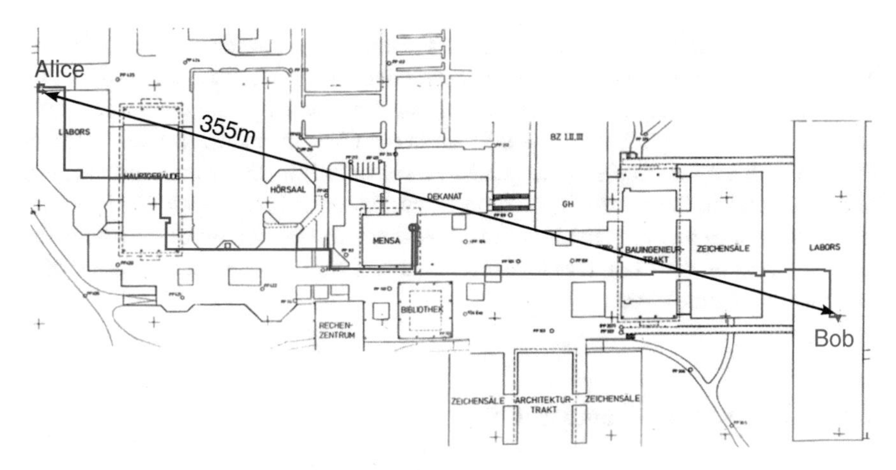

Fig. 26.1 Two 500 m long optical fibers connected the source location in an underground installation room near the center of the Campus Technik in Innsbruck to the two measurement stations labelled "Alice" and "Bob" at the west and east ends of the campus. The extra length of optical fiber stayed on spools at the source location.

each of the two end stations and the distance between the end stations. We argued later, that the fiber on the spool, and the fact that the fiber can hardly ever be installed in a straight line but has to follow corridors and cable trays merely reduce the effective speed of the photons, which is not a problem per se in tests of Bell's inequality. Eventually, and with a lot of help but also many an argument with our administration we managed to have our fibers installed as outlined in Fig. 26.1, with walkable underground connections between all three stations. I then estimated the distances from the campus plans and arrived at about 400 m end-to-end separation, which is the number quoted in Ref. [1]. A later measurement carried out using tape measures and a somewhat haphazard vectorial addition of all the little segments yielded an actual separation of only 355 m. This didn't change any of the conclusions from the experiment, because even the smaller distance is much bigger than the relevant times (multiplied with the speed of light). Still it was a lesson learned about having to estimate ones uncertainties accurately.

Basis Switching

The main limiting factor for the distance was originally thought to be the device that would switch between the two required polarization bases for the Bell inequality measurement. As it turned out there were very fast devices available, with switching times of less than one nanosecond, however, these could only be switched at low repetition rates of a few kHz due to the huge dissipated power. Another subtle detail was that in order for random switching to work, we could not use a resonant device but needed a wideband one. Therefore we settled for lower speed ($\approx$10 ns) electro-optic modulators that we could run continuously. Even at this reduced speed the switching of the device would only require a distance of a few ten meters. In the end it boils down to analyzing the whole measurement chain to determine how long the measurement takes from the generation of random number to the recording of the photon detection. In particular, it is difficult to say when the measurement ends. Is it finished with a classical electrical pulse in a coaxial cable, the data in the memory of the instrument, the controlling computer, or only when it is saved to the PCs hard drive? For our experiment the first three possible cuts would have been outside the other side's light cone, but not the recording on the hard drive, which occurred on a much too slow time scale.

Random Number Generation

One of the earliest discussions we had was about how to generate the random bits. It was clear to us that it would have to be physical random bits and not just pseudorandom ones, in order to go significantly beyond Alain Aspect's experiments. Because of Anton's experience in radioactivity and neutron physics we considered a radioactive source of randomness. We quickly realized, though, that to get a few events within the estimated few hundred nanosecond time interval, we would need a source with an activity in excess of 10 MBq, much beyond what would have been permissible and possible to handle. Also, using the light of distant stars appeared to be too cumbersome so that eventually we settled on random number generation using an LED and a beam-splitter [13]. This idea of using quantum randomness to generate physical random numbers has arguably created the most commercially successful quantum technology so far and has also developed into a field of its own with questions of certifiability and device independence [14].

Data Acquisition

Perhaps the biggest change compared to all earlier experiments was the idea to collect data at each end of the experiment, i.e. *Alice* and *Bob* locally and perform the correlation analysis later. From our laboratory experiments we knew that in order to avoid too much background, we would need a coincidence time window, within which an event at Alice and Bob would be considered simultaneous, of a few nanoseconds. At the expected source rates we envisioned collection times of at least a few minutes in order to achieve statistical significance, in which we would simply record the arrival time of every photon on each side with a resolution of better than 1 ns. We figured that to gather enough data and thus statistical significance we would need to run the experiment at least for a few minutes at a time. Devices capable of recording precise detection times ("time-tagging") continuously for long or indefinite periods were difficult to find. Because we had two detectors, two channels were a minimum and in addition we wanted to record the actual basis modulator voltage. Eventually we only got two channels and to record one bit indicating the two possible bases had to be somehow encoded into the data stream. This encoding proceeded by generating a double pulse for every "1" setting of the modulator bit and a single pulse for every "0". Unfortunately this encoding also made our problem of synchronizing all 16 possible combinations of detector and basis on both sides even worse, thus making the experiment a little vulnerable to the coincidence time loophole for extremely short coincidence windows. Measuring accurate arrival times, for say 100 s demands a relative clock stability of 1 ns/100 s $= 10^{-11}$, which is achievable with a good quartz oscillator. However, the independent data collection at the two points also requires that the absolute frequencies of the two oscillators are the same with better than 10^{-11} accuracy, which may be possible with syntonized oven controlled quartz oscillators, but is easier achievable with Rubidium-based atomic clocks. One could also use GPS-derived clocks, which are affordable nowadays, but weren't at the time.

Experiment Control

One of the most important encounters was my first meeting with Alain Aspect when he visited Innsbruck in 1996. His main message was "automate everything!", meaning I should make sure that all subsystems of the experiment could run independently and unsupervised. Thus I set up a system of three computers, on at the source site and one each at Alice's and Bob's stations. For lack of established tools I had to write my own protocols for controlling the data acquisition at the end stations in a synchronized but independent way and to have an alignment mode that would allow me to align the source, and in particular the fiber polarization controllers with live measurement data from the end stations. Yet, for lack of time and other resources I couldn't follow Alain's advice completely. For example the optical alignment of the source or the modulators had to be done by hand and, owing to the unsteady environmental conditions in our makeshift laboratories in boiler rooms and other quarters, quite frequently too. Ideally the experiment was thus operated by three people (Thomas Jennewein, Christoph Simon and myself). In addition we operators quite often had to walk from one point to the other. Not very far, but too far to retrieve a forgotten screwdriver. Eventually though it was possible that a single person collected data and finally the best data set came about on a very quiet Easter Sunday, April 12, 1998.

Results and Response

It was a very nice experience to show this campus-spanning experiment to several very important people, among them His Holiness the 14th Dalai Lama, who was very curious and excited to "see" entangled photons, but who did not like the randomness they come with. Abner Shimony took a walk with me along the whole setup from end to end and was very enthusiastic about the experiment and the clear separation between the two observers it achieved for the first time. He said: “I didn't think that within my lifetime one could do an experiment like this one!” All of these encounters with giants of the field made it clear that we had broken through another barrier. We then decided to make raw data from the experiment available online. The data is still available nowadays, albeit not on the same website, which doesn't exist anymore. It turned out that several if not many people had downloaded the data and performed their own analyses, just as intended. At some point, though, someone asked me at a conference what I thought about someone else's published criticism of my experiment. I had to answer that I was not even aware of the publication. After this incident we removed the public link and henceforth interested researchers have to ask for access and agree to alert us before they publish anything. As an advantage all data sets are available through my new website, not just the one. This ties into the debate about open access to research raw data, which for some parts of the life sciences is already established, but is still in its infancy for physics.

Several peculiarities were found through a variety of analyses. People who analyzed the data include Guillaume Adenier [15], Peter Bierhorst [16], James H. Bigelow [17], Hans De Raedt [18], Michel Fodje, Donald Graft [19], Alejandro Hnilo [20], Emanuel Knill, Marian Kupczynski, Franz Kohl, Peter Morgan [21], Dan Shanahan, and Sandro Sozzo. The most interesting anomalies can be traced back to timing problems. Various channels not being perfectly temporally aligned and electrical pulse echoes can produce such signatures. In particular the “nonlocality” found by Adenier and Khrennikov [15] comes about by taking coincidences with a narrow window, which has the effect of rejecting events dependent on both detection times, so that even the detection result on one side appears to depend on the setting of the other side. Needless to say, when analyzed without coincidence filtering there is no action at a distance, which would indeed be very spooky.

Conclusions

This experiment was the one of the first in a series of long-distance experiments in our groups and others culminating in the beautiful series of experiments done between some of the Canary islands [22] and the first long-distance Greenberger-Horne-Zeilinger experiment [23]. As a consequence new loopholes were discovered and closed that we had not foreseen [24, 25]. As the experiments got bigger and bigger so did the teams and collaborations required to run them. Still at this point at least the terrestrial ones can be run by a local collaboration. Clearly, in going to space, e.g. on satellites this will no longer be possible. In addition, convincing the funding and space agencies to support quantum missions is a fulltime job in itself. Clearly many of us in the foundations community are convinced that doing Bell-type experiments over distances that permit humans to do a local basis choice is very important. In addition, though, the same line of experiments can also be used to explore the interface between quantum physics and relativity, in particular in communication scenarios with relativistic observers. It appears that in several countries and through international collaborations space-borne long-distance experiments on entanglement are now under way. The best result would be if these mammoth tests would produce as many new questions as they will answer.

References

1. G. Weihs, T. Jennewein, C. Simon, H. Weinfurter, A. Zeilinger, Violation of Bell's inequality under strict einstein locality conditions. Phys. Rev. Lett. **81**, 5039 (1998). doi:10.1103/PhysRevLett.81.5039
2. A. Aspect, P. Grangier, G. Roger, Experimental test of realistic local theories via Bell's theorem. Phys. Rev. Lett. **47**, 460 (1981). doi:10.1103/PhysRevLett.47.460
3. A. Aspect, J. Dalibard, G. Roger, Experimental test of Bell's inequalities using time-varying analyzers. Phys. Rev. Lett. **49**, 1804 (1982). doi:10.1103/PhysRevLett.49.1804
4. A. Aspect, P. Grangier, G. Roger, Experimental realization of einstein-podolsky-rosen-bohm gedankenexperiment: a new violation of bell's inequalities. Phys. Rev. Lett. **49**, 91 (1982). doi:10.1103/PhysRevLett.49.91
5. A. Einstein, B. Podolsky, N. Rosen, Can quantum-mechanical description of physical reality be considered complete? Phys. Rev. **47**, 777 (1935). doi:10.1103/PhysRev.47.777
6. B. Hensen et al., Experimental loophole-free violation of a Bell inequality using entangled electron spins separated by 1.3 km. Nature **526**, 682 (2015). doi:10.1038/nature15759
7. M. Giustina, M.A.M. Versteegh, S. Wengerowsky, J. Handsteiner, A. Hochrainer, K. Phelan, F. Steinlechner, J. Kofler, J.-A. Larsson, C. Abellán, W. Amaya, V. Pruneri, M.W. Mitchell, J. Beyer, T. Gerrits, A. E. Lita, L.K. Shalm, S.W. Nam, T. Scheidl, R. Ursin, B. Wittmann, and A. Zeilinger, Significant-loophole-free test of Bell's theorem with entangled photons. Phys. Rev. Lett. **115**, 250401 (2015), doi:10.1103/PhysRevLett.115.250401
8. L.K. Shalm, E. Meyer-Scott, B.G. Christensen, P. Bierhorst, M.A. Wayne, M.J. Stevens, T. Gerrits, S. Glancy, D.R. Hamel, M.S. Allman, K.J. Coakley, S.D. Dyer, C. Hodge, A.E. Lita, V.B. Verma, C. Lambrocco, E. Tortorici, A.L. Migdall, Y. Zhang, D.R. Kumor, W.H. Farr, F. Marsili, M.D. Shaw, J.A. Stern, C. Abellán, W. Amaya, V. Pruneri, T. Jennewein, M.W. Mitchell, P.G. Kwiat, J.C. Bienfang, R.P. Mirin, E. Knill, and S.W. Nam, Strong loophole-free test of local realism. Phys. Rev. Lett. **115**, 250402 (2015). doi:10.1103/PhysRevLett.115.250402
9. A. Zeilinger, C.G. Shull, M. Horne, K.D. Finkelstein, Effective mass of neutrons diffracting in crystals. Phys. Rev. Lett. **57**, 3089 (1986). doi:10.1103/PhysRevLett.57.3089
10. G. Weihs, M. Reck, H. Weinfurter, A. Zeilinger, All-fiber three-path Mach-Zehnder interferometer. Opt. Lett. **21**, 302 (1996). doi:10.1364/ol.21.000302
11. P.G. Kwiat, K. Mattle, H. Weinfurter, A. Zeilinger, A. Sergienko, Y. Shih, New high-intensity source of polarization-entangled photon pairs. Phys. Rev. Lett. **75**, 4337 (1995). doi:10.1103/PhysRevLett.75.4337
12. N. Gisin, J.-P. Von der Weid, J.-P. Pellaux, Polarization mode dispersion of short and long single-mode fibers. J. Lightwave Technol. **9**, 821 (1991). doi:10.1109/50.85780
13. T. Jennewein, U. Achleitner, G. Weihs, H. Weinfurter, A. Zeilinger, A fast and compact quantum random number generator. Rev. Sci. Instrum. **71**, 1675 (2000). doi:10.1063/1.1150518
14. S. Pironio, A. Acin, S. Massar, A.B. de la Giroday, D.N. Matsukevich, P. Maunz, S. Olmschenk, D. Hayes, L. Luo, T.A. Manning, C. Monroe, Random numbers certified by Bell's theorem. Nature **464**, 1021 (2010). doi:10.1038/nature09008
15. G. Adenier, A. Khrennikov, Is the fair sampling assumption supported by epr experiments? J. Phys. B. (2006). arXiv:quant-ph/0606122
16. P. Bierhorst, in *A New Loophole in Recent Bell Test Experiments* (2013). arXiv:1311.4488

17. J.H. Bigelow, in *A Close Look at the EPR Data of Weihs et al* (2009). arXiv:0906.5093
18. H. De Raedt, K. De Raedt, K. Michielsen, K. Keimpema, S. Miyashita, Event-based computer simulation model of aspect-type experiments strictly satisfying einstein's locality conditions. J. Phys. Soc. Jpn. **76**, 104005 (2007). doi:10.1143/JPSJ.76.104005
19. D.A. Graft, A local realist account of Weihs. Jennewein, Simon, Weinfurther, and Zeilinger EPRB experiment, Physics Essays **26**, 214 (2013). arXiv:1301.1670
20. A.A. Hnilo, M.G. Kovalsky, G. Santiago, Low dimension dynamics in the EPRB experiment with random variable analyzers. Found. Phys. **37** (2007). doi:10.1007/s10701-006-9091-7
21. P. Morgan, in *A graphical presentation of signal delays in the datasets of Weihs et al.* (2012). arXiv:1207.5775
22. X.-S. Ma, T. Herbst, T. Scheidl, D. Wang, S. Kropatschek, W. Naylor, B. Wittmann, A. Mech, J. Kofler, E. Anisimova, V. Makarov, T. Jennewein, R. Ursin, A. Zeilinger, Quantum teleportation over 143 kilometres using active feed-forward. Nature **489**, 269 (2012). doi:10.1038/nature11472
23. C. Erven, N. Ng, N. Gigov, R. Laflamme, S. Wehner, G. Weihs, An experimental implementation of oblivious transfer in the noisy storage model. Nature Commun. **5** (2014). doi:10.1038/ncomms4418
24. R.D. Gill, G. Weihs, A. Zeilinger, M. Żukowski, No time loophole in Bell's theorem: the Hess-Philipp model is nonlocal. Proc. Natl. Acad. Sci. **99**, 14632 (2002). doi:10.1073/pnas.182536499
25. T. Scheidl, R. Ursin, J. Kofler, S. Ramelow, X.-S. Ma, T. Herbst, L. Ratschbacher, A. Fedrizzi, N.K. Langford, T. Jennewein, A. Zeilinger, Violation of local realism with freedom of choice. Proc. Nat. Acad. Sci. **107**, 19708 (2010). doi:10.1073/pnas.1002780107

Chapter 27
Quantum Information Experiments with Free-Space Channels

Yuan Cao, Qiang Zhang, Cheng-Zhi Peng and Jian-Wei Pan

Abstract Satellite based quantum communication is believed to be a feasible way to achieve global unconditional secure network. Here, we present several free space experiments towards this direction. We also show that the technology developed in these experiments provides a platform to test the foundation of quantum theory.

Introduction

In order to test Bell inequality, quantum state generation, transmission, manipulation and detection technology has been developed over the past decades. These techniques, later on, find immediate applications in information science, including secure communication, quantum teleportation, and quantum computation.

Generally speaking, there are mainly two types of quantum channels, fiber links and free-space links for quantum information processing. The distance of current fiber-based experiments are limited to hundreds of kilometers due to the exponentially channel losses and decoherence [1–3]. The quantum repeater scheme [4–6] presents an efficient way to generate highly entangled states between long distant locations, hence providing a novel solution to the photon loss and decoherence problem. However the realization of quantum repeaters remains experimentally challenging [7]. Meanwhile, for free-space links, the photon losses and decoherence are almost negligible in outer space, therefore the satellite-based quantum communication offers a promising way to run large-scale quantum communication schemes. It has become a research highlight that studying the feasibility of satellite-based global-

Y. Cao · Q. Zhang · C.-Z. Peng · J.-W. Pan (✉)
Shanghai Branch, National Laboratory for Physical Sciences at Microscale
and Department of Modern Physics, University of Science and Technology of China,
Shanghai 201315, China
e-mail: pan@ustc.edu.cn

Y. Cao · Q. Zhang · C.-Z. Peng · J.-W. Pan
CAS Centre for Excellence and Synergetic Innovation Centre in Quantum Information
and Quantum Physics, University of Science and Technology of China,
Hefei, Anhui 230026, China

R. Bertlmann and A. Zeilinger (eds.), *Quantum [Un]Speakables II*,
The Frontiers Collection, DOI 10.1007/978-3-319-38987-5_27

scale quantum communication by using free-space channels near earth atmosphere. Along this line, the free-space quantum communication systems have been successfully demonstrated over large-scale distances [8–11], or with more practical and compact systems [12–15].

On the other hand, the technology developed in quantum communication experiment makes more and more gendanken experiments possible. For example, satellite-based quantum technology provides a platform to test the Bell inequality in the outer space. One can also perform the experimental probes of quantum phenomena at large length scales, at which gravitational effects will play a significant role, by utilizing the future satellites in the free-space channels [16]. Furthermore, by sending astronauts and experimental setups in the space, testing Bell inequality with free will can be implemented.

Here, we will review a series of experimental works on quantum information with the free-space channel in recent decade by our group. It is organized as follows. In section "Test of Bell Inequality over Free-Space Channels", two quantum entanglement distribution and Bell inequality test experiments are introduced. In section "Quantum Teleportation in Free-Space Channels", quantum teleportation over 100 km free-space channel is reviewed. In section "Quantum Cryptography in Free-Space Channels", we will summarize our efforts on quantum cryptography in long distance free-space channels. We finally present our experiment of testing the speed of nonlocal correlations in section "Testing the Speed of Nonlocal Correlations".

Test of Bell Inequality over Free-Space Channels

There are two main kinds of loopholes in Bell inequality test experiments, the locality loophole [17] and detection loophole [18]. In order to close the locality loophole, the correlation measurement events must be space like separate.

In 2005, we report free-space distribution of entangled photon pairs over a noisy ground atmosphere of 13 km [8] and observe a violation of Bell inequality of 2.45 ± 0.09 [19]. Meanwhile, considering the effective thickness of the aerosphere is on the order of 5–10 km (i.e., the whole aerosphere is equivalent to 5–10 km ground atmosphere), it's the first time to experimentally demonstrate the entanglement can still survive after both entangled photons have passed through the noisy ground atmosphere with a distance beyond the effective thickness of the aerosphere.

The entangled photons at the sender are collected into two single-mode fibers, which are connected to the two sending telescopes, respectively. Because of the disturbance of the atmosphere, the size and position of the received beam vary randomly, causing a reduction of the collecting efficiency. To solve this problem, we have used the two sending telescopes to expand the beam diameter to about 12 cm for long-distance propagation. Moreover, at each receiver a similar telescope is used to receive the entangled photons. After being focused, entangled photons are coupled into 62.5 μm multi-mode fibers and finally sent to single-photon detectors. With

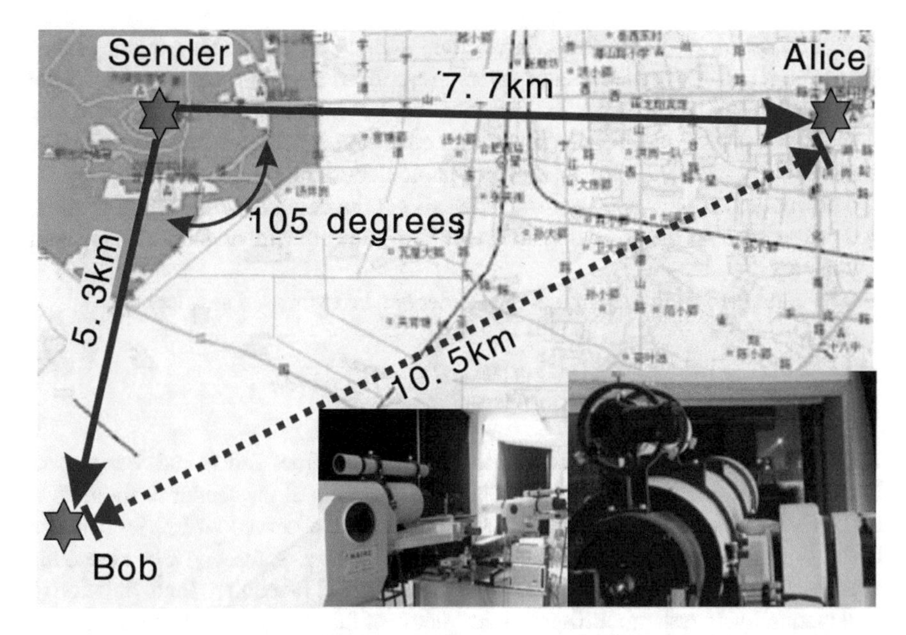

Fig. 27.1 Schematic diagram of locations in the experiment. The source of entangled photons is located at the foot of a high television tower, on the top of Dashu Mountain. Alice is located on the west campus of USTC, and Bob is located at Feixi, a county of Hefei city. Photons from the sender to the receivers experience a noisy city environment. Therefore, strongly influenced by the air pollution and noisy background lights, even after rain with less air pollution, the background count rate can reach about 30 000 per second at night without using interference filters.

these efforts, we manage to keep the transmission system working stably for a couple of hours. For example, in the right photograph of Fig. 27.1 we can see at the Alice side a bright and stable adjusting laser beam from the sender. Since the distances from the sender to Alice and to Bob are not equal, the two entangled photons will arrive at each receiver at different times. The air disturbance will cause this time difference to vary randomly, resulting in a time difference shake ΔT. To coincide the detected events at the two receivers, we have to make sure that the coincident time window is wider than ΔT. However, when we widen the coincident time window to get the adequate true coincident events, the accidental coincident count rate also increases and thus results in a reduction of the visibility. In our experiment, we utilized the method of laser pulse synchronization to achieve time coincidence between the two receivers. At the sender, Q-switched laser pulses with a wavelength of 532 nm are separated into two parts, and then sent to the receivers, experiencing the same optical path as the entangled photons. At each receiver, we measure the time difference between the signal of the single-photon event and the signal of the corresponding synchronous laser pulse for the subsequent coincidence via the classical communication link. Considering other ingredients causing time shake, we set the time window to 20 ns in our experiment.

Finally, to minimize the background count rate, the experiment is performed during night, and 2.8 nm interference filters are utilized at each receiver to block the noisy background light. With the filters added, the average background count rate is about 400 per second. When the weather condition is perfect with a considerably high visibility (>15 km), the total single-photon count rate is about 40000 per second at Bob and about 18000 per second at Alice; the coincident count rate is about 300 per second. At the normal visibility (10 km), the coincident count rate is about 150 per second.

The entangled state prepared at the sender can be expressed as follows:

$$|\Psi^-\rangle = \frac{1}{\sqrt{2}}(|H\rangle_A|H\rangle_B - |V\rangle_A|V\rangle_B), \tag{1}$$

where photon A is sent to Alice, photon B is sent to Bob, and H and V represent horizontal and vertical polarization. The local visibility at the sender is about 98 % in the H/V basis, and 94 % in the +45/−45 basis. The observed visibilities between two separated receivers are 94 % and 89 % in the two bases, respectively. Hence the average visibility reaches 91 %, which is well beyond 71 % required for a violation of Bell inequality. In order to further test the quality of the entangled state, we measured the Clauser-Horne-Shimony-Holt (CHSH) inequality which is one type of the Bell inequalities [19]. The polarization correlation coefficient is defined as follows:

$$E(\phi_A, \phi_B) = \frac{N_{++} + N_{--} - N_{+-}N_{-+}}{N_{++} + N_{--} + N_{+-}N_{-+}}, \tag{2}$$

where $N_{ij}(\phi_A, \phi_B)$ are the coincidences between the i channel of polarizer of Alice set at angle ϕ_A and the j channel of the polarizer of Bob set at angle ϕ_B. In the CHSH inequality, parameter S is defined as

$$S = |E(\phi_A, \phi_B) - E(\phi_A, \phi'_B) + E(\phi'_A, \phi_B) + E(\phi'_A, \phi'_B)|, \tag{3}$$

In the local realistic view, no matter what angles ϕ_A and ϕ_B are set to, parameter S should be below 2. But in view of the quantum mechanics, S will get to the maximal value $2\sqrt{2}$ when the polarization angles are set to $(\phi'_A, \phi'_A, \phi_B, \phi'_B) = (0°, 45°, 22.5°, 67.5°)$.

In this experiment, the whole measurement process was completed in 20 s. He measured result of parameter S is 2.45 ± 0.09, with a violation of the CHSH inequality by 5 standard deviations. This result firmly ascertains that entanglement has been built between the two distant receivers.

Although compared to the previous experiments our experimental results might seem to be only a modest step forward, the implication is profound. First, our experiment demonstrated for the first time that entanglement can still survive after penetrating the effective thickness of the aerosphere by showing a violation of the Bell inequality with space-like separated observers. Obviously, the strong violation of the Bell inequality is sufficient to guarantee the absolute security of the quantum cryptography scheme, hence closing the eavesdropping loophole. Second, the link efficiency of entangled photon pairs achieved in our experiment is about a few percent, which is well beyond the threshold required for satellite-based free-space quantum communication. Meanwhile, the methods developed in the present experiment to establish a high stable transmission channel and achieve synchronization between two distant receivers provide the necessary technology for future experimental investigations of global quantum cryptography and quantum teleportation in free space.

We further extend the entanglement distribution distance to 100 km at Qinghai Lake in 2012, in which violation of the Clauser-Horne-Shimony-Holt inequality is observed without the locality loophole.

In this experiment, we put the entanglement source close to the middle of the free-space channel, an island in the middle of Qinghai Lake (Haixin, 36°51′38.7″N, 100°8′15.2″E, Fig. 27.2a). In order to show a two-link entanglement distribution between two sites, Alice and Bob cannot see each other directly (Fig. 27.2a). Char-

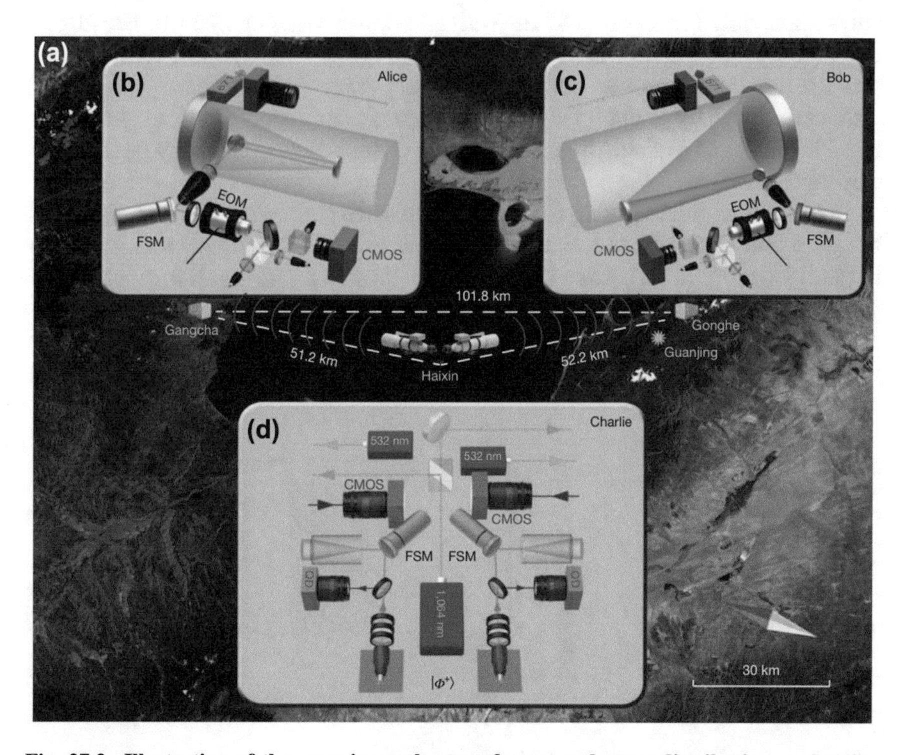

Fig. 27.2 Illustration of the experimental set-up for entanglement distribution. **a**, Satellite image of experiment site. **b**, The receiver, Alice, collects the photon sent by Charlie utilizing a 600 mm Cassegrain telescope. **c**, The receiver, Bob, collects the photon sent by Charlie utilizing a 400 mm off-axis reflecting telescope. **d**, The Charlie's site. The entangled photon pairs are created by Charlie at the centre island of Qinghai Lake (Haixin) and then distributed to Alice and Bob.

lie first prepares the entangled photon pairs in the state $|\Phi^+\rangle$, which are then sent to Alice and Bob via two telescopes each mounted on a two-dimensional rotatable platform (Fig. 27.2b–d). The distances between Charlie and the two receivers are 51.2 km (Alice) and 52.2 km (Bob), and the distance between Alice and Bob is 101.8 km.

Furthermore, this experiment closed the locality loophole. The entangled photon pairs were distributed along two opposite directions to Alice and Bob: these parties are separated by 101.8 km, a distance that takes 340 μs for light to travel, and the path difference of the two links is within 1 km, which results in a 3 μs delay between the two measurement events. Thus, the two measurement events on Alice's and Bob's sites are space-like separated. Meanwhile the two receivers used fast EOMs to switch between the two possible polarization bases. The two EOMs were controlled by two independent quantum random number generators, each of which generates a random number every 20 μs (less than 340 μs). Thus the measurement-setting choices are also space-like separated. Hence, the locality loophole is closed.

Finally, we obtained 208 coincidences during an effective time of 32,000 s. By comparison with the counts of our entanglement source, we found that the channel attenuation varied from 66 to 85 dB with an average value of 79.5 dB. For 20 cm-aperture satellite optics at an orbit height of 600 km and 1 m-aperture receiving optics, the total loss for a two-downlink channel between a satellite and two ground stations is typically about 75 dB. The measured correlation functions (shown in Fig. 27.3) resulted in $S = 2.51 \pm 0.21$, which violates Bell's inequality by 2.4 standard deviations.

In contrast to previous long-distance free-space experiments with entangled photon pairs using only one-link channels [20], our two-link experiment requires tracking and synchronization between three different locations. Our two-link experiment—most comparable with satellite-to-ground quantum entanglement distribution—has achieved a distance between two receivers that is an order of magnitude larger than in previous experiments.

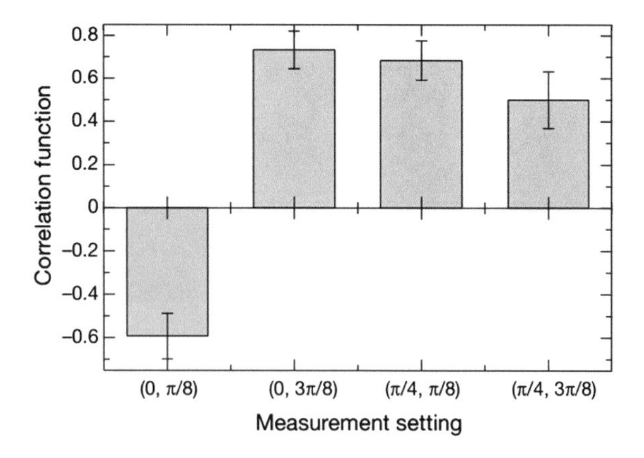

Fig. 27.3 Correlation functions of a CHSH-type Bell's inequality for entanglement distribution. The measurement setting (φ_A, φ_B) represents the measured polarization bases of photons by Alice and Bob, respectively. Error bars, statistical errors (±1 *s.d.*).

Quantum Teleportation in Free-Space Channels

Quantum teleportation lies at the heart of a number of quantum protocols, finding particular use in quantum repeaters, quantum relays and so on, and is central to the practical realization of quantum communication [4, 21]. Since its initial proposal by Bennett and colleagues [22], quantum teleportation has triggered significant research activity and become a focus in the field of quantum-information science.

An optical free-space link is highly desirable for extending the transfer distance, because of its low atmospheric absorption for certain ranges of wavelength. In 2010, by following the Rome scheme [23], which allows a full Bell-state measurement, we reported free-space implementation of quantum teleportation over 16 km, as shown in Fig. 27.4.

In this experiment, we have followed the Rome scheme to achieve quantum teleportation in free space over a distance of 16 km. At that time, this is the longest reported distance over which photonic teleportation has been achieved, more than 20 times longer than the previous implementation for a fibre channel [24, 25]. Various techniques have been developed for accomplishing this goal, including real-time feedback control of the high-stability interferometer for single-photon BSM, active feed-forward manipulation of the single-photon state for reconstruction of the initial teleported qubit, novel design of telescopes tailored for teleportation experiments,

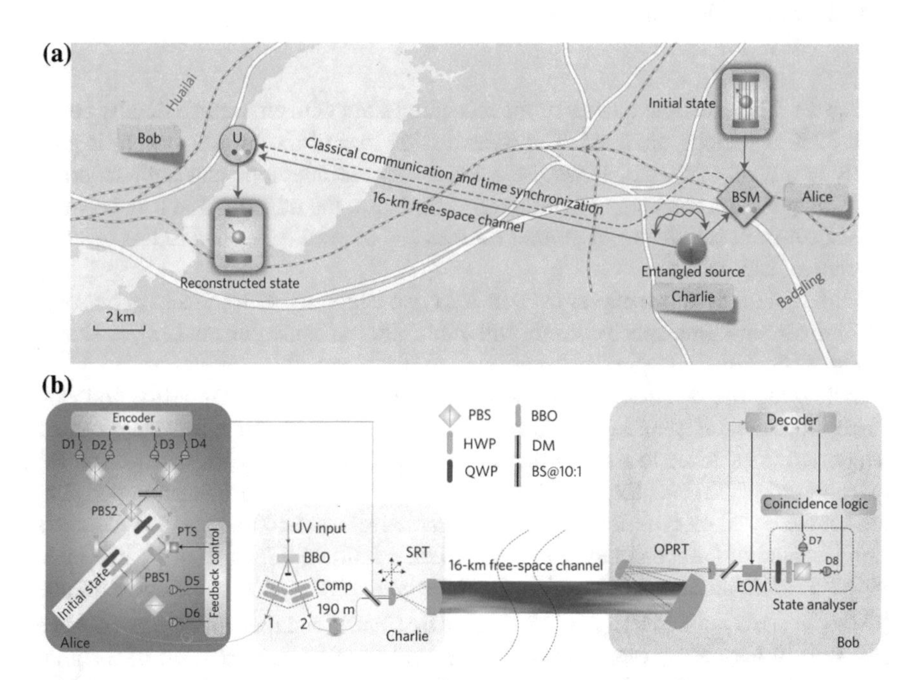

Fig. 27.4 Experimental quantum teleportation in free space. **a**, A birds-eye view of the 16 km free-space quantum teleportation experiment. **b**, Sketch of the experimental system.

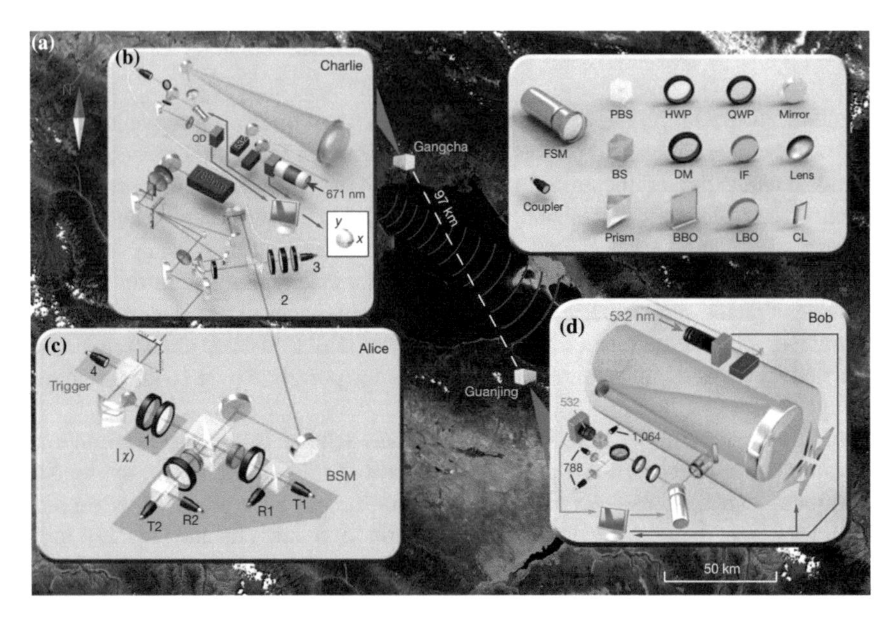

Fig. 27.5 Bird's-eye view and schematic diagram for free-space quantum teleportation. **a**, Satellite image of experiment site. **b**, Entanglement generation and distribution on Charlies side. **c**, Initial state preparation and Bell-state measurement (BSM) on Alice's side. **d**, Receiving system and polarization analysis on Bob's side.

and so on. The excellent quality of the recovered state with an average fidelity better than 89 % is thereby obtained. The transmission loss of the overall system is generally at a level of 30 dB. If we use a large-aperture telescope and high-accuracy ATP (acquisition, tracking and pointing) techniques, the transmission loss between low-Earth-orbit satellites and ground stations can be well controlled to this level by theoretical estimation.

Different to the above experiment, in 2011, we teleported independent qubits over a 97 km one-link free-space channel with multi-photon entanglement [22], as shown in Fig. 27.5.

Following the original quantum teleportation scheme [22, 26], Alice and Bob share an entangled photon pair distributed by Charlie. An unknown state can be teleported from Alice to Bob by performing a joint Bell-state measurement on the two photons with Alice. Experimentally, we start with an ultra-bright entangled photon source based on type-II spontaneous parametric down-conversion. On Charlie's side (located at Gangcha next to Qinghai Lake; latitude $37°16'42.4''N$, longitude $99°52'59.8''E$; altitude 3,262 m; Fig. 27.5), an entangled photon pair 2 and 3 in state $|\Phi^{+}\rangle_{23} = (|HH\rangle_{23} + |VV\rangle_{23})/\sqrt{2}$ is created by Charlie and then distributed to Alice and Bob, Where $H(V)$ represents the horizontal (vertical) polarization of the photonic state (see Fig. 27.5b). An average twofold coincidence rate of $4.4 \times 10^{5}\text{s}^{-1}$ for the entangled photon source was observed.

To prepare the unknown state to be teleported, Alice uses an ultraviolet laser to pump a collinear β-barium borate (BBO) crystal, which emits photon pairs in $|HV\rangle_{14}$ along the pumping direction (see Fig. 27.5c). After filtering out the pumping laser, a polarized beam splitter (PBS) splits the photon pair. A twofold coincidence rate of $6.5 \times 10^5 \mathrm{s}^{-1}$ was observed. A half-wave plate (HWP) and a quarter-wave plate (QWP) are applied to photon 1 to prepare the initial state $|\chi\rangle_1 = \alpha|H\rangle + \beta|V\rangle$, when triggered by photon 4. Alice then performs a joint Bell-state measurement on photons 1 and 2 by interfering them on a PBS and performing polarization analysis on the two outputs. The subsequent coincidence measurements can identify the $|\Phi^{\pm}\rangle$ Bell states in our experiment. In the joint Bell-state measurement, the observed visibility of interference on the PBS was 0.6. Finally, we observed a fourfold coincidence of $2 \times 10^3 \mathrm{s}^{-1}$ locally. Such a brightness supports successful quantum teleportation over a high channel loss, which can be greater than 50 dB.

Charlie sends photon 3 (by means of a compact transmitting system) through a 97 km free-space channel to Bob (Fig. 27.5b). A 127 mm $f/7.5$ (that is, aperture 127 mm, focal length 952.5 mm) extra-low dispersion apochromatic refractor telescope is used as an optical transmitting antenna. For near-diffraction-limited far-field divergence angles, we have designed our systems to reduce chromatic and spherical aberrations substantially. The divergence angle of our compact quantum transmitter is about 20 μrad. On the other side of Qinghai Lake, Bob received photon 3 in a 400 mm-diameter off-axis reflecting telescope. An integrated measurement system, consisting of an HWP, a QWP and a PBS, is assembled at the telescopes exit for state analysis. Passing through two band-pass filters (full-width at half-maximum bandwidth $\Delta_{FWHM} = 80$ nm) and one narrow-band interference filter ($\Delta_{FWHM} = 10$ nm) used to reduce background noise, the photons are coupled in multi-mode fibres and then detected by the single-photon counting modules (SPCMs) with ultra-low dark counts ($< 20\ \mathrm{s}^{-1}$). The noise that we observed, including the dark counts and ambient counts, was in total about 160 to 300 s^{-1}, depending on the position of the Moon.

In this work, we experimentally realized free-space quantum teleportation for independent qubits over a 35–53 dB-loss one-link channel. We obtained 1, 171 coincidences during an effective time of 14,400 s. Six distinct polarization states, namely, $|H\rangle$, $|V\rangle$, $|\pm\rangle = (|H\rangle \pm |V\rangle)/\sqrt{2}$, $|R\rangle = (|H\rangle + i|V\rangle)/\sqrt{2}$ and $|L\rangle = (|H\rangle - i|V\rangle)/\sqrt{2}$ were teleported. The experimental fidelities for the six teleported states range from 76 to 89 %, all well beyond the classical limit of $2/3$ [27], with an overall average fidelity of 80 %. In comparison with previous multi-photon experiments, we have enhanced the transmission distance by two orders of magnitude to 97 km.

Our experiment therefore confirms the feasibility of teleportation-based quantum communications for satellite-ground applications. Our results show that even with a high-loss ground-to-satellite uplink channel, quantum teleportation can be realized. Furthermore, our APT system can be used to track an arbitrarily moving object with high frequency and high accuracy, which is essential for future satellite-based ultra-long-distance quantum communication. We believe our experiment will help fundamental tests of the laws of quantum mechanics on a global scale to be achieved.

Quantum Cryptography in Free-Space Channels

Quantum key distribution (QKD) is one of the most research domain of quantum communication. QKD provides the only intrinsically unconditional secure method for communication based on the fundamental principles of quantum mechanics. There are two types of QKD schemes: one is the prepare-and-measure scheme, such as the Bennett-Brassard-1984 (BB84) protocol [28] and the Bennett-1992 protocol [29]; the other is the entanglement-based scheme, such as the Ekert-1991 protocol [30] and the Bennett-Brassard-Mermin-1992 (BBM92) protocol [31]. Since the first QKD experimental demonstration in the early 1990s [32], QKD has rapidly developed towards the stage for real-life applications.

Compared with fibre-based demonstrations, free-space links could provide the most appealing solution for communication over much larger distances. Despite significant efforts, most realizations to date rely on stationary sites. Experimental verifications are therefore extremely crucial for applications to a typical low Earth orbit satellite. To achieve direct and full-scale verifications of our setup, we have carried out three independent experiments with a decoy-state QKD system, and overcome all conditions at Qinghai Lake and reported these experimental results in 2013 [14]. The system is operated on a moving platform (using a turntable), on a floating platform (using a hot-air balloon), and with a high-loss channel to demonstrate performances under conditions of rapid motion, attitude change, vibration, random movement of satellites, and a high-loss regime. The experiments address wide ranges of all leading parameters relevant to low Earth orbit satellites. These results pave the way towards ground-satellite QKD and a global quantum communication network.

We implemented direct and comprehensive verifications for establishing successful quantum cryptography communication via satellites in three independent experiments conducted at night to address every aspect of the above mentioned crucial issues (as shown in Figs. 27.6 and 27.7). First, quantum communication experiments using a turntable (the site at 40 km in Fig. 27.6) and a hot-air balloon (the site at 20 km in Fig. 27.6) were performed to simulate a platform in a rapidly moving orbit as well as the vibration, random motion and attitude change associated with a satellite in such orbit. Second, we illustrate the generation of secret keys for a 96 Km free-space channel with loss of ~50 dB, which is more severe than the 30–50 dB loss associated with links between ground stations and a low Earth orbit satellite (LEOS) [33, 34]. Many key technologies, which can be directly used in satellite-ground QKD, were developed and improved, such as a high-speed and compact QKD source based on a decoy scheme, an acquisition, tracking and pointing (ATP) system, and tailored optical transmitters and receivers integrated to be lightweight and portable terminals. The secure distances achieved here are significantly greater than the effective thickness of the atmosphere (equivalent to ~8–10 km of ground atmosphere). Our verification environment has not only incorporated all possible motion modes, but also applied more extreme situations in relation to vibration, random movement and attitude change by using a hot-air balloon. Accordingly, our implementations, for the first time, provide comprehensive and direct verifications for secure key exchanges

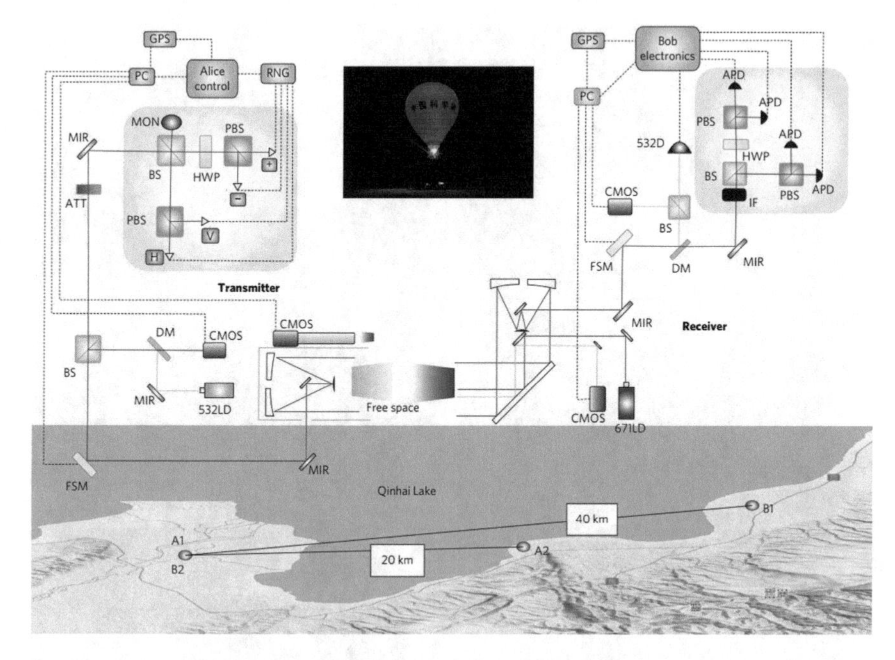

Fig. 27.6 Schematic of experimental set-up. The signal states and decoy states emitted from the polarization-encoding module pass through a reflecting Cassegrain telescope before transmitting to the receiver side. Once received by another telescope, they are directed to the detection module for polarization analysis. The 532 nm and 671 nm beacon light passes through the same channel for tracking, and the 532 nm light also acts as a synchronization signal. A1 (A2) and B1 (B2) represent the transmitter and receiver sites for the moving (floating) platform experiments, respectively. Red lines represent light at 850 nm, green lines and violet lines represent the tracking beacon light of receiver and transmitter, respectively, and dotted lines represent electric cable. PBS, polarizing beam splitters; BS, beam splitter; HWP, half-wave plate; MON, monitor window; MIR, mirror; ATT, attenuator; DM, dichroic mirror; 532LD, 532 nm laser; FSM, fast steering mirror; 671LD, 671 nm laser; 532D, 532 nm detector; IF, interference filter; APD, avalanche photodiode. Inset: rising and erupting hot-air balloon in the floating platform experiment.

via fast-moving platforms such as satellites or aircraft, and bring global-scale quantum communication closer to fruition.

Aiming to the future satellite-based quantum communication, we spent great efforts on ground testing the feasibility. In 2013, we reported a direct experimental demonstration of the satellite-ground transmission of a quasi-single-photon source [35] cooperating with Shanghai Astronomical Observatory. In this experiment, single photons (~0.85 photon per pulse) are generated by reflecting weak laser pulses back to earth with a cube-corner retro-reflector on the satellite CHAMP, collected by a 600 mm diameter telescope at the ground station, and finally detected by single-photon counting modules after 400 km free-space link transmission. With the help of high accuracy time synchronization, narrow receiver field-of-view and high-repetition-rate pulses, a signal-to-noise ratio (SNR) of better than 16:1 is obtained, which is sufficient for a secure quantum key distribution. This experimental results

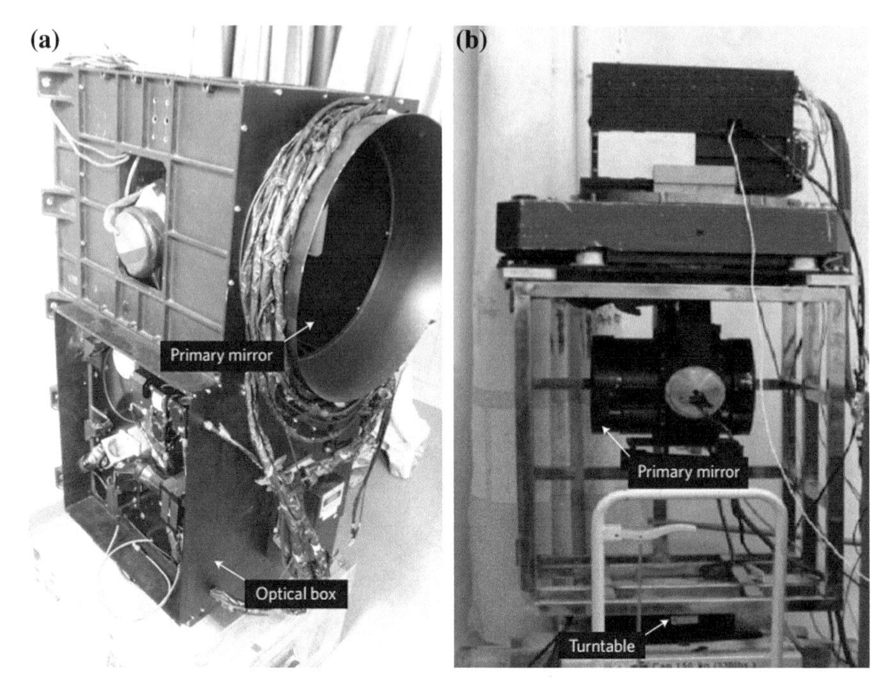

Fig. 27.7 Photographs of the QKD receiver and transmitter terminals. **a**, The receiver terminal has approximate dimensions of 500 mm × 500 mm × 900 mm and is used in experiments involving the simulation of satellite orbiting and the floating platform provided by a hot-air balloon. It contains the receiver telescope and an optical box. **b**, The transmitter terminal is mounted on a turntable and has approximate dimensions of 500 mm × 450 mm × 600 mm. This is used for performing the experimental simulation of satellite orbiting. When the turntable carries out a complex motion, the transmitter terminal will move accordingly. Therefore, simulation can be achieved with an angular velocity and angular acceleration larger than that in a typical LEOS.

demonstrate the feasibility of realizing satellite-ground QKD with current technology. In the same year, we report another experiment of free-space entanglement-based quantum key distribution, implementing the biased basis protocol between two sites which are 15.3 km apart. Photon pairs from a polarization-entangled source are distributed through two 7.8 km free-space optical links [13]. This experiment suggests that the efficient BB84 protocol [36] is indeed an easy and effective way to increase the key rate for QKD and can be used in our future plan of quantum science satellite.

Quantum physics allows for unconditionally secure communication between parties that trust each other. However, when the parties do not trust each other such as in the bit commitment scenario, quantum physics is not enough to guarantee security unless extra assumptions are made. Unconditionally secure bit commitment only becomes feasible when quantum physics is combined with relativistic causality constraints. We experimentally implemented a quantum bit commitment protocol with relativistic constraints that offers unconditional security at Shanghai in 2013, as

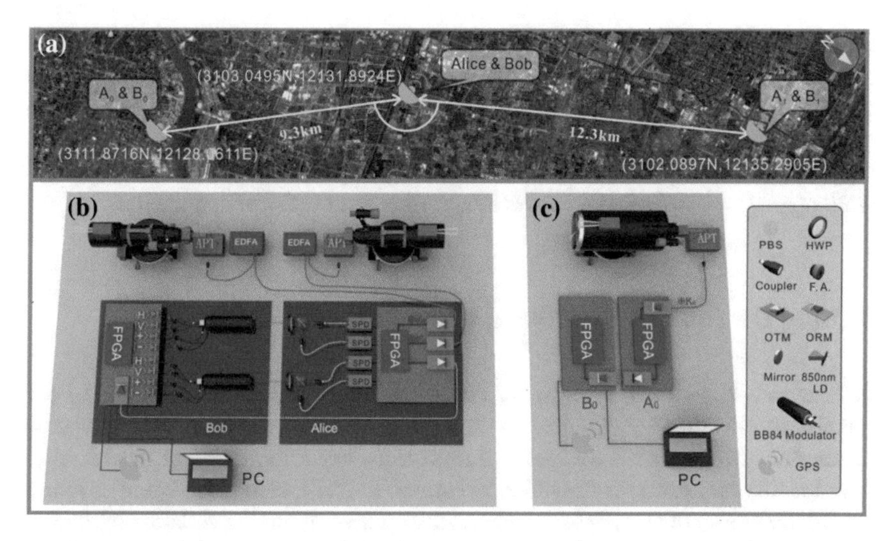

Fig. 27.8 **a** Diagram of the geographical distribution of the parties. The distance between Alice and Bob, and between A_0 and B_0 (and A_1 and B_1) is less than one meter. Alice communicates with her agents using sending and receiving telescopes. Each site has a global position system (GPS) for synchronization. **b** Diagram of Bobs and Alices setup. Triggered by a GPS signal, Bob attenuates and encodes laser pulses with a BB84 module, which is composed of two Wollason polarization prisms and a beam splitter, and sends them to Alice. Alices commitment setup consists of a half wave plate (HWP), a polarization beam splitter (PBS) and silicon avalanched photo-diode single photon detectors (SPDs). A field programmable gate array (FPGA) board is used to record the detected signals and communicate with Bob and Alices agents. Alice encodes, amplifies and sends the measurement results to her agents through telescopes. The four BB84 polarization states are denoted as $|H\rangle$, $|V\rangle$, $|+\rangle$, and $|-\rangle$. F. A.: fixed attenuator. OTM: optical transmission module. ORM: optical receiving module. EDFA: erbium-doped fiber amplifiers. APT: acquiring, pointing and tracking. **b** Diagram of B_0's and A_0's setup. The one of B_1 and A_1 is identical. $A_{(0)}$'s FPGA board receives and decrypts the detection information, stores it until the last classical signal is received, and then sends the measurement results to B_0. Bob's agents record the timing of these signals and send the results to Bob.

shown in Fig. 27.8. The commitment is made through quantum measurements in two quantum key distribution systems in which the results are transmitted via free-space optical communication to two agents separated with more than 20 km. The security of the protocol relies on the properties of quantum information and relativity theory. In each run of the experiment, a bit is successfully committed with less than 5.68×10^{-2} cheating probability. This demonstrates the experimental feasibility of quantum communication with relativistic constraints.

Our experiment, for the first time, provides unconditionally secure bit commitment between two mutually mistrustful parties, which is also a building block for many cryptographic primitives, including coin tossing [37, 38], zero-knowledge proofs [39, 40], oblivious transfer [41, 42] and secure two-party computation [43]. In future applications, QKD would be the preferable choice to distribute secret keys between Alice and her agents. However, to cover long distances in this scenario it would be necessary to employ either satellite-based links or quantum repeaters. While both quantum mechanics and relativity have changed our understanding of the Universe, our experiment together with that of Lunghi et al. [44] show for the first time that when we combine them, we can solve a fundamental problem with practical applications, and for which there is no solution using only one of them on their own. Our work opens a promising new research field with technological applications.

Testing the Speed of Nonlocal Correlations

In their well-known paper, Einstein, Podolsky, and Rosen called the nonlocal correlation in quantum entanglement a "spooky action at a distance." If the spooky action does exist, what is its speed? All previous experiments along this direction have locality and freedom-of-choice loopholes. In 2012, using free-space channels, we strictly closed the loopholes by observing a 12 h continuous violation of the Bell inequality and concluded that the lower bound speed of spooky action was 4 orders of magnitude of the speed of light if Earth's speed in any inertial reference frame was less than 10^3 time the speed of light.

We distributed entangled photon pairs over two sites that were 16 km apart via a free-space optical link, and implemented a space-like Bell test [17, 19] to close the locality loophole. Meanwhile, we utilized fast electro-optic modulators (EOMs) to address the freedom-of-choice. As almost all the photonic Bell experiments, we utilized the fair sampling assumption to address the detection loophole [18]. This assumption is justified by J. S. Bell's comments [45], "...it is hard for me to believe that quantum mechanics works so nicely for inefficient practical setups and is yet going to fail badly when sufficient refinements are made. Of more importance, in my opinion, is the complete absence of the vital time factor in existing experiments. The analyzers are not rotated during the flight of the particles."

Our two sites are east-west oriented sites at the same latitude as Eberhard proposed [46]. Takeing advantage of Earth's rotation, the configuration of our experiment allowed us to determine, for any hypothetically privileged frame, the lower bound of the spooky action speed by distributing polarization entangled photon pairs over two exactly east-west oriented sites and observed a 12 h continuous spacelike Bell inequality violation. It should be noted that, with only two parties one may always hold the opinion that a real fast enough spooky action could explain quantum correlations violating Bell inequality without leading to superluminal communication. Recently, some researchers found that this no longer holds when more parties are involved [47] (Fig. 27.9).

In this experiment, the developed fast time tagging, long-distance laser tracking, and synchronizing technology in the experiment can also find immediate applications in satellite-Earth quantum communication, multiparty quantum communication, and tests of the space-like GHZ theorem.

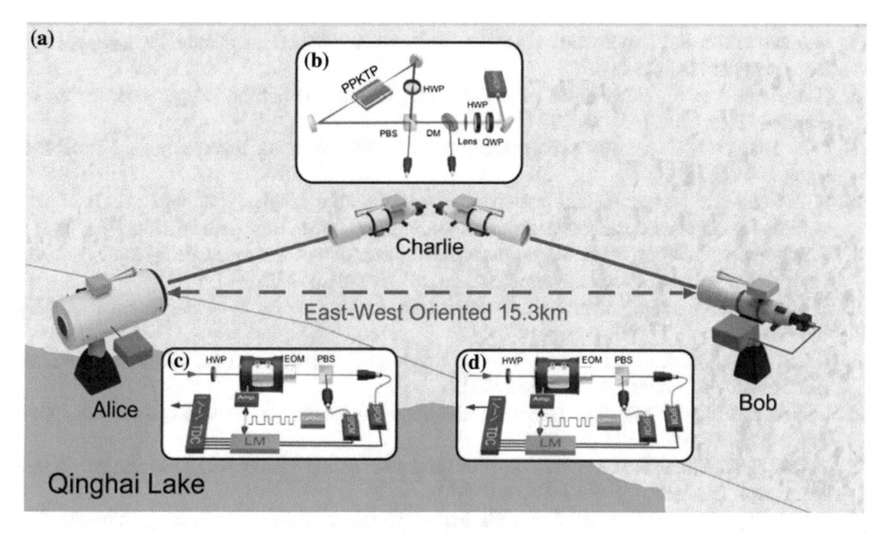

Fig. 27.9 Diagram of experimental setup for testing the speed of spooky action. **a**, Birds-eye view of the experiment. We generated entangled photon pairs in the sending point and utilized an integrated sending system to distribute them into two east-west oriented receiving points A and B. Receiving site A used an off-axis reflection telescope with 400 mm diameter, and Bused a refraction telescope with 127 mm diameter. Both receiving sites A and B had polarization analyzing units (the orange box). **b**, Entanglement generation setup. A pair of 45° mirrors and a dual-wavelength PBS formed the Sagnac interferometer. A 22.5° dual-wavelength HWP was used to swap the horizontal and vertical polarization. The DM was used to distinguish the pump light and the signal light. The entangled photons were coupled to SMFs through aspherical lens. **c**, Active polarization analyzing unit in receiving site A. A HWP, an EOM, a PBS, and two multimode fiber-coupled single photon counting modules (SPCMs) constitute an active polarization analyzing unit. A quantum random number generator (QRNG) and an amplifier were used to drive the EOM. The signals from the SPCMs combined with the logic module were sent to a time digital converter (TDC). **d**, Active polarization analyzing unit in receiving site B.

Acknowledgments This work has been supported by the National Fundamental Research Program (under Grant No. 2011CB921300 and 2013CB336800), the "Strategic Priority Research Program" of the Chinese Academy of Sciences (under Grant No. XDA04020000), the NNSF of China and the Chinese Academy of Sciences.

References

1. H. Takesur et al., Quantum key distribution over a 40-dB channel loss using superconducting single-photon detectors. Nat. Photon **1**, 343–348 (2007)
2. D. Stucki et al., High rate, long-distance quantum key distribution over 250 km of ultra low loss fibres. New J. Phys. **11**, 075003 (2009)
3. Y. Liu et al., Decoy-state quantum key distribution with polarized photons over 200 km. Opt. Express **18**, 8587–8594 (2009)
4. H.-J. Briegel et al., Quantum repeaters: the role of imperfect local operations in quantum communication. Phys. Rev. Lett. **81**, 5932 (1998)

5. M. Zukowski et al., “Event-ready-detectors” Bell experiment via entanglement swapping. Phys. Rev. Lett. **71**, 4287 (1993)
6. C.H. Bennett et al., Purification of noisy entanglement and faithful teleportation via noisy channels. Phys. Rev. Lett. **76**, 722 (1996)
7. Z.-S. Yuan et al., Experimental demonstration of a BDCZ quantum repeater node. Nature **454**, 1098–1101 (2008)
8. C.-Z. Peng et al., Experimental free-space distribution of entangled photon pairs over 13 km: towards satellite-based global quantum communication. Phys. Rev. Lett. **94**, 150501 (2005)
9. T. Schmitt-Manderbach et al., Experimental demonstration of free-space decoy-state quantum key distribution over 144 km. Phys. Rev. Lett. **98**, 010504 (2007)
10. R. Ursin et al., Entanglement-based quantum communication over 144 km. Nat. Phys. **3**, 481–486 (2007)
11. J. Yin et al., Quantum teleportation and entanglement distribution over 100-kilometre free-space channels. Nature **488**, 185–188 (2012)
12. C. Erven et al., Entangled quantum key distribution over two free-space optical links. Opt. Express **16**, 16840–16853 (2008)
13. Y. Cao et al., Entanglement-based quantum key distribution with biased basis choice via free space. Opt. Express **21**, 27260–27268 (2013)
14. J.-Y. Wang et al., Direct and full-scale experimental verifications towards ground-satellite quantum key distribution. Nat. Photon. **7**, 387–393 (2013)
15. S. Nauerth et al., Air-to-ground quantum communication. Nat. Photon. **7**, 382–386 (2013)
16. D. Rideout et al., Fundamental quantum optics experiments conceivable with satellites—reaching relativistic distances and velocities. Class. Q. Grav. **29**, 224011 (2012)
17. J.S. Bell, On the Einstein Podolsky Rosen paradox. Physics **1**, 195 (1964)
18. D. Salart et al., Reply to the “Comment on: Testing the speed of ‘spooky action at a distance”’. arXiv:0810.4607
19. J. Clauser et al., Proposed experiment to test local hidden-variable theories. Phys. Rev. Lett. **23**, 880 (1969)
20. A. Fedrizzi et al., High-fidelity transmission of entanglement over a high-loss free-space channel. Nat. Phys. **5**, 389–392 (2009)
21. N. Gisin et al., Quantum cryptography. Rev. Mod. Phys. **74**, 145–195 (2002)
22. C.H. Bennett et al., Teleporting an unknown quantum state via dual classic and Einstein-Podolsky-Rosen channels. Phys. Rev. Lett. **70**, 1895–1899 (1993)
23. D. Boschi et al., Experimental realization of teleporting an unknown pure quantum state via dual classical Einstein-Podolsky-Rosen channels. Phys. Rev. Lett. **80**, 1121–1125 (1998)
24. R. Ursin et al., Quantum teleportation across the Danube. Nature **430**, 849 (2004)
25. Y.-A. Chen et al., Memory-built-in quantum teleportation with photonic and atomic qubits. *Nat. Phys.* **4**, 43.1–43.14 (2002)
26. D. Bouwmeester et al., Experimental quantum teleportation. Nature **390**, 575–579 (1997)
27. S. Popescu, Teleportation versus Bell’s inequalities. What is nonlocality? Phys. Rev. Lett. **72**, 797–799 (1994)
28. C.H. Bennett, G. Brassard, Quantum cryptography: public key distribution and coin tossing, In *Proceedings of the IEEE International Conference on Computers, Systems and Signal Processing* (New York, 1985), p. 175
29. C.H. Bennett, Quantum cryptography using any two nonorthogonal states. Phys. Rev. Lett. **68**, 3121–3124 (1992)
30. A.K. Ekert, Quantum cryptography based on Bell’s theorem. Phys. Rev. Lett. **67**, 661–663 (1991)
31. C.H. Bennett et al., Quantum cryptography without Bell’s theorem. Phys. Rev. Lett. **68**, 557–559 (1992)
32. C.H. Bennett et al., Experimental quantum cryptography. J. Cryptol. **5**, 3–28 (1992)
33. P. Villoresi et al., Experimental verification of the feasibility of a quantum channel between space and Earth. New J. Phys. **10**, 033038 (2008)

34. C. Bonato et al., Feasibility of satellite quantum key distribution. New J. Phys. **11**, 045017 (2009)
35. J. Yin et al., Experimental quasi-single-photon transmission from satellite to earth. Opt. Express **21**, 20032–20040 (2013)
36. H.-K. Lo et al., Efficient quantum key distribution scheme and a proof of its unconditional security. J. Cryptol. **18**, 133–165 (2005)
37. M. Blum, in *Digest of Papers Spring COMPCON'82. High Technology in the Information Industry* (IEEE Computer Society, New York, 1982), p. 133
38. G. Brassard et al., in *Proceedings of the 10th Annual International Cryptology Conference, Advances in Cryptology—CRYPTO'9*, vol. 537 (Springer, New York, 1991), p. 49
39. S. Goldwasser et al., in *Proceedings of the 17th Annual ACM Symposium on Theory of Computing, STOC'85* (ACM, New York, 1985), p. 291
40. O. Goldreich et al., in *Proceedings of the 27th Annual Symposium on Foundations of Computer Science, FOCS'86* (IEEE Computer Society, New York, 1986), p. 174
41. C.H. Bennett et al., in *Proceedings of the 11th Annual International Cryptology Conference, Advances in Cryptology—CRYPTO'91*, vol. 576 (Springer, 1991), p. 351
42. D. Unruh, in *Proceedings of the 29th Annual International Conference on the Theory and Applications of Cryptographic Techniques, Advances in Cryptology—EUROCRYPT 2010*, vol. 6110 (Springer, New York, 2010), p. 486
43. J. Kilian, in *Proceedings of the 20th Annual ACM Symposium on Theory of Computing, STOC'88* (ACM, New York, 1988), p. 20
44. T. Lunghi et al., Phys. Rev. Lett. **111**, 180504 (2013)
45. J.S. Bell, *Speakable and Unspeakable in Quantum Mechanics* (Cambridge University Press, Cambridge, England, 1987), p. 139
46. P.H. Eberhard, *Quantum Theory and Pictures of Reality* (Springer, Berlin, 1989)
47. J.-D. Bancal, Quantum non-locality based on finite-speed causal influences leads to superluminal signalling. Nat. Phys. **8**, 867–870 (2012)

Chapter 28
Bell's Theorem, Bell Inequalities, and the "Probability Normalization Loophole"

John F. Clauser

Abstract Fifty years ago in 1964, John Bell [6], showed that deterministic local hidden-variables theories are incompatible with quantum mechanics for idealized systems. Inspired by his paper, Clauser, Horne, Shimony and Holt (CHSH) [12] in 1969 provided the first experimentally testable Bell Inequality and proposed an experiment to test it. That experiment was first performed in 1972 by Freedman and Clauser [20]. In 1974 Clauser and Horne (CH) [13] first showed that all physical theories consistent with "Local Realism" are constrained by an experimentally testable loophole-free Bell Inequality—the CH inequality. These theories were further clarified in 1976–1977 in "An Exchange on Local Beables", a series of papers by Bell, Shimony, Horne, and Clauser [8] and by Clauser and Shimony (CS) [15] in their 1978 review article. In 2013, nearly fifty years after Bell's original 1964 paper [6], two groups, Giustina et al. [24] and Christensen et al. [11] have finally tested the loophole-free CH inequality. Clauser and Shimony (CS) [15] also showed that the CHSH inequality is testable in a loophole-free manner by using a "heralded" source. It was first tested this way by Rowe et al. [35] in 2001, and more convincingly in 2008 by Matsukevich et al. [33]. To violate a Bell Inequality and thereby to disprove Local Realism, one must experimentally examine a two component entangled-state system, in a configuration that is analogous to a *Gedankenexperiment* first proposed by Bohm [9] in 1951. To be used, the configuration must generate a normalized coincidence rate with a large amplitude sinusoidal dependence upon adjustable apparatus settings. Proper normalization of this amplitude is critical for the avoidance of counterexamples and loopholes that can possibly invalidate the test. The earliest tests used the CHSH inequality without source heralding. The first method for normalizing coincidence rates without heralding was proposed by CHSH [12] in 1969. It consists of an experimental protocol in which coincidence rates measured with polarizers removed are used to

A talk presented at Quantum [Un] Speakables II, 50 Years of Bell's Theorem 19–22 June 2014, University of Vienna & Austrian Academy of Sciences.

J.F. Clauser (✉)
817 Hawthorne Drive, Walnut Creek, CA 94596, USA
e-mail: john@jfcbat.com

R. Bertlmann and A. Zeilinger (eds.), *Quantum [Un]Speakables II*,
The Frontiers Collection, DOI 10.1007/978-3-319-38987-5_28

normalize coincidence rates measured with polarizers inserted. Very high transmission polarizers are required when using this method. Highly reasonable and very weak supplementary assumptions by CHSH and by CH allow this protocol to work in a nearly loophole free manner. A second method for normalizing coincidence rates was offered by Garuccio and Rapisarda [22] in 1981. As will be discussed below, it allows experiments to be done more easily, but at a significant cost to the generality of their results. It was first used in the experiment by Aspect, Grangier, and Roge [3] in 1982. It uses "ternary-result" apparatuses and allows the use of highly absorbing polarizers, which would not work with other normalization methods. It normalizes using a sum of coincidence rates. Gerhardt et al. [23] in 2011 theoretically and experimentally demonstrated counterexamples for tests that use this normalization method. Their experiments thus obviate the validity of their counterexamples, and further indicate that very high transmission polarizers are necessary for convincing tests to be performed.

Introduction to Bell's Theorem and the Bell Inequalities

Bell's Theorem is formulated in terms of a set of individually named inequalities, each with increasing generality and scope. These inequalities are collectively referred to as the "Bell Inequalities". They surprisingly follow from very simple natural assumptions concerning the nature of reality. These assumptions, along with their associated consequences via Bell's Theorem, then constitute a minimal framework for a whole class of theories originally named "Objective Local Theories" by Clauser and Horne (CH) [13] in 1974, and subsequently renamed "Local Realism" by Clauser and Shimony (CS) [15] in 1978. The assumptions underlying Local Realism are so simple and natural that one of this conference's organizers, Anton Zeilinger, recently commented that if Bell's Theorem had been discovered before quantum mechanics, it would have been promoted to be considered a law of nature on its own, whereupon the subsequently discovered quantum mechanics must obviously be wrong! The assumptions underlying Local Realism are reviewed in the Appendix.

The essence of Bell's Theorem is that theories based on Local Realism cannot give the same prediction for certain "entangled-state" two-component systems as does the theory of quantum mechanics. Thus, these two opposing theories are experimentally distinguishable from each other. It is then the task for experimental physicists to determine which of these two incompatible theories correctly describes the world in which we live. To refute Local Realism (and/or to refute quantum mechanics) experimentally, one can perform an experiment whose quantum mechanical predictions violate a Bell Inequality. The first experiment to do so was that by Freedman and Clauser [20] in 1972. Their results were then the first to violate the CHSH inequality (but still leave open the normalization and locality loopholes). Freedman and Clauser's experimental results have been overwhelmingly confirmed by many other experiments, some of which are discussed and tabulated in the section "Some Experimental Results". The experiment by Aspect, Dailbard and Roger [4] again violated the CHSH inequality, and was first to close the locality loophole, but still leave open the normalization loophole. The normalization-loophole-free heralded-source CHSH inequality was first tested by Rowe et al. [35] in 2001, and more convincingly in 2008 by Matsukevich et al. [33], with both experiments still leaving open the locality loophole. Giustina et al. [24] and Christensen et al. [11] have finally tested the normalization-loophole-free CH inequality in 2013. Experiments are currently in progress to finally close both the normalization loophole and the locality loophole simultaneously in a single experiment.

The particular entangled-state two-component systems referred to above were used in a *Gedankenexperiment* that was first envisaged by David Bohm [9] in 1951. That entangled-state system was used by John Bell [6] in his now-famous 1964 paper. Bohm's *Gedankenexperiment* is described below in the section "Bohm's 1951 *Gedankenexperiment* and Its Relation to Bell's Theorem". Bohm's arrangement then provides a prototype configuration for Bell's Theorem experiments.

The generality and scope of Bell's Theorem has evolved since its discovery. John Bell's 1964 paper [6] introduced the first Bell-Inequality and showed that no Local Hidden Variables Theory (LHVT) can give the quantum mechanical prediction for Bohm's idealized *Gedankenexperiment*. In that paper, however, Bell made no reference to the experimental status, or even to the experimental testability of his result. Moreover, Bell's 1964 [6] Inequality applied only to idealized systems. To bridge the gap between theory and the real world in which we live, Clauser, Horne, Shimony, and Holt (CHSH) [12] in 1969 introduced the second Bell Inequality—the CHSH inequality—and showed that it holds for deterministic LHVT's that govern realizable systems. More importantly, unlike Bell's 1964 result, the CHSH inequality is experimentally testable. CHSH also were the first to propose an actual experiment to test the Theorem's predictions.

The extension of Bell's Theorem to include Objective Local Theories (and Local Realism) was made by Clauser and Horne (CH) [13] in 1974, and these theories were renamed Local Realism by Clauser and Shimony [15] in 1978. Clauser and Horne therein introduced what is now commonly referred to as the CH inequality. It is experimentally testable and is loophole free, and is what we herein refer to as an "R-inequality" (see below). A further discussion clarifying the meaning and scope of Bell's Theorem followed CH in "An Exchange on Local Beables" [8]—a series of papers by Bell, Shimony, Horne and Clauser. A review of the various proofs and interpretations of Bell's Theorem, the various Bell Inequalities, and the available modalities for experimental testing is given by the Clauser and Shimony (CS) [15] review article.

To violate a Bell inequality and thereby to disprove Local Realism, one must experimentally examine a two component entangled-state system, in a configuration that is analogous to Bohm's *Gedankenexperiment*. The experiment is done with two widely separated apparatuses. To be used for a test, the configuration must generate normalized coincidence rates at these two apparatuses with a measured large amplitude sinusoidal dependence upon the two adjustable apparatus settings. Proper normalization of the coincidence rates is critical for the avoidance a loophole.

A loophole exists when a counterexample exists that invalidates the experimental test. Loopholes sometimes arise when technology limits just how closely one can approach the ideal experiment specified by a Bell Inequality. Various experimental tricks are then generally used, along with associated supplementary assumptions to plug these loopholes. These added assumptions generally do not rely on either locality or realism (or quantum mechanics), although it is highly desirable that they at least be consistent with these theories. To evaluate the assumptions, one may examine how reasonable a supplementary assumption is, along with how contrived the associated counterexample is. Such assumptions thus become the weak point in any argument claiming an experimental disproof of Local Realism. The obvious question is always offered—are you testing the fundamental assumptions behind Local Realism, or are you just testing the supplementary assumption(s)? Fortunately, recent experiments closing the remaining loopholes are now rendering this last question moot.

The first identified loophole is the so-called "locality loophole". Curiously, it was first noted by Bohm and Aharonov [10] in 1957, prior to Bell's 1964 paper [6]. Under the locality loophole, a hypothetical collusion between the two separated apparatuses can possibly occur, whereby the apparatuses communicate their settings to each other. Such communication can then possibly can account for the strange quantum-mechanical predictions associated with the entanglement of widely separated particles. This possibility was promoted further by Bell in his 1964 paper [6]. Bohm and Aharonov had suggested that a rapid change of the two apparatuses of Bohm's *Gedankenexperiment,* while the entangled-state particles are in flight, can thereby exclude any such collusion. While this locality-loophole counterexample may seem somewhat contrived, it has become particularly important to close it when Bell's Inequality experimental results are under attack by malevolent efforts, as may occur when the experimental outcomes are used for quantum communication and cryptography, for which malevolent forces are well known to exist (e.g. by eavesdroppers). The first experiment to close this loophole was performed in 1982 by Aspect, Dalibard, and Roger [4].

The second identified loophole is what we herein call the "normalization loophole". It occurs when the measured large amplitude sinusoidal dependence on adjustable apparatus settings is less than that required for an actual violation of a Bell Inequality. The so-called "detection loophole" is one of several examples of the normalization loophole. The detection (normalization) loophole commonly occurs when low detection efficiency reduces the measured amplitude of the coincidence-rate variation to below that needed for a violation of a Bell Inequality.

Sometimes the normalization loophole occurs without it even being recognized. Indeed, the transition from Bell's 1964 inequality [6] to the CHSH inequality [12] involved closure of the first example of a normalization loophole, wherein Bell assumed and indeed required a perfect apparatus correlation. (See the sections "One Possible Cause for the Normalization Loophole" and "Bell's 1964 E-Inequality for Idealized Binary Result Apparatuses" below.) In general, the normalization loophole can be closed only with highly precise apparatus, and with careful count-rate normalization. It has only been closed recently. Closure of both the locality loophole and the normalization loophole simultaneously in a single experiment has not yet been done, but experiments now in progress promise to do so soon.

Much of the remainder of this paper addresses the normalization loophole. There are two known routes to plug it—either violate the CHSH inequality using a heralded source, as was first suggested by Clauser and Shimony (CS) [15] in 1978, or directly violate the CH inequality [13]. Both routes have met significant technical difficulties, and both routes require highly efficient detection schemes and highly efficient high-transmission analyzers (polarizers).

In the section "Bohm's 1951 *Gedankenexperiment* and Its Relation to Bell's Theorem", we describe Bohm's 1951 *Gedankenexperiment* [9]. It provides a basic prototype for Bell's Theorem experiments. In the section "One Possible Cause for the Normalization Loophole", we discuss the origin of the normalization loophole. The loophole's nature depends on the nature of the associated Bell Inequality being tested. In deriving a Bell Inequality, there are at least two ways to proceed. One way is to start directly from observed quantities, such as the number of observed particle detections (per unit time), then to calculate probabilities for them, and finally to derive an inequality constraining them that is consistent with the requirements for locality and realism. This path was followed by CH. It yields the CH Inequality, and the experiments that it constrains are then inherently free from the normalization-loophole. It is described in the section "Normalization-Loophole Free Clauser-Horne (CH) R-Inequality for Binary-Result Apparatuses". The CH inequality uses binary-result apparatuses. The 1978 review article by Clauser and Shimony (CS) [15] describes a variety of alternative methods for deriving the CH inequality. The CH inequality directly constrains observed count rates, and is thus, what we herein call an R-inequality, as an abbreviation for count-Rate-inequality. An R-inequality directly compares one linear combination of measured count rates with another.

A second method for deriving a Bell Inequality is that originally followed by Bell and by CHSH. It requires one to first define "result values" (as discussed in the section "Result Values and Expectation-Value Inequalities (E-Inequalities)"). That method then provides inequalities that constrain the expectation values for the various observed results. We call these "E-inequalities",[1] as an abbreviation for Expectation-value-inequalities. The first such E-inequality was derived by Bell in his original 1964 paper [6]. It is discussed in the section "Bell's 1964 E-Inequality for Idealized Binary Result Apparatuses". The second E-inequality was that by CHSH [12]. It is discussed in the section "Clauser Horne Shimony Holt (CHSH) E-Inequality for Real Binary Result Apparatuses". Unfortunately, an E-inequality is not directly testable, unless it is first converted to an R-inequality. Care must be exercised when one is performing the conversion in order to avoid unnecessarily introducing a normalization loophole.

[1]Historical Note: Both Bell [6] and CHSH [12] use the symbol P rather than E for the expectation value of the product of the binary result values A and B.Subsequent works generally now use the symbol E.

Care must also be exercised in recognizing whether or not one is using binary-result or ternary-result apparatuses. (See the section "One Possible Cause for the Normalization Loophole") Three methods have been used for doing the conversion. One method, described in section "Clauser Horne Shimony Holt (CHSH) R-Inequalities for Real Binary-Result Apparatuses via the CHSH Polarizer-Removal Protocol", is to use the CHSH polarizer-removal protocol to get a CHSH R-inequality. It requires the use of binary-result apparatuses, and it was used for all of the earliest tests of the CHSH E-and R-inequalities. The protocol consists of measuring coincidence rates with polarizers removed as well as measuring coincidence rates with polarizers inserted. The former measurements are used to normalize the latter. Very high transmission polarizers are required for this method to work. Highly reasonable supplementary assumptions by CHSH and by CH allow this protocol to provide a very reasonable argument for avoidance of the loophole. Clauser and Horne (CH) [13], however, do provide an *ad hoc* somewhat contrived counterexample, discussed in the section "The CH Counterexample". Thus, a residual normalization loophole remains with this method, despite the high plausibility of their associated supplementary assumption and the rather contrived nature of that counterexample.

A second method for converting an E-inequality into an R-inequality is described in the section "CHSH R-Inequality with Heralding". It uses a heralded source, and it was first explicitly suggested by Clauser and Shimony (CS) [15] in 1978. It then allows one to use the CHSH E-inequality directly to get a loophole-free CHSH R–inequality. It can use either binary- or ternary-result apparatuses.

A third method for normalizing coincidence rates was first proposed in 1981 by Garuccio and Rapisarda (GR) [22], and was first used in 1982 in the experiment by Aspect, Grangier, Roger (AGR) [3]. (See the section "Garuccio and Rapisarda/Aspect Grangier Roger R-Inequalities for Real Ternary-Result Apparatuses".) It uses "ternary-result" apparatuses only. Notably, it allows the use of highly absorbing polarizers, whereby a violation of an associated R-inequality is much easier to achieve experimentally. It normalizes the coincidence rates using a sum of these coincidence rates, and ignores unobserved (and unobservable) events. Unlike the CHSH polarizer-removal protocol, no polarizers are removed using this method, and no additional normalizing data need be taken. It also requires a much stronger supplementary assumption than that required by the CHSH polarizer-removal protocol. The GR/AGR supplementary assumption is now commonly (and gratuitously) referred to as the "fair-sampling assumption". Gerhardt et al. [23] provide a convincing experimental demonstration of the ease by which it can be countered, especially by malevolent efforts, as may occur in "security related scenarios" and quantum cryptography. Despite the need for these strong supplementary assumptions, GR/AGR normalization has been used by many experiments, presumably because of it's ease of experimental implementation. Its use has become sufficiently common that it is often cited (incorrectly) as an integral necessary part of the CHSH E-inequality, despite strident protestations made by this author at the first Quantum [Un]speakables conference (Clauser [16]). It is not!

We conclude in the section "Some Experimental Results" with a description and tabulation of various experimental results that test the predictions made by these various Bell Inequalities.

Bohm's 1951 *Gedankenexperiment* and Its Relation to Bell's Theorem

Figure 28.1 shows Bohms's 1951 *Gedankenexperiment* [9]. It provides the prototype for Bell's-Theorem discussions. It consists of a highly idealized pair of binary-result apparatuses interacting with a quantum mechanically entangled two-particle system. In this *Gedankenexperiment*, a spin-zero particle decays into a pair of spin-entangled spin-½ particles. Each of these particles, in turn, flies into an associated rotatable Stern-Gerlach analyzer, where it then follows one of two trajectories, and is detected by one of two associated detectors. For this system, Bohm [9] and Bell [6] both assume that the following highly idealized requirements hold:

(a) The initial state of the pair is a 100 % pure spin-singlet. (ψ = singlet = $\uparrow\downarrow$—$\downarrow\uparrow$ in any coordinate system)
(b) Both particles enter the collimators.
(c) The system's collimation is perfect and the propagation is loss-free.
(d) The propagation and spin-state selection are depolarization-free, and
(e) Both detectors have 100 % efficiency.

It is important to note, in passing, that these idealized specifications are, in general, impossible to realize in practice.

We define the result values (see the section "Result Values and Expectation-Value Inequalities (E-Inequalities)") at each apparatus A $\equiv \pm 1$, and B $\equiv \pm 1$, respectively. With the above idealized specifications, for an ensemble of decaying spin-zero particles, the quantum mechanical predictions for the probabilities of the four possible outcomes are

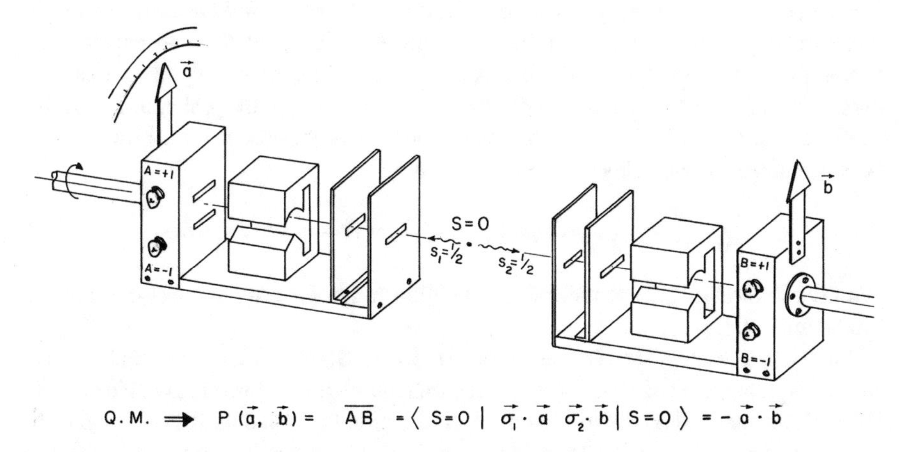

Fig. 28.1 Bohm's [9] *Gedankenexperiment*, that provides the prototype for Bell's-Theorem discussions.

$$\begin{aligned}
\text{prob}_{\text{QM}}(\text{A}=1,\text{B}=1) &= 1/2\sin^2(\text{ang}(\mathbf{a},\mathbf{b})/2),\\
\text{prob}_{\text{QM}}(\text{A}=-1,\text{B}=-1) &= 1/2\sin^2(\text{ang}(\mathbf{a},\mathbf{b})/2),\\
\text{prob}_{\text{QM}}(\text{A}=1,\text{B}=-1) &= 1/2\cos^2(\text{ang}(\mathbf{a},\mathbf{b})/2),\\
\text{prob}_{\text{QM}}(\text{A}=-1,\text{B}=-1) &= 1/2\cos^2(\text{ang}(\mathbf{a},\mathbf{b})/2),
\end{aligned} \tag{1}$$

where ang(**a**, **b**) is the angle between the two Stern-Gerlach analyzer orientations, and prob_{QM} (A = i, B = j) is the probability (as predicted by quantum mechanics) that apparatus A will yield the result i, and that apparatus B will yield the result j.

It is the large amplitude sinusoidal dependence of Eqs. (1) that is at the heart of Bell's Theorem, and it was Bell's genius to first note that this dependence cannot be obtained by any local hidden-variables theory, but instead can only be obtained by quantum mechanics! Bell thus discovered that the large amplitude sinusoidal dependence in Eqs. (1) is strictly peculiar to quantum-mechanical entangled-state systems. He further discovered that virtually any reasonable attempt to model the behavior of Bohm's *Gedankenexperiment* via hidden-variables gave instead, a strange non-sinusoidal result, and/or a low amplitude result that is very different from that given by Eqs. (1).

Bell's observation thus became the inspiration for experimentalists, who, in turn, wondered if nature really behaves the way quantum mechanics strangely predicts here. Relaxation of the ideal specifications for this *Gedankenexperiment*, in turn, reduces the amplitude of this sinusoidal dependence, whereupon a normalization loophole can result when the relaxation goes too far. In practice, very little relaxation from the ideal can be tolerated.

One Possible Cause for the Normalization Loophole

An important but frequently overlooked feature of Bohm's *Gedankenexperiment* is that each apparatus provides the binary result, ±1. Thus, for the two apparatuses and a given pair of spin-entangled particles, there are then only four possible outcomes, and four associated probabilities. For any set of probabilities to be sensible, and for Bell's Theorem to obtain, the sum of these four probabilities must be normalized to one. That is, we must have

$$\textstyle\sum_{\text{i}=\pm1}\sum_{\text{j}=\pm1}\text{prob}(\text{A}(\mathbf{a})=\text{i},\text{B}(\mathbf{b})=\text{j})=1,\ \text{for all } \mathbf{a},\mathbf{b}. \tag{2}$$

We note here that this normalization condition holds for the quantum mechanical predictions of Eqs. (1).

One can measure the various prob(A(**a**) = i, B(**b**) = j) experimentally from event frequencies. To do so, one needs the total number of i, j events, N(A(**a**) = i, B (**b**) = j), normalized by the total number N of emitted-pair events. Then, if and only if all events are properly accounted for, the above normalization condition becomes

$$\sum_{i=\pm 1} \sum_{j=\pm 1} N(A(\mathbf{a})=i, B(\mathbf{b})=j) = N, \text{ for all } \mathbf{a}, \mathbf{b}. \tag{3}$$

We further note that if there are missing non-zero terms in these summations, then the normalization condition of Eq. (3) does not hold.

Now consider any actual realization of Bohm's *Gedankenexperiment*. There, we really have ternary-result rather than binary-result apparatuses. In practice, one or both of the particles will fail to enter the collimators. Additionally, any real detector will sometimes fail to detect a particle entering it, and/or will sometimes falsely detect a particle, even when one is not present (a "dark-count"). As a result, there will be un-paired detections at the two apparatuses and/or totally missing paired detections. Correspondingly, for any realization of Bohm's *Gedankenexperiment* we really have the possible outcomes for each apparatus as being one of three possibilities: +1, −1, and No-detection (with no result-value, as yet, assigned to this possibility). For binary-result apparatuses, there are 4 nonzero terms in the above double summations. For ternary-result apparatuses, however, there are 9 nonzero terms. Unfortunately, at most, only 8 of those 9 can be observed by the two apparatuses, since the 9th term is a probability of nothing happening at both apparatuses.

Of course the value of the 9th unobserved term can be determined via an *a priori* knowledge of N by using Eq. (3). This latter possibility is now commonly referred to as "heralding", wherein the source apparatus signals (heralds) that a particle pair has been emitted and is ready for analysis and detection. The possible use of said heralding measurements was first noted by Clauser and Shimony (CS) [15], and was therein given the name "event-ready detectors". (The modern term "heralding" had not yet been invented in 1978.) It is discussed in the section "CHSH R-Inequality with Heralding".

A simpler alternative to the use of heralding was offered by CH [13]. They avoid a need for knowing the value of N by producing a Bell Inequality that only involves ratios of the various $N(A(\mathbf{a}) = i, B(\mathbf{b}) = j)$, whereupon the unknown value of N cancels out! Worries about unobserved particles may seem unimportant until one recognizes that in the earliest realizations of Bohm's *Gedankenexperiment*, the overwhelming majority of emitted pairs were, in fact, wholly or partially unobserved. The ratio of paired (coincident events) to unpaired detections ("singles events") detections was about 10^{-3}, while the ratio of the number of paired (coincident events) to the number of emitted particle pairs was about 10^{-6}. Only now, nearly 5 decades later have experiments evolved to the point where these event rates are all of comparable orders of magnitude.

Normalization-Loophole Free Clauser-Horne (CH) R-Inequality for Binary-Result Apparatuses

Clauser and Horne (CH) [13] start from an experimental arrangement that is slightly different from that of Fig, 28.1. It is shown in Fig. 28.2. It is configured to automatically enforce the above-noted need for binary results. A source at the center

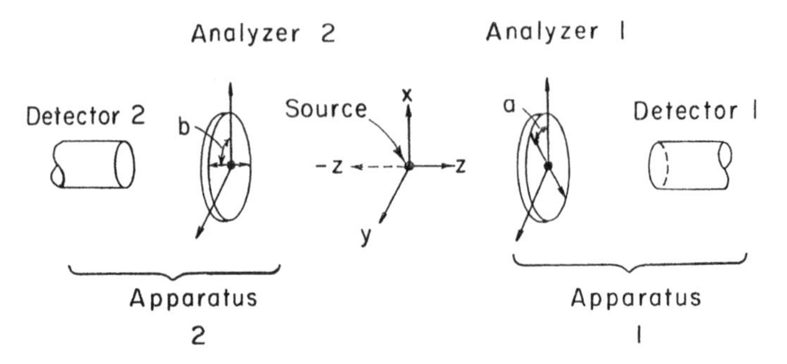

Fig. 28.2 Clauser and Horne's configuration for Bell's inequality experiments using binary-result apparatuses. Figure from Clauser and Shimony [14].

emits pairs of particles. The source is viewed by two apparatuses, named 1 and 2. Each apparatus consists of an adjustable attenuating analyzer and an associated single detector. A particle of the pair can pass through one of the analyzers, wherein after it is or is not detected by the associated detector. By the design of the experimental layout, only binary events, i.e. detection or non-detection events, can occur at each apparatus.

The CH derivation of Bell's Theorem and of the CH inequality starts by using directly observed event rates. In the apparatus of Fig. 28.2, one measures an individual detector's detection rate, and also simultaneously measures the ("coincident") paired detection rate of the two detectors. During a long period of time, t, the source emits say N of the two particle systems of interest. For this period, denote by $N_1(\mathbf{a})$ and $N_2(\mathbf{b})$ the number of detections at detectors 1 and 2 respectively, and by $N_{12}(\mathbf{a}, \mathbf{b})$ the number of nearly simultaneous (coincident) detections at the two detectors. From these numbers of detections, when sufficiently large, one may correspondingly define (measure) the ensemble probabilities

$$\begin{aligned} p_1(\mathbf{a}) &= N_1(\mathbf{a})/N, \\ p_2(\mathbf{b}) &= N_2(\mathbf{b})/N, \\ p_{12}(\mathbf{a},\mathbf{b}) &= N_{12}(\mathbf{a},\mathbf{b})/N. \end{aligned} \tag{4}$$

Here, p_{12} is the probability of joint (coincident) detections by both detectors; p_1 is the probability of a detection by detector 1, independently of what happens at detector 2; and p_2 is the probability of a detection by detector 2, independently of what happens at detector 1.

CH showed that the probabilities associated with correlated particle pairs that are described by any Local Realistic Theory (i.e. one that describes pairs of localized objects—see the Appendix), are constrained by the following inequality:

$$-1 \le p_{12}(\mathbf{a},\mathbf{b}) - p_{12}(\mathbf{a},\mathbf{b}') + p_{12}(\mathbf{a}',\mathbf{b}) + p_{12}(\mathbf{a}',\mathbf{b}') - p_1(\mathbf{a}') - p_2(\mathbf{b}) \le 0. \tag{5}$$

The left-hand inequality requires careful normalization of these probabilities (i.e. it requires one to know N), but the right-hand inequality does not! It is independent of N. Via Eqs. (4) one then has

$$-N \leq N_{12}(\mathbf{a},\mathbf{b}) - N_{12}(\mathbf{a},\mathbf{b}') + N_{12}(\mathbf{a}',\mathbf{b}) + N_{12}(\mathbf{a}',\mathbf{b}') - N_1(\mathbf{a}') - N_2(\mathbf{b}) \leq 0. \quad (6)$$

Dividing by t finally gives the associated CH R-inequality

$$-R_{Source} \leq R_{12}(\mathbf{a},\mathbf{b}) - R_{12}(\mathbf{a},\mathbf{b}') + R_{12}(\mathbf{a}',\mathbf{b}) + R_{12}(\mathbf{a}',\mathbf{b}') - r_1(\mathbf{a}') - r_2(\mathbf{b}) \leq 0, \quad (7)$$

where, $R_{12}(\mathbf{a},\mathbf{b})$ is the coincidence detection rate for the two detectors, and $r_1(\mathbf{a}')$ and $r_2(\mathbf{b})$ are respectively the individual singles detection rate at detectors 1 and 2. The quantity R_{Source} is the source rate, which may be used if it has been measured via heralding. (See the section "CHSH R-Inequality with Heralding".) For experiments where a heralded source is not employed, one may rewrite the right-hand side of Ineq. (7) as

$$[R_{12}(\mathbf{a},\mathbf{b}) - R_{12}(\mathbf{a},\mathbf{b}') + R_{12}(\mathbf{a}',\mathbf{b}) + R_{12}(\mathbf{a}',\mathbf{b}')]/[r_1(\mathbf{a}') + r_2(\mathbf{b})] \leq 1. \quad (7')$$

Here, the singles rates r_1 and r_2, are used to normalize the sum of four R_{12} coincidence rates. The minus sign preceding the second R_{12} term in the numerator may be permuted among any of the four terms and Ineq. (7') still holds.

Inequalities (5)–(7') are known as the CH inequalities. The CH R-inequality (7') is noteworthy in that it gets its normalization by using only the number of singles events at the two detectors. So doing, it provides a Bell Inequality that does not rely on the value of N (or R_{Source}), that is usually difficult to measure. Thereby Ineq. (7') is self-normalizing. The good news associated with the CH inequality is that the influence of N (or R_{Source}) vanishes, and the inequality is normalization-loopholefree. The bad news is that from an experimental viewpoint, the CH inequality is very difficult to violate. For low detector efficiencies and/or for small solid-angle collection efficiencies, the singles rates r_1 and r_2 are typically very much larger than R_{12}, (by a factor of about 10^3 for the typical cascade-photon experiments), Ineq. (7') is then automatically satisfied, whereupon no normalization-loophole free experiment can then be done.

For these (and for all other Bell Inequalities), it is generally necessary to perform a sequence of different experiments and compare their different results. In the present case there are four required experiments, at each of the respective analyzer orientation pairs, (**a**,**b**), (**a**,**b'**), (**a'**,**b**), and (**a'**,**b'**). Additionally, measurements must also be taken of the normalizing singles count rates r_1 and r_2 at angles **a'** and **b**, respectively, although these measurements are usually already obtained simultaneously during the coincidence rate measurements.

The experimental difficulties associated with designing an experiment to violate Ineq. (7') were lessened significantly by an observation made by Eberhard [17] in 1993. When the quantum state of the particle pair is maximally entangled, the singles rates $r_1(\mathbf{a}')$ and $r_2(\mathbf{b})$ in Ineq. (7') are constant, (i.e. independent of **a**' and **b**). On the other hand, Eberhard noted that there is no need for the normalizing denominator in Ineq. (7') to be constant. If the quantum state of the particle pair is not maximally entangled, then at least one of these two singles rates may be made small, and Ineq. (7') is more readily violated. Eberhard's observation has led to the recent experimentally observed violations of the CH inequality by Giustina et al. [24] and by Christensen et al. [11].

Clauser and Shimony [15] show in their review article that the methods of proof of Bell's Theorem used by Wigner [41], Bell [7], Belinfante [5], and Holt [26] can all lead to the CH inequality. CS further note that the CHSH inequality is a special case of the CH inequality, and that Bell's 1964 inequality [6] in turn is a special case of the CHSH inequality.

Result Values and Expectation-Value Inequalities (E-Inequalities)

Bell's original 1964 inequality and the CHSH inequality are both in a form that constrains expectation values for observed "result values". Herein, we refer to these as E-inequalities. The expectation values are calculated using previously defined "result values". What is a "result value", and why is it needed? To some extent, for Bell's Theorem, result values are only a historical artifact, and, as noted above in the section "Normalization-Loophole Free Clauser-Horne (CH) R-Inequality for Binary-Result Apparatuses", Bell Inequalities can be derived without invoking them. One may ask, where did they come from?

Bohm was discussing the quantum theory of the measurement process and the Einstein, Podolsky, and Rosen [19] paradox when he introduced his 1951 *Gedankenexperiment* of Fig. 28.1. In such a discussion, it is presumably necessary to assert that something is actually being "measured". Whatever is being measured, then should have a "result value" that is to be determined by the "measurement". The assumptions underlying Bell's Theorem, on the other hand, do not depend on whether or not something is being measured. In fully general LHVT's or Local Realistic Theories, one really does not have the faintest idea about what one is doing on a microscopic level when one performs a given experiment. Indeed, a dispute over what is happening internally in such an experiment (or indeed, if anything is happening at all on a microscopic level, as Bohr insisted) is at the very heart of any fully general theory that is an alternative to or consistent with quantum mechanics. Recall that the Copenhagen interpretation of quantum mechanics asserts that there is no possible explanation of the microphysics of Bohm's *Gedankenexperiment*, whereupon it would seem to be highly presumptuous to assert that one knows what one is measuring! Correspondingly, for a discussion of Bell's Theorem, the result values A, B $\equiv \pm 1$ chosen by Bohm and Bell are perfectly arbitrary. Mostly, they simply provide names like Tom, Dick, Harry, +1, 0, spin-up, top, beauty, and +ħ/2, etc. for the possible experimental outcomes. Indeed, the CH inequality which governs Local Realism dispenses with the use of result values altogether!

For a discussion of Bell's Theorem, however, CHSH and Bell did find some utility in defining result-values. If the result values have integer numerical values, then they may be used as indices in summations. Result values also allow one to construct expectation values for these results, which then lead to E-inequalities. Additionally, expectation values are readily calculated using quantum mechanics via matrix elements. Then, as long as care is exercised in converting said E-inequality into a useful R-inequality, result values serve their purpose.

The expectation value of the product of the binary-result values for A and B is also known as the correlation function of these two values. For Bohm's 1951 *Gedankenexperiment*, it is given by

$$E(\mathbf{a},\mathbf{b}) \equiv <AB> = \sum\nolimits_{i,j=\pm 1} A(\mathbf{a})B(\mathbf{b})\ \mathrm{prob}(A(\mathbf{a})=i, B(\mathbf{b})=j). \tag{8}$$

Using Eqs. (1), one can calculate the quantum mechanical prediction for this correlation function for Bohm's idealized *Gedankenexperiment* of Fig. 28.1 as

$$E_{QM}(\mathbf{a},\mathbf{b}) = -\mathbf{a}\cdot\mathbf{b}. \tag{9}$$

In order to measure E(**a**, **b**) for use in the CHSH inequality, one must measure the various prob(A(**a**) = i, B(**b**) = j), for all i and j in a ternary result experiment, i.e. a realization of Bohm's 1951 *Gedankenexperiment*. To do so using event frequencies, one needs (for sufficiently large N)

$$\mathrm{prob}(A(\mathbf{a})=i, B(\mathbf{b})=j) = N(A(\mathbf{a})=i, B(\mathbf{b})=j)/\sum\nolimits_{\mathrm{all}\ i}\sum\nolimits_{\mathrm{all}\ j} N(A(\mathbf{a})=i, B(\mathbf{b})=j) \tag{10}$$

where the double summation in the denominator must be taken over all possible values of i and j.

As we have noted above, this cannot be done without a knowledge of N. The possible routes are then

1. Ignore details of a realization of Bohm's *Gedankenexperiment*, and thereby ignore the normalization loophole, as was done by Bell [6], 1964 (See the section "Bell's 1964 E-Inequality for Idealized Binary Result Apparatuses").
2. Use the CH experimental configuration of Fig. 28.2 and define "Detection" and "NoDetection" to be the binary results needed. This option only works with very high detection efficiency, and was used first experimentally by Giustina et al. [24] and by Christensen et al. [11].
3. Use the Bell experimental configuration described in the section "CHSH R-Inequality With Heralding" and measure N via heralding, as was done experimentally by Rowe et al. [35] and by Matsukevitch et al. [33].

4. Use the CH experimental configuration of Fig. 28.2 and the polarizer-removal protocol given by CHSH, along with the associated CH supplementary assumption. This option was used by the earliest Bell's Inequality experimental tests, the first of which was by Freedman and Clauser [20].
5. Bypass the normalization loophole by employing a supplementary assumption that is much stronger than that needed by the polarizer-removal protocol. This method is herein referred to as GR/AGR normalization (See the section "Garuccio and Rapisarda/Aspect Grangier Roger R-Inequalities for Real Ternary-Result Apparatuses"). It was first used experimentally by Aspect, Grangier and Roger (AGR) [3] in 1982.

Bell's 1964 E-Inequality for Idealized Binary Result Apparatuses

Bell's original 1964 paper [6] considered Bohm's idealized *Gedankenexperiment*, and assumed the idealized specifications as listed above in the section "Bohm's 1951 *Gedankenexperiment* and Its Relation to Bell's Theorem". Given these specifications, he noted (indeed he required) that the quantum mechanical prediction for parallel analyzers has the value

$$\mathrm{E}(\mathbf{a},\mathbf{a}) = -1. \tag{11}$$

Bell assumed Eq. (11) to hold exactly for at least one value of **a**. This assumption thus requires that for that value of **a**, the *Gedankenexperiment* must exhibit a perfect correlation. Using this assumption, he first notes that determinism follows directly from it for the *Gedankenexperiment*. In addition, using these assumptions, he goes on to show that no Local Hidden Variables Theory (LHVT) can give the quantum mechanical prediction for Bohm's idealized *Gedankenexperiment*. He does so by showing that the inequality

$$1 + \mathrm{E}_{\mathrm{LHVT}}(\mathbf{b},\mathbf{c}) \geq |\mathrm{E}_{\mathrm{LHVT}}(\mathbf{a},\mathbf{b}) - \mathrm{E}_{\mathrm{LHVT}}(\mathbf{a},\mathbf{c})|, \tag{12}$$

holds for any LHVT. Surprisingly, he discovers that the quantum-mechanical prediction for this system given by Eq. (9) does not satisfy Ineq. (12) for a significant range of **a** and **b**.

Unfortunately, Bell's mathematical analysis applies only to totally idealized systems, as per the discussion of the section "Bohm's 1951 *Gedankenexperiment* and Its Relation to Bell's Theorem", above. With even an infinitesimal departure from the perfect system described in that section, his mathematical arguments fail.

Here, it is perhaps appropriate to invoke Ben Franklin's famous observation:

"The only predictions that can be made with certainty are for death and taxes."

From a physicist's standpoint, Ben Franklin's observation can be paraphrased (interpreted) to mean

> If a theoretical argument relies on predictions that must obtain with certainty, but said predictions can never obtain in reality, then that argument applies only to a vanishing subset of reality that never occurs, whereupon the argument applies to nothing at all, and thus may be disregarded.

CHSH noted that Bell's argument requires Eq. (11) to hold exactly for at least one value of **a**. However, since Eq. (11) is not equivalent to either death or taxes, then it never obtains for any realizable (or real) systems. Correspondingly, Bell's argument (as it stands) does not apply to any realizable (or real) systems. Unfortunately, without the constraint by Eq. (11), for at least one value of **a**, Bell's mathematical argument fails. Fortunately, despite the above paraphrasing of Ben Franklin, Bell's mathematical reliance on Eq. (11) by no means implies that Bell's result may be disregarded. CHSH show that Eq. (11) is not a necessary requirement for a useful (but different) Bell Inequality to be derived via an alternative mathematical argument.

Clauser Horne Shimony Holt (CHSH) E-Inequality for Real Binary Result Apparatuses

CHSH first showed that for systems that do not comply with the unrealizable specifications outlined in the section "Bohm's 1951 *Gedankenexperiment* and Its Relation to Bell's Theorem" for Bohm's *Gedankenexperiment*, and especially for systems that do not comply with the unrealizable restriction of Eq. (11), then an alternative inequality can be written that does apply to realizable systems. CHSH show that all deterministic local hidden variable theories are constrained by the alternative E-inequality for real binary result apparatuses,

$$|\mathrm{E}(\mathbf{a},\mathbf{b}) - \mathrm{E}(\mathbf{a},\mathbf{c})| \leq 2 - \mathrm{E}(\mathbf{b'},\mathbf{b}) - \mathrm{E}(\mathbf{b'},\mathbf{c}). \tag{13}$$

Shimony [37] further pointed out that Ineq. (13) can be rewritten as

$$-2 \leq \mathrm{E}(\mathbf{a},\mathbf{b}) - \mathrm{E}(\mathbf{a},\mathbf{b'}) + \mathrm{E}(\mathbf{a'},\mathbf{b}) + \mathrm{E}(\mathbf{a'},\mathbf{b'}) \leq 2, \tag{14}$$

wherein the single minus sign may be permuted among the four terms. Unlike Bell's original inequality (12), the CHSH Ineq. (14) applies to realizable systems. The LHVT subscript has been dropped because CH subsequently showed that Ineq. (14) also holds for the more general theories of Local Realism.

Clauser Horne Shimony Holt (CHSH) R-Inequalities for Real Binary-Result Apparatuses via the CHSH Polarizer-Removal Protocol

Since Bell [6] did not address either the experimental status or the testability of his results, Clauser Horne Shimony Holt (CHSH) [12] in 1969 pursued those issues. Armed with Ineq. (14), they first noted that there were no existing experimental data available for comparison with Ineq. (14). They then proposed a real experiment to fill this gap and to actually test Ineq. (14). Their proposed experiment was for a modification of an experiment that had been performed two years earlier by Kocher and Commins [28].

In the CHSH-proposed experiment, two polarization-entangled photons are emitted by an atomic-cascade decay. High efficiency linear polarizers are then used to analyze the entangled photons. These polarizers then replace the Stern Gerlach analyzers of Bohm's *Gedankenexperiment*. To maintain binary-result apparatuses, CHSH use the apparatus configuration of Fig. 28.2. They first propose using the associated result values Detection = +1, NoDetection = −1. Unfortunately, with the technology and detector efficiencies available in 1969, it was still not possible to violate Ineq. (14) using those definitions. Undeterred, CHSH introduce the polarizer-removal protocol. Under this protocol, coincidence rates are measured with both polarizers in place as a function of the polarizer orientations. Additionally, coincidence rates are measured with one polarizer, or the other, or both polarizers removed. The needed multiplicity of experiments, i.e. experiments with polarizer(s) removed and with polarizers inserted at the various needed orientations, are all performed in such a manner that the source rate and the effective detector acceptance solid-angles remain unchanged among them. CHSH modify the result definitions from Detection/ No Detection to Passage/ NoPassage of the photons through the polarizers. CHSH then offer the following supplementary assumption:

> Given the emergence of a pair of photons from the associated pair of analyzers, we assume that the joint detection probability is independent of the analyzer orientations ***a** and **b**.*

Using the polarizer-removal protocol and the CHSH (or CH—see below) supplementary assumption, one can then write for the correlation function

$$\mathrm{E}(\mathbf{a},\mathbf{b}) = 1 + 4\left[\mathrm{R}(\mathbf{a},\mathbf{b}) - 2\mathrm{R}(\mathbf{a},\infty) - 2\mathrm{R}(\infty,\mathbf{b})\right]/\mathrm{R}(\infty,\infty), \tag{15}$$

where the symbol, ∞, denotes the exceptional case when a polarizer has been removed. The associated measured coincidence detection rates may be abbreviated as

$$\begin{aligned} \mathrm{R}(\mathbf{a},\infty) &\equiv \mathrm{R}_1(\mathbf{a}), \\ \mathrm{R}(\infty,\mathbf{b}) &\equiv \mathrm{R}_2(\mathbf{b}), \\ \mathrm{R}(\infty,\infty) &\equiv \mathrm{R}_0. \end{aligned} \tag{16}$$

Further, assuming that $R_1(\mathbf{a}) = R_1$, and that $R_2(\mathbf{b}) = R_2$ are both measured to be constant and respectively independent of **a** and **b**, then CHSH show that (13)–(16) can be combined to yield the CHSH R-inequality

$$|R(\mathbf{a},\mathbf{b}) - R(\mathbf{a},\mathbf{c})| + R(\mathbf{b}',\mathbf{b}) + R(\mathbf{b}',\mathbf{c}) - \mathbf{R_1} - \mathbf{R_2} \leq \mathbf{0}. \tag{17}$$

Also, when **a** and **b** are the scalar angles a and b, as shown in Fig. 28.2, and when it is experimentally demonstrated that the coincidence rate R(**a**,**b**) only depends on the angle $\Phi \equiv \text{ang}(\mathbf{a},\mathbf{b})$ between the polarizers, as per

$$R(\mathbf{a},\mathbf{b}) = R(\Phi), \tag{18}$$

then Ineq. (17) can be written as

$$-R_0 \leq 3R(\Phi) - R(3\Phi) - R_1 - R_2 \leq 0. \tag{19}$$

Freedman [18] further noted that if one takes the optimal value $\Phi = \pi/8$ for maximal violation of (19) by cascade-photon experiments, then a particularly compact form of a Bell R-inequality results, as per

$$|R(\pi/8) - R(3\pi/8)|/R_0 \leq 1/4. \tag{20}$$

Here, the coincidence rate with polarizers in place, R (Φ) is normalized by the coincidence rate with polarizers removed, R_0. Significant utility is provided by this compact form in that only three independent coincidence rates need be measured for it to be tested, although it is still necessary to verify the required rotational invariance of R(Φ). Such rotational invariance can be assured, however, by simply averaging the R(Φ) measurement over common rotations of the pair of analyzers.

Freedman's version of the CHSH inequality is noteworthy in that it provides a somewhat graphic measure of the minimum sinusoidal amplitude variation of the normalized coincidence rate that is needed to violate a Bell Inequality. It also graphically indicates that if R(Φ) in Ineq. (20) is normalized by a coincidence rate other than R_0, say by a smaller rate, then a larger violation occurs, and a violation may then occur where it otherwise would not. On the other hand, if the magnitude of R(Φ) is significantly diminished, say by even modest absorption by the polarizers, then no violation of Ineq. (20) can occur, and the experimental configuration is insufficient to test a Bell Inequality. Thus, violation or no-violation of the CHSH inequality critically depends on the count-rate's normalization.

The experimental requirements for a violation of Ineqs. (17), (19) or (20) are highly demanding upon the required polarizer quality. Those requirements are specified quantitatively by CHSH for their proposed experiment. That experiment was first performed by Freedman and Clauser in [20]. They found that the only available polarizers (in 1972) meeting the requirements for very low absorption were the pile-of-plates variety. Most other early experiments testing the CHSH inequality followed their example, and also used pile-of-plates polarizers.

Clauser and Horne (CH) [13] significantly improve upon the CHSH supplementary assumption. They thus provide a much weaker supplementary assumption that leads to the same result as that by CHSH. CH call their improved supplementary assumption the "no-enhancement" assumption. It is the following:

> We assume that the presence of an analyzer does not somehow enhance (increase) a particle's probability of detection, relative to the probability of its detection with the polarizer absent.

Under the CH no-enhancement assumption, and with the polarizer-removal protocol, the CH Ineq. (5), discussed in the section "Normalization-Loophole Free Clauser-Horne (CH) R-Inequality for Binary-Result Apparatuses" above, reduces to the CHSH R-inequality predictions Ineqs. (17)–(20), whereupon the Freedman-Clauser experiment [20] refutes Local Realism (but is still, of course, subject to the locality loophole).

The CH Counterexample

Clauser and Horne [13], provide an *ad hoc* counterexample that employs enhancement and that can predict the Freedman-Clauser [20] results. Thus, despite the high plausibility of the associated CH supplementary assumption, their counterexample shows that the no-enhancement assumption (or some other assumption) is still needed for experiments that use the polarizer–removal protocol in order to evade the normalization loophole. Under the CH counterexample, the normalization loophole is carefully exploited, in a somewhat pathological manner. Low detection efficiency may be present in the system for a variety of reasons. The CH counterexample carefully exploits the low efficiency produced by absorbing polarizers and other losses to "collude" with the detectors to generate an anomalous violation of the CHSH R-inequality by a Local Realistic theory. Under the CH counterexample, some photons passing through a polarizer have diminished detection probability, i.e. their detection probability is attenuated. Other photons passing through the polarizer have increased detection probability, i.e. their detection probability is "enhanced", or supercharged. Recall that the CH supplementary assumption (see the section "Clauser Horne Shimony Holt (CHSH) R-Inequalities for Real Binary-Result Apparatuses via the CHSH Polarizer-Removal Protocol") specifies that this latter enhancement process does not occur. In the CH counterexample, the polarizer and detector collude with each other to create an anomalous violation. While such "collusion" seems pathological, it should be noted that similar collusion was taken seriously when one considered tests of the locality loophole, as mentioned above. A major difference here is that the malevolent force providing the collusion, as mentioned above, now must be nature, herself, rather than that by a determined cryptography eavesdropper.

Marshal et al. [31] offer a counterexample that is vaguely similar to the CH counterexample. While the CH counterexample generates an exactly sinusoidal variation of the coincidence rate as a function of polarizer orientation, the Marshal-Santos-Selleri counterexample generates a non-sinusoidal variation that would have been readily apparent in the experimental data, which they claim "*fits the existing data as closely as the quantum-mechanical model.*" Scrutiny of their model's prediction, however, reveals their claim to be false, at least for the "existing data" from the Freedman Clauser [20] experiment.

CHSH R-Inequality with Heralding

Bell in his 1971 paper [7] continued to use Bohm's ternary-result *Gedankenexperiment* that he used in his 1964 paper [6]. In this second paper, he acknowledged the CHSH assertion that one must account for the failure of one or both of the apparatuses to detect a particle. To handle this situation, he first showed that the CHSH E-inequality obtains as long as the result-values A and B are defined such that $|A| \leq 1$ and $|B| \leq 1$. Correspondingly, he proposed the definitions,

$$A, B \equiv \pm 1 \text{ Detections, and } 0 \equiv \text{No Detection,} \tag{20}$$

for use with ternary-result apparatuses. Clauser and Horne [13], in their Appendix B, however, show that Bell's scheme will not work in an actual experimental context, because it requires measuring events where nothing happens. It eventually became clear (private communication between CH and Bell) that Bell was tacitly assuming that "it was already known (by some unspecified means) that a particle pair was emitted into the associated detector entrance solid angles", whereupon accountability of the unobserved events is then possible. Clauser and Shimony [15] thus clarified Bell's [7] proposed scheme by depicting "Bell's configuration", as shown in Fig. 28.3, and contrasting it with the "CH configuration" of

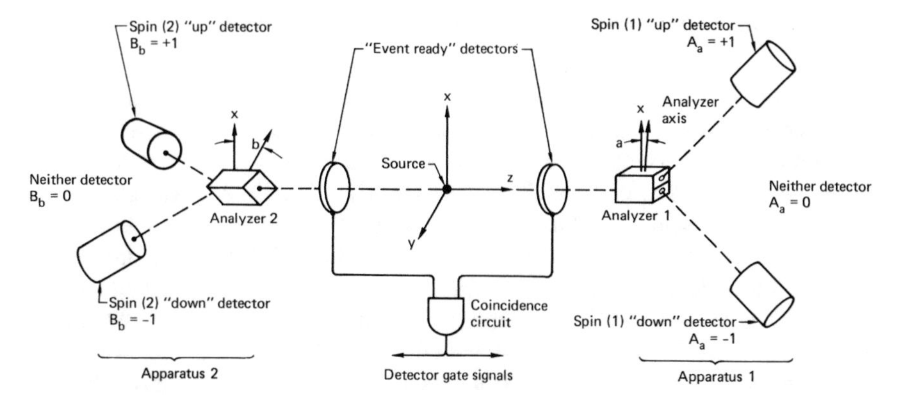

Fig. 28.3 Bell' configuration for Bell's inequality experiments using ternary-result apparatuses and source heralding. Figure from Clauser and Shimony [14].

Fig. 28.2. Thus explicitly clarified, Bell's configuration includes "event-ready" detectors as a means for specifically knowing that a particle pair has been emitted into the associated detector entrance solid angles. Subsequently, the term "heralding" has been coined to describe this process.

In order to use heralding and Bell's [7] result-value definitions in the CHSH E-inequality (14) with an experiment that uses ternary-result apparatuses, one may rewrite Eq. (8) for ternary-result apparatuses as

$$\mathrm{E}(\mathbf{a},\mathbf{b}) \equiv \ <\mathrm{AB}> \ = \textstyle\sum_{i=-1,0,1} \sum_{j=-1,0,1} \mathrm{A}(\mathbf{a})\mathrm{B}(\mathbf{b})\ \mathrm{prob}(\mathrm{A}(\mathbf{a}) = \mathrm{i}, \mathrm{B}(\mathbf{b}) = \mathrm{j}). \tag{21}$$

The probabilities appearing in Eq. (21) may now be measured using observed count rates via

$$\mathrm{prob}(\mathrm{A}(\mathbf{a}) = \mathrm{i}, \mathrm{B}(\mathbf{b}) = \mathrm{j}) = \mathrm{R}(\mathrm{A}(\mathbf{a}) = \mathrm{i}, \mathrm{B}(\mathbf{b}) = \mathrm{j})/\mathrm{R}_{\mathrm{Source}}, \tag{22}$$

where $\mathrm{R}_{\mathrm{Source}}$ is the source rate, as measured via heralding. To be sure, some of the needed detection rates in Eqs. (21) and (22) are unmeasured, i.e. those for $i = 0$ and/or $j = 0$, and thus are not known. However, since their contributions to Eq. (21) have zero for their associated coefficients in Eq. (21), their unknown values are of no importance. The expectation values needed for the CHSH E-inequality (14) are now fully determined by observed detection rates, and are given by

$$\mathrm{E}(\mathbf{a},\mathbf{b}) = [\mathrm{R}_{+,+}(\mathbf{a},\mathbf{b}) + \mathrm{R}_{-,-}(\mathbf{a},\mathbf{b}) - \mathrm{R}_{+,-}(\mathbf{a},\mathbf{b}) - \mathrm{R}_{-,-}(\mathbf{a},\mathbf{b})]/\mathrm{R}_{\mathrm{Source}}, \tag{23}$$

where the following shorthand notation is used: $\mathrm{R}_{i,\,j} \equiv \mathrm{R}(\mathrm{A}(\mathbf{a}) = \mathrm{i}, \mathrm{B}(\mathbf{b}) = \mathrm{j})$.

Garuccio and Rapisarda/Aspect Grangier Roger R-Inequalities for Real Ternary-Result Apparatuses

Garuccio and Rapisarda (GR) [22] in 1981 proposed a new method for normalizing coincidence rates for "ternary-result" apparatuses without the associated requirement for heralding. Thereby they provide a new R-inequality, that was first tested experimentally by Aspect, Grangier, Roger (AGR) [3] in 1982. These efforts proceeded despite known difficulties for ternary-result apparatuses, as found earlier by CH [13] and by CS [16]. Recall that Bell had originally proposed the use of ternary-result apparatuses, first, inadvertently, in 1964, and again, advertently, in 1971 [7] (see the section "CHSH R-Inequality with Heralding"). In his 1971 paper, Bell thus proposed using the ternary-result values (+1, −1, and 0), wherein 0 represents unobserved No-Detection events in the CHSH E-inequality. Unfortunately, he did not offer an associated testable R-inequality. He tacitly assumed that source heralding was being used, although he did not specifically say so. However,

CH [13] (see their Appendix B) reanalyzed Bell's 1971 proposal [7] and found that, no loophole-free testable R-inequality can be generated for such systems, because they require the measurement of a probability of nothing happening, which cannot be done without knowing the associated probability of something happening. Knowing the latter then requires knowing the source rate, R_{Source}, (e.g. via a heralded source). CH show that, alternatively, one may convert each ternary-result apparatus into a binary-result apparatus, "at the very beginning of the derivation", and then use the CH inequality. CH pointed out that such conversion can be accomplished, for example, by combining the No-Detection "0" channels with say the –1 Detection channels at each apparatus, and then by looking only at the +1 Detection channels. Such a conversion is tantamount to simply ignoring the –1 detections. CH show that by doing so, one can then generate a CH inequality (using Bell's method of proof) that involves only (+1,+1) detections. Actually, depending upon which pair of channels one chooses to ignore, one can alternatively generate four independent CH inequalities for each of the (+1,+1), (−1, −1), (−1,+1), and (+1, −1) channels. Of course, such a conversion destroys any symmetry between the (+1,+1) channels and the partly unobserved (−1 & 0, −1 & 0) channels. But, of course, neither locality, nor realism, nor quantum mechanics have any need or requirement for experiment symmetry.

CH inequalities generated by ignoring channels were still not testable using the technology available in 1982, i.e. by using low quantum-efficiency photomultiplier tube detectors and atomic-cascade decay entangled-photon sources. With 1982 technology, some additional experimental protocol and set of supplementary assumptions was thus still needed for testing ternary-result apparatus experiments.

Despite these known difficulties for ternary-result apparatus experiments, Aspect, Grangier, Roger (AGR) [3] in 1982 attempted to build a ternary-result apparatus for testing a Bell Inequality by using a supplementary assumption proposed earlier by Garuccio and Rapisarda [22]. As a starting point, GR/AGR use the CHSH E-inequality (14),

$$-2 \leq \mathrm{E}(\mathbf{a}, \mathbf{b}) - \mathrm{E}(\mathbf{a}, \mathbf{b}') + \mathrm{E}(\mathbf{a}', \mathbf{b}) + \mathrm{E}(\mathbf{a}', \mathbf{b}') \leq 2. \tag{14}$$

Their experiment uses the Bell's configuration of Fig. 28.2, except that it does not include the "event-ready detectors" used for heralding by that configuration.

In the section "One Possible Cause for the Normalization Loophole" above, we note that a ternary-result apparatus experiment requires one to know the values of nine independent coincidence rates, wherein their sum can be used for probability normalization. In the AGR experiment, however, only four coincidence rates are measured. Using Bell's [7] ternary-result values, (+1, −1, and 0), where 0 represents unobserved No-Detection events, the four coincidence rates measured by AGR are

$R_{++}(\mathbf{a}, \mathbf{b}) \equiv (A = +1, B = +1)$ observed coincidence rate,
$R_{--}(\mathbf{a}, \mathbf{b}) \equiv (A = -1, B = -1)$ observed coincidence rate,
$R_{+-}(\mathbf{a}, \mathbf{b}) \equiv (A = +1, B = -1)$ observed coincidence rate,
$R_{-+}(\mathbf{a}, \mathbf{b}) \equiv (A = -1, B = +1)$ observed coincidence rate.

The other five (needed) coincidence rates that AGR do not measure are:

$R_{+0}(\mathbf{a}, \mathbf{b}) \equiv (A = +1, B = 0)$ unobserved "coincidence" rate,
$R_{0+}(\mathbf{a}, \mathbf{b}) \equiv (A = 0, B = +1)$ unobserved "coincidence" rate,
$R_{-0}(\mathbf{a}, \mathbf{b}) \equiv (A = -1, B = 0)$ unobserved "coincidence" rate,
$R_{0-}(\mathbf{a}, \mathbf{b}) \equiv (A = 0, B = -1)$ unobserved "coincidence" rate,
$R_{00}(\mathbf{a}, \mathbf{b}) \equiv (A = 0, B = 0)$ unobserved "coincidence" rate.

While the first four of these last five could, in fact, have been determined from the singles rates, that last one of these five, i.e. the (NoDetection, NoDetection) coincidence rate cannot be measured.

As per the discussion of the section "Result Values and Expectation-Value Inequalities (E-Inequalities)" above, the appropriate expectation value for use in the CHSH E-inequality (14) is

$$E_{CHSH}(\mathbf{a},\mathbf{b}) = \frac{R_{++} + R_{--} - R_{+-} - R_{-+} + 0 \times [R_{+0} + R_{-0} + R_{0+} + R_{0-} + R_{00}]}{R_{++} + R_{--} + R_{+-} + R_{-+} + [R_{+0} + R_{-0} + R_{0+} + R_{0-} + R_{00}]}$$
$$= \frac{R_{++} + R_{--} - R_{+-} - R_{-+}}{R_{++} + R_{--} + R_{+-} + R_{-+} + [R_{+0} + R_{-0} + R_{0+} + R_{0-} + R_{00}]}. \tag{24}$$

To make their results testable, AGR use the GR normalization scheme and arbitrarily set equal to zero the term in square brackets in the normalizing denominator that includes the five unobserved rates. So doing, they define

$$E_{GR/AGR}(\mathbf{a},\mathbf{b}) \equiv \frac{R_{++}(\mathbf{a},\mathbf{b}) + R_{--}(\mathbf{a},\mathbf{b}) - R_{+-}(\mathbf{a},\mathbf{b}) - R_{-+}(\mathbf{a},\mathbf{b})}{R_{++}(\mathbf{a},\mathbf{b}) + R_{--}(\mathbf{a},\mathbf{b}) + R_{+-}(\mathbf{a},\mathbf{b}) + R_{-+}(\mathbf{a},\mathbf{b})}. \tag{25}$$

The CHSH E-inequality (14) and GR/AGR's definition (25) are then combined to produce a new GR/AGR R-inequality, which was then violated by their experimental data.

We note, however, that the deleted term in square brackets in the denominator of Eq. (24) is very much larger than the remaining terms by a factor of about 10^6, when it is used with an experiment that uses an atomic cascade photon source and photomultiplier tube detectors, as did the AGR [3] experiment. Correspondingly, omission of that very large term by GR/AGR deserves careful scrutiny, especially since it represents a dramatic renormalization (by a factor of about 10^6). Without the term's deletion, it would be impossible for $E_{GR/AGR}(\mathbf{a},\mathbf{b})$ as given by Eq. (25), to violate (14).

GR and AGR state that

> …we assume that the ensemble of actually detected pairs is a faithful example of all emitted pairs. ….

The GR/AGR assumption is now commonly and "gratuitously" referred to as the "fair-sampling assumption". This author's use of the description "gratuitous" will now be justified.

Let us examine the implications of GR/AGR's so-called "fair-sampling assumption". An important issue is whether or not it is consistent with the requirements and/or the assumptions underlying either quantum mechanics and Local Realism. In particular, one must first consider whether or not this supplementary assumption tacitlyor explicitly assumes/requires that all detected events have the same *a priori* probability of detection.

The GR/AGR assumption takes a particle-centric view of wave-particle duality / wave-particle ambiguity, which asserts an equivalence of the wave and particle viewpoints in quantum mechanics, but leaves vague and/or ambiguous the mechanism providing said equivalence. In a wave-centric viewpoint, the inverse square law for detected flux versus source-to-detector distance is due to diminished wave amplitude with increasing distance, and an associated diminished detection probability with said diminished amplitude. However, in a particle-centric view, it is due instead to a diminished geometrically diluted particle flux. In a particle centric view, particle detection probability is always constant, given a particle's presence at a detector. On the other hand, in a wave-centric view, particle detection probability is variable and depends on wave amplitude. Wave amplitude at each detector, in turn, may depend on the associated analyzer's orientation.

Consider first an assumption underlying quantum mechanics. Under Born's rule for calculating probabilities, a particle's detection probability is proportional to the absolute square of its probability amplitude. Particles that pass through polarization analyzers at differing orientations will have different transmitted probability amplitudes. Correspondingly, not all particles arriving at a detector will have the same probability amplitude or the same a priori detection probability. Thus, under a fundamental requirement by quantum mechanics, different events must be allowed to have different *a priori* detection probabilities. Any reasonable supplementary assumption used for a Bell's Inequality test must correspondingly allow for and be consistent with this possibility. GR/AGR instead assume particle detection probability is always constant, and *a fortiori* exclude theories with variable detection probability (including quantum mechanics).

Next consider a straightforward local realistic theory in which a photon is modeled simply as a short-pulse (or wave packet) of classical electromagnetic radiation. Under this theory, for example, one may assume that the semiclassical model for the photoelectric effect proposed by Lamb and Scully [30] holds. Then a pulse with a large classical amplitude will have a higher probability of generating a photoelectron, and an associated detectable output pulse of electric current from a photomultiplier tube, than will one with a small classical amplitude. Again, under Local Realism, different events must be allowed to have different *a priori* detection probabilities.

We thus see that properly testing both quantum mechanics and Local Realism requires one to allow for a variable detection probability of the detected particles by the particle detectors. Now, consider the implications of this requirement. Naturally, the set of detected particle events will have a preponderance of events with a higher *a priori* detection probability than will the set of undetected-particle events, which, in turn, will be dominated by events with a lower *a priori* detection probability. Correspondingly, the ensemble of detected pairs is clearly not "a faithful example of all emitted pairs" as GR/AGR assert. Instead, it provides an ensemble that is significantly biased in favor of events with a high *a priori* detection probability.

One may view the detection process as a competition for detection among the particles at the detectors. The GR/AGR assumption implies that the winners (the detected particles) in the competition for detection were no more capable of "winning the competition" (being detected) than were the losers (the undetected particles) in the competition. Viewed in this light, one may ask, is the GR/AGR assumption truly reasonable, and is the sampling truly "fair"? For comparison, note that in atmost any sports competition, it would be hard to find a winner who didn't truly believe that he/she did not "fairly" win the event because of superior ability (i.e. a higher *a priori* probability of winning) rather than by simple luck. Such a sports competitor would thus strongly disagree with the GR/AGR assumption as being "fair" and reasonable. Correspondingly, an assumption that all of the "winners" in a competition for being detected by a photomultiplier tube were equally capable of winning the competition, is equivalent to saying that the detected particles form a representative subset of all of the emitted particles, as far as their probability for being detected is concerned.

The GR/AGR assumption is thus a very strong assumption, indeed! It even seems to violate the fundamental premises underlying both Local Realism and quantum mechanics. Correspondingly, its application to the testing of not only the above very reasonable Local Realistic model, but also to the testing of quantum mechanics itself appears to be highly dubious. Clearly then, the appellation "fair sampling assumption" must be considered gratuitous.

It should also be noted that the GR/AGR assumption is not equivalent to the very much weaker CH no-enhancement assumption, as coupled to the CHSH polarizer-removal protocol. By contrast, the expectation that different particles may have different *a priori* detection probabilities is explicit in the CH no-enhancement assumption. CH simply assume that passage of a photon through a (presumably) attenuating polarizer does not somehow enhance its *a priori* detection probability. Moreover, polarizer absorption has a very strong effect when one is using the CHSH R-inequality (17), (19) or (20) that results from the CHSH polarizer-removal protocol and the no-enhancement assumption. In such a case, when the polarizer absorption is even modestly too large, no inequality violation occurs. It also has a very strong effect on the viability of the CH counterexample. On the other hand, photon absorption by a polarizer has no effect at all on the resulting numerical value obtained from using Eq. (25) in (14), and correspondingly has no effect on whether or not a violation occurs.[2]

[2]In defense of the AGR experiment's polarizer parameters, their parameters do appear to meet the CHSH transmission requirements, although they are not required to do so in order to violate the

As suggested in the introduction, in order to evaluate a supplementary assumption, one may also compare how reasonable it is with how contrived an associated counterexample is. Given the GR/AGR normalization method's relative insensitivity to polarizer absorption, it is not surprising that one can readily build counterexamples that do not violate the CHSH E-inequality when the polarizer-removal protocol and no-enhancement assumption is used, but do violate the CHSH E-inequality when the GR/AGR assumption and protocol are used. (See Clauser [16].)

Further evidence of just how vulnerable the GR/AGR assumption is to counterexamples was given by Gerhardt et al. [23], who provide both a theoretical and a convincing experimental demonstration of the ease by which an actual experiment can be countered, especially in "security related scenarios". It should be noted that Gerhardt et al.'s demonstrated violations of a Bell Inequality all use GR/AGR normalization. Gerhardt et al., however, mistakenly attribute their counterexample's existence to a loophole in what they mistakenly refer to as the CHSH inequality, which instead is really the GR/AGR R-inequality produced by a combining (14) and (25) above. It should be emphatically noted that Gerhardt et al. do not produce an experimental (or theoretical) counterexample that employs the CHSH polarizer-removal protocol. However, they do provide a convincing experimental demonstration of the ease by which schemes that use GR/AGR normalization can be countered, especially by malevolent efforts (by "Eve"), as may occur in "security related scenarios" and quantum cryptography.

GR/AGR normalization and the associated gratuitously named "fair-sampling" assumption and the GR/AGR-inequality have nonetheless been used by many experiments, despite the associated proliferation of counterexamples and above noted shortcomings. (See the section "Some Experimental Results"). One possible reason for their popularity presumably is their relative ease of experimental implementation. Beyond allowing the use of strongly absorbing polarizers, no polarizers are removed under this method, and no additional normalizing data need be taken. Data collection is thereby expedited. The use of GR/AGR normalization also avoids a further difficulty associated with ternary-result apparatuses, for which polarizer removal is not readily possible. For such apparatuses, polarizer removal necessarily disturbs the collimation geometry of at least one of the channels, whereupon the polarizer-removal protocol then cannot be used.

Ursin et al. [39] in their test of a Bell's Inequality go even further than AGR by using a passive non-polarizing beam-splitter to precede a pair of ternary-result apparatuses on each side of their experiment. Each composite apparatus then has five possible results. For each ternary-result apparatus that follows the beam-splitter, Ursin et al. use a modified GR/AGR normalization scheme, where the denominator only includes coincidences associated with that ternary-result

(Footnote 2 continued)

CHSH E-inequality using Eq. (25). Subsequent experiments that use GR/AGR normalization, however, do not always meet these requirements.

apparatus. The normalization for one ternary result apparatus on one side of the experiment thus ignores coincidences occurring in the other ternary-result apparatus on the same side of the experiment. An additional facility is gained here from the use of GR/AGR normalization. Given that all four apparatus orientations needed for an evaluation of the GR/AGR R-inequality can be taken in parallel, GR/AGR normalization then allows greatly expedited data collection and a single experimental run with no required apparatus changes.

It should be noted that there is considerable confusion and misinformation in the literature on what constitutes the "CHSH inequality". Many writers mistakenly appear to believe that GR/AGR normalization is an integral part of the CHSH inequality, and fail to distinguish the CHSH E-inequality from the CHSH R-inequalities. For example, Gerhardt et al. [23] say that the CHSH E-inequality's use necessarily requires the use of the GR/AGR normalization scheme and its associated R-inequality. Giustina et al. [24] mistakenly claim that

> ...[separated apparatuses named] Alice and Bob ... each require two detectors for testing a Clauser-*Horne-Shimony-Holt inequality*.

The sections "Result Values and Expectation-Value Inequalities (E-Inequalities)"–"Garuccio and Rapisarda/Aspect Grangier Roger R-Inequalities for Real Ternary-Result Apparatuses" above all show the falsity of these claims.

Finally, it should also be noted that Christensen et al. [11], mistakenly claim that

> ...all previous experiments have had to make fair-sampling assumptions that the collected photons are typical of those emitted (this assumption is demonstrably false (Marshal et al. [31]) for many of the pioneering experiments using atomic cascades (Freedman and Clauser [20], and Aspect Dalibard and Roger [4]) and has been intentionally exploited to fake Bell violations in recent experiments (Gerhardt et al. [23]) ....

Their erroneous statement clearly does not apply to the two cascade photon experiments they quote, notably to that by Freedman and Clauser [20] and that by Aspect et al. [4], which both use the CHSH polarizer-removal protocol and CH no-enhancement assumption, that, in turn, is not demonstrably false.

Some Experimental Results

Table 28.1 lists chronologically some of the experimental tests to date of the various Bell Inequality predictions. (I apologize to the authors of references omitted from this table.) The experiment is identified in columns 1 and 2. The entangled systems and source are given in column 3, along with whether or not locality was tested. The number of apparatus channels (binary or ternary) is shown in column 4. The Bell Inequality that was tested is shown in column 5, along with the associated normalization protocol that was used. Whether an accidental background rate was subtracted is indicated in column 6, and the magnitude of the observed inequality violation is shown in column 7.

Table 28.1 Bell's inequalities—experimental tests (incomplete list). See the section "Some Experimental Results" for a discussion of these experiments

Year	Author(s); Institution(s)	System, excitation method (locality tested?)	# Channels per apparatus	Bell Inequality tested, normalization method	BkgSubtr? Violation
1972	Freedman, Clauser; U. Cal., Berkeley	Ca cascade photons, UV discharge lamp	1Det. + No Cnt. (binary)	CHSH & Freedman $\neq$'s, pol removal + CH no Enh.	No, 6.3σ
1973	Holt, Pipkin (unpub.); Harvard U.	Hg cascade photons, e-beam	1Det. + No Cnt. (binary)	CHSH-Freedman $\neq$, pol removal + CH no Enh.	Yes, None
1976	Clauser; U. Cal., Berkeley	Hg cascade photons, e-beam	1Det. + No Cnt. (binary)	CHSH-Freedman $\neq$, pol removal + CH no Enh.	Yes, 4.1σ
1976	Fry, Thompson; Texas A&M U.	Hg cascade photons, e-beam + laser	1 Det. + No Cnt. (binary)	CHSH-Freedman $\neq$, pol removal + CH no Enh.	Yes, 3.3σ
1981	Aspect, Grangier, Roger; U. Paris-Sud	Ca cascade photons, 2 lasers	1 Det. + No Cnt. (binary)	CHSH-Freedman $\neq$, pol removal + CH no Enh.	Yes, 13σ
1982	Aspect, Grangier, Roger; U. Paris-Sud	Ca cascade photons, 2 lasers	2 Det. + No Cnt. (ternary)	CHSH $\neq$ norm. by GR/AGR sum of 4 coinc. rates	Yes, 5σ
1982	Aspect, Dalibard, Roger; U. Paris-Sud	Ca cascade photons, 2 lasers (test of locality)	1 or 1 (active sw) + No Cnt. (binary)	CHSH $\neq$ pol removal + CH no Enhancement	Yes, 5.1σ
1987	Shih, Alley; U. Maryland	PDC (parametric down-conversion) photons	1 Det. + No Cnt. (binary)	CHSH (Freedman) $\neq$ pol removal + CH no Enh.	3σ
1988	Ou, Mandel; U. Rochester	PDC photons	1 Det. + No Cnt. (binary)	CHSH $\neq$ pol removal + CH no Enhancement	Yes, 5.8σ
1995	Kwiat et al. (6 authors); U. Innsbruck et al.	Type II PDC photons	1 Det. + No Cnt. (binary)	CHSH $\neq$ norm. by GR/AGR sum of 4 coinc. rates	No, 102σ
1998	Weihs et al. (5 authors); U. Innsbruck	Type II PDC photons (test of locality > 400 m)	2 Det. + No Cnt. (ternary-active sw)	CHSH $\neq$ norm. by GR/AGR sum of 4 coinc. rates	No, 37σ
1998	Tittel et al. (4 authors); U. Geneva	PDC photons, Franson Energy-time entanglement (test of locality > 10 km)	2 Det. + No Cnt. (ternary via beam-splitter)	CHSH $\neq$ norm. by GR/AGR sum of 4 coinc. rates	No, 16σ Yes, 25σ

(continued)

Table 28.1 (continued)

Year	Author(s); Institution(s)	System, excitation method (locality tested?)	# Channels per apparatus	Bell Inequality tested, normalization method	BkgSubtr? Violation
2001	Rowe et al. (7 authors); NIST	Induced entanglement of 2 Be ions (spatially unresolved)	2 Det. (binary via active switch)	CHSH $\neq$, norm. by # heralded events	No, 8.3σ
2006	Matsukevich et al. (6 authors); GeorgiaTech	Induced entanglement of 2 Rb clouds	2 Det. + No Cnt. (ternary)	CHSH $\neq$, norm. by GR/AGR sum of 4 coinc. rates	No, 5.3σ
2007	Ursin et al. (18 authors); U. Vienna et al.	PDC photons (test of locality through 144 km of air)	4 Det. + No Cnt. (fivefold via passive switch/beam-splitter)	CHSH $\neq$ norm. by GR/AGR selected sum of 4 (out of 16) coinc. rates	No, 14σ
2008	Matsukevich et al. (5 authors); U. Maryland	Induced entanglement of trapped Yb ion & photon of 2 trapped Yb ions	2 actively selected Det. (binary)	CHSH $\neq$ norm. by # heralded events	No, 27σ No, 3.1σ
2009	Ansmann et al. (13 auth.); U. Cal., Santa Barbara	Induced entanglement of Josepson phase qubits	1 Det. (binary via active switch)	CHSH $\neq$ norm. by # heralded events	No, 244σ
2012	Hofmann et al. (7 authors); Ludwig-Maximilians U. et al.	Induced entanglement of 2 trapped Rb atoms	2 actively selected Det. (binary)	CHSH $\neq$ norm. by # heralded events	No, 2.1σ
2013	Giustina et al. (12 authors); U. Vienna et al.	PDC photons (non-maximally entangled)	1 Det. + No Cnt. (binary)	CH $\neq$ norm. by # heralded events & singles rates	No, 69σ
2013	Christensen et al. (14 authors); U. Illinois Urbana-Champaign et al.	PDC photons (non- maximally entangled)	1 Det. + No Cnt. (binary)	CH $\neq$ norm. by # heralded events & singles rates	No, 7.5σ

All of these experiments except one—that by Holt and Pipkin [27], (see also Holt [26])—agree with the associated predictions by quantum mechanics. Clauser [14] repeated the Holt and Pipkin experiment with only a few minor changes and obtained the opposite results, i.e. results in agreement with quantum mechanics. The earliest experimental tests by Freedman and Clauser [20], Clauser [14], Fry and Thompson [21], Aspect, Grangier, and Roger [2], and Aspect, Dalibard, and Roger [4] all used photons emitted by an atomic cascade, and also used the CHSH polarizer-removal protocol along with the CH no-enhancement assumption. GR/AGR normalization via Eq. (25) of the CHSH E-inequality (14) was used in the experiments by Aspect, Grangier, and Roger [3], Kwiat et al. [29], Weihs et al. [40], Tittel et al. [38], Ursin et al. [39], and Matsukevich et al. [32].

The experiments by Aspect, Dalibard, and Roger [4], Weihs et al. [40], Tittel et al. [38] and Ursin et al. [39] all changed the analyzers while the entangled-state photons were in flight, thereby providing a direct realization of Bohm and Aharonov's [10] locality-test. The experiment by Tittel et al. [38] is noteworthy in that the photons use energy-time entanglement, rather than polarization-state entanglement.

The Fry and Thompson [21] experiment was the first to use tunable laser excitation of the source atomic cascades, thereby providing a dramatic boost in count rates over previous experiments. The use of parametric down conversion in a crystal as a source of entangled-state photons was first offered in 1988 by Shih and Alley [36], and by Ou and Mandel [34]. It provides a further dramatic boost to count rates when compared to those emitted by atomic cascade decays. Kwiat et al. [29] further enhanced count rates via the use of Type II parametric down conversion.

The experiment by Rowe et al. [31] was the first to violate the "heralded" CHSH inequality using a heralded source, and thereby to avoid the normalization loophole. However, in their experiment, light from the two entangled-state Beryllium ions is commingled indistinguishably in a single detector. By contrast, the basic locality postulates associated with Bell's Theorem prototype configurations call instead for a pair of widely separated independent detectors with no worry about their possible intercommunication. Unfortunately, the two ions in the Rowe et al. experiment were seemingly in intimate communication with each other, and even share the same probe laser light that was used to determine their excitation states. Correspondingly, it is not clear if there were any interfering interference effects (classical, quantum mechanical, or otherwise) from unresolved light emitted by both, ions. Interference effects were indeed observed earlier in a similar experiment by Eichmann et al. [18]. Rowe et al. do note, however that the ions' separation was wide enough that associated Young's fringes average out. However, this fact does not rule out some other perhaps non-quantum-mechanical and/or non-classical interfering interference effect. Recall that Bell's Inequality tests seek to determine whether or not quantum-mechanics is correct, and/or even whether or not any of the physics generally assumed to govern the formation of interference fringes is correct. Thus, given the level of generality required for such tests, such claims of independence are not fully reassuring, and such assumed physics cannot be relied upon here.

The experiment by Ansmann et al. [1] entangled a pair of Josephson phase qubits to violate the heralded CHSH inequality. Since the entangled qubits were only 3.1 mm separated, intimately coupled, and indistinguishably probed, this experiment is subject to similar criticisms to those regarding Rowe et al.

More convincing violations of the CHSH inequality using a heralded source with well separated apparatuses were subsequently reported by Matsukevich et al. in [33] in 2008, and by Hofmann et al. [25] in 2012. The experiment by Matsukevich et al. [33] entangled a pair of Yb^+ remotely trapped ions. The experiment by Hofmann et al. [25] entangled a pair of widely-separated remotely trapped Rubidium atoms.

Finally, the experiment by Giustina et al. [24] was the first to directly violate the CH inequality. It was followed very shortly by a similar experiment by Christensen et al. [11]. Closure of both the normalization loophole and the locality loophole simultaneously in a single experiment has not yet been done.

Appendix: Local Realism

Local Realism was first explicitly defined by Clauser and Horne (CH) [13] in 1974, and further clarified in a series of papers by Bell et al. [8] in 1976–1977 and by Clauser and Shimony [15] in 1978. CH originally called the theories governed by it, "Objective Local Theories". Clauser and Shimony renamed these theories "Local Realism". Local Realism is the combination of the philosophy of realism with the principle of locality. The locality principle is based on special relativity. It asserts that nature does not allow the propagation of information faster than light to thereby influence the results of experiments. Without locality, one must contend with paradoxical causal loops, as are now popular in science fiction thrillers involving time travel. Upholding locality is effectively a denial of the reality of causal loops. Equivalently, it is the assertion that history is single valued. Realism is a philosophical view, according to which external reality is assumed to exist and have definite properties, whether or not they are observed by someone. Bell's Theorem, and the experimental predictions made by the associated CHSH and CH Bell's Inequalities, along with the associated experimental tests of these predictions, show that any theory that combines Realism and locality, must be in observed disagreement with these experiments. Consequently, it can now be asserted with reasonable confidence that either the thesis of Realism or that of locality (or perhaps even both) must be abandoned.

Another way of describing what we mean by Realism here is to say that it specifies that nature consists of "objects", i.e. stuff with "objective reality". Realism assumes that objects exist and have inherent properties on their own. It does not require that these properties fully determine the results of an experiment locally performed on said object. Instead, in a possibly non-deterministic world, it simply allows the properties of an object to influence the probabilities of experiments being performed on it. There is also nothing in our specification that prohibits an act of observation or measurement of an object from influencing, perturbing and/or even destroying said properties of the object.

Realism thus assumes that an object's properties determine minimally the probabilities of the results of experiments locally performed on it. Realism, under the additional constraint of locality, i.e. Local Realism, then assumes that the results of said experiments do not depend on other actions performed far away by someone else, especially when those actions are performed outside of the light-cone of the local experiment.

Properties, as referred to here, are what John Bell called Beables, and what Einstein et al. [19] called "elements of reality". The properties of an object constitute a description of the stuff that is "really there" in nature, independently of our observation of it. When we perform a "measurement" of these properties, we don't really need to know what we are actually doing, or what we are really measuring. What we are assuming is that what is "really there" somehow influences what we observe, even if said influence is inherently stochastic and/or perhaps irreproducible from one measurement to the next.

Recall that Einstein et al. [19] attempted to define an object's properties as something that one can measure, but they further required that the measurement result be predictable with certainty. However, given Ben Franklin's observation that the only predictions that are certain in life are for death and taxes (see the section "Bell's 1964 E-Inequality for Idealized Binary Result Apparatuses"), said definition becomes meaningless, because it describes nothing that can ever occur in reality, (unless, of course, said properties are equivalent to death and taxes). Our definition is very much looser and requires no predictions with certainty.

Precisely how does one define an object with such extreme generality? For the purposes of Local Realism and its tests via Bell's Theorem, a purely operational definition of an object suffices. An object (or collection of objects) is stuff with properties that one can put inside a box, wherein one can then perform measurements inside said box and get results whose values are presumably influenced by the object's properties. What then is a box? A box is defined as a closed three-dimensional Gaussian surface,[3] inside of which one can perform said measurements of said properties. For Local Realism, such a box becomes a four-dimensional Gaussian surface consisting of the backward light cone (extending to $t = -\infty$) enveloping a three dimensional box, that contains the object(s) being measured, at the time that they are being measured.

Familiar examples of "classical" objects that can be put into boxes are galaxies, stars, airplanes, shoes, trapped clouds of atoms, single trapped atoms, electrons, y-polarized photons, a single bit of information, etc. All of these can be put into a box and have their properties (e.g. color, mass, charge, etc.) measured. Or can they? Via Bell's Theorem experiments, one may ask—are there examples of objects that

[3]Gauss showed that a "Gaussian surface" is one that divides all of space into two disjoint volumes, wherein one of these volumes may be called the inside, and the other the outside.

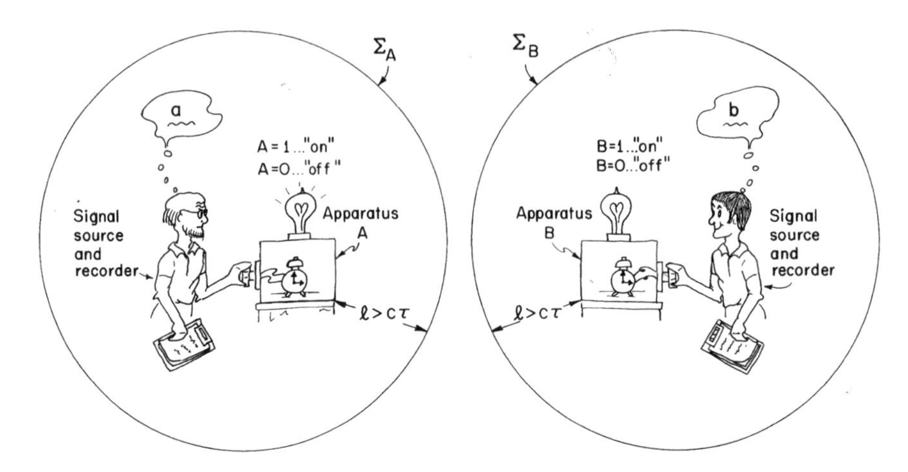

Fig. 28.4 Worst-case set of required elements for a Bell's Inequality experiment. This figure was first presented by the author at the 1976 International "Ettore Majoranna" Conference in Erice, Sicily on "Experimental Quantum Mechanics". The conference was organized by John Bell, Bernard d'Espagnat, and Antonino Zichichi. With present-day jargon, the characters labeled "Signal source and recorder" might now be named Alice and Bob.

cannot be put inside such boxes?[4] If so, such objects cannot be described by Local Realism. Furthermore, if there are parts of nature that cannot be described by Local Realism, then Local Realism must be discarded as a description of all of nature. Sadly, (for Local Realism advocates[5]) the individual particles comprising a quantum-mechanically entangled pair of particles are parts of nature that cannot be described by Local Realism.

Figure 28.4 shows the worst-case set of required elements for a fully loophole-free Bell's Inequality experimental test. Two objects and associated binary-result apparatuses are each contained in associated boxes that are space-like separated at the time of the measurement events. The apparatuses measure quantum-mechanically entangled pairs of particles. The boxes are labeled Σ_A. and Σ_B in the Figure. Each box contains a signal recorder and signal source. Each signal source generates via the free-will of an observer an appropriate apparatus parameter setting. The two settings are respectively called **a** and **b**. Each box contains a clock that permits synchronized measurements in the two boxes of the object pairs, that were emitted in the past and that have propagated into these boxes at subliminal speed for measurement by the apparatuses.

[4]The fact that the simplest possible object—a single bit of information—cannot be put into a "box", in turn gives rise to the field of quantum information. It also calls into question a claim often made by general relativists that information is always contained within a given spatial volume and cannot be destroyed.

[5]John Bell and I have both confessed to being former advocates of Local Realism.

An important issue discussed in "An Exchange on Local Beables", (Bell, Shimony, Horne, and Clauser [8]), is that the apparatus parameters **a** and **b** must be generated independently, for example by the presumed free will of the observers, and are not to be counted as part of the objective reality being measured.

To derive a Bell Inequality, one then needs to assume the following requirements for a Local-Realistic theory:

(1) The probability of obtaining the measured result A in box Σ_A may depend on all of the stuff (objects) that are inside the box at the time of the measurement, including any stuff that may have propagated into the box at a velocity less than or equal to the speed of light since the beginning of time.
(2) The probability of obtaining the measured result A in box Σ_A may depend on the freely chosen apparatus parameter **a**.
(3) Locality, however, prohibits the probability of obtaining the measured result A in box Σ_A from depending on the apparatus parameter **b** that was freely chosen in the space-like separated box Σ_B.
(4) Locality, similarly, prohibits the probability of obtaining the measured result A in box Σ_A from depending on the result B, as measured in box Σ_B, which, of course, is allowed to depend on the parameter **b**.
(5) Similar reciprocal permissions and prohibitions like (1)–(4) govern the probability of obtaining the measured result B in box Σ_B.

Surprisingly, that's all you need to derive the CH (and CHSH) inequality and thereby to constrain and test Local Realism!

References

1. M. Ansmann et al., Science **461**, 504–506 (2009)
2. A. Aspect, P. Grangier, G. Roger, Phys. Rev. Lett. **47**, 460 (1981)
3. A. Aspect, P. Grangier, G. Roger, Phys. Rev. Lett. **49**, 91 (1982)
4. A. Aspect, J. Dalibard, G. Roger, Phys. Rev. Lett. **49**, 1804–1807 (1982)
5. F. Belinfante, *A Survey of Hidden Variables Theories* (Pergamon, Oxford, 1973)
6. J. Bell, Physics **1**, 195 (1964)
7. J. Bell in *Foundations of Quantum Mechanics, Proceedings of the International School of Physics "Enrico Fermi"*, ed. by B. d'Espagnat (Academic Press, New York, 1971)
8. J. Bell, A. Shimony, M. Horne, J. Clauser (1976–1977) Epistemological Letters (Association Ferdinand Gonseth, Institut de la Methode, Case Postale 1081, CH-2501, Bienne.) republished as *An Exchange on Local Beables* in Dialectica, **39**, 85–110 (1985)
9. D. Bohm, *Quantum Theory* (Prentis Hall, Englewood Cliffs, NJ, 1951)
10. D. Bohm, Y. Aharonov, Phys. Rev. **108**, 1070 (1957)
11. B. Christensen et al., Phys. Rev. Lett. **111**, 130406–130409 (2013)
12. J. Clauser, M. Horne, A. Shimony, R. Holt, Phys. Rev. Lett. **23**, 880–884 (1969)
13. J. Clauser, M. Horne, Phys. Rev. D **10**, 526–535 (1974)
14. J. Clauser, Phys. Rev. Lett. **36**, 1223 (1976)
15. J. Clauser, A. Shimony, Rep. Prog. Phys. **41**, 1881–1927 (1978)

16. J. Clauser in *Quantum [Un]speakables, From Bell to Quantum Information, Proceedings of the 1st Quantum [Un]speakables Conference*, ed. by R. Bertlmann, A. Zeilinger (Springer, Berlin, 2002), pp 61–96
17. P. Eberhard, Phys. Rev. A **47**, R747–R750 (1993)
18. U. Eichmann et al., Phys. Rev. Lett. **70**, 2359–2362 (1993)
19. A. Einstein, B. Podolsky, N. Rosen, Phys. Rev. **47**, 777–780 (1935)
20. S. Freedman, J. Clauser, Phys. Rev. Lett. **28**, 938–941 (1972)
21. E. Fry, R. Thompson, Phys. Rev. Lett. **37**, 465 (1976)
22. A. Garuccio, V. Rapisarda, 1981. Nuovo Cimento **A65**, 269 (1981)
23. I. Gerhardt et al., Phys. Rev. Lett. **107**, 170404 (2011)
24. M. Giustina et al., Nature **497**, 227–230 (2013)
25. J. Hofmann et al., Science **337**, 72–75 (2012)
26. R. Holt, Ph. D. thesis, Harvard Univ. (1973)
27. R. Holt, F. Pipkin, unpublished preprint, Harvard University. (1973)
28. C. Kocher, E. Commins, Phys. Rev. Lett. **18**, 575 (1967)
29. P. Kwiat et al., Phys. Rev. Lett. **75**, 4337–4341 (1995)
30. W. Lamb, M. Scully, in *Polarization: Matiere at Rayonnement*, Société Française *de Physique*, Presses Universitaires de France, Paris (1968)
31. T. Marshall, E. Santos, F. Selleri, Phys. Lett. A **98**, 5–9 (1983)
32. D. Matsukevich et al., Phys. Rev. Lett. **96**, 030405 (2006)
33. D. Matsukevich et al., Phys. Rev. Lett. **100**, 150404 (2008)
34. Z. Ou, L. Mandel, Phys. Rev. Lett. **61**, 50–53 (1988)
35. M. Rowe et al., Nature **409**, 791–794 (2001)
36. Y. Shih, C. Alley, Phys. Rev. Lett. **61**, 2921–2924 (1988)
37. A. Shimony in *Foundations of Quantum Mechanics, Proceedings of the International School of Physics "Enrico Fermi"*, ed. by B. d'Espagnat (Academic Press, New York, 1971)
38. W. Tittel et al., Phys. Rev. Lett. **81**, 3563–3566 (1998)
39. R. Ursin et al., Nat. Phys. **3**, 481–486 (2007)
40. G. Weihs et al., Phys. Rev. Lett. **81**, 5039–5043 (1998)
41. E. Wigner, Am. J. Phys. **38**, 1005–1009 (1970)

Chapter 29
On Loopholes and Experiments

Marissa Giustina

It is particularly striking that the concept of local realism would allow itself to be tested in an experiment. Thus in a conference celebrating the theoretical breakthrough of Bell's inequality, it seems fitting to consider the recent progress toward such a laboratory realization. This essay will follow my brief presentation at the 2014 [Un]Speakables conference, and takes two parts. The first part is an overview and discussion of the "most important" loopholes. The second part outlines our recent experiment violating Bell's inequality with entangled photons, continuing with an outline of what improvements are needed for such an experiment to close all aforementioned loopholes simultaneously. The pace of this review is significantly slower than that of the talk. Those who understand Bell's inequality and the various loopholes can comfortably skip the first section; those well acquainted with photonic tests of Bell's inequality will likely learn little from reading the second section. However, to those with only limited background in the field, my hope is that you will find in this note a concise introduction to the motivations and challenges inherent in the construction of a "definitive" Bell test using photons.

M. Giustina (✉)
Institute for Quantum Optics and Quantum Information (IQOQI),
Austrian Academy of Sciences, 1090 Vienna, Austria
e-mail: marissa.giustina@univie.ac.at

M. Giustina
Vienna Center for Quantum Science and Technology (VCQ), 1090 Vienna, Austria

M. Giustina
Faculty of Physics, University of Vienna, 1090 Vienna, Austria

R. Bertlmann and A. Zeilinger (eds.), *Quantum [Un]Speakables II*,
The Frontiers Collection, DOI 10.1007/978-3-319-38987-5_29

On Loopholes

Introduction

Intuition is, for a physicist, perhaps the most prized of senses. To develop a functioning theory about a system, the physicist draws on some fundamental intuition about what he observes in nature and attempts to codify this as simply and generally as possible with the aid of mathematics. He considers what elements of nature might be relevant to his description. To improve the theory, the physicist conceives of a physical situation at the edges of his theory's credibility, and builds an experiment there. Armed with valuable new observations he carefully revises his understanding and perhaps with it, the theory. The cycle continues.

Quantum mechanics may be called a lot of things, but "intuitive" is not among them. With the advent of quantum mechanics came a fundamental shift in our approach to physics that seemed to undermine the very specificity to which physics as a discipline had previously owed such success. The concept of a "complete description"[1] of a system changed dramatically. Precise and decisive statements about any degree of freedom for any element of any system were replaced by probability distributions and ensemble descriptions with fundamentally limited specificity and precision.

The concept of an ensemble is familiar from statistical physics, but the statistical description should not be confused with the quantum ensemble description. In a statistical theory of gases, for instance, the probability distribution covers an underlying reality about the state of each individual gas molecule, and even while using the statistical theory, one continues to envisage individual distinguishable molecules, each with its own individual properties. For example, the velocity distribution gives the fraction of molecules that possess each possible value of "velocity", and is in the end an elegant way to think or talk about many individually describable molecules at once. The approach of quantum mechanics is quite different, for it does not permit any element to be considered individually. In the quantum mechanical ensemble, individual elements are considered truly "indistinguishable", and each is completely described only by the probability distribution of the whole ensemble. This is quite a strong statement especially in light of the observation that the same measurement performed on multiple (indistinguishable) elements of the quantum ensemble may give a distribution of different outcomes. It means that although different measurements yield different outcomes, there *existed no information* before the measurement by which one element of the ensemble could be differentiated from any other element (otherwise the description of each element could not have been called complete—

[1]Einstein, Podolsky, and Rosen (EPR) [1] define completeness in the following way: In a complete theory, "*every element of the physical reality must have a counterpart in the physical theory*". They further define, "*If without in any way disturbing a system we can predict with certainty (i.e. with probability equal to one) the value of a physical quantity, then there exists an element of physical reality corresponding to this physical quantity.*".

it should have included the differentiating information). How does quantum theory justify such an odd codification of nature?

Actually, these two approaches to the concept of a probability distribution seem operationally interchangeable: Both offer the same (good) prediction of what outcomes we will observe if we perform a given measurement on a randomly chosen collection of ensemble elements. Thus we would be inclined to retain our intuition, namely that a truly complete description of each element should include information about which particle will yield which measurement result. We would do away with the indistinguishability of quantum mechanics, augmenting the quantum mechanical description with so-called *hidden variables* that are deterministically (or at least probabilistically) responsible for the individual outcomes of each possible measurement. Only then would we consider the quantum description to be truly complete.

As it turns out, unless we allow some kind of *nonlocal* behavior, a so-called "action at a distance" (physical influences exceeding the speed of light), this is impossible. No theory of local hidden variables can reproduce the predictions of quantum mechanics.

Bell's Inequality

John Bell's famous theorem [2] puts a limit on the amount of measurement outcome correlation that can be predicted for a pair of independent measurements by any theory of local hidden variables. To summarize the above: In such a theory, physical influences are limited to light speed and measurement outcomes are defined prior to and independent of measurement. (Note that the inequality may also be derived for stochastic local hidden variables [3], where the outcomes could be defined probabilistically.) This worldview is often called "local realism." Bell imagined many pairs of separated spin-1/2 particles, all identically prepared, and considered the results of measurements done pairwise with Stern-Gerlach magnets at different rotations (two different rotational settings were considered per side), recording for each particle whether it was deflected up or down by the apparatus. When locality is enforced, that is, when the information about the *local* rotational setting and measurement outcome from one side is not accessible to the *distant* measurement party, and when one assumes that the individual choices of rotational setting for any given measured pair are in no way correlated with any of the hypothetical hidden variables, Bell's inequality defines a limit on the amount of correlation that can be observed across the paired measurement results in any local realist world. For certain prepared states of the particles and certain sets of rotational settings, the quantum mechanical prediction exceeds this limit!

In this way, Bell succeeded to establish a divide between the predictions of local realism and those of quantum mechanics. It became clear that any attempts to recover the quantum mechanical theory on a local realist foundation were futile. The inequality itself, a purely theoretical statement, sheds little light on the "nature of reality". However, it provides clear hints about the sort of experiment one should perform

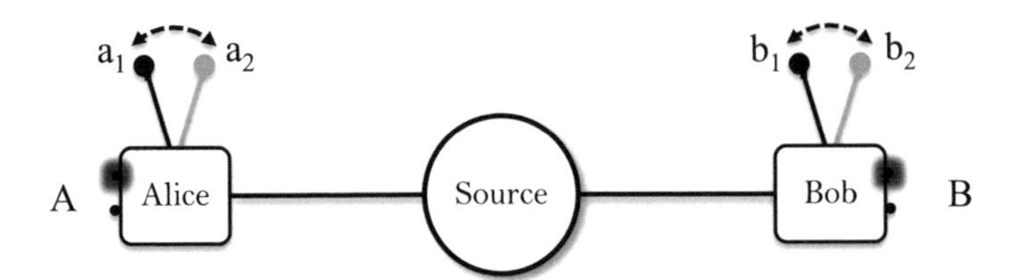

Fig. 29.1 Idea of the Bell experiment: a source of entangled particles dispatches a pair, sending one particle each to two observers Alice and Bob. Independently from the rest of the experiment, Alice and Bob choose a measurement setting with which to measure their particle (for Alice, a_1 or a_2; for Bob, b_1 or b_2), and record one of two possible measurement outcomes (Alice's and Bob's outcomes are denoted A and B). Later, they compile their results to determine a list of which (joint) measurement settings yielded which (joint) results. The Bell inequality enables them to determine whether their observations were consistent with local realism.

(see Fig. 29.1). Although the inequality as originally presented was not suitable for testing, it was later rewritten in a form accessible to experiment by Clauser, Horne, Shimony and Holt (CHSH) [4, 5]. Gradually, a blueprint was developed for an experiment testing the validity of local realism: a pair of entangled spin-1/2 particles (or photons, as the case may be) would be dispatched in opposite directions and measured in two different rotational (polarization) settings. This would be repeated many times and the results plugged into the inequality. Any experiment satisfying the inequality could be explained by a local realist theory, while an experiment violating the inequality shows that "reality", at least to some extent in some laboratory, is at odds with local realism.

It is a common misconception that such an experiment "tests quantum mechanics". The Bell experiment tests local realism, not quantum mechanics; whether the data agree with quantum mechanics is a separate issue entirely. Indeed, quantum mechanics predicts a violation of the Bell inequality and, due to its general success as a theory, can be a useful tool for designing an experiment that hopes to observe such a violation in the lab. If a carefully considered quantum mechanical model of the experiment is in good agreement with the observations, that further highlights the usefulness of quantum mechanics as a theory. Still, one cannot consider this a test confirming quantum mechanics—according to the scientific method it is not possible to confirm any theory with finality. In the end, an observed Bell violation must not be seen as a confirmation of quantum mechanics but rather a refutation of local realism.

Loopholes

When the first experimental violations of the inequality were reported [6], some philosophers and physicists alike were startled. Local realism should not be so hastily abandoned! Surely it would be unreasonable for serious physicists to discard the intuitive local realist worldview in favor of something fraught with quirks and allegations of fundamental uncertainty. Such measures should be taken only if no alternative remains!

And so the blueprint was revisited. Of course, the process of mapping a gedanken-experiment into a laboratory involves some inevitable assumptions. Real laboratories can approximate gedankenlaboratories only so well. Perhaps, by exploiting one of these additional assumptions, local realism could after all explain the inequality violation in any particular experiment. A variety of common experimental assumptions have been identified, along with corresponding strategies for using local realism to exploit them and simulate the violation of Bell's inequality. The latter are known as "loopholes."

As the field has developed, many loopholes have been identified and analyzed. Some can be closed. Some cannot be closed. Some can be closed only part way. Some are subsets of other loopholes. Perhaps some have not yet been discovered. I will briefly discuss four of the "most important" loopholes to date and their related assumptions.

Locality

When assuming that the measurement outcome on one side is not affected by the outcome or setting choice on the other side, one invokes the locality assumption. Ideally, the local measurement outcome should depend only on the local measurement setting and information brought with the entangled particle (including any hypothetical local hidden variables). If a signal at or below the speed of light could carry distant setting or outcome information to influence the local measurement, this would exploit the locality loophole. More specifically: if information about the distant outcome could be transmitted, this violates "outcome independence"; if information about the distant measurement setting could be transmitted, this violates "setting independence".

The standard approach to closing the locality loophole is to space-like separate the events that must remain independent. That is, the experiment must be constructed in space-time so that no speed-of-light signal could carry out the forbidden information transfer. In this case, to ensure outcome independence, the two measurement events must be space-like separated (see Fig. 29.2). To ensure setting independence as well, the local measurement must in addition be space-like separated from the choice and implementation of the distant setting. The act of localizing a setting choice in space-time is non-trivial, but I will come to this later. For a more detailed discussion of this loophole, see [7].

Freedom of Choice

One assumption in the derivation of the Bell inequality is that the choice of measurement settings is statistically independent from any hidden variables. If this were not

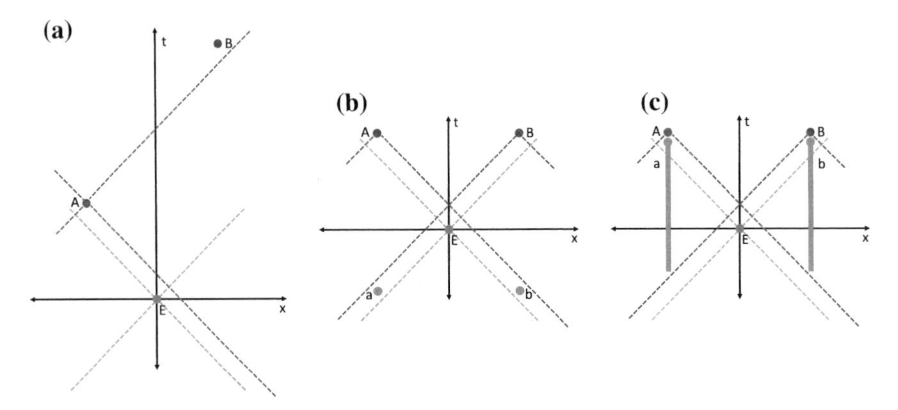

Fig. 29.2 A poll of the audience augmented with a fair sampling assumption (see third loophole) suggested that by this point in the [Un]Speakables conference, a startling half of participants preferred to look at animated flying balls rather than space-time diagrams. As this is a printed essay, I am limited to space-time diagrams here, which are to be interpreted with (one-dimensional) space on the x-axis and time on the y-axis. The *dotted lines* at a 45° angle to the axes indicate the speed of light. The emission event at **E** marks the birth of the entangled pair. The range of **E**'s physical influence (both influence *on* **E** by the universe, *below* the x-axis, and *by* **E** on the universe, *above* the x-axis) is denoted by the faintly shaded triangular region of spacetime, also known as **E**'s *light-cone*, and extends arbitrarily far into the past and future, with only a limited depiction in the figure. **(a)** indicates Alice's measurement performed at **A** and Bob's at **B**, situated a distance apart with the emission event between them in space. Since Alice and Bob need the particles emitted at **E** in order to make their measurements, **A** and **B** will always be within the future lightcone of **E**. However, in this configuration, the measurement at **A** takes place so far in advance of the measurement at **B** that outcome independence is not fulfilled: **B** lies within **A**'s forward lightcone. **(b)** resolves this issue by making measurements **A** and **B** at the same time at distant locations–we say they are *space-like separated* because neither occupies the other's light cone. However, we now consider as well the process of choosing the measurement settings, denoted by **a** for Alice's setting choice and **b** for Bob's. If these choices occur too far in the past, Alice's choice **a** could be available to Bob's measurement **B** (and vice versa), violating setting independence. **(c)** shows a configuration (including a range of possibilities for the setting choices) in which both measurement and setting independence are fulfilled, and thus the locality loophole is closed. **(a)** Violation of outcome independence. **(b)** Violation of setting independence. **(c)** Closing the locality loophole.

the case, one could imagine a source producing pre-determined states, tailor-made for the particular settings that are about to be measured, or vice-versa.

To close this loophole, one would like to space-like separate both setting choices from the production of hypothetical hidden variables (see Fig. 29.3). Since neither a setting choice nor a hidden variable has a well-defined birth place/time, this loophole can be closed only within particular assumptions. (Since the locality loophole also requires space-like separation of setting choices, similar assumptions are required for its closure as well.) With a reasonable model for the experiment and a reasonable model for the setting generators, one can identify which events should be space-like separated.

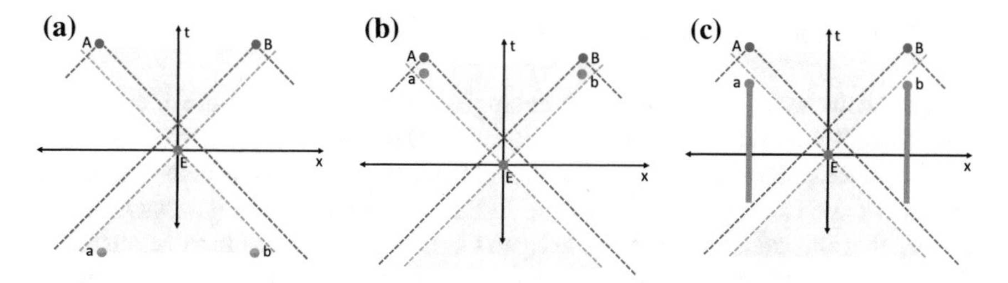

Fig. 29.3 These figures relate to the freedom of choice loophole. We assume that any hypothetical hidden variables are born with the entangled pair at **E**. Then if the measurement choices **a** and **b** are determined within the past lightcone of **E**, as in **(a)**, these choices could influence the emission to produce a state tailor-made for the settings to be measured. (This configuration also violates setting independence as discussed above, thus being unsuitable in multiple ways.) **(b)** indicates another way to violate the freedom of choice loophole: the settings are chosen in the future lightcone of **E**. Finally, the configuration depicted in **(c)** closes both the locality and freedom of choice loopholes. **(a)** Violation of freedom of choice (and setting independence). **(b)** Violation of freedom of choice. **(c)** Closing locality and freedom of choice.

One can make the argument that this loophole is unimportant since it cannot be decisively closed anyway. However, it can be decisively left open! A conscientious experimentalist might thus argue that he or she would be remiss not to address the loophole as far as it is possible. For a more detailed discussion of this loophole, see [8].

Fair Sampling

Experiments on entangled photon pairs are plagued by losses. This often forces experimentalists to draw a conclusion about the statistics of an ensemble (of pairs) based on measurements of only part of that ensemble. In doing so, they assume that the statistics of the full ensemble are accurately represented by the sub-ensemble. This so-called "fair sampling assumption" opens another loophole[2]: If the decision of whether or not a given photon will be detected is allocated to a hidden variable, it is possible for a local hidden variable model to exploit this assumption and fake a Bell violation [9–11]. To close the loophole, an experimentalist can either confirm a high detection efficiency or use a version of the inequality—such as those derived by Clauser and Horne (CH) and by Eberhard [5, 12]—that does not rely on the fair-sampling assumption. Such inequalities will be more difficult to violate due to the high requisite efficiency, however, any violation then automatically ensures closure of the loophole with no need for any further (efficiency) characterization.

---

[2]Note that this loophole is sometimes called the "detector" loophole. Since it is not clear what element of the detector is responsible for the vulnerability, and since even perfectly efficient detectors could be involved in unfair sampling when they are connected to lossy experiments, I will stick to the term "fair sampling" for both the assumption and the loophole in this essay.

Fair Coincidences Assumption/Coincidence-Time Loophole

In continuous wave experiments on entangled photons, photons are identified as pairs based on closeness in their arrival times. Detector clicks occurring within a given "coincidence window" are considered "coincident" and thus belong to a pair. The use of such an assignment protocol in Bell tests carries with it an assumption, namely that the statistics of identified pairs sufficiently represents the statistics of all detected pairs if they had been correctly identified. Choosing the wrong coincidence window can result in the misidentification of pairs, resulting in a discrepancy between the number of pairs for which both photons were detected and the number of pairs actually identified by the algorithm in use. A hypothetical setting-dependent timing uncertainty (jitter) in the detector would be one way to exploit the loophole, which can be closed by using an inequality that does not rely on the assumption of fair coincidences. For more detail on the loophole, see the seminal paper on it [13]; for more detail on the loophole as it relates to the CH inequality, see [14].

Other Loopholes, Other Thoughts

In addition to these loopholes there exist others, some of which could be categorized as a variant of one listed here, some of which cannot be closed at all. For example, a fully deterministic worldview postulates that all events are deterministically traceable arbitrarily far back in time, which means that no relevance can be attributed to any space-like separation. Such a worldview is also inaccessible to the scientific method.

The following comment is motivated by the term loophole itself. When I first met the term "loophole", I rejected it as bad terminology. It seemed much more reasonable to speak of assumptions than of loopholes, and it seemed each loophole has a corresponding assumption anyway. When discussing a given Bell test, I thought it would be much clearer to discuss what assumptions *were* made rather than starting a list of assumptions that might have been made but weren't. However, I overlooked something in this thinking: there may be more than one approach to closing a given loophole. In general, one can:

- Ensure that the corresponding assumption has not been exploited in the experiment (this does not mean confirming that the assumption is valid–although in principle that would work too)
- Avoid making the corresponding assumption in the first place.

Consider the fair-sampling loophole for an informative example. In massive systems (atoms, ions, superconducting qubits, NV-centers in diamond), the experimenters can be certain of the fate of every pair in their ensemble. They can be certain that the experiment was not vulnerable to the fair-sampling assumption, and can use analysis methods that make a fair-sampling assumption, since the assumption cannot be violated anyway. Photon-based experiments, on the other hand, struggle to achieve sufficient detection efficiency. They use the second approach: by violating the

CH inequality, which can be derived without making the fair-sampling assumption in the first place (and is thus harder to violate), these experiments can also close the loophole. Thus the term "loophole" offers a linguistic efficiency that should not be underestimated!

Considering the four loopholes, one can organize them into two general groups based on how they may be closed. For locality and freedom-of-choice, there is no hope to construct a Bell inequality without these assumptions, as these are a fundamental part of Bell's derivation. Thus to close the corresponding loopholes, it is necessary to design and build the setup carefully, space-like separating the relevant events to ensure that these loopholes are not exploited in the experiment. Furthermore, since both of these loopholes involve space-like separation of setting choices, they are closable only under some further assumption that gives meaning to the space-time layout employed. In contrast, the closure of the fair-sampling and coincidence-time loopholes can be ensured by using the proper data analysis. A skeptic needn't trust that the experimentalists have properly characterized the efficiency of their experiment if the data violate the CH inequality.

Loopholes represent the biggest impediments to a "definitive" test of local realism. Of course, the idea of a "definitive" test is subjective. Some local realist explanations, such as a fully deterministic worldview, cannot be tested at all. What one person views as an important loophole might strike another as inconsequential or unreasonable. Furthermore, there is no way to ensure that a hypothetical experiment simultaneously closing all loopholes listed here would not be vulnerable to some as-of-yet undiscovered loophole. Thus we must be satisfied with statements about the set of local realist theories excluded by a particular test.

On Experiments

With all of this in mind, what kind of experiment could one design to test local realism as completely as possible?

Accessibility, both to the theorist and to the experimentalist, drives the appeal of the photon as a test medium. Bohm's 1951 gedankenexperiment [15] involving a separated pair of spin-1/2 particles analyzed by rotating Stern-Gerlach magnets can, with minimal alteration, be rewritten to consider pairs of polarization-entangled photons analyzed by rotating polarizers. Thus, photon pairs are appealing because they are easy to think about. For physicists accustomed to considering two-level spin-1/2 particles, the polarized photon is easy to digest: the familiar function of rotating Stern-Gerlach magnets can be intuitively mapped to rotatable polarizers.

Furthermore, their (relative) experimental simplicity made pairs of entangled photons the first system in which a Bell experiment was feasible at all. Already in 1967, Kocher and Commins [16] were able to observe polarization correlations between visible photon pairs emitted from calcium atoms. Unlike atoms, superconducting qubits, and most other massive quantum systems, which involve complicated traps and/or cryogenic temperatures, photons can be created in any laboratory; simple glass or plastic polarizers and well-understood photomultiplier tubes suffice for analysis and detection. For this reason, the first Bell experiment [6] and indeed an entire generation of early Bell experiments were carried out on pairs of entangled photons.

As the experiments developed and began to address the open loopholes, photons continued to display advantages. Since photons travel at the speed of light, achieving the space-like separation necessary to address the locality and freedom-of-choice loopholes was a logical goal for photonic Bell experiments of ever-increasing complexity [7, 8, 17]. Unfortunately, as easy as photons are to send, they are even easier to lose, and until recently there was no hope for a photon-based experiment free of the fair-sampling assumption. Here, massive systems (which are much more difficult to lose) could easily surpass the photon, and a number of experiments ([18–21], among others) showed Bell violations in which every entangled pair could be counted.

Modern photon pair sources based on spontaneous parametric downconversion (SPDC) in nonlinear optical crystals bear little resemblance to the atomic cascade sources of the early Bell tests, whose emission was comparatively dim and scattered in all directions. State-of-the-art sources, in which photons from a pump laser are converted into pairs of lower-energy entangled photons, can now be designed so efficiently that in 2013, the first photonic experiments to close the fair-sampling loophole were published [22, 23]. These experiments, in contrast to their counterparts in massive systems, used the CH/Eberhard inequality [5, 12], which can be derived without the fair-sampling assumption, to ensure the loophole was closed. Eberhard's critical improvement on the CH inequality was the observation that by a careful selection of state and measurement parameters, the detection efficiency necessary for violation can be as low as 2/3 (in the absence of background), which is just on the edge of achievable for state-of-the-art photon pair sources and detectors. In addition, neither 2013 experiment was vulnerable to the fair-coincidence assumption [14]. This made the photon the first system for which all major loopholes have been closed–albeit in different experiments.

Despite the progress in photon-based Bell tests, a definitive test of local realism remains an open challenge. The 2013 experiments ensured no space-like separation in their measurements or setting choices; the task of separating the paired photons over long distances brings an additional loss, which would have made the requirements on the photon detection efficiency even more stringent. In the remainder of this essay, I will briefly expound on our experiment [22] in order to consider in detail what is still required to make the experiment "loophole-free."

A schematic of our experiment, borrowed from [22], is displayed in Fig. 29.4. The block labeled "Source" depicts the triangular Sagnac interferometer with the ppKTP (periodically poled potassium titanyl phosphate) downconversion crystal on the hypotenuse, which is designed to convert a horizontally polarized blue photon at 405 nm into a pair of orthogonally polarized infrared photons at 810 nm. Ultraviolet (UV) light from a continuous wave laser enters the loop at a polarizing beam-splitter cube, where it is directed based on its polarization to traverse the loop in the clockwise direction (for vertical or "V" polarized light), the counter-clockwise direction (for horizontal or "H" polarized light), or both directions (for light of any other

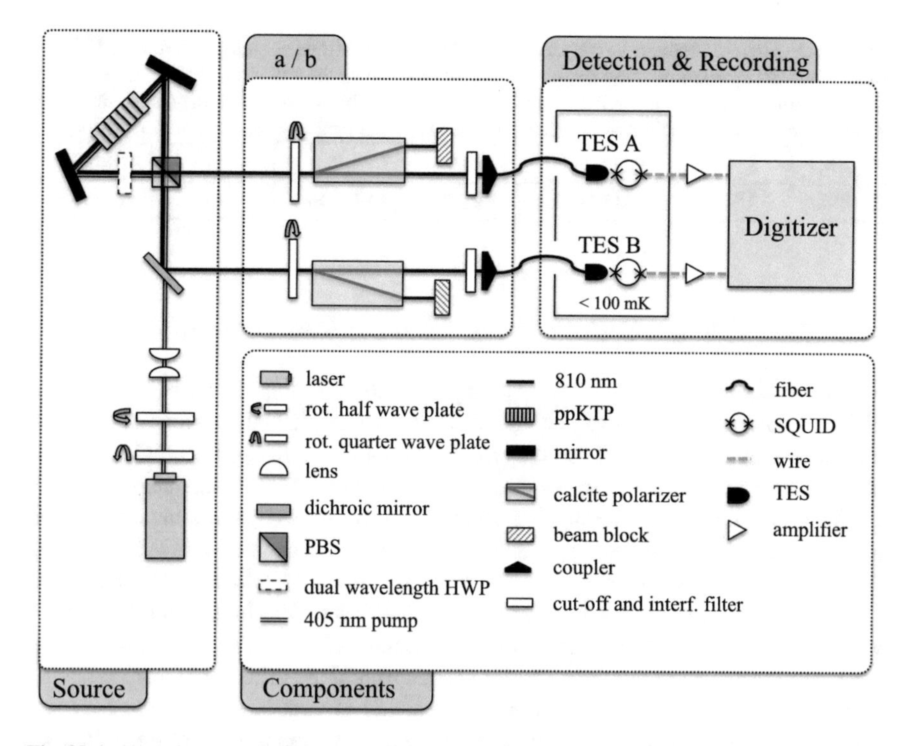

Fig. 29.4 Experimental construction used for our Bell violation without the fair-sampling assumption (figure borrowed from [22]). See text for more information.

polarization, which can be expressed as a mixture of H and V polarizations). When a clockwise-traveling UV photon undergoes downconversion, the infrared photons are emitted such that the top arm contains a V photon, and the bottom arm an H photon. However, when a counter-clockwise-traveling UV photon undergoes downconversion, the infrared photon polarizations are the other way around, with H in the top arm and V in the bottom arm. By pumping in both directions simultaneously, it is possible to overlap the (V_{top}, H_{bottom}) with the (H_{top}, V_{bottom}) cases. Then, for any given pair of photons entering the box labeled "a / b", it is not possible to know which photon will have what polarization, but in any case, the photons in the two arms will have opposite polarizations. The polarization is defined for the pair but not for the individual photons, and we call the photons entangled.

Equipped with a source of polarization-entangled photons, we can consider the rest of the experiment, in which the photons are passed through polarizers and detected. We define $C(i,j)$ as the number of identified coincidences when the polarizers are set to setting i on Alice's side with $i \in \{a_1, a_2\}$ and j on Bob's side with $j \in \{b_1, b_2\}$. Similarly, $S^A(i)$ and $S^B(j)$ represent the number of singles counts, that is, the total number of clicks identified respectively in Alice's and Bob's detectors, for the given polarizer setting. To evaluate the Eberhard inequality [12],

Table 29.1 Recorded counts aggregated over 300 s of recorded data per setting, as reported in [22]. See text for an explanation of the labels. The **bolded** terms help the violation (contribute negatively to the parameter J) while the `fixed width` terms hinder the violation (contribute positively to J). A non-negative value for J is in agreement with local realism; here we observe a violation.

$C(a_1b_1)$	$S^A(a_1)$	$C(a_1b_2)$	$S^B(b_1)$	$C(a_2b_1)$	$C(a_2b_2)$	J (total)
1069306	`1522865`	**1152595**	`1693718`	**1191146**	`69749`	−126715

$$J = -C(a_1b_1) + S^A(a_1) - C(a_1b_2) + S^B(b_1) - C(a_2b_1) + C(a_2b_2) \geq 0 \quad (1)$$

we must collect data at four different joint polarization settings, corresponding to two different polarization settings per side. This was accomplished by using a half-wave plate in a rotatable mount followed by a calcite polarizer and a highly efficient single-photon detector. In particular, we used fiber-coupled transition-edge sensors (TES) to detect the photons. These superconducting calorimetric detectors must be cooled to below 100 mK, have timing jitter on the order of 100 ns, and produce an analog signal which must be interpreted, rather than a series of "clicks" [24]. However, the detectors also boast the highest known efficiencies for single-photons and are virtually free of dark counts [25]. Thus, TES detectors are indispensable to photon-based Bell tests where efficiency and low background are critical parameters.

We recorded a total of three hundred seconds of data in each of the four setting combinations, and identified coincidences by a standard "moving window" approach: a coincidence is counted when two clicks are separated by less than a certain amount of time. These counts can be used to evaluate the Eberhard inequality (1), as shown in Table 29.1, which yielded a clear violation of the inequality. (For more information, please see [22, 26].) The experiment also violated two other inequalities, variations on the CH/Eberhard inequality that were derived without the fair-coincidence assumption [14]. One variation involved analyzing the data in discrete time blocks, where the block, rather than the identified photon pair, becomes the experimental unit. The other approach was similar to the "moving window" analysis but with one window enlarged to the sum of the other three.

This experiment indeed closed the fair-sampling and coincidence-time loopholes, but it did not address locality or freedom-of-choice. In fact, the two detectors sat next to each other in the same cryostat, and the choice of measurement settings was determined very far in advance. I have been asked on several occasions what obstacles prevent us from carrying out a loophole-free test immediately; furthermore, there seems to exist a misconception that merely separating the detectors to a great distance would be a sufficient upgrade. Unfortunately, this is not the case (Fig. 29.5).

There are three main issues that must be resolved in order to make this experiment into a "definitive" test. First, the polarizers and detectors must be physically separated to a distance at which space-like separation of measurements can be enforced. For the TES detectors, this implies the procurement of a second cryogenic system and a second set of readout and recording electronics. (For the first experiment, both detectors could be kept in the same cryostat and monitored with two different channels

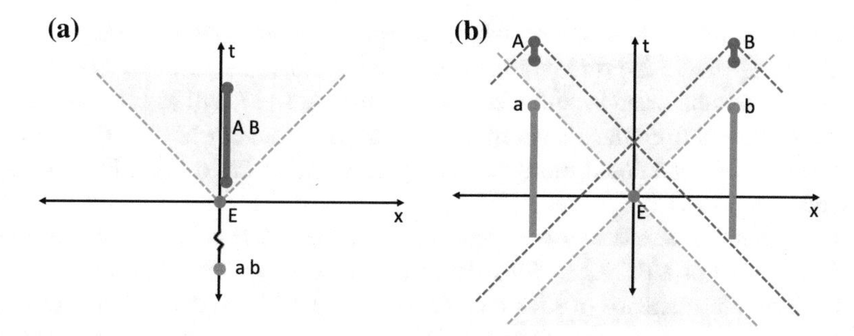

Fig. 29.5 Space-time diagrams representing **(a)** Experimental situation in experiments [22], **(b)** ideal situation for a "definitive" Bell test.

of the same electronics.) In principle, one would like a very large separation, since more spatial separation of the measurements corresponds to less stringent requirements on their syncrhonization. However, the farther the photons are separated, the larger the losses due to transit (for instance, optical fiber has an intrinsic loss per length; efficient free-space links over long distances would require vibrational stability on each end and ever-larger lenses to collect the diverging beam). Since the TES detectors are fiber-coupled, one instance of fiber coupling will be required in any case—and fiber coupling brings more intrinsic losses. All losses associated with the requisite separation will either compromise the strength of the violation or must be compensated by improvements in the pair source.

A second issue, related to the space-like separation of the measurements, is the time duration of the measurement. The TES detectors, like any photon detector, have an associated timing uncertainty known as *jitter*. This means that a photon arriving at a given time may be registered by the detector as arriving at a different time; the width of this spread quantifies the jitter. To ensure that a local detection event is not privy to information about the opposite setting, the detection must be confined to a finite time window. To keep this time window (and thus the requisite distance for space-time separation) as small as possible, it is desirable to reduce the large timing jitter of the TES detectors as much as possible.

Finally, it is necessary to ensure that the setting choice is isolated, not only from the distant choice and measurement, but also from the (hidden variable) source itself. For this, it is common to employ so-called "Quantum Random Number Generators" (QRNG). An ideal number generator creates a brand new random variable within a prescribed time window. This value should have no connection to any other variables in the experiment and no memory of any past experimental events (including the history of the number generator itself). In fact, this behavior is not at all represented by the descriptions "quantum" and "random," which are neither necessary nor sufficient for our needs. A sluggish random number that makes its way to all experimental parties before being used for a setting choice cannot be considered space-like separated from the rest of the experiment—regardless of its possibly quantum origins. Although quantum processes may provide a good source of random variables, there is no inherent reason why a classical process couldn't be equally well suited. Finally, there is no requirement that the numbers be perfectly "random"—even biased numbers can be sufficient for our purposes [27, 28] as long as they are free of influences from the distant measurement party and any hidden variables. So in fact what we need could more accurately be described as a "Fast Free Number Generator" (FFNG).

Unfortunately, there is no way to determine whether a particular number generator fulfills the ideal of a FFNG. A fully deterministic or superdeterministic worldview forbids even the existence of such a device. We would like to demand that the generated values are independent from any hidden variables—but hidden variables are by definition inaccessible to our characterization. Must we abandon the endeavor? Certainly not! For while it is impossible to certify any number generator as fully independent, it is relatively easy to identify some number generators as unsuitable. A conscientious experimentalist cannot space-like separate the setting choices beyond the shadow of a doubt, but he or she would be remiss to use a setting generator known to produce old numbers or numbers that are correlated with known past events in a predictable way.[3] In other words, one cannot close the loophole with certainty, but one can leave it open with certainty. The latter should be avoided.

Once the setting choice is determined by some appropriate number generator, it must be implemented quickly. Electro-optic modulators that behave like switchable half-wave plates can enable a particular polarization rotation on the order of nanoseconds. After this (switchable) rotation comes a fixed polarizer and finally the detector. However, there is a non-trivial loss associated with the use of such switches, which have only been well developed for use in free-space. Light in fiber must be coupled out into free-space, through the polarization switches, and back into fiber (since the detectors are fiber-coupled). Recall that any added loss must be overcome by further improvements in the source in order for the experiment to remain feasible.

A final change, unessential but desirable, would be to modify the setup from continuous operation to a pulsed experiment. Such a change would have two benefits. Firstly, it would allow the coincidence-time loophole to be closed implicitly, since the experimental "timeslot" unit would be built into the experiment. In addition, confining the emission event to a known window smaller than the detector jitter reduces the total time window needed to identify a coincidence (see Fig. 29.6). With the possible arrival interval confined absolutely in time, one will find a click (if it is there) by searching in a region defined by the detector jitter around this possible arrival time. If, however, the search for a coincident photon on one side is triggered by a detection on the other, then a larger time region must be searched, since the jitter from both detectors could contribute to the total time offset between the "coincident" clicks. A shorter measurement duration translates to less physical separation required to space-like separate **A** and **B**. This is particularly relevant because loss increases with separation distance.

---

[3]A more detailed evaluation of some common photon-based random number generators can be found in [28].

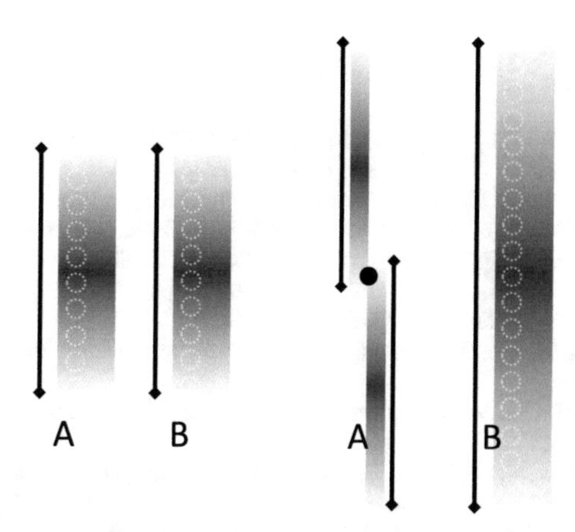

Fig. 29.6 This figure illustrates how confining the photon emission to a known time can reduce the required size of the coincidence window. Time runs vertically. In the situation on the left, the emission is assumed to be confined to a single instant in time. As a result of detector jitter, the registered clicks could fall anywhere along the black lines, with a distribution indicated by the grey stripe. Thus the total time in which to search for a coincidence can be determined based on knowledge of the emission time and the detector jitter. The situation on the right represents continuous emission. The detection of a photon on Alice's side triggers the search for a coincident photon on Bob's side. However, since we do not know how the detector jitter affected Alice's timestamp, it is necessary to consider a much larger window on Bob's side in order to find a possible coincident photon.

In total, an experimentalist starting with the laboratory situation described in [22] now sees the following formidable to-do list:

- Separate Alice's and Bob's detectors into two distant cryostats
- Configure a low-loss optical link between the source and the remote measurement stations
- Lower the jitter in TES detectors
- Find and employ suitable number generators, based on processes that are as fast and free as possible
- Implement fast-switching polarization rotations at each remote measurement station
- Compensate for the losses introduced by the above steps, probably by making further improvements in the source
- (Introduce a pulsed source)
- Carefully synchronize the experiment

Assuming the Bell violations are still observed in these extreme conditions, the first such tests will in fact launch a next generation of Bell tests. Photons are not the only system in which such a Bell test seems to be just on the horizon. We can hope that ever more loopholes will be addressed in a variety of systems—photonic and massive. Improved sources of randomness can further push the limit to which local realism can explain our experiments [29]. Eventually we may be forced to accept a reality in which, at least sometimes, local realism can be violated.

References

1. A. Einstein, B. Podolsky, N. Rosen, Can quantum-mechanical description of physical reality be considered complete? Phys. Rev. **47**, 777–780 (1935)
2. J. Bell, On the Einstein Pedolsky Rosen paradox. Physics **1**, 3 (1964)
3. J. Bell, Introduction to the hidden-variable question, in *Speakable and Unspeakable in Quantum Mechanics* (Cambridge University Press, 2004), pp. 29–39
4. J.F. Clauser, M.A. Horne, A. Shimony, R.A. Holt, Proposed experiment to test local hidden-variable theories. Phys. Rev. Lett. **23**, 880–884 (1969)
5. J.F. Clauser, M.A. Horne, Experimental consequences of objective local theories. Phys. Rev. D **10**, 2 (1974)
6. S.J. Freedman, J.F. Clauser, Experimental test of local hidden-variable theories. Phys. Rev. Lett. **28**, 938–941 (1972)
7. G. Weihs, T. Jennewein, C. Simon, H. Weinfurter, A. Zeilinger, Violation of Bell's inequality under strict Einstein locality conditions. Phys. Rev. Lett. **81**, 5039–5043 (1998)
8. T. Scheidl, R. Ursin, J. Kofler, S. Ramelow, X.-S. Ma, T. Herbst, L. Ratschbacher, A. Fedrizzi, N.K. Langford, T. Jennewein, A. Zeilinger, Violation of local realism with freedom of choice. Proc. Natl. Acad. Sci. USA **107**, 19708–19713 (2010)
9. F. Selleri, A. Zeilinger, Local deterministic description of Einstein-Podolsky-Rosen experiments. Found. Phys. **18**, 1141–1158 (1988)
10. N. Gisin, B. Gisin, A local hidden variable model of quantum correlation exploiting the detection loophole. Phys. Lett. A **260**(5), 323–327 (1999)
11. P. Pearle, Hidden-variable example based upon data rejection. Phys. Rev. D **2**, 1418–1425 (1970)
12. P. Eberhard, Background level and counter efficiencies required for a loophole-free Einstein-Podolsky-Rosen experiment. Phys. Rev. A **47**, 2 (1993)
13. J.-Å. Larsson, R.D. Gill, Bell's inequality and the coincidence-time loophole. EPL (Europhys. Lett.) **67**(5), 707 (2004)
14. J.-Å. Larsson, M. Giustina, J. Kofler, B. Wittmann, R. Ursin, S. Ramelow, Bell-inequality violation with entangled photons, free of the coincidence-time loophole. Phys. Rev. A **90**, 032107 (2014)
15. D. Bohm, *Quantum Theory* (Prentice Hall, New York, 1951)
16. C.A. Kocher, E.D. Commins, Polarization correlation of photons emitted in an atomic cascade. Phys. Rev. Lett. **18**, 575–577 (1967)
17. A. Aspect, J. Dalibard, G. Roger, Experimental test of Bell's inequalities using time-varying analyzers. Phys. Rev. Lett. **49**, 1804–1807 (1982)
18. M. Rowe, D. Kielpinski, V. Meyer, C. Sackett, W. Itano, C. Monroe, D.J. Wineland, Experimental violation of a Bell's inequality with efficient detection. Lett. Nat. **409** (2001)
19. D. Matsukevich, P. Maunz, D. Moehring, S. Olmschenk, C. Monroe, Bell inequality violation with two remote atomic qubits. Phys. Rev. Lett. **100**(15), 150404 (2008)
20. M. Ansmann, H. Wang, R.C. Bialczak, M. Hofheinz, E. Lucero, M. Neeley, A.D. O'Connell, D. Sank, M. Weides, J. Wenner, A. Cleland, J.M. Martinis, Violation of Bell's inequality in Josephson phase qubits. Lett. Nat. **461** (2009)
21. J. Hofmann, M. Krug, N. Ortegel, L. Gérard, M. Weber, W. Rosenfeld, H. Weinfurter, Heralded entanglement between widely separated atoms. Science **337**(6090), 72–75 (2012)
22. M. Giustina, A. Mech, S. Ramelow, B. Wittmann, J. Kofler, J. Beyer, A. Lita, B. Calkins, T. Gerrits, S.W. Nam, R. Ursin, A. Zeilinger, Bell violation using entangled photons without the fair-sampling assumption. Nature **497**, 227–230 (2013)
23. B.G. Christensen, K.T. McCusker, J.B. Altepeter, B. Calkins, T. Gerrits, A.E. Lita, A. Miller, L.K. Shalm, Y. Zhang, S.W. Nam, N. Brunner, C.C.W. Lim, N. Gisin, P.G. Kwiat, Detection-loophole-free test of quantum nonlocality, and applications. Phys. Rev. Lett. **111**, 130406 (2013)
24. A.E. Lita, A.J. Miller, S.W. Nam, Counting near-infrared single-photons with 95 % efficiency. Opt. Express **16**, 5 (2008)

25. A.J. Miller, S.W. Nam, J.M. Martinis, A.V. Sergienko, Demonstration of a low-noise nearinfrared photon counter with multiphoton discrimination. Appl. Phys. Lett. **83**, 791–793 (2003)
26. J. Kofler, S. Ramelow, M. Giustina, A. Zeilinger, On 'Bell violation using entangled photons without the fair-sampling assumption' (2013). arXiv:1307.6475v1[quant-ph]
27. R.D. Gill, Time, finite statistics, and Bell's fifth position, in *Proceedings of "Foundations of Probability and Physics—2"*. Series Mathematical Modelling in Physics and Engineering, Engineering, Economy and Cognitive Science, vol. 5 (Växjö Univ. Press, 2003), pp. 179–206
28. J. Kofler, M. Giustina, Requirements for a loophole-free Bell test using imperfect setting generators (2014). arXiv:1411.4787
29. J. Gallicchio, A.S. Friedman, D.I. Kaiser, Testing Bell's inequality with cosmic photons: closing the setting-independence loophole. Phys. Rev. Lett. **112**, 110405 (2014)

Chapter 30
New Dimensions for Entangled Photons: The Role of Information

Anton Zeilinger

Introduction

This paper follows very closely the talk given at *Quantum (Un)Speakables II: 50 Years of Bell's Theorem*. It will be argued that information plays a central role in interpreting and understanding real experiments.

The relation of quantum theory to physical reality has been at the core of discussions since the very first beginnings of quantum mechanics. Just to emphasize the level of technical development achieved, Fig. 30.1 shows a telescope. It is the OGS (Optical Ground Station) operated by the European Space Agency on the Canary Island of Tenerife. This instrument was built for developing and testing optical communication with satellites. In our experiments, we utilize it for testing optical quantum communication over large earthbound distances, specifically ca. 150 km between the Canary Islands of La Palma and Tenerife. In that image, one can see a laser beam which serves as a guiding beacon for continuously adjusting the sending and the receiving telescopes onto each other in order to reduce the effect of atmospheric fluctuations.

This setup has served as a workhorse for many experiments on entanglement, some of which test the ideas of reality put forward by Einstein, Podolsky and Rosen in 1935 [1]. It is actually quite instructive to investigate the number of citations the Einstein-Podolsky-Rosen paper received (see Fig. 30.2). This paper came out in 1935. In the

A. Zeilinger (✉)
Vienna Center for Quantum Science and Technology (VCQ), 1090 Vienna, Austria
e-mail: Anton.Zeilinger@univie.ac.at

A. Zeilinger
Institute for Quantum Optics and Quantum Information (IQOQI), Austrian Academy of Sciences, 1090 Vienna, Austria

A. Zeilinger
Faculty of Physics, University of Vienna, 1090 Vienna, Austria

R. Bertlmann and A. Zeilinger (eds.), *Quantum [Un]Speakables II*, The Frontiers Collection, DOI 10.1007/978-3-319-38987-5_30

Fig. 30.1 Optical Ground Station OGS on the Canary Island of Tenerife. The laser beam shown is used as a guiding beacon for the quantum optical communication link to the island of La Palma. (*Photo* Daniel Padrón).

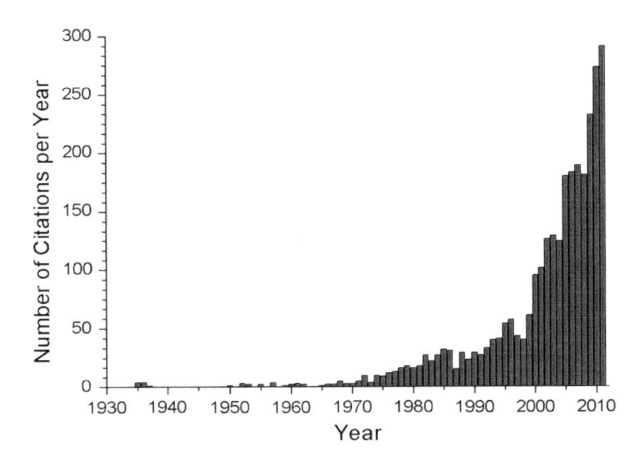

Fig. 30.2 Number of annual citations of the Einstein-Podolsky-Rosen paper.

beginning, it had very few citations. I would like to remark that these citations were not so bad. Two of them were by Schrödinger [2, 3] and one by Bohr [4]. So the paper immediately attracted attention by leading figures in quantum science. Yet, after that, there were no citations until the beginning of the 1950s. I would like to remark as a comment on today's situation regarding the ways in which careers are often decided that this paper would not have gotten Einstein a permanent position because of its low citation numbers. But then, citations took off for two reasons. Slowly, in the 1950s, the field of the foundations of quantum mechanics started again after it had lain dormant

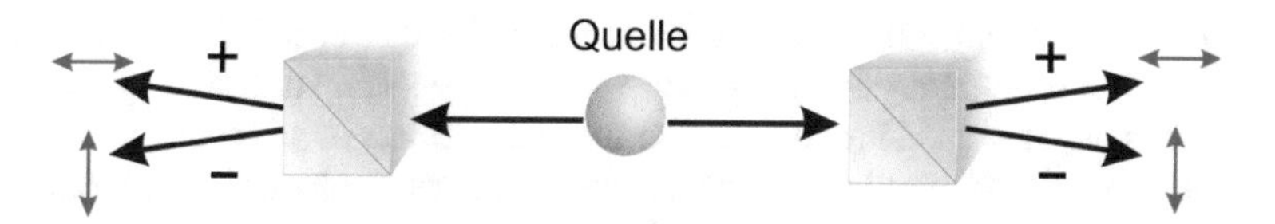

Fig. 30.3 Typical experimental set-up for measurements on polarization-entangled photon pairs. The source (Quelle) is assumed to emit an entangled pair, say, in the state of Eq. 1. Each photon path is a two-channel polarizer which may be rotated around the incident beam direction. Then, one always measures perfect correlations in either horizontal or vertical polarization when the polarizers are oriented along the same direction. A violation of Bell's Inequality arises for skewed angles between two-channel polarizers. (*Image* Thomas Jennewein)

for a significant time. And most importantly, in 1964, John Bell showed [5] that the fundamental concepts of reality and locality, as proposed by Einstein, Podolsky and Rosen, are in conflict with some predictions of quantum mechanics (for an explicit formulation of these conditions, see paper [6], which builds on [7]). The other significant increase in the number of citations occurred around 2000, when entanglement and Bell's theorem proved to be essential to basic concepts and new ideas in quantum information science. I should mention that, as was often the case in the history of physics, this development towards future applications was not foreseen by any of the early pioneers. Both theoretical and experimental work on the foundations of quantum mechanics was originally motivated by nothing but curiosity, and without even the slightest indication of applications on the horizon.

We will now discuss the specific situation of two entangled photons emitted by a source (Quelle) in the state Φ_+ (Fig. 30.3).

$$|\Phi_+\rangle = \frac{1}{\sqrt{2}}(|H\rangle|H\rangle + |V\rangle|V\rangle) \tag{1}$$

This is a maximally entangled state, where measurements on the two photons always show the same linear polarization, horizontal H or vertical V, independent of the measurement basis chosen, that is, along any direction as long as it is the same for both photons. Such states today are called "Bell states".

Albert Einstein most succinctly analyzed the situation in "Remarks to the Essays Appearing in this Collective Volume", his reply to the articles in the famous volume *Albert Einstein—Philosopher-Scientist* [8]. This volume, which has already been mentioned by Bertlmann in an earlier chapter of the current volume [9], contains a great collection of interesting papers relating to many aspects of Einstein's work. For example, it contains the famous Gödel paper, where he shows for the first time that there could be closed time-like curves in the Universe [10]. Many open questions that were raised in these papers, including some in mathematics, have not been resolved to date.

Einstein in his reply basically argues that when performing a measurement on photon 1, say the one on the left-hand side in Fig. 30.3, we can freely decide along which direction we measure, for example its linear polarization. If one decides to measure the polarization along x, then, Einstein argues, one definitely gets an x polarization eigenstate for photon 2 on the right-hand side. It is certainly unknown and quantum mechanically maximally uncertain whether it will be a +x or a −x eigenstate, but it definitely is an eigenstate to the x basis and not to the y-basis. On the other hand, if we decide to measure the polarization of photon 1 along the y direction, we will definitely get a y eigenstate for the other photon. Again, either +y or −y (z is the propagation direction). Therefore, the choice of measurement basis on one photon decides what kinds of eigenstates are permitted for the other photon, and this is a definite property of the system upon the first measurement. Therefore, as Einstein said, "the quantum state cannot describe the 'real factual situation'" [8]. In my opinion, this statement is absolutely correct. The question is, what is its significance? In my eyes, it can be interpreted that the quantum state is not about a reality existing prior to or independent of measurement, but it is the representation of information about possible future measurement results. It allows the maximal set of, in general probabilistic, predictions of future measurement results.

The essence of Bell's theorem is that for certain correlations between measurements on both photons, the predictions of quantum mechanics are in conflict with the philosophical position of local realism, as exposed by EPR. This is signified by Bell's inequality.

$$E_{11} + E_{12} + E_{21} - E_{22} \leq 2 \tag{2}$$

where E_{ab} describes the correlation between measurement results when photon 1 is measured along direction a and photon 2 is measured along direction b. Without going into detail, the implication is that the sum of correlations, as expressed in Eq. 2, between measurements on both sides is limited to 2 in a local realist point of view. But for quantum mechanics, the sum on the left-hand side can be as high as $2\sqrt{2}$ for two qubits, a qubit being any quantum mechanical two-state system.

Many contributions to the current conference refer to specific reformulations of this inequality and its variants, but the essential point is always the same, as outlined above.

The Role of Information as Underlined by Entanglement Swapping

The role of information in the understanding of quantum mechanics can most clearly be discussed for a variation of entanglement swapping [11] as proposed by Peres [12]. We have two EPR sources (Fig. 30.4), each one producing an entangled pair of photons. One of the photons of each source, photon 1 and photon 4, is measured locally. In some polarization bases, each basis is chosen arbitrarily and independently by Alice or Bob respectively. The results are recorded and printed out as shown. Photons 2 and 3, one from each source, are then directed to Victor, who can decide whether he wants to project these two photons onto an entangled state or whether he wants to measure their polarizations separately. These decisions can be made by Victor randomly and at any time before or after the outer two photons have been registered.

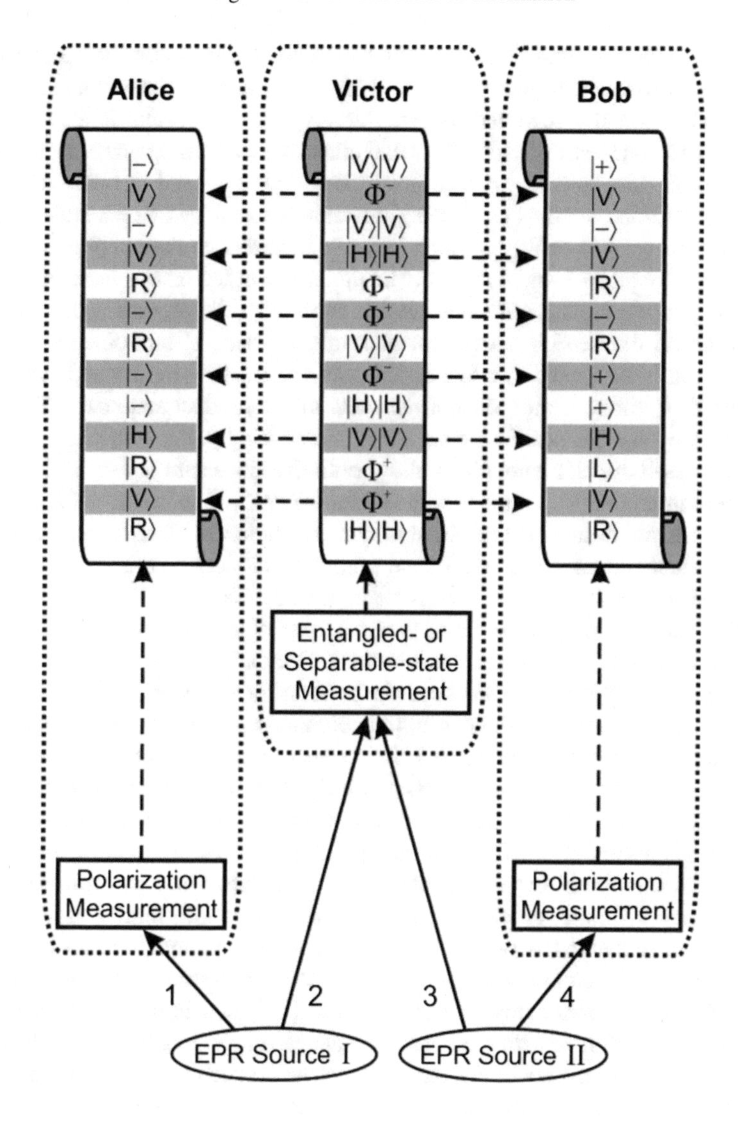

Fig. 30.4 The concept of delayed-choice entanglement swapping. Two entangled pairs—photons 1 and 2 and photons 3 and 4—are produced in the state $|\Psi^-\rangle_{12}\otimes|\Psi^-\rangle_{34}$ in the EPR sources I and II respectively. At first, Alice and Bob perform polarization measurements on photons 1 and 4, choosing freely the polarization analysis basis among three mutually unbiased bases (*horizontal/vertical* $|H\rangle/|V\rangle$, *right-circular/left-circular* $|R\rangle/|L\rangle$, *plus/minus* $|+\rangle/|-\rangle$, and record the outcomes. Photons 2 and 3 are sent to Victor, who then subjects them to either an entangled-state measurement or a separable-state measurement. He projects them randomly into one of two possible maximally entangled Bell states ($|\Phi^+\rangle_{23}$ or $|\Phi^-\rangle_{23}$) or one of two separable states ($|HH\rangle_{23}$ or $|VV\rangle_{23}$). Victor records the outcome and keeps it to himself. This procedure projects photons 1 and 4 onto a corresponding entangled ($|\Phi^+\rangle_{14}$ or $|\Phi^-\rangle_{14}$) or separable state ($|VV\rangle_{14}$ or $|HH\rangle_{14}$) respectively. According to Victor's choice and his results, he can sort Alice's and Bob's already recorded data into subsets and can verify that each subset behaves as if it consisted of either entangled or separable pairs of distant photons, which have neither communicated nor interacted in the past. From [16].

We consider the case that when Victor makes his decision and performs the respective measurement, photons 1 and 4 have long been registered and the results have been classically recorded. A possible sequence of results is also shown in Fig. 30.4. The interesting point now is the following. Depending on the measurement that Victor decides to perform, the already long ago registered records for photon 1 and 4 can be interpreted as confirming the entanglement by using a suitable entanglement witness or as being consistent with the factorizable state of photons 1 and 4. But, clearly, photons 1 and 4 can only be either entangled or in a factorizable state.

Depending on the kind of measurement chosen by Victor, entangled or not, and depending on the specific measurement result, the sets of data obtained by Alice and Bob can be separated into four independent data sets in the entangled state case and into four different independent data sets in the product state case. These four data sets are very different in the two cases. Interestingly, the data sets obtained by Alice and Bob are rich enough to allow both divisions into subsets. Therefore, a consistent interpretation of the recorded data emerges. This is sometimes seen as a puzzle, as it might appear that the decision by Victor whether to project onto an entangled state or onto a product state might modify the results already obtained and classically recorded. Yet this is certainly not the case. The data have been recorded and are not changed. But their interpretation is missing until Victor has made his decision and done the measurement. Only after that measurement, Alice's and Bob's already long ago recorded data obtain their proper interpretation. Peres points out very clearly that there is nothing contradictory about the situation. It is all consistent with quantum mechanics. In a sense, one might say that the data are somehow the most primitive and basic concept in a quantum measurement. Their interpretation can certainly depend on future actions, as is the case here.

In a way to see the situation from the point of view of information, one can say that the information in the four-photon state is limited. The four photons are only able to carry a limited amount of information. If this information is information implying entanglement between 1–2 and 3–4, then it cannot also be information implying entanglement between 1–4 and 2–3. It can never be both at the same time. The well-known monogamy of entanglement [13] thus is a consequence of the limited amount of information a quantum state can carry. It is interesting to see that the original quantum state of all four photons can be decomposed into sums of four product states of maximally entangled Bell states,

$$|\Psi\rangle_{1234} = |\Psi_-\rangle_{12} \otimes |\Psi_-\rangle_{34} = \frac{1}{2}\left(|\Psi_+\rangle_{14} \otimes |\Psi_+\rangle_{23} - |\Psi_-\rangle_{14} \otimes |\Psi_-\rangle_{23} - |\Phi_+\rangle_{14} \otimes |\Phi_+\rangle_{23} + |\Phi_-\rangle_{14} \otimes |\Phi_+\rangle_{23}\right), \tag{3}$$

where

$$\begin{aligned} |\Psi_{+,-}\rangle_{ab} &= \frac{1}{\sqrt{2}} |\mathrm{H}\rangle_a |\mathrm{V}\rangle_b \pm |\mathrm{V}\rangle_a |\mathrm{H}\rangle_b, \\ |\Phi_{+,-}\rangle_{ab} &= \frac{1}{\sqrt{2}} |\mathrm{H}\rangle_a |\mathrm{H}\rangle_b \pm |\mathrm{V}\rangle_a |\mathrm{V}\rangle_b. \end{aligned} \tag{4}$$

Entanglement swapping can also be seen as teleportation of an entangled state. It is either the teleportation of the state of photon 2 over to photon 4 or the teleportation of photon 3 over to photon 1.

The first experimental realization was done in Vienna [14, 15]. In these experiments, the situation was static, and no special precaution was taken to perform the Bell measurement after the independent photons had been registered. More recently, the proposal of Peres was realized [16] by measuring one photon from each entangled pair separately and sending the other photon from each pair into a rapidly tunable Mach-Zehnder interferometer (Fig. 30.5).

When the internal phase of the interferometer is set to an integer multiple of π, registration at the final detectors allows to conclude from which source each of the photons 2 and 3 came. And thus, in that case, they are projected onto a product state. In contrast, when the internal phase is set to, say, $\pi/2$, no conclusion is possible about the source of the incoming photons 2 and 3. Thus, they are projected onto a maximally entangled state. Consequently, also the earlier registered photons 1 and 4 are projected either into a product state or into a maximally entangled state. It should also be mentioned that the registration, and even the setting of the internal phase, was done at a time after the other two photons had been registered already. When photons 2 and 3 were projected onto an entangled state, an entanglement witness confirmed the existence of entanglement between photons 1 and 4. This witness was basically a measurement of the correlations in three mutually unbiased bases. In the case of a projection of photons 2 and 3 onto a product state, polarization correlation of photons 1 and 4 was only observed in one basis, confirming that there was no entanglement between them.

To analyze the situation from a conceptual point of view, we realize that the individual polarization registration events of photons 1 and 4 are classical long before the complete measurement, including photons 2 and 3, is finished. But only the complete measurement on photons 2 and 3 allows to determine the quantum state for all four photons, and therefore allows to fully interpret the situation. It is therefore suggestive that the events are a more fundamental reality than the quantum states which are only determined in the end.

The Quantum State as Representation of Information

Let us now come back to the famous double-slit experiment [17]. The message of that experiment is again related to the notion of information. Interference occurs if, and only if, no path information exists anywhere in the Universe about which of the two paths the particle took. It is irrelevant whether an observer takes note of that information. The quantum state then is just a representation of possible future measurement results or, following Schrödinger [2], an expectation catalogue (Fig. 30.6).

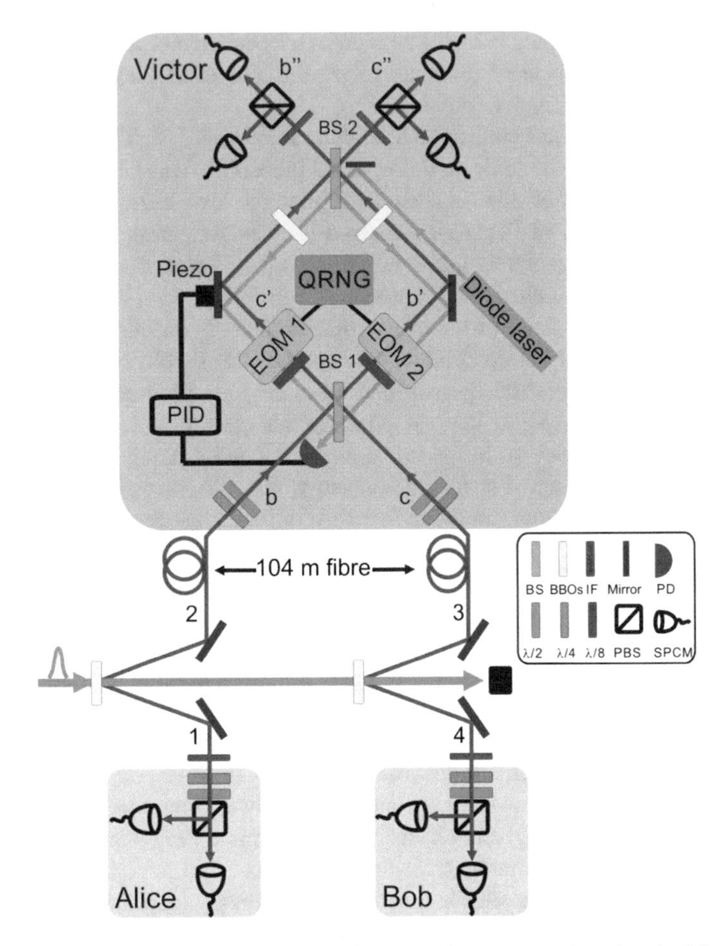

Fig. 30.5 Experimental set-up of delayed-choice entanglement swapping. A pulsed laser beam with a central wavelength of 404 nm, a pulse duration of 180 fs, and a repetition rate of 80 MHz successively passes through two BBO crystals to generate two polarization-entangled photon pairs (photons 1 and 2 and photons 3 and 4) through type-II spontaneous parametric down-conversion. Photons 1 and 4 are directly subjected to the polarization measurements performed by Alice and Bob (*green blocks*). Photons 2 and 3 are each delayed with a 104 m single-mode fiber and then coherently overlapped at the input beam splitter BS1 of a tunable Mach-Zehnder interferometer acting as a switchable interferometric Bell-state analyzer (*purple block*). Depending on the internal phase of the interferometer, it operates either as a Bell-state measurement device or it does not superpose the amplitudes of the two photons coming from the two down-conversion crystals. In the first case, photons 2 and 3 are projected onto an entangled state. In the second case, each photon's polarization is measured individually by Victor. The choice between the two measurements is made by a fast quantum random number generator at a time after the twin photons 1 and 4 have been measured. From [16].

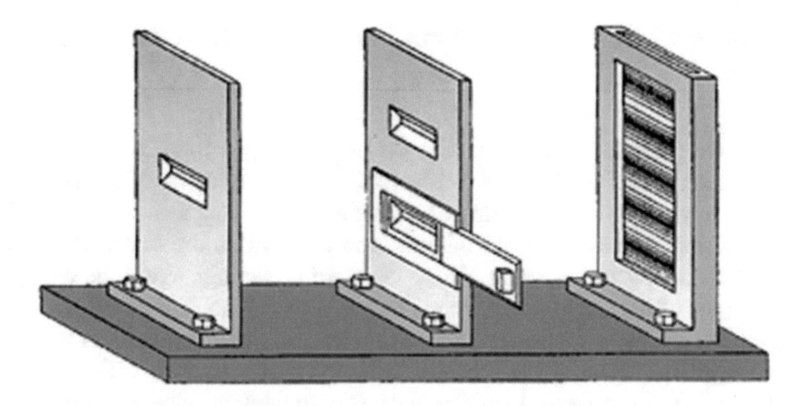

Fig. 30.6 Young's double-slit experiment as drawn by Niels Bohr in [17].

What is meant by information here is the possibility to obtain knowledge. Here, it is the possibility to obtain knowledge about the path taken. As long as that possibility exists, there cannot be any interference. All quantum eraser experiments [18] and their delayed-choice variants [19] indeed work by erasing just this possibility of obtaining knowledge about the path, i.e. the particle-like property. Finally, we remark that this knowledge could be obtained by any observer! It is irrelevant whether it is a conscious observer or not.

At about the same time when Bell discovered his theorem, Wigner expressed his position very clearly [20]:

> … it is impossible to give a satisfactory description of atomic phenomena (*i.e. by quantum mechanics*)[1] without reference to the consciousness. … the "reduction of the wave packet" … takes place when the result of an observation enters the consciousness of the observer … or, to be even more painfully precise, my own consciousness.

It is clear that this position differs significantly from Bell's view as expressed above. In contrast, Heisenberg wrote in a letter to Renninger on February 2 1960 [21]:

> The act of recording, on the other hand, which leads to the reduction of the state, is not a physical, but rather, so to say, a mathematical process. With the sudden change of our knowledge also the mathematical representation of our knowledge undergoes of course a sudden change.

I interpret Heisenberg's position—and in that respect, I share his point of view—in the way that there is no measurement problem. All that happens is a change of our representation of information, i.e. the quantum state. Concluding, I might suggest that it is most natural to change the representation of information, that is, the quantum state, when our knowledge changes because of new information obtained by measurement.

---

[1]Cursive text in brackets inserted by the author of the current essay.

Information and Quantum Imaging with Undetected Photons

Recently, we performed an experiment [22] which brings forth the role of information in a very succinct way. Let us first recall that scattering experiments are very central for physics. In any scattering experiment, the essential idea is the following: Some radiation interacts with the sample, and one detects the scattered radiation (certainly including transmission, that is, forward scattering). From the change of the properties of the scattered radiation compared to the incident radiation one can make inferences about properties of the scatterer. An important point is that in all such experiments hitherto, the scattered radiation that interacted with the sample must be registered somehow, by a film, a detector or another suitable device. This is even true for holography, where the scattered radiation is superposed with coherent radiation. But it is not true in the experiment [22] which we will now discuss.

The experiment builds on a beautiful paper by Zou et al. [23]. The paper acknowledges Jeff Ou for the suggestion of aligning NL1 and NL2 as to make the idler trajectories coincide. In our review of multi-particle interference in Physics Today in 1993 [24], Greenberger, Horne and myself called it a 'mind-boggling' experiment.

The salient features of the experiment are (Fig. 30.7): Two nonlinear crystals are pumped by the same pump. In each of the two crystals, photon pairs can be produced. Let us assume that the pump intensity is such that the production of two photon pairs is negligible. Consider therefore the case that one pair is produced either in crystal NL1 or in crystal NL2. Quantum mechanically speaking, the photon pair is produced in a superposition of both possibilities. The outgoing modes in each crystal are called idler i and signal s. If now the two modes of the signal photon are brought back together, there should be no interference in general, because the idler could be used to find out by which of the two crystals the pair was produced, thereby revealing path information.

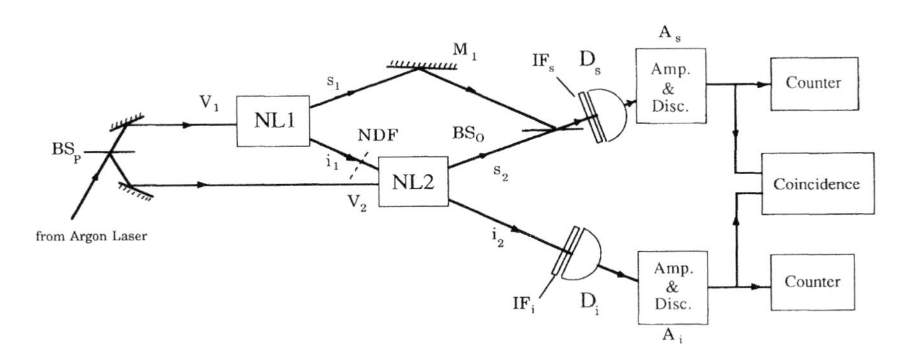

Fig. 30.7 Outline of the interference experiment of Zou, Wang and Mandel. From [23].

The central idea of the paper was to make the path information carried by the idler photon disappear. This was done with a very special trick, such that the mode of the idler coming from the first crystal after passage through the second one is identical to the idler mode coming directly from the second crystal. That way, no information about which crystal produced the pair can be found out by looking at the idler. Thus, no path information for its twin signal photon exists, and interference at the beam splitter may occur. We note that this situation is very different from a quantum eraser, which could for example be done by combining the two idler modes also at a beam splitter. In such a case, the idler would have to be detected to finally erase the source information. In the new situation, the detection of the idler photon is not necessary at all. Most importantly, the fringes for the signal occur not in coincidence, but as singles, in contrast to the usual ghost imaging.

I would suggest calling this phenomenon "coherence by identity". This is because the two signals for the idler are coherent for the very reason that the two modes of the idler are completely identical. Strictly speaking, it is actually wrong to talk about two modes for the idler. There is only one mode, which emerges from both crystals. Again, it is the absolute unavailability of information, and thus the impossibility to obtain path knowledge, which makes coherence and interference possible.

The experiment has various interesting, maybe even counter-intuitive properties. This can be seen, for example (Fig. 30.8), when inserting a phase-shifter into the idler mode between NL1 and NL2. The important point is that that phase shift cannot be attributed to the idler mode alone, even as it is introduced by the phase-shifter in the idler beam. Rather, it is a nonlocal phase shift experienced by the total two-photon state emerging from crystal NL1. It can therefore be revealed as the phase shift between the signal modes c and e meeting at the beam splitter BS2. Furthermore, if the object has some absorptive properties, full or partial, this provides path information between the signal modes c and e, which results in a reduction of the outgoing interference contrast in modes g and h.

We now discuss two imaging experiments which also underline salient features of the experimental set-up. Before discussing both experiments, we should realize once more that the idler in that experiment is not registered at all. It is just left to pass away freely. An experimental situation where the path information is directly relevant for obtaining the image occurs when an absorptive mass is placed into the idler beam. The object is a cardboard cut-out of the image of a cat. Light can freely pass through the cat opening, while outside, light is absorbed. This object is placed into the idler beam between the two crystals of Fig. 30.8. The image (Fig. 30.9) on the top left clearly shows that in both outgoing beams of the signal photon after the beam splitter, the cat image appears. We will now discuss briefly the image itself and how it relates to information. On the left-hand side, the cat is seen as bright, namely constructive, interference of the two signal beams. On the right-hand side, the cat is dark—destructive interference of the two signal beams. This is easily understood. The beam path was adjusted in such a way that for one of the two signal beams, constructive interference results, and for the other one, destructive interference arises. Outside of the cat images, we have a shadow region of the same brightness in both cases. This simply results from the fact that the idler mode

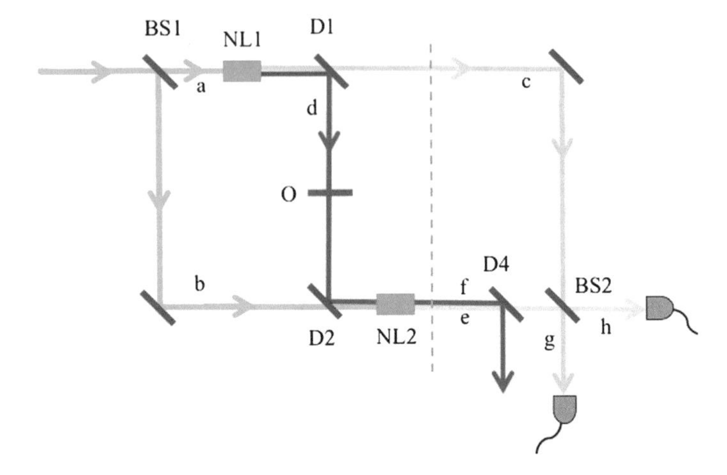

Fig. 30.8 Schematic of the experiment of quantum imaging with undetected photons. Laser light (*green*) splits at beam splitter *BS1* into modes *a* and *b*. Beam *a* pumps nonlinear crystal *NL1*, where collinear down-conversion may produce a pair of photons of different wavelengths called signal (*yellow*) and idler (*red*). After passing through the object *O*, the idler reflects at dichroic mirror *D2* to align with the idler produced in *NL2*, such that the final emerging idler *f* does not contain any information about which crystal produced the photon pair. As a consequence, signals *c* and *e* combined at beam splitter *BS2* interfere. Consequently, the superposition signal beams *g* and *h* reveal idler transmission properties of object *O*. From [22].

between crystal 1 and 2 is absorbed. Therefore, the idler emerging from the experiment could easily be used to find out that the pair was created in the second crystal. Likewise, we could consider the absorbing cardboard as a detector which, if it fires, tells us that the pair was created in the first crystal. Therefore, in that situation, one has path information outside the cat openings, and thus, no interference for the signal photon can arise. That is, the outside shady regions simply represent incoherent summations or incoherent mixtures of the two modes of the signal photon coming via the two paths. Thus, the existence or nonexistence of path information is clearly responsible for the production of the image. And, to stress that point again, that information is carried—or not—by the idler photon, which need not at all be deleted or erased.

A different physical argument leads to the imaging of phase objects (Fig. 30.10). There, an object introducing a phase shift of π was inserted into the idler mode between the two crystals. As discussed above, that phase cannot be considered to be carried by the idler beam alone. In fact, because the mode of the idler after emerging from the second crystal is the same whether the idler was created in either crystal, that phase cannot manifest itself in the emerging idler beam at all. Rather, it now manifests itself as a phase difference between the two signal modes superposed at the beam splitter. Again, the picture here is a consequence of firstly the fact that nowhere in the image, path information is available about where the pair was created, and secondly, it is a consequence of the fact that in a product state, the phase shift cannot be assigned to either member of the product.

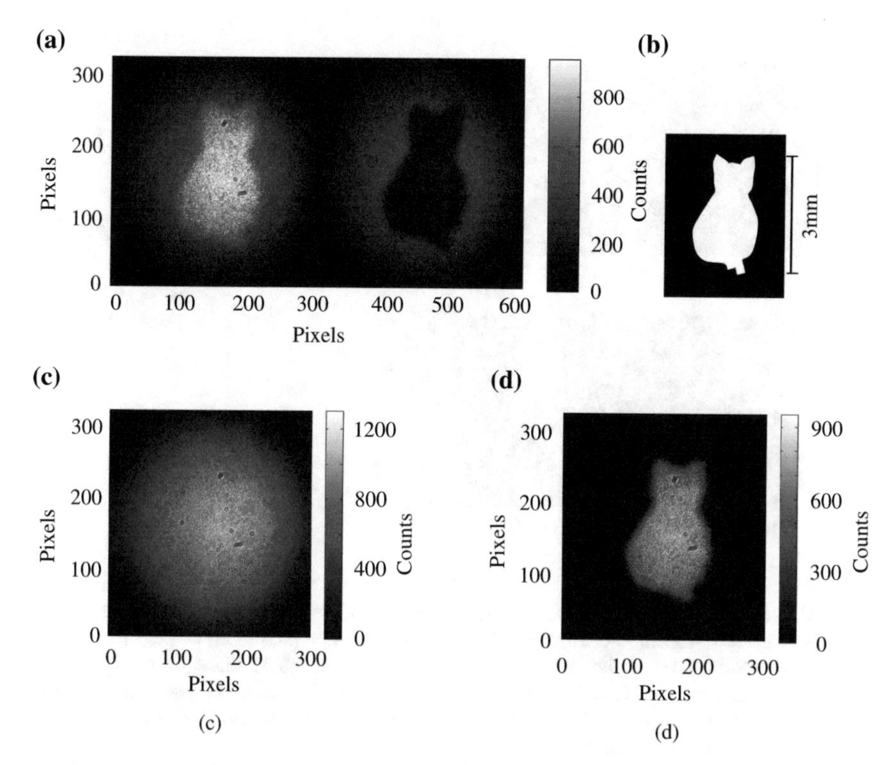

Fig. 30.9 Which-path imaging. **a** Inside the cat, constructive and destructive interference are observed at the two outputs of BS2 when we placed the cardboard cut-out shown in (**b**) into the idler path between NL1 and NL2. Outside the cat opening, the idler photon from NL1 is blocked and therefore the corresponding signal photon does not interfere, resulting in an unstructured background in both images in (**a**). **c** The sum of the outputs gives the intensity profile of the signal beam, not showing any effect of the absorption. **d** The subtraction of the outputs leads to an enhancement of the interference contrast, as outside of the cut-out, the two backgrounds completely subtract. The image arises because outside of the cut-out area, path information is available, while inside, this is not the case. From [22].

Finally, a small remark on entanglement might be in order. It is not irrelevant that the state used in the experiment is a high-dimensionally entangled state of superposed pairs created at different locations transversely to the pump beams.

Conclusion

I have given a number of examples which underline the role of information in quantum mechanics. This is part of an emerging view where it appears that information has a much more fundamental role in quantum physics than realized hitherto in general. I am very confident that while John Bell would not have liked

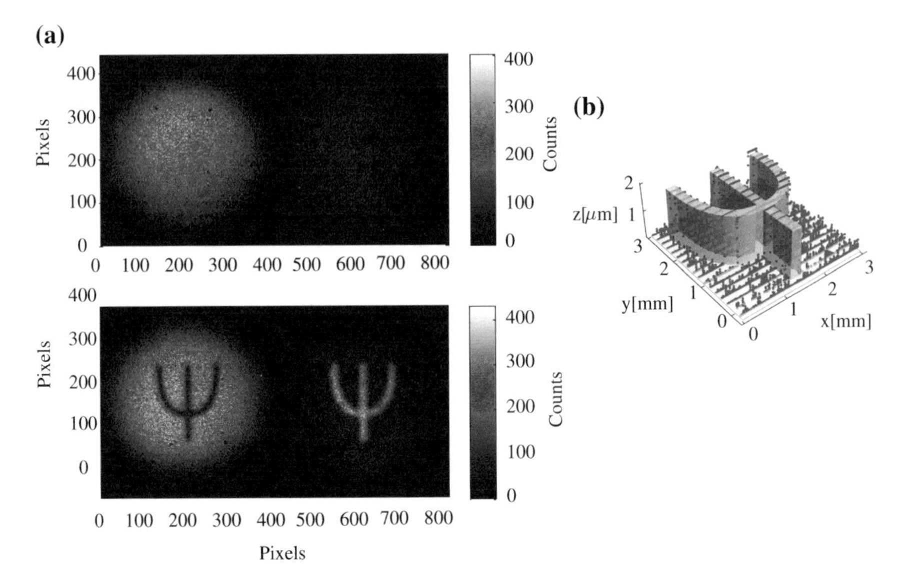

Fig. 30.10 Phase imaging of a phase step. The object **b** is etched into silicon such that it provides a π phase shift at 1515 nm, the wave length of the idler, or a 2π phase step at 820 nm, the wave length of the signal. The top picture (*top* of **a**) was taken with the object placed in the signal beam after the nonlinear crystal NL1. No image results. Yet, when the object is placed into the idler beam between NL1 and NL2, a very well defined phase image is obtained in the emerging signal beam (*bottom* of (**a**)). From [22].

Fig. 30.11 Photograph taken in Amherst, Massachusetts, in 1990. From *right to left*; John Bell, Mary Bell, AZ and Arthur Zajonc. In the *lower left corner*, you can see a beer, held by Mike Horne. (*Photo* Kurt Gottfried).

Fig. 30.12 The *Quantum [Un]Speakables* conference series. (Graphic design *Quantum [Un] Speakables* I: Julia Petschinka; graphic design *Quantum [Un]Speakables* II: Raoul Krischanitz).

the emerging increasing role of measurement or information in the interpretation of quantum mechanics (see e.g. [25]), he would certainly have loved to see the modern experiments. Such experiments also include quantum interference for higher and higher quantum numbers, as for example witnessed in the entanglement of orbital angular momentum states [26]. Since these experiments were performed, the quantum numbers for entanglement have been raised to beyond 10.000 $\hbar$.

In the end, I would like to emphasize my deep appreciation and admiration of John Bell. I am very grateful for having met him personally, and having had the opportunity to discuss with him a couple of times our views of the foundations of quantum mechanics, even if they differed significantly (Fig. 30.11).

As you know, this is the second conference on *Quantum [Un]Speakables*. The first conference took place in 2000, to commemorate the tenth anniversary of John Bell's death. The present conference, in 2014, celebrates 50 years of Bell's theorem. It is therefore natural to see this as the start of a longer series. The next conference, again in 14 years' time, will therefore take place in 2028, and we will commemorate the 100th birthday of John Stewart Bell (Fig. 30.12).

Acknowledgments This work was supported by the Austrian Science Fund (FWF) with SFB F40 (FoQuS).

References

1. A. Einstein, B. Podolsky, N. Rosen, Can quantum-mechanical description of physical reality be considered complete? Phys. Rev. **47**, 777 (1935)
2. E. Schrödinger, *Die gegenwärtige Situation in der Quantenmechanik*, Naturwissenschaften **23**, 823–828 (1935). (Translation: E. Schrödinger, The present situation in quantum mechanics. Proc. Am. Philos. Soc. **124**, 323–38 (1980))

3. E. Schrödinger, Discussion of probability relations between separated systems. Proc. Camb. Philo. Soc. **31**, 555–563 (1935)
4. N. Bohr, Can quantum-mechanical description of physical reality be considered complete? Phys. Rev. **48**, 696 (1935)
5. J. Bell, On the einstein podolsky rosen paradox. Physics **1**, 195–200 (1964)
6. D.M. Greenberger, M.A. Horne, A. Shimony, A. Zeilinger, Bell's theorem without inequalities. Am. J. Phys. **58**, 1131–1143 (1990)
7. D. Greenberger, M.A. Horne, A. Zeilinger, Going beyond Bell's Theorem, in *Bell's Theorem, Quantum Theory, and Conceptions of the Universe*, ed. by M. Kafatos (Kluwer Academic, 1989), pp. 73–76
8. A. Einstein, Remarks to the essays appearing in this collective volume, in *Albert Einstein, Philosopher-Scientist,* ed. by P.A. Schilpp (Tudor, New York, 1949), pp. 663–688
9. R. Bertlmann, Bell's universe: a personal recollection, *in Quantum [Un]Speakables II*, ed. by R. Bertlmann and A. Zeilinger (2015, this volume)
10. K. Gödel, A remark about the relationship between relativity theory and idealistic philosophy, in *Albert Einstein, Philosopher-Scientist*, ed. by P.A. Schilpp (Tudor, New York, 1949), pp. 555–562
11. M. Zukowski, A. Zeilinger, M.A. Horne, A.K. Ekert, "Event-ready-detectors" Bell experiment via entanglement swapping. Phys. Rev. Lett. **71**, 4287 (1993)
12. A. Peres, Delayed choice for entanglement swapping. J. Mod. Opt. **47**, 139–143 (2000)
13. V. Coffman, J. Kundu, W.K. Wooters, Distributed entanglement. Phys. Rev. A **61**, 052306 (2000)
14. J.W. Pan, M. Daniell, S. Gasparoni, G. Weihs, A. Zeilinger, Experimental demonstration of four-photon entanglement and high-fidelity teleportation. Phys. Rev. Lett. **86**, 4435 (2001)
15. T. Jennewein, G. Weihs, J.W. Pan, A. Zeilinger, Experimental nonlocality proof of quantum teleportation and entanglement swapping. Phys. Rev. Lett. **88**, 017903 (2001)
16. X.S. Ma, S. Zotter, J. Kofler, R. Ursin, T. Jennewein, Č. Brukner, A. Zeilinger, Experimental delayed-choice entanglement swapping. Nat. Phys. **8**, 479–484 (2012)
17. N. Bohr, Discussions with Einstein on epistemological problems in atomic physics, in *Albert Einstein: Philosopher-Scientist*, eds. P.A. Schilpp (Tudor, New York, 1949), pp. 32–66
18. M.O. Scully, K. Drühl, Quantum eraser: a proposed photon correlation experiment concerning observation and "delayed choice" in quantum mechanics. Phys. Rev. A **25**, 2208 (1982)
19. X.S. Ma, J. Kofler, A. Zeilinger, Delayed-choice gedanken experiments and their realizations. Rev. Mod. Phys. **88**, 015005 (2016)
20. E.P. Wigner, Two kinds of reality. Monist **48**, 248 (1964)
21. W. Heisenberg, Letter to Renninger, in *The Philosophy of Quantum Mechanics*, ed. M. Jammer (Wiley, New York, 1974)
22. G.B. Lemos, V. Borish, G.D. Cole, S. Ramelow, R. Lapkiewicz, A. Zeilinger, Quantum imaging with undetected photons. Nature **512**, 409–412 (2014)
23. X.Y. Zou, L.J. Wang, L. Mandel, Induced coherence and indistinguishability in optical interference. Phys. Rev. Lett. **67**, 318 (1991)
24. D.M. Greenberger, M.A. Horne, A. Zeilinger, Multiparticle interferometry and the superposition principle. Phys. Today **46**, 22–29 (1993)
25. J. Bell, Against measurement. Phys. World **8**, 33–40 (1990)
26. R. Fickler, R. Lapkiewicz, M. Huber, M.P. Lavery, M.J. Padgett, A. Zeilinger, Interface between path and orbital angular momentum entanglement for high-dimensional photonic quantum information. Nat. Commun. **5**, 4502 (2014)

Appendix A
Contributions to the Conference

This collection of essays draws on contributions to the conference "Quantum [Un]Speakables II: 50 Years of Bell's Theorem", which was organized by the editors of this volume and took place June 19–22, 2014.

The contributions at the conference were:

Invited talks[1]
In chronological order.

Address to the conference participants
Mary Bell
European Organization for Nuclear Research (CERN), Geneva, Switzerland

Putting the scientist into the science
N. David Mermin
Cornell University, NY, USA

Xtreme nonlocality
Bradley G. Christensen and **Paul G. Kwiat**
University of Illinois at Urbana-Champaign, IL, USA

Quantum correlations in Newtonian space and time: faster than light communication or nonlocality
Nicolas Gisin
University of Geneva, Switzerland
Nicolas Gisin was unable to present his talk in person due to family reasons. Valerio Scarani presented the talk in his place.

[1]Invited speakers are marked in bold.

R. Bertlmann and A. Zeilinger (eds.), *Quantum [Un]Speakables II*,
The Frontiers Collection, DOI 10.1007/978-3-319-38987-5

Can quantum-mechanical description of causal relations be considered complete?
Caslav Brukner
Austrian Academy of Sciences, Austria

Designing Bell inequalities via Tsirelson bounds
Michael Epping, Hermann Kampermann and **Dagmar Bruß**,
Heinrich Heine University Düsseldorf, Germany

A GHZ experiment under strict Einstein locality conditions
Gregor Weihs
University of Innsbruck, Austria

On closing loopholes in Bell experiments
Sven Ramelow
Austrian Academy of Sciences, Austria

Search for hidden variables in neutron experiments
Helmut Rauch
Vienna University of Technology, Austria

Testing Bell's theorem in high energy physics
Beatrix Hiesmayr
University of Vienna, Austria

Quantum optics experiments using satellites
Rupert Ursin
Austrian Academy of Sciences, Austria

A full state in a single number
Valerio Scarani
National University of Singapore, Singapore

Physics above and below the Bell horizon: re-examining quantum foundations and glimpsing the post-quantum world via photonics
Andrew White
University of Queensland, Australia

Superconducting Xmon qubits with gate fidelity at the surface code threshold
John M. Martinis
University of California at Santa Barbara, CA, USA

The maximally entangled set of multipartite quantum states
Cornelia Spee[1], Julio Iñigo de Vicente Majúa[2] and **Barbara Kraus**[1]
[1]University of Innsbruck, Austria
[2]Charles III University of Madrid, Spain

Some Bell's theorem test loopholes added in the last 36 years
John F. Clauser
J. F. Clauser and Associates, CA, USA

Entanglement in massive systems: what do we learn?
Markus Aspelmeyer
University of Vienna, Austria

On spatial entanglement wave functions
Michael Horne
Stonehill College, MA, USA

Whose *information? Information about* what?
Jeffrey Bub
University of Maryland, MD, USA

Magic moments with John Bell: collaboration and friendship
Reinhold Bertlmann
University of Vienna, Austria

Public lecture: *From Einstein's intuition to quantum bits: a new quantum age*
Alain Aspect
École supérieure d'optique, France

Analyzing multiparticle quantum states: problems and solutions
Otfried Gühne[1], Matthias Kleinmann[2] and Tobias Moroder[1]
[1]University of Siegen, Germany
[2]University of the Basque Country, Spain

Quantum non-locality: a resource for information processing
Antonio Acín[1,2] and Miguel Navascués[3]
[1]The Institute of Photonic Sciences (ICFO), Spain
[2]Catalan Institution for Research and Advanced Studies (ICREA), Spain
[3]University of Bilkent, Turkey

On causal explanations of quantum correlations
Robert Spekkens
Perimeter Institute for Theoretical Physics, ON, Canada

Non-locality?—It ain't necessarily so
Marek Zukowski
University of Gdańsk, Poland

My struggle to face up to un-reality
Terence Rudolph
Imperial College London, UK

Heralded entanglement between distant atoms. Towards a loophole free test of Bell's inequality?
Harald Weinfurter
Ludwig Maximilian University of Munich, Germany

Quantum correlations: where, how and why
Adán Cabello
University of Sevilla, Spain

Bell violation with entangled photons, free of the fair-sampling assumption
Marissa Giustina[1,2], Alexandra Mech[1,2], Sven Ramelow[1,2], Bernhard Wittmann[1,2], Johannes Kofler[1,3], Jörn Beyer[4], Adriana Lita[5], Brice Calkins[5], Thomas Gerrits[5], Sae Woo Nam[5], Rupert Ursin[1] and Anton Zeilinger[1,2]
[1]Austrian Academy of Sciences, Austria
[2]University of Vienna, Austria
[3]Max Planck Institute of Quantum Optics (MPQ), Germany
[4]Physikalisch-Technische Bundesanstalt, Germany
[5]National Institute of Standards and Technology (NIST), MD, USA

The dynamical roles played by mass and proper time in physics
Daniel Greenberger
City College of New York, NY, USA

Causation and the two Bell's theorems of John Bell
Howard Wiseman[1] and Eric G. Cavalcanti[2]
[1]Griffith University, Australia
[2]University of Sydney, Australia

The freedom of choice assumption and its implications
Renato Renner
Swiss Federal Institute of Technology (ETH), Switzerland

Bell violation with entangled photons, free of the coincidence-time loophole
Jan-Åke Larsson
Linköping University, Sweden

John Bell and quantum information theory
Andrew Whitaker
Queen's University Belfast, UK

Steering, and maybe why Einstein didn't go all the way to Bell's argument
Reinhard F. Werner
Leibniz University of Hanover, Germany

Quantum mechanics in a new key
Simon B. Kochen
Princeton University, NJ, USA

New dimensions for entangled photons
Anton Zeilinger
Austrian Academy of Sciences, Austria

Poster presentations[2]

Quantum circuits cannot control unknown operations
Mateus Araújo
University of Vienna, Austria

Probing macroscopic realism via Ramsey correlations measurements
Ali Asadian[1], Caslav Brukner[2] and Peter Rabl[1]
[1]Vienna University of Technology, Austria
[2]Austrian Academy of Sciences, Austria

Non-locality of the multipartite W state upon loosing parties
Jean-Daniel Bancal[1], **Tomer Jack Barnea**[2], Nicolas Brunner[3], Péter Diviánszky[3], Nicolas Gisin[2], Tamás Vértesi[3] and Yeong-Cherng Liang[4]
[1]National University of Singapore, Singapore
[2]University of Geneva, Switzerland
[3]Hungarian Academy of Sciences, Hungary
[4]Swiss Federal Institute of Technology (ETH), Switzerland

Tripartite quantum state violating the hidden influence constraints
Tomer Jack Barnea[1], Jean-Daniel Bancal[2], Yeong-Cherng Liang[1,3] and Nicolas Gisin[1]
[1]University of Geneva, Switzerland
[2]National University of Singapore, Singapore
[3]Swiss Federal Institute of Technology (ETH), Switzerland

Quantifying the non-locality of experimental qutrits without Bell inequalities
Bänz Bessire[1], Marcel Pfaffhauser[2], Christof Bernhard[1], Alberto Montina[2], André Stefanov[1] and Stefan Wolf[2]
[1]University of Bern, Switzerland
[2]University of Lugano, Switzerland

[2]Presenting authors are marked in bold.

Compressed simulation of evolutions of the XY-model
Walter L. Boyajian, Valentin Murg and Barbara Kraus
University of Innsbruck, Austria

Fast and efficient detection of atomic states for a conclusive test of Bell's inequality
Daniel Burchardt, Norbert Ortegel, Kai Redeker, Robert Garthoff, Markus Rau, Michael Krug, Markus Weber, Wenjamin Rosenfeld and Harald Weinfurter
Ludwig Maximilian University of Munich, Germany
and Max-Planck-Institute for Quantum Optics, Germany

Detecting entanglement and nonlocality in path-entangled states
Valentina Caprara Vivoli[1], Jean-Daniel Bancal[2], Charles Ci Wen Lim[1], Nicolas Sangouard[1], Pavel Sekatski[3] and Nicolas Gisin[1]
[1]University of Geneva, Switzerland
[2]National University of Singapore, Singapore
[3]University of Innsbruck, Austria

Analysis of entanglement photons generated by four-wave mixing in a nonlinear ring resonator
Thitanan Chittha[1], Surasak Chiangga[1] and Till D. Frank[2]
[1]Kasetsart University, Thailand
[2]University of Connecticut, CT, USA

On the spatial locations of quantum systems
John V. Corbett
Macquarie University, Australia

Entanglement properties of hypergraph states
Marti Cuquet[1], Otfried Gühne[2], Frank E.S. Steinhoff[2], Tobias Moroder[2], Matteo A.C. Rossi[3], D. Bruß[4], B. Kraus[1] and C. Macchiavello[5]
[1]University of Innsbruck, Austria
[2]University of Siegen, Germany
[3]University of Milan, Italy
[4]Heinrich Heine University Düsseldorf, Germany
[5]University of Pavia, Italy

Multipartite quantum correlations—indefinite causal order and delocalized nonlocality
Florian Curchod[1], Yeong-Cherng Liang[2,1] and Nicolas Gisin[1]
[1]University of Geneva, Switzerland
[2]Swiss Federal Institute of Technology (ETH), Switzerland

Qudit based on the orbital angular momentum of light: an experimental tool for fundamental quantum mechanics
Vincenzo D'Ambrosio[1], Adán Cabello[2] and Fabio Sciarrino[1]
[1]Sapienza University of Rome, Italy
[2]University of Sevilla, Spain

Entanglement and nonlocality are inequivalent for any number of particles
Remigiusz Augusiak, **Maciej Demianowicz**, Jordi Tura and Antonio Acín
The Institute of Photonic Sciences (ICFO), Spain

Time domain matter-wave interferometry with nanoparticles
Nadine Dörre, Philipp Haslinger, Philipp Geyer, Jonas Rodewald and Markus Arndt
University of Vienna, Austria

Multisetting Bell inequalities for N spin-1 systems avoiding the Kochen-Specker contradiction
Arijit Dutta, Marcin Wiesniak and Marek Zukowski
University of Gdańsk, Poland

A Kolmogorovian account of probabilistic contextuality
Ehtibar Dzhafarov[1] and Janne Kujala[2]
[1]Purdue University, IN, USA
[2]University of Jyväskylä, Finland

Quantum interference experiments with complex molecules
Sandra Eibenberger, Joseph Cotter, Xiaxi Cheng, Lukas Mairhofer and Markus Arndt
University of Vienna, Austria

A construction of tripartite Bell inequalities
Michael Epping, Hermann Kampermann and Dagmar Bruß
Heinrich Heine University Düsseldorf, Germany

A glimpse into the post-quantum world: simulating stronger-than quantum correlations in photonic experiments
Martin Ringbauer, Dominic Berry, Andrew G. White and **Alessandro Fedrizzi**
University of Queensland, Australia

Indefinite causal order in multipartite scenarios
Adrien Feix
University of Vienna, Austria

EPR paradox and quantum steering in a multimode optomechanical system
Qiongyi He[1,2] and **Zbigniew Ficek**[3]
[1]Peking University, PR China
[2]Collaborative Innovation Center of Quantum Matter, PR China
[3]The National Centre for Mathematics and Physics (KACST), Saudi Arabia

Interface between path and OAM entanglement for high-dimensional photonic quantum information
Robert Fickler[1,2], Radek Lapkiewicz[1,2], Marcus Huber[3,4], Martin Lavery[5], Miles Padgett[5] and Anton Zeilinger[1,2]
[1]University of Vienna, Austria
[2]Austrian Academy of Sciences, Austria
[3]Autonomous University of Barcelona, Spain
[4]The Institute of Photonic Sciences (ICFO), Spain
[5]University of Glasgow, UK

Quantum entanglement of complex photon polarization patterns in vector beams
Robert Fickler, Radek Lapkiewicz Sven Ramelow and Anton Zeilinger
Austrian Academy of Sciences, Austria

Are past events still real if we forget them?
Daniela Frauchiger and Renato Renner
Swiss Federal Institute of Technology (ETH), Switzerland

Implementing quantum control for unknown subroutines
Nicolai Friis, Vedran Dunjko, Wolfgang Dür and Hans J. Briegel
Austrian Academy of Sciences, Austria
and University of Innsbruck, Austria

Classical microwaves as a universal model system for quantum contextuality
Diego Frustaglia, Jose-Pablo Baltanas, María-C Velazquez, Armando Fernandez, Vicente Losada, Manuel-Jose Freire and Adán Cabello
University of Sevilla, Spain

Violations of entropic Bell inequalities with coarse-grained quadrature measurements for continuous-variable states
Zeng-Bing Chen, **Yao Fu** and Yu-Kang Zhao
University of Science and Technology of China, PR China

(In)definite causal order with n *parties*
Christina Giarmatzi[1] and Ognyan Oreshkov[2]
[1]École polytechnique, France
[2]Université libre de Bruxelles, Belgium

Genuinely multipartite entangled states and orthogonal arrays
Dardo Goyeneche[1] and Karol Życzkowski[2]
[1]University of Concepción, Chile
[2]Jagiellonian University, Poland

Experimental test of a four-party GHZ-theorem
Chiara Greganti
University of Vienna, Austria

Almost quantum correlations
Miguel Navascués[1,2], **Yelena Guryanova**[1], Matty J. Hoban[3] and Antonio Acín[3,4]
[1]University of Bristol, UK
[2]Autonomous University of Barcelona, Spain
[3]The Institute of Photonic Sciences (ICFO), Spain
[4]Catalan Institution for Research and Advanced Studies (ICREA), Spain

Error-disturbance uncertainty relation studied in successive spin-measurements
Yuji Hasegawa
Vienna University of Technology, Austria

Dynamics of genuine three-mode quantum steering in an optomechanical system
Meng Wang[1,2], **Qiongyi He**[1,2] and Zbigniew Ficek[3]
[1]Peking University, PR China
[2]Collaborative Innovation Center of Quantum Matter, PR China
[3]The National Centre for Mathematics and Physics (KACST), Saudi Arabia

Towards a loophole-free Bell test with spin qubits in diamond
Bas Hensen, Hannes Bernien, Wolfgang Pfaff, Machiel S. Blok, Lucio Robledo, Tim H. Taminiau and Ronald Hanson
Delft University of Technology, Netherlands

Entanglement swapping over a 143 km free-space link
Thomas Herbst, Bernhard Wittmann, Rupert Ursin, Anton Zeilinger, Thomas Scheidl, Matthias Fink and Johannes Handsteiner
Austrian Academy of Sciences, Austria

Quantum entanglement and teleportation in pulsed cavity optomechanics
Sebastian G. Hofer[1,2], Witlef Wieczorek[1], Markus Aspelmeyer[1] and Klemens Hammerer[2]
[1]University of Vienna, Austria
[2]Leibniz University of Hanover, Germany

Negative probabilities as deterministic relations between observable properties: Why quantum physics must violate inequalities
Holger F. Hofmann
University of Hiroshima, Japan

Experimental evaluation of error and disturbance of Ozawa's inequality in two-level systems
Masataka Iinuma, Yutaro Suzuki, Ryuji Kinoshita and Holger F. Hofmann
University of Hiroshima, Japan

Towards the creation and detection of momentum entangled atom pairs
Michael Keller, Mateusz Kotyrba, Maximilian Ebner and Anton Zeilinger
Austrian Academy of Sciences, Austria

CHSH inequality: Quantum probabilities as conditional probabilities
Andrei Khrennikov
Linnaeus University, Sweden

Sequences of projective measurements in generalized probabilistic theories
Matthias Kleinmann
University of Siegen, Germany

Quantum theory as a causal inference: a nonlinear noncommutative approach
Ryszard Kostecki
Perimeter Institute for Theoretical Physics, ON, Canada

A (100 × 100)-dimensional entangled quantum system
Mario Krenn[1], Marcus Huber[2,3,4], Robert Fickler[1], Radek Lapkiewicz[1], Sven Ramelow[1] and Anton Zeilinger[1]
[1]Austrian Academy of Sciences, Austria
[2]University of Bristol, U.K.
[3]The Institute of Photonic Sciences (ICFO), Spain
[4]University of Barcelona, Spain

Cavity cooling of free nanoparticles in high vacuum
Stefan Kuhn, Peter Asenbaum, Stefan Nimmrichter, Ugur Sezer and Markus Arndt
University of Vienna, Austria

Randomness extraction from quantum systems with different levels of trust in the working of the devices
Yun Zhi Law, Jean-Daniel Bancal and Valerio Scarani
National University of Singapore, Singapore

The Colbeck-Renner claim on the completeness of quantum mechanics as a theorem

Klaas Landsman[1] and **Gijs J. Leegwater**[2]
[1]Radboud University Nijmegen, Netherlands
[2]Erasmus University Rotterdam, Netherlands

Ψ-epistemic models are exponentially bad at explaining the distinguishability of quantum states

Matthew Leifer
Perimeter Institute for Theoretical Physics, ON, Canada

Quantum imaging with undetected photons

Gabriela B. Lemos, Victoria Borish, Garrett D. Cole, Sven Ramelow, Radek Lapkiewicz and Anton Zeilinger
Austrian Academy of Sciences, Austria

Simultaneously testing the Kochen-Specker and Bell theorems

Gustavo Cañas[1], Sebastián Etcheverry[1], Esteban S. Gómez[1], Carlos Saavedra[1], Guilherme B. Xavier[1], **Gustavo Lima**[1] and Adán Cabello[2]
[1]University of Concepción, Chile
[2]University of Sevilla, Spain

Quantum mechanics with single atoms and photons

Philipp Müller, Pascal Eich, Stephan Kucera, José Brito, Christoph Kurz, Michael Schug and Jürgen Eschner
Saarland University, Germany

Causal inference in quantum networks

Jacques Pienaar and Caslav Brukner
Austrian Academy of Sciences, Austria

Macroscopic quantum systems and gravitational phenomena

Igor Pikovski
University of Vienna, Austria

Beyond local causality: causation and correlation after Bell

Matthew Pusey
Perimeter Institute for Theoretical Physics, ON, Canada

Entanglement control in a quantum dot-microcavity system by an external magnetic field applied
Luisa Fernanda Ramírez Ochoa[1], Herbert Vinck-Posada[1], Luis E. Cano[2], Paulo S.S. Guimaraes[2] and Boris A. Rodríguez[3]
[1]National University of Colombia, Colombia
[2]Federal University of Minas Gerais, Brazil
[3]University of Antioquia, Colombia

Algorithmic synthetic unity
Allan F. Randall
York University, UK

Equivalence between adaptive and non-adaptive nonlocality distillation protocols
Jibran Rashid and Stefan Wolf
University of Lugano, Switzerland

Tailored two-photon correlation and fair-sampling: a cautionary tale
Jacquiline Romero[1], Daniel Giovannini[1], Daniel S. Tasca[1], Steve M. Barnett[2] and Miles J. Padgett[1]
[1]University of Glasgow, UK
[2]University of Strathclyde, UK

Heralded entanglement of single neutral atoms for a conclusive test of Bell's inequality
Wenjamin Rosenfeld, Daniel Burchardt, Norbert Ortegel, Kai Redeker, Robert Garthoff, Julian Hofmann, Markus Weber and Harald Weinfurter
Ludwig Maximilian University of Munich, Germany
and Max-Planck-Institute for Quantum Optics, Germany

Bell inequalities from 1964 to 2014: a compendium
Denis Rosset[1], Jean-Daniel Bancal[2] and Nicolas Gisin[1]
[1]University of Geneva, Switzerland
[2]National University of Singapore, Singapore

Greenberger-Horne-Zeilinger theorem for N *qudits*
Junghee Ryu[1], Changhyoup Lee[3], Marek Żukowski[1] and Jinhyoung Lee[2,4]
[1]University of Gdańsk, Poland
[2]Hanyang University, Korea
[3]National University of Singapore, Singapore
[4]Seoul National University, Korea

Noise tolerant entanglement verification in an untrusted quantum network
Dylan J. Saunders, Anthony J. Bennett and Geoff J. Pryde
Griffith University, Australia

Quantum communication with satellites, its preparatory terrestrial free-space demonstrations and future missions
Thomas Scheidl, Rupert Ursin and Anton Zeilinger
Austrian Academy of Sciences, Austria

Optimal LOCC conversion of 3-qubit states
Katharina Schwaiger
University of Innsbruck, Austria

On detecting the quantum correlations in the early universe
Yutaka Shikano[1], Yusuke Hayashi[2] and Masaaki Hashimoto[2]
[1]Tokyo Institute of Technology, Japan
[2]Kyushu University, Japan

Internal structure of the Heisenberg and Robertson-Schrödinger uncertainty relations
Lubomír Skála
Charles University in Prague, Czech Republic

Improved quantum metrology using quantum error-correction
Wolfgang Dür[1], **Michalis Skotiniotis**[1], Florian Fröwis[1,2] and Barbara Kraus[1]
[1]University of Innsbruck, Austria
[2]University of Geneva, Switzerland

Axiomatic approach for the function bound of all Bell's inequalities
Wonmin Son
National University of Singapore, Singapore

Remote entanglement preparation
Cornelia Spee[1], Julio Iñigo de Vicente Majúa[2] and Barbara Kraus[1]
[1]University of Innsbruck, Austria
[2]Charles III University of Madrid, Spain

Experimental reconstruction of complex joint probabilities for arbitrary photon polarization via sequential measurements of non-commuting observables
Yutaro Suzuki, Masataka Iinuma, Ryuji Kinoshita and Holger F. Hofmann
University of Hiroshima, Japan

Bit commitment based on Bell's theorem
Marcelo Terra Cunha
Federal University of Minas Gerais, Brazil

Robust test of Bell's inequality with amplified NOON states
Falk Töppel[1,2,3] and Magdalena Stobínska[4,5]
[1]Max Planck Institute for the Science of Light, Germany
[2]Friedrich-Alexander University Erlangen-Nürnberg, Germany
[3]Erlangen Graduate School in Advanced Optical Technologies, Germany
[4]University of Gdańsk, Poland
[5]Polish Academy of Sciences, Poland

Past of a quantum particle: speakable after all!
Lev Vaidman
Tel Aviv University, Israel

Single pairs of time-bin entangled photons
Marijn A. M. Versteegh[1,2,3], Michael E. Reimer[1], Aafke A. van den Berg[1], Gediminas Juska[4], Valeria Dimastrodonato[4], Agnieszka Gocalinska[4], Emanuele Pelucchi[4] and Val Zwiller[1]
[1]Delft University of Technology, Netherlands
[2]Austrian Academy of Sciences, Austria
[3]University of Vienna, Austria
[4]University College Cork, Ireland

Exploiting Bell's inequality to extend the device-independent quantum key distribution
Giuseppe Vallone, Alberto Dall'Arche, Marco Tomasin and **Paolo Villoresi**
University of Padova, Italy

Loophole-free Einstein Podolsky Rosen experiment via quantum steering
Bernhard Wittmann[1,2], Sven Ramelow[1,2], Fabian Steinlechner[2], Nathan K. Langford[2], Nicolas Brunner[3], Howard M. Wiseman[4], Rupert Ursin[2] and Anton Zeilinger[1,2]
[1]University of Vienna, Austria
[2]Austrian Academy of Sciences, Austria
[3]University of Bristol, UK
[4]Griffith University, Australia

Genuine energy-time entanglement-based quantum key distribution over installed telecom fibers
Gonzalo Carvacho[1], Gabriel Saavedra[1], Alvaro Cuevas[2], Jaime Cariñe[2], Miguel Figueroa[2], Adán Cabello[2], Paolo Mataloni[3], Gustavo Lima[3] and **Guilherme B. Xavier**[3]
[1]University of Concepción, Chile
[2]University of Sevilla, Spain
[3]Sapienza University of Rome, Italy

Experimental methods of detecting steering for arbitrary dimensional states
Yu-Lin Zheng, Yi-Zheng Zhen, Nai-Le Liu, Kai Chen, Zeng-Bing Chen and Jian-Wei Pan
University of Science and Technology of China, PR China

Violation of Bell inequalities with time
Magdalena Zych
Austrian Academy of Sciences, Austria

Printed in the United States
By Bookmasters